Hybrid and Hierarchical Composite Materials

Chang-Soo Kim • Charles Randow • Tomoko Sano
Editors

Hybrid and Hierarchical Composite Materials

Springer

Editors
Chang-Soo Kim
Materials Science and Engineering
University of Wisconsin-Milwaukee
Milwaukee
Wisconsin
USA

Tomoko Sano
US Army Research Laboratory
Aberdeen Proving Ground
Maryland
USA

Charles Randow
US Army Research Laboratory
Aberdeen Proving Ground
Maryland
USA

ISBN 978-3-319-35050-9 ISBN 978-3-319-12868-9 (eBook)
DOI 10.1007/978-3-319-12868-9

Springer Cham Dordrecht Heidelberg New York London

Softcover reprint of the hardcover 1st edition 2015

Printed on acid-free paper

Springer is part of Springer Science+Business Media (www.springer.com)

Contents

Contributors

Khelendra Agrawal Biomaterials and Multiscale Mechanics Lab, Department of Metallurgical and Materials Engineering, Indian Institute of Technology Roorkee, Roorkee, India

Michael R. Bockstaller Department of Materials Science and Engineering, Carnegie Mellon University, Pittsburgh, PA, USA

Golda L. Chakkalakal Department of Chemical and Biomedical Engineering, FAMU-FSU College of Engineering, Tallahassee, FL, USA

Randall M. Erb Department of Mechanical and Industrial Engineering, Northeastern University, Boston, MA, USA

Nikhil Gupta Composite Materials and Mechanics Laboratory, Mechanical and Aerospace Engineering Department, New York University Polytechnic School of Engineering, Brooklyn, NY, USA

Chang-Soo Kim Materials Science and Engineering, University of Wisconsin-Milwaukee, Milwaukee, WI, USA

Theodora Krasia-Christoforou Department of Mechanical and Manufacturing Engineering, University of Cyprus, Nicosia, Cyprus

Manoj Kumar R Biomaterials and Multiscale Mechanics Lab, Department of Metallurgical and Materials Engineering, Indian Institute of Technology Roorkee, Roorkee, India

Debrupa Lahiri Biomaterials and Multiscale Mechanics Lab, Department of Metallurgical and Materials Engineering, Indian Institute of Technology Roorkee, Roorkee, India

Rafael Libanori Complex Materials, Department of Materials, ETH Zurich, Zurich, Switzerland

Lele Peng Materials Science and Engineering Program and Department of Mechanical Engineering, The University of Texas at Austin, Austin, TX, USA

Shashank Priya Bio-inspired Materials and Devices Laboratory (BMDL), Center for Energy Harvesting Materials and Systems (CEHMS), Virginia Tech, Blacksburg, VA, USA

Subramanian Ramakrishnan Department of Chemical and Biomedical Engineering, FAMU-FSU College of Engineering, Tallahassee, FL, USA

Charles L. Randow US Army Research Laboratory, Aberdeen, MD, USA

Tomoko Sano US Army Research Laboratory, Aberdeen, MD, USA

Vasanth Chakravarthy Shunmugasamy Composite Materials and Mechanics Laboratory, Mechanical and Aerospace Engineering Department, New York University Polytechnic School of Engineering, Brooklyn, NY, USA

Meisha L. Shofner School of Materials Science and Engineering, Georgia Institute of Technology, Atlanta, Georgia, USA

André R. Studart Complex Materials, Department of Materials, ETH Zurich, Zurich, Switzerland

Chongchen Xiang Composite Materials and Mechanics Laboratory, Mechanical and Aerospace Engineering Department, New York University Polytechnic School of Engineering, Brooklyn, NY, USA

Yongke Yan Bio-inspired Materials and Devices Laboratory (BMDL), Center for Energy Harvesting Materials and Systems (CEHMS), Virginia Tech, Blacksburg, VA, USA

Guihua Yu Materials Science and Engineering Program and Department of Mechanical Engineering, The University of Texas at Austin, Austin, TX, USA

Yu Zhao Materials Science and Engineering Program and Department of Mechanical Engineering, The University of Texas at Austin, Austin, TX, USA

Chapter 1
Introduction

Tomoko Sano, Charles L. Randow and Chang-Soo Kim

Abstract In the past decade, hybrid and hierarchical composites have come into focus for their ability to be property-specific solutions to design problems that cannot be solved by homogeneous materials. Significant advances have been especially made in the more established areas of organic–inorganic hybrids, and bioinspired and biomimetic hierarchical composites have led the development in the hierarchical composite research areas. Because of this, numerous review articles and books have been written on each of these topics (Buehler et al., Eur J Inorg Chem 2012(32):5091–5420, 2012; Sanchez et al., Chem Soc Rev 40(2):453–1152, 2011; Studart, Adv Mater 24(37):5024–5044, 2012; Chen and Pugno, J Mech Behav Biomed Mater 19:3–33, 2013; Fratzl and Weinkamer, Prog Mater Sci 52(8):1263–1334, 2007; Qian et al., J Mater Chem 20(23):4751, 2010; Kickelbick, Hybrid materials: synethesis, characterization, and applications, Wiley-VCH Verlag, Weinheim, 2007; Gomez-Romero and Sanchez, Functional Hybrid Materials, Wiley-VCH Verlag, Weinheim, 2004; Su et al., Hierarchically structured porous materials: from nanoscience to catalysis, separation, optics, energy, and life science, Wiley-VCH Verlag, Weinheim, 2011; Zhang and Wei, Advanced hierarchical nanostructured materials, Wiley-VCH Verlag, Weinheim, 2014) with some focused on specific material class or application. However, few have focused on the nanoscale contributions and linkages of these composites. The goal of this book is to cover a broad range of current nano-lengthscale hybrid and hierarchical composite research, including the processing, properties, modeling, and applications of these composites, for a technical audience interested in learning more about these subject areas.

T. Sano (✉)
US Army Research Laboratory, RDRL-WMM-F, Aberdeen Proving Ground, Aberdeen, MD 21005, USA
e-mail: tomoko.sano.civ@mail.mil

C. L. Randow
US Army Research Laboratory, RDRL-WMP-D Aberdeen Proving Ground, Aberdeen, MD 21005, USA

C.-S. Kim
Materials Science and Engineering, University of Wisconsin-Milwaukee, 3200 N Cramer St., Milwaukee, WI 53211, USA

C.-S. Kim et al. (eds.), *Hybrid and Hierarchical Composite Materials*,
DOI 10.1007/978-3-319-12868-9_1

1.1 Introduction

There are a variety of definitions used in the field of hybrid and hierarchical composites. In this work, material class definitions, in a manner similar to Ashby (Ashby and Brechet 2003), are used in defining hybrid composites and hierarchical composites. We will specifically consider the following material classes in this work: organic polymers, inorganic polymers, molecular arrangements, and elemental groups (e.g., carbon nanotubes, metal oxides), metals, and ceramics and clays.

A hybrid composite is a system composed of (at least) two different material classes with length scales for at least one of the materials on the order of one micron or less. A hierarchical composite is a composite with an additional, smaller length scale component (e.g., a composite of a composite), which is a composite that consists of three or more critical length scales. Some hierarchical composites could be considered a hybrid composite with an additional component that adds another level of functionality. We will be considering hybrid and hierarchical composites with at least a single length scale on the order of one micron or less (e.g., a hierarchical composite with nanograin reinforcement). Although metal matrix composites (MMCs) are hybrid composites, we will not consider MMCs in this work due to the large volume of work published in this subject area. Figure 1.1 schematically shows the hybrid and hierarchical systems, a subset of all composites, that will be covered in the following chapters.

With nanotechnology in mind, this book will uniquely address a broad spectrum of areas of interest in both hybrid composites and hierarchical composites. These areas of interest include recent development of processing technologies, structural designs, modern computational simulation techniques, and the relationships among processing, structure, property, and performance.

Fig. 1.1 A schematic of the hybrid and hierarchical composite subjects covered in this book

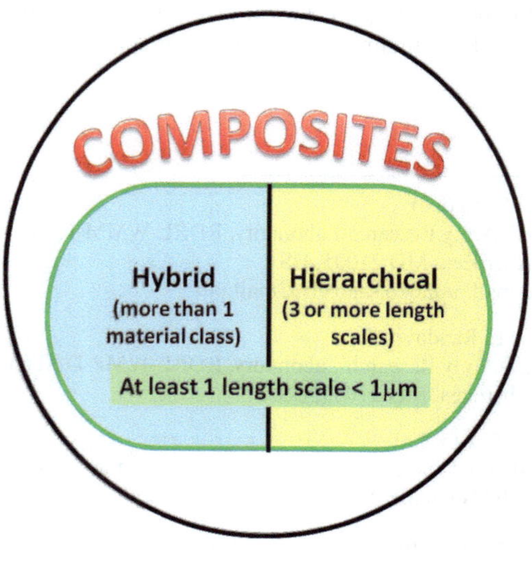

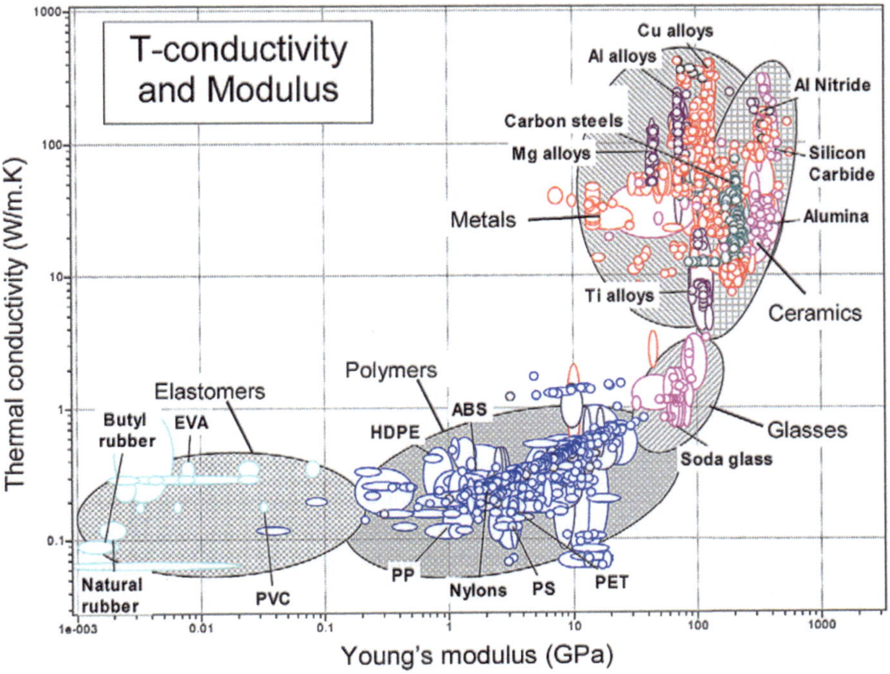

Fig. 1.2 A material property chart indicating the thermal conductivity and Young's modulus for a variety of materials. (Reprinted from Ashby and Brechet (2003), with permission from Elsevier)

1.2 Hybrid Composites

Hybrid composites are composites that combine two or more different material classes, such as an inorganic ceramic and an organic polymer. There are many classifications of hybrids. Gomez-Romero and Sanchez (2004) as well as Kickelbick (2007) distinguish hybrids with weak interactions such as van der Waals and hydrogen bonding or weak electrostatic interactions as class I. They classify class II as hybrids with strong chemical bonding between the two components. Ashby and Bréchet (2003) address hybrids from a functional classification, and Nanko (2009) classifies hybrids into three categories: structurally hybridized materials (composites), materials hybridized in chemical bond, and functionally hybridized materials. The differences in classifications depend on one's point of view. A polymer chemist could view hybrids differently than a design engineer.

Regardless of terminology used, the purpose of hybridizing is to obtain combined properties, such that the hybrid is an improvement compared to each of the parent materials. For example, a solution to an engineering problem might require a new material with a higher Young's modulus and a lower thermal conductivity. Figure 1.2 presents a typical property chart to show the relationship between the thermal conductivity and Young's modulus for a variety of materials (Ashby and

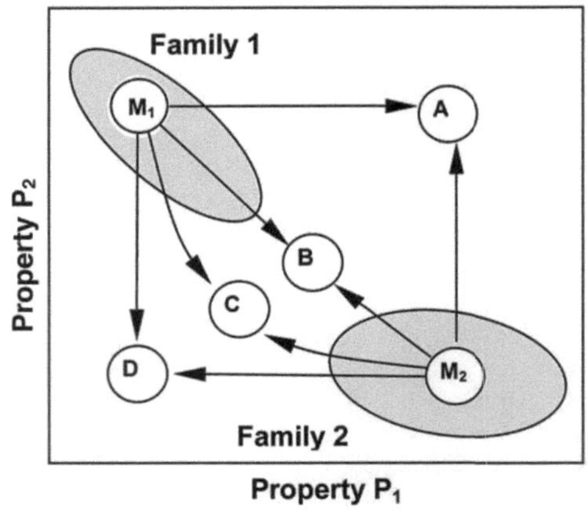

Fig. 1.3 Hybridization property map scenarios. (Reprinted from Ashby and Brechet (2003), with permission from Elsevier)

Brechet 2003). Examining Fig. 1.2, one might decide to combine a polymer with a metal to obtain a hybrid composite with these desired properties. However, the processing technique, the volume percentages, the choice of catalysts, etc., can play an important role in the hybrid's homogeneity, interfaces, bonding, and the resulting structure and properties. An idealization of such a process is shown in Fig. 1.3, where the combination of Material 1 (M_1) and Material 2 (M_2) results in the hybrid A, with its improved synergistic properties. In the figure, hybrid B would be a combination that follows the "rule of mixtures," hybrid C is the combination that has one improved property, and hybrid D would be the worst case scenario with no benefit to hybridization.

In the following hybrid chapters, the various successful combinations of hybrid composites will be introduced and discussed in detail. Chapter 2 covers the synthesis of organic–inorganic polymer hybrids and provides an overview of their applications. In Chap. 3, the emerging field and potential applications of particle brushes, or polymer-tethered inorganic nanoparticle hybrids, are discussed. Chapter 4 details the engineering of multiferroic and magnetoelectric hybrids, as well as the properties, processing, and modeling and simulation of these hybrid systems. In Chap. 5, the last chapter in the hybrid composites area, the types, processing, properties, and modeling work of clay–polymer nanocomposites will be discussed.

1.3 Hierarchical Composites

The simplest form of a composite consists of a two-phase structure. As described earlier, a hierarchical composite is a composite with an additional, smaller length scale component that is part of one of the phases. For example, a structural hierarchical composite might contain a structural component possessing its own unique substructure (Lakes 1993). As another example, one of the most notable biological hierarchical composite is bone. As shown in Fig. 1.4, at the macroscopic scale, bone is made up of osteons, which consist of haversion canals that surround nerves and

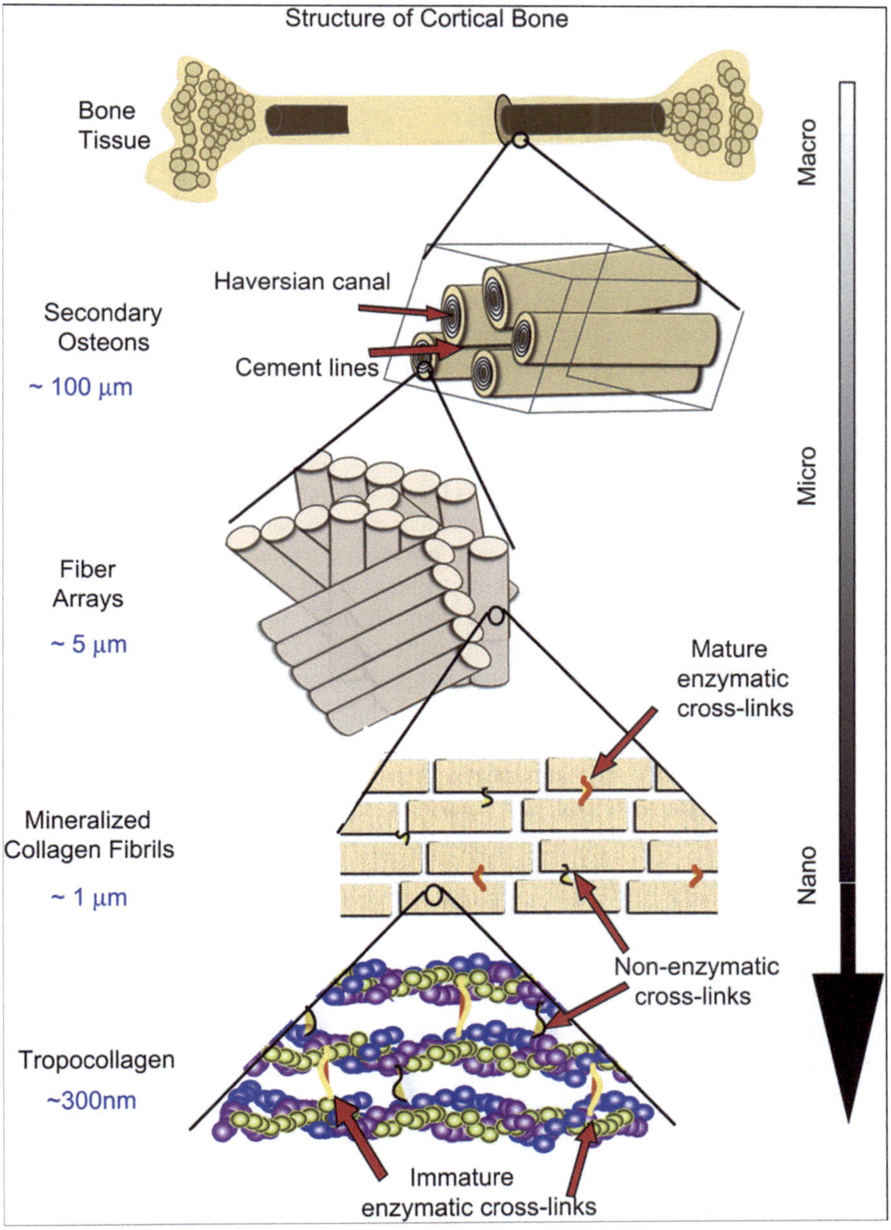

Fig. 1.4 The hierarchical structure of bone at multiple length scales. (Reprinted from Barth et al. (2011), with permission from Elsevier)

blood vessels, and cylindrical lamellae. The lamellae are made up of mineralized collagen fibers. The mineralized collagen fibers are bundles of mineralized fibrils, which in turn are made up of tropocollagen molecules (Barth et al. 2011; Vaz et al. 2011). Each of the hierarchical components and their materials play a role in the strengthening and toughening mechanisms at multiple length scales, and contribute to the anisotropic properties of bone. By natural evolution, biological systems have combined a variety of materials with specific properties to create hierarchical composite solutions for survival.

These naturally occurring hierarchical materials have inspired the research and generation of engineered or synthetic hierarchical composites made with different classes of materials for a variety of applications (Chen and Pugno 2013; Fratzl and Weinkamer 2007; Studart 2012; Su et al. 2011; Vaz et al. 2011; Zhang and Wei 2014; Dunlop and Fratzl 2013; Launey and Munch 2009; Fratzl 2007). Hence, the goal for many synthetic hierarchical composites is to obtain better performing materials by intentionally creating material complexity by designing composites with multi-length scale property control. For example, structural hierarchical composites have been shown to provide both stiffness and damping characteristics, which are generally competing properties (Lakes, 2002).

Hierarchical composite research has broadened with the increased interest in creating hierarchical composite solutions for various applications. The multi-length scale design and creation of such hierarchical composites include advanced characterization of the material components, processing techniques that optimize the structure–property relationships, and modeling and simulations. In the second half of this book, these application-specific hierarchical composite systems and their design metrics will be described, starting with medical applications of hierarchical composites in Chap. 6. The hierarchical composite material types, their properties, and specific medical applications of the hierarchical composites are described. In Chap. 7, electrochemical hierarchical composites used for energy applications are discussed. These include mechanisms of energy storage, hierarchical composite use in batteries, energy conversion, and sensors. Chapter 8 covers bioinspired structural hierarchical composites, providing examples from nature and showing how researchers have applied a design methodology for hierarchical composites. Elements of these methodologies include reinforcements, hierarchical porosity, and hierarchical topography. In Chap. 9, various morphologies of hierarchical composites with carbon nanotube additions, their properties, and processing methods are described. Depending on the incorporation method, the carbon nanotubes can play several different roles in the matrix, such as changing the polymer mobility, or improving the mechanical properties.

1.4 Concluding Remarks

The following chapters will reveal that hybrid and hierarchical composite research has grown and will continue to grow for some time. There are new formulations, processing methods, and structure–property relationships to optimize for numerous applications. We hope this book educates the reader in some of the exciting research currently being conducted in these fields, and inspires them for further inquiry.

References

Ashby MF, Brechet YJ (2003) Designing hybrid materials. Acta Mater 51:5801

Barth HD, Zimmermann EA, Schaible E, Tang SY, Alliston T, Ritchie RO (2011) Characterization of the effects of x-ray irradiation on the hierarchical structure and mechanical properties of human cortical bone. Biomaterials 32:8892–8904

Buehler M, Rabu P, Taubert A (eds) (2012) Organic-Inorganic hybrid materials: design and applications (cluster issue). Eur J Inorg Chem 2012(32):5091–5420

Chen Q, Pugno NM (2013) Bio-mimetic mechanisms of natural hierarchical materials: a review. J Mech Behav Biomed Mater 19:3–33

Dunlop JW, Fratzl P (2013) Multilevel architectures in natural materials. Scripta Materialia 68:8–12

Fratzl P (2007) Biomimetic materials research: what can we really learn from nature's structural materials? J R Soc Interface 4:637–642

Fratzl P, Weinkamer R (2007) Nature's hierarchical materials. Prog Mater Sci 52(8):1263–1334

Gomez-Romero P, Sanchez C (2004) In: Gomez-Romero P, Sanchez C (eds) Functional hybrid materials. Wiley-VCH, Weinheim

Kickelbick G (2007) Hybrid materials: synthesis, characterization, and applications. Wiley-VCH, Weinheim

Lakes R (1993) Materials with structural hierarchy. Nature 361:511–515

Lakes RS (2002) High damping composite materials: effect of strucural heirarchy. J Compos Mater 36(3):287–296

Launey ME, Munch E (2009). Designing highly toughened hybrid composites through nature-inspired herarchical complexity. Acta Materialia 57:2919–2932

Nanko M (2009) Definitions and categories of hybrid materials. Adv Technol Mat Mat Proc J 11(1):1–8

Qian H, Greenhalgh ES, Shaffer MS, Bismark A (2010) Carbon nanotube-based hierarchical composites: a review. (4762, ed) J Mater Chem 20(23):4751

Sanchez C, Shea JK, Katagawa S (eds) (2011) Hybrid materials themed issue. Chem Soc Rev 40(2):453–1152

Studart AR (2012) Towards high-performance bioinspired composites. Adv Mater 24(37):5024–5044

Su B-L, Sanchez C, Yang X-Y (eds) (2011) Hierarchically structured porous materials: from nanoscience to catalysis, separation, optics, energy, and life science. Wiley-VCH Verlag, Weinheim

Vaz MF, Canhao H, Fonseca JE (2011) Analysis of natural and man-made materials. In: Tesinova P (ed) Bone: a composite natural material, advanced in composite materials. InTech

Zhang Q, Wei F (eds) (2014) Advanced hierarchical nanostructured materials. Wiley-VCH, Weinheim

Part I
Hybrid Composites

Chapter 2
Organic–Inorganic Polymer Hybrids: Synthetic Strategies and Applications

Theodora Krasia-Christoforou

Abstract Hybrid materials generated via the combination of functional polymers with inorganic nanostructured compounds, with the latter exhibiting size-dependent physical and chemical properties, have become a major area of research and technological development owing to the remarkable properties and multifunctionalities deriving from their nanocomposite/nanohybrid structure. In this chapter, the different fabrication routes for generating organic–inorganic polymer hybrid materials are discussed. Those include blending processes, sol–gel methods, emulsion polymerization and photopolymerization, metallosupramolecular and coordination approaches, intercalation, microwave-assisted and electrochemical synthesis, synthetic routes based on surface grafting, and finally self-assembly and block copolymer-mediated synthetic strategies. The existing versatility in materials' design in organic–inorganic polymer hybrids, in respect to the structural, compositional, and architectural characteristics, creates new prospects for many applications in very diverse areas. In the second section of this chapter, the applicability of organic–inorganic polymer hybrids in various fields including biomedicine, sensing, environmental remediation, energy, construction, automotive and coating technologies, catalysis, and optoelectronics is reviewed.

2.1 Introduction

The modern society is facing a number of challenges related to health, sustainability, and environmental protection. Inevitably, the necessity of designing and developing new functional materials that could improve the quality of life and lead to the development of advanced technologies is of paramount importance.

During the past decades, polymer science has strongly been entering the field of organic–inorganic hybrids, aiming to combine the properties of polymers with those of inorganic components. The former include ease of processing, lightweight, flexibility, extremely high versatility in materials' design in regard to the structural

T. Krasia-Christoforou (✉)
Department of Mechanical and Manufacturing Engineering, University of Cyprus,
75, Kallipoleos Avenue, P.O. Box 20537, 1678 Nicosia, Cyprus
e-mail: krasia@ucy.ac.cy

C.-S. Kim et al. (eds.), *Hybrid and Hierarchical Composite Materials*,
DOI 10.1007/978-3-319-12868-9_2

and architectural characteristics and consequently in materials' properties. Inorganic materials exhibit superior thermal and mechanical behavior as well as unique optical, electrical, catalytic, and magnetic properties, especially when those are characterized by nanoscale (10^{-9} m) dimensions.

The tremendous increase in the number of publications, reports, and patents appearing during the past years on the design, synthesis, properties, and applications of organic–inorganic polymer hybrids demonstrates the high interest of the scientific community and of the industrial sector toward this rapidly emerging field, as shown by the number of publications from 2000 onwards on "polymer hybrids" and "polymer composites" (Source: Web of Science, June 2014) exceeding 85,000.

The advancements in the synthetic methods employed for developing such materials as well as in the characterization techniques and computational tools have enabled the fine-tuning of the materials' structural and compositional characteristics resulting—in many cases in a predictable manner—in the desirable behavior, property, or response.

Organic–inorganic polymer hybrids based on combinations of polymers, with metals, ceramics, or both, have been prepared by a variety of synthetic methods and have been exploited in different applications. The use of an appropriate polymer coating provides solubility and high colloidal stability of the inorganic nanocomponents in organic or aqueous solvents, thus facilitating their low-temperature, low-cost processing, whereas it enables their applicability in certain areas requiring the use of nanoparticles with long-term colloidal stability in solution (e.g., magnetic nanoparticles destined for use as contrast enhancement agents in magnetic resonance imaging (MRI) or ceramic-based nanocontainers employed as drug delivery systems). Most importantly, the flexibility in materials' design and the high versatility in existing synthetic methodologies that can be employed for the preparation of a targeted polymer-based hybrid system enables the development of novel multifunctional materials with new characteristics and functions that derive from the combination of the properties of the organic and the inorganic components, as well as from additional synergistic effects that allow the tailoring and fine-tuning of the materials' properties.

The present chapter aims in providing a general overview of the different synthetic methods developed toward the preparation of organic–inorganic polymer-based nanohybrids/nanocomposites and of their applications in various fields including biomedicine, sensing, energy, catalysis, optoelectronics, environmental remediation, and finally in construction, automotive, and coating technologies.

2.2 Synthetic Strategies for the Preparation of Organic–Inorganic Polymer Hybrids

Numerous experimental methodologies have been developed for synthesizing polymer–inorganic hybrids. In this section, the different synthetic approaches used toward this purpose are reviewed (Luan et al. 2012; Jeon and Baek 2010). First, blending methods are discussed, followed by the sol–gel synthetic approach that has been used for the synthesis of polymer–metal oxide nanocomposites focusing mostly on polymer–silica systems. Subsequently, emulsion polymerization is presented with particular emphasis in miniemulsion processes that have been proved to be highly promising and effective in the preparation of polymer-based hybrid materials. Other synthetic strategies are also briefly discussed such as metallosupramolecular polymerization, photopolymerization, intercalation, electrochemical synthesis, and microwave irradiation. The surface grafting methods are presented in more detail including the "grafting from," "grafting to," and the "grafting through" approaches. Finally, self-assembly techniques and block copolymer-mediated synthetic methods are discussed.

2.2.1 Blending

2.2.1.1 Solution and Melt Blending

One of the simplest methods for producing polymer–inorganic nanocomposites is solution blending. In this method, a polymer solution is generated first by dissolving the polymeric material in a good solvent followed by the introduction of the inorganic nanoparticulates. The dispersion of the latter is accomplished via mechanical stirring or ultrasonication. The resulting polymer-based nanocomposites can then be easily processed to generate solid-state nanocomposites upon solvent evaporation. Although this method for preparing polymer-based hybrids is simple and of low cost, it suffers from nanoparticle aggregation due to the strong tendency of the inorganic nanoparticles to agglomerate so as to reduce their surface energy. Such nanoparticle agglomeration phenomena lead to materials characterized with high inhomogeneity and inferior physical, chemical, and mechanical properties compared to their counterparts, resulting from the ineffective mixing of the organic and the inorganic components.

For this reason, and to ensure good intermixing between materials with different cohesion energies, the inorganic nanoparticles are usually introduced in the polymer solution after being functionalized with either organic molecules or macromolecules to ensure good solubility and mixing at a molecular level with the polymer matrix as schematically depicted in Fig. 2.1.

Melt blending does not require the use of organic solvents, thus rendering it more environmentally friendly compared to the solution blending method. Moreover, it is performed under similar processing conditions as those commonly used for the parent polymers, i.e., thermal treatment and shearing at temperatures above the polymer-softening point. During this process, the polymer is present in its melt state when mixed with the inorganic nanofillers. Melt blending has been commonly introduced in the synthesis of polymer-based nanocomposites. Many examples describe the synthesis of polymer–silica systems by using the melt-blending technique, but there are also numerous literature reports describing the fabrication of polymer-based hybrids in which the polymers are combined with different types of

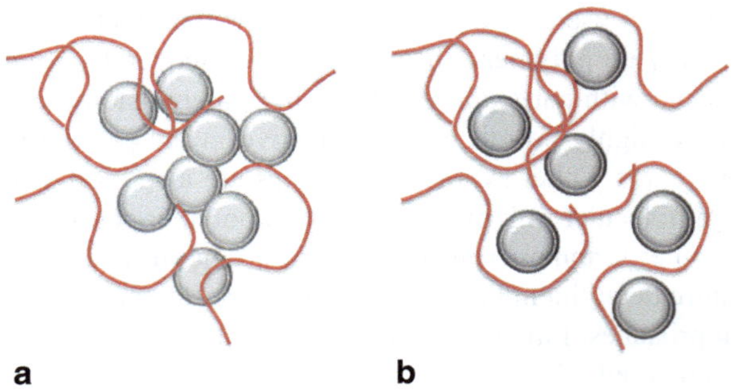

Fig. 2.1 Polymer–inorganic nanocomposite material in which **a** nanoparticle agglomeration takes place and **b** nanoparticles are well dispersed within the polymer matrix

inorganic nanoparticles including nanoclays (Albdiry et al. 2013), silica (Lai et al. 2013; Reddy and Das 2005), metal nanoparticles (MNP) (Xia et al. 2006), semiconducting quantum dots (QD), (Minhao Wong et al. 2012) and Al_2O_3 nanoparticles (Psarras et al. 2011) to mention only a few.

2.2.1.2 Powder Blending

Solid-state powder processing has been also used for the synthesis of polymer–inorganic hybrids. The use of solid-state processes is advantageous from the aspect that they do not require any solvents (i.e., environmental issues arising from solvent removal and disposal do not exist) or polymer heat treatment, and they overcome the processing problems associated with the high nanoparticle loading, observed in the case of solution and in particular in the case of melt processing due to the tremendous increase in the viscosity of the melt rendering the whole process infeasible. Moreover, polymers with poor solubility in organic solvents or with high melting temperatures cannot be co-processed with inorganic nanoparticles using the aforementioned techniques to obtain polymer–inorganic hybrid materials.

High-energy ball milling (HEBM; Castrillo et al. 2007) has been used for synthesizing polymer–inorganic hybrid materials. During the HEBM process, a series of energy-transfer events takes place promoting intermixing between polymers and inorganic nanoparticles. Existing agglomerates of the filler particles break up into smaller (nano) particles during the process resulting in their homogeneous dispersion inside the polymer matrix. This in turn leads to a composite material where the properties of the individual components are retained, whereas depending on the degree of interaction between the filler particles and the polymer matrix, new properties may arise due to synergistic interfacial effects.

Polymers have been combined with silica (Olmos et al. 2012; Pantaleon and Gonzalez-Benito 2010; Gonzalez-Benito and Gonzalez-Gaitano 2008), barium titanate (Olmos et al. 2013; Serra-Gomez et al. 2012), TiO_2 (Ouyang et al. 2012), nanoclay minerals (Vertuccio et al. 2009; Mangiacapra et al. 2006), magnetoresponsive MNP (Giri 1997; Bao and Jiang 2005; Zhu et al. 2006; 2008), and nickel-ferrite nanopowder (Azhdar et al. 2008) employing cryogenic HEBM.

As in the case of melt and solution blending, the mixing efficacy between polymers and inorganics can be further assisted by functionalization of the inorganic surface with organic (macro) molecules.

2.2.2 Sol–Gel Synthesis

The synthesis of polymer-based organic–inorganic hybrids by the so-called sol–gel method dates back more than two decades. Nevertheless, this is still a highly emerging field since newly developed sol–gel synthetic approaches result in the generation of advanced polymer-based hybrid materials enabling at the same time structural and configurational control (Pandey and Mishra 2011). Such materials combining organic polymers with inorganic solids at the nanoscale, exhibiting properties not only of their counterparts but also new ones arising by the hybridization process, are highly promising in optical, mechanical, biological, electrical, catalytic, and other applications (Ogoshi and Chujo 2005).

2.2.2.1 Basics of the Sol–Gel Method

The sol–gel method makes use of metal alkoxides that undergo two stepwise reactions: hydrolysis and condensation. These reactions result in the generation of an inorganic network. A general reaction scheme of the sol–gel process is provided in Fig. 2.2.

The incorporation of organic macromolecules within the inorganic network generated via sol–gel including silica, alumina, titania, germania, and vanadia is feasible due to the mild reaction conditions employed (Chujo and Saegusa 1992; Zou et al. 2008; Mark 1996). As a consequence, numerous organic (polymer)–inorganic (metal oxide) hybrids have been prepared by this method. However, most studies describe the synthesis of silica-based polymer hybrids due to the well-known chemistry of silica, the high stability of the Si–O bond, as well as the availability of the starting materials, whereas the high reactivity of other metal oxides in sol–gel processes results in aggregation phenomena and inhomogeneity issues.

A principal requirement for obtaining materials characterized by high homogeneity and transparency, is the molecular- or nanodispersion of the organic counterpart within the inorganic matrix. This is realized via the development of specific interactions between the organic and the inorganic elements that can be either physical or chemical bonding. Depending on the type of interactions developed, the polymer-based hybrids prepared by the sol–gel approach can be divided into two classes as schematically presented in Fig. 2.3.

Class I includes the systems where only physical interactions take place between the organic and the inorganic components, including hydrogen bonding, ionic, aromatic (π–π interactions), etc. In this case, both the polymer and the starting inorganic "monomer" must possess specific functionalities that would enable the devel-

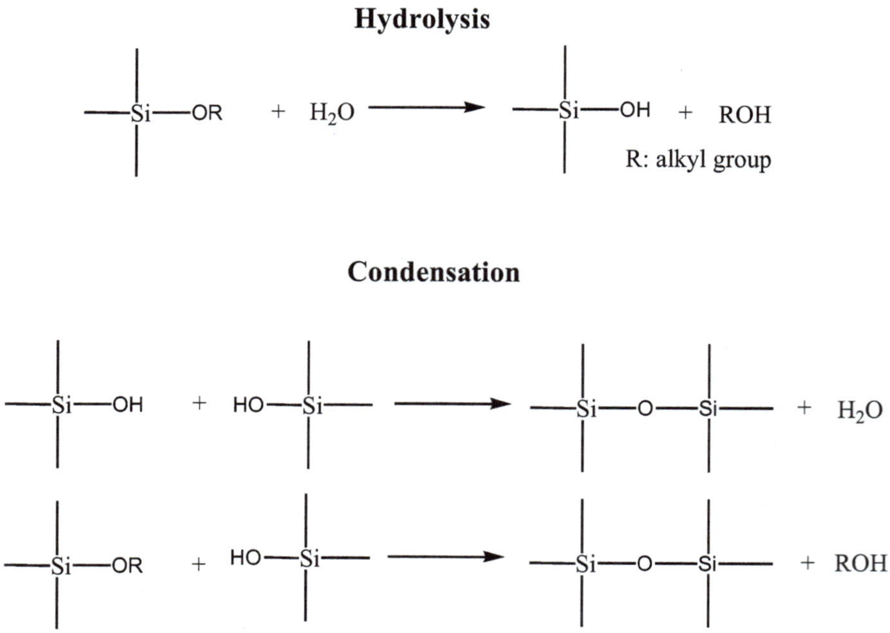

Fig. 2.2 The sol–gel reaction process involving hydrolysis and subsequent condensation steps

opment of such interactions. For example, polymers possessing H-bonding functionalities, including polyamides (Sarwar et al. 2009), polyurethanes (Zhang et al. 2011a), polycarbonates (Lionti et al. 2013), as well as biopolymers (Coradin et al. 2006), have been successfully employed in the synthesis of *Class I* polymer-based hybrids by sol–gel. Similarly, the inorganic precursors can be selected so as to promote such interactions with the polymer chain. For example, the incorporation of silicate monomers containing aromatic moieties facilitate the generation of homogeneous polymer–silica hybrids with polymers containing aromatic groups (Tamaki et al. 1998). In contrast, when the inorganic precursors used lack such moieties, the resulting hybrid materials are characterized by insufficient homogeneity due to the immiscibility, and consequently phase separation between the organic and inorganic components occurs.

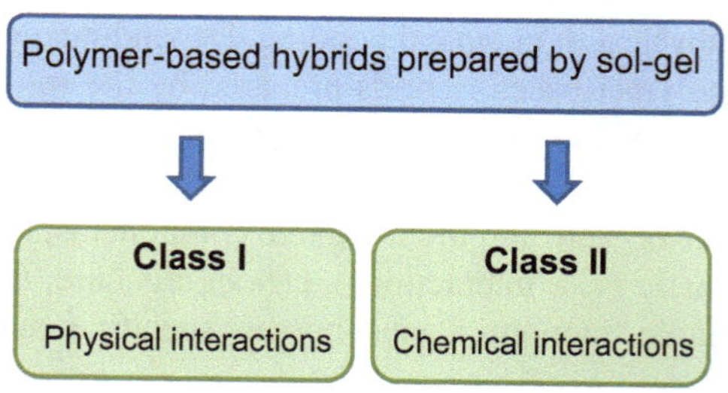

Fig. 2.3 Classification of polymer-based hybrids prepared by the sol–gel approach in *Class* I and *Class* II materials

In *Class II*, the different phases are strongly held together by chemical bonding. Different synthetic approaches were used toward this purpose with the most popular ones being the use of alkoxysilane (–Si(OR)$_3$)-functionalized polymers (Huang et al. 2004; Sardon et al. 2010; McDowell et al. 2010) and of telechelic (possessing two identical reactive end groups) polymers with functional moieties that enable co-condensation reactions to take place (Julián et al. 2004; Uilk et al. 2003; Chujo et al. 1993).

Alternatively, for achieving miscibility between noncompatible polymer–metal oxide systems, different strategies have been used including the incorporation of compatibilizers. These are usually macromolecules exhibiting good miscibility properties with both, the inorganic metal oxide network and the polymer that is incompatible with the inorganic counterpart. Alternatively, macrocycles assisting the dispersion of polymer chains via the host–guest interactions developed between the polymer and the hydrophobic cavity of the macrocycle have been employed (Ogoshi and Chujo 2005).

2.2.2.2 In Situ Preparation of Polymer-Based Hybrids via Sol–Gel

(i) H_2O-catalyzed sol–gel/polymerization

The use of a monomeric unit combining an organic polymerizable group with a metal alkoxide moiety, allows for the simultaneous formation of the polymer chain, and of the inorganic network (Fig. 2.4). Depending on the type of the polymerizable group employed, different polymerization methods have been used for the in-situ formation of polymer–metal oxide hybrids including ring-opening polymerization (ROP) and free-radical polymerization (Ogoshi and Chujo 2005).

(ii) Nonhydrolytic sol–gel/polymerization

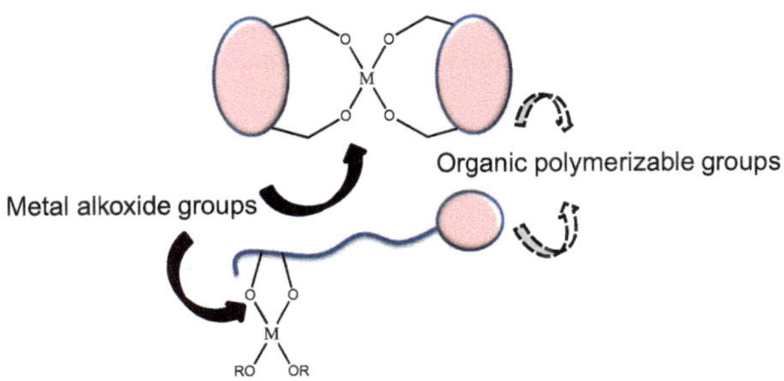

Fig. 2.4 Two different structures of hybrid "monomers" containing both metal oxide and organic polymerizable moieties, employed for the in-situ preparation of polymer–metal oxide hybrid materials by H_2O-catalysed sol–gel polymerization. The *dashed white* arrows show the organic polymerizable groups (e.g., vinyl monomers or monomers susceptible to ROP), whereas the *black* arrows point to the metal–alkoxide moieties employed for the generation of the inorganic gel.

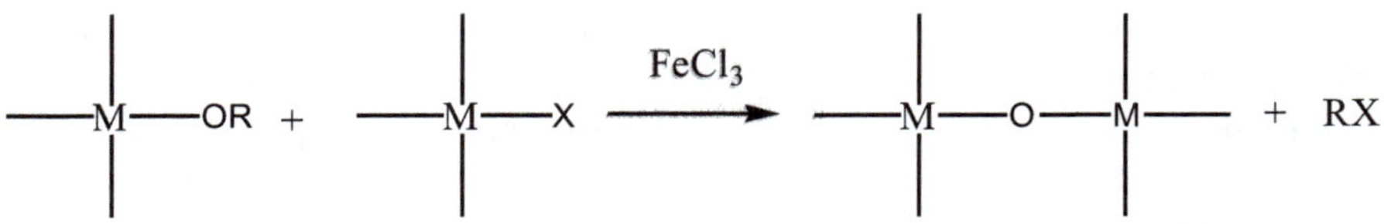

Fig. 2.5 General reaction scheme of the NHSG process. *NHSG* stands for nonhydrolytic sol–gel

In 2001, the first report appeared describing the sol–gel synthesis of a polymer–metal oxide network, obtained in the absence of solvents (i.e., the organic monomer was polymerized in the bulk simultaneously with the formation of the inorganic metal oxide network), and without the necessity of using water as a catalyst (Hay and Raval 2001). This synthetic approach involves a ligand exchange reaction that takes place in the presence of metal halides (e.g., $FeCl_3$) or organic molecules with oxygen donor moieties such as an alcohol, ether, etc., as seen in Fig. 2.5.

For example, nonhydrolytic sol–gel (NHSG) has been used for synthesizing silica–dimethylsiloxane hybrids where the NHSG synthesis of silica took place simultaneously with the ROP of hexamethylcyclotrisiloxane in the presence of $FeCl_3$ introduced as a catalyst (Apperley et al. 2002). In other cases, the NHSG process is used to prepare the colloidal metal oxide first, which is then mixed with preformed polymer chains to yield the desired nanocomposite (Song et al. 2008). Alternatively, the polymer/metal oxide nanocomposites are prepared by an in situ formation of the metal oxide through NHSG, starting from a mixture of the corresponding metal oxide precursor, an oxygen donor, and preformed polymer chains (Morselli et al. 2011).

(iii) The interpenetrating polymer network approach

The interpenetrating polymer network (IPN) process involves the simultaneous cross-linking processes of the organic and the inorganic components resulting in the formation of an interpenetrated organic–inorganic hybrid network structure as schematically depicted in Fig. 2.6 (Wen and Wilkes 1996).

Thermo-reversible IPN polymer hybrids consisting of organic polymer chains and silica gel were prepared by acid-catalyzed sol–gel reaction of tetramethoxysilane (TEOS) to generate the silica network. The reaction took place in the presence

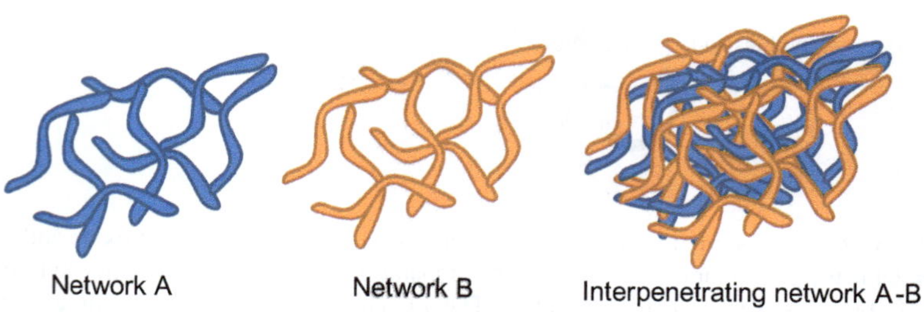

Fig. 2.6 Schematic presentation of an interpenetrating network consisting of two different networks (*Network A* inorganic; *Network B* organic) strongly interpenetrating each other

of poly(oxazoline) polymers bearing maleimide and furan groups that reacted together via a Diels–Alder reaction process to generate the organic network (Imai et al. 2000). In another example, bipyridyl-containing polystyrene and poly(ethylene oxide) led to network formation in the presence of a ruthenium salt. This coordination cross-linking process was carried out simultaneously with the generation of silica gel via a sol–gel process starting from TEOS (Tomoki et al. 2001).

2.2.2.3 Structurally Defined Polymer-Based Hybrids by Sol–Gel

To date, there is considerable attention on the development of new synthetic routes that enable the fabrication of polymer–metal oxide hybrid materials with controlled particle size, morphology, and porosity. After the first report back in 1992 on the synthesis of mesoporous silica (MCM41) by following a surfactant-templating procedure (Beck et al. 1992), the main focus has been on the development of mesoporous metal oxides exhibiting ordered structures at the 1–50-nm scale (Gu and Schueth 2014; Soler-Illia and Azzaroni 2011) using different types of surfactants (Antonietti 2001; Palmqvist 2003; Holmberg et al. 2005; El-Safty 2011), block copolymers (Soler-Illia et al. 2003; Deng et al. 2013; Orilall and Wiesner 2011), surfactant/block copolymer mixtures (Hellweg 2011; Ma et al. 2003), and macromolecular dendrimers (Tsiourvas et al. 2013) as structure-directing mediators.

The different pathways developed so far for producing mesoporous materials are presented schematically in Fig. 2.7. Those involve the direct precipitation, the true liquid-crystal templating (TLCT), the evaporation-induced self-assembly (EISA) and exotemplating routes (Soler-Illia and Azzaroni 2011). Briefly, the precipitation process involves the assembly of the inorganic precursors and the organic template occurring via hydrolysis/condensation of the inorganic precursors followed by template removal, resulting in the desired mesoporous material. In the TLCT templating, a liquid crystalline mesophase is generated first, followed by infiltration of the inorganic precursor within this mesophase, and subsequent creation of the inorganic network around the liquid crystalline nanodomains. A popular, solution-based route to synthesize oriented mesoporous metal oxide materials is based on the combination of sol–gel chemistry and dip or spin coating, using organic templating agents such as low molecular weight surfactants or amphiphilic block copolymers (ABCs). This method is called EISA (Brinker 2004).

The periodic structure, porosity, and degree of orientation of the resulting mesoporous films depend strongly on the template used. Triblock copolymers of the Pluronic family (poly(ethylene oxide)–poly(propylene oxide)–poly(ethylene oxide), PEO–PPO–PEO), or polyethylene oxide-based surfactants, such as the Brij family (CH_3–$(CH_2)_{10-16}$–$(O$–$C_2H_4)_{1-25}$–OH), have been found particularly useful as templates in EISA processes (Liu 2009; Chu et al. 2014; Herregods et al. 2013; Ma et al. 2010; Henderson et al. 2006) by acting as structure-directing agents. In such processes, the inorganic precursors are mixed together with the organic amphiphile in solution.

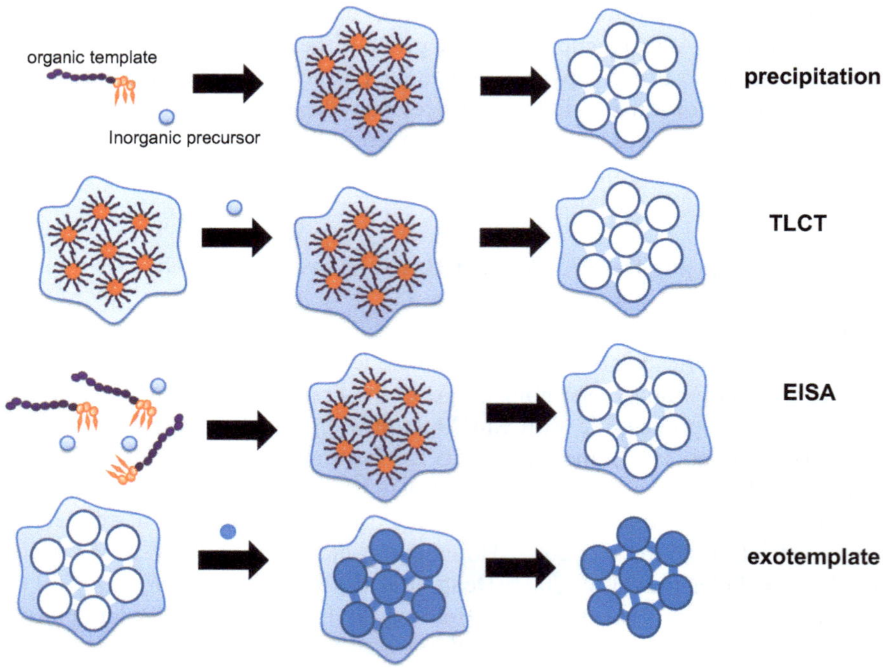

Fig. 2.7 Synthetic pathways leading to the generation of mesoporous inorganic materials: The direct precipitation, the true liquid crystal templating (*TLCT*), the evaporation-induced self-assembly (*EISA*), and the exotemplating route

This is followed by solvent evaporation resulting in an increase in solution concentration of the templating reagent that eventually exceeds the critical micellar concentration (cmc) resulting in its self-assembly into spherical, cylindrical, or lamellae nanostructures. Subsequently, the inorganic material solidifies around the organic self-assembled nanomorphologies resulting in well-ordered mesoporous hybrids. Upon calcination at higher temperatures, the organic structure-directing reagents can be removed resulting in the generation of well-ordered inorganic nanostructures. Finally, in the exotemplating route, a hard mesoporous matrix is used as a template. In this template, the desired inorganic precursors are loaded via infiltration, followed by gelation, and subsequent removal of the exotemplate. The structural characteristics of the final product imitate the structure of the initial mesoporous template in a "complementary" manner.

Different synthetic strategies have been proposed for the development of polymer-based mesoporous hybrids. These include (a) polymerization (chemical, electrochemical) of monomeric species incorporated within the pores of the mesoporous template either from the liquid or from the gaseous phase, (b) infiltration of preformed polymers inside the pores of the template, (c) in situ growth of macromolecules with linear or dendronized architectures and (d) use of structure-directing agents with polymerizable functionalities.

Moreover, synthetic efforts have been also focusing on the generation of polymer-based mesoporous hybrids in which the location of the macromolecules can be precisely controlled, i.e., functional polymeric materials can be selectively placed on specific locations (for example, either on the inner or on the outer surfaces) of the mesoporous framework (De Juan and Ruiz-Hitzky 2000). More details can be found in an excellent recent review by Soler-Illia and Azzaroni, focusing on the development of multifunctional hybrids derived by the combination of ordered mesoporous materials and macromolecular building blocks (Soler-Illia and Azzaroni 2011).

2.2.3 *Emulsion Polymerization*

Emulsion (Asua 2004; Chern 2006; Bourgeat-Lami and Lansalot 2010), dispersion (Oh 2008), microemulsion (Antonietti et al. 1995; Pavel 2004), and miniemulsion (Capek and Chern 2001; Hu et al. 2011; Solans et al. 2005; Schork et al. 2005) polymerization belong to the general class of heterogeneous polymerization methods.

In a conventional emulsion polymerization process, a hydrophobic monomer is dispersed in an aqueous solution containing surfactant molecules at concentrations exceeding the cmc. As a consequence, micelles are generated in solution that serve as nucleation centers for the formation of polymer particles, via the incorporation of polymerization radicals within these micellar aggregates (heterogeneous nucleation mechanism). However, the existence of undesired phenomena, such as homogeneous nucleation translated into the generation of oligomeric species into the aqueous phase followed by precipitation during the traditional emulsion polymerization, is unavoidable resulting in poor control over the morphology, size, and size distribution of the resulting polymeric spheres.

For overcoming such problems, miniemulsion polymerization has been used during the past years for obtaining uniform polymer particles with controllable sizes (Landfester 2009). In the miniemulsion process, an amphiphilic molecule (surfactant) is combined with a costabilizer resulting in the generation of very small, stable, and uniform droplets under high shear conditions, distributed in the continuous phase. Typically, the droplet dimensions range from 10–500 nm. Such nanodroplets may serve as nanoreactors for the polymerization of different monomers (Fig. 2.8).

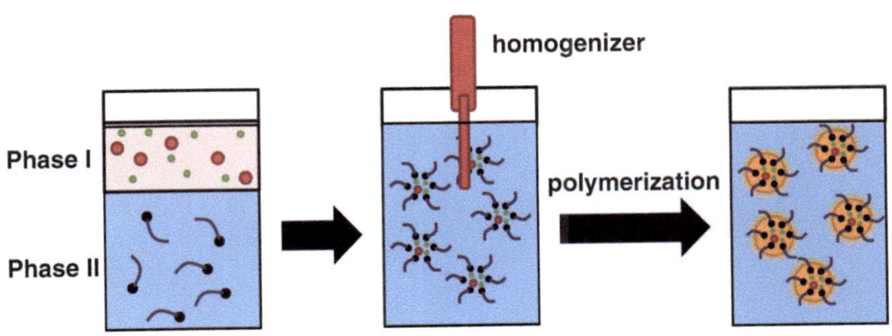

Fig. 2.8 Schematic presentation of the miniemulsion polymerization

In contrast to conventional emulsion polymerization, in miniemulsion, the polymerization is not limited to free-radical processes, but it expands to the generation of polymeric nanoparticles via anionic, cationic, as well as polycondensation reactions.

In particular, the use of miniemulsion processes in the synthesis of polymer–inorganic nanocomposites has attracted considerable attention in recent years, since polymer-based hybrids with controlled dimensions, structure, and morphology (core shell, raspberry like, multicore shell) can be prepared. Very nice reviews exist on the use of miniemulsion polymerization as a promising and highly efficient synthetic route toward the preparation of polymer–inorganic composites including magnetoactive, clay-based, silica- and titania-based polymer nanocomposites (Landfester 2003; Faucheu et al. 2010; Weiss and Landfester 2010).

2.2.4 Metallosupramolecular and Coordination Approaches

The incorporation of metallic centers within the backbone of a synthetic organic polymer can be realized by means of supramolecular approaches. Supramolecular chemistry based on noncovalent interactions, including metal–ligand, ionic, and hydrogen bonding, has been extensively used for the fabrication of hybrid materials consisting of organic and inorganic components, thus expanding their properties and applicability in comparison to their pristine (organic or inorganic) counterparts. Inspired by the pioneer work of I. Manners (Manners 1996), many researchers have been focusing on the synthesis of metallopolymers employing metallosupramolecular polymerization processes. Various metallic centers have been incorporated within such hybrid materials including transition metals, main-group elements, and lanthanides (Whittell et al. 2011).

Linear polymer–metal hybrid chains can be obtained by combining metal ions with small molecules containing two ligating functionalities as schematically shown in Fig. 2.9a (Schmatloch et al. 2003). Other types of coordination processes can also be carried out resulting in polymer–metal ion hybrids including the coupling between homopolymers of different chemical compositions enabling the preparation of block copolymer–metal hybrids (Fig. 2.9b; Gohy et al. 2003), and the use of polymers possessing the metal-ligating moieties as side chains (Fig. 2.9c; Schubert and Eschbaumer 2002; Tew et al. 2006).

Besides the synthesis of linear metallosupramolecular systems, the generation of other more complex architectures is also possible. These include branched (Yu et al. 2013) and network structures (Beck and Rowan 2003) and metallosupramolecular polymers with cyclic architectures (Zhang et al. 2013b).

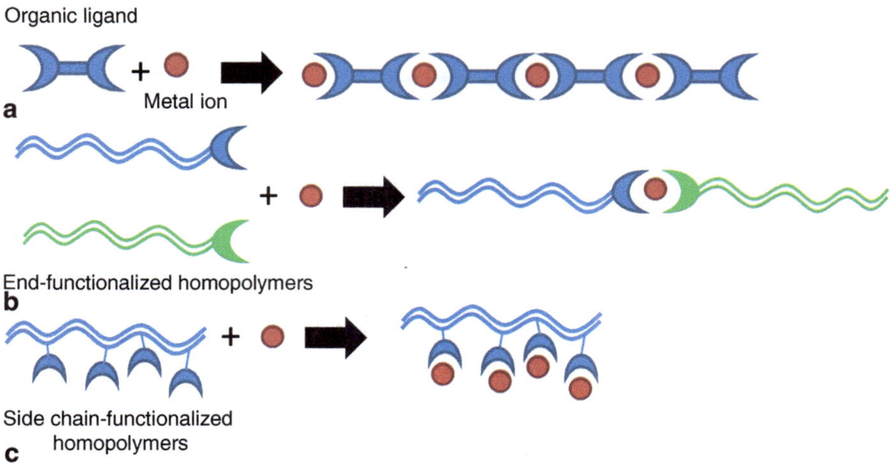

Fig. 2.9 **a** Metallosupramolecular polymers generated via self-assembly of metal ions, and small organic molecules possessing two metal-binding groups, **b** Schematic presentation of linear metallosupramolecular block copolymers comprised of two homopolymer chains of different chemical compositions possessing end-group metal-binding functionalities and metal ion centers, **c** Metallosupramolecular polymers in which side-chain metal–ligand interactions take place

2.2.5 Photopolymerization

The generation of polymeric species upon ultraviolet (UV) irradiation has been realized by introducing suitable photoinitiators in monomers or oligomers (Decker 1998). The interaction of the photoinitiators with UV light results in the generation of radicals or cations promoting polymer formation. If this photocrosslinking polymerization process takes place in the presence of inorganic nanofillers, polymer–inorganic nanocomposites are generated (Yusuf Yagci 2012). Organic–inorganic polymer nanocomposites prepared by photoinitiation polymerization processes are of high interest in coating technologies, and advances in this field enabled the fabrication of polymer-based nanocomposite coatings with improved physicochemical properties.

Common systems employed in such processes include, among others, acrylates (Sangermano et al. 2009; Chibac et al. 2012), methacrylates (Sangermano et al. 2011) epoxy- (Kahraman et al. 2009) and vinyl ether-based systems (Sangermano et al. 2007) combined with metal oxide nanoparticles including silica, titania and alumina (Sangermano et al. 2011), clays (Ceccia et al. 2008), and MNPs (Ijeri et al. 2010).

As an example, direct-monomer photopolymerization was employed for the synthesis of polymer/clay nanocomposites by using intercalated ion-containing photoinitiators capable of being anchored onto the clay surfaces, as schematically presented in Fig. 2.10 (Nese et al. 2006).

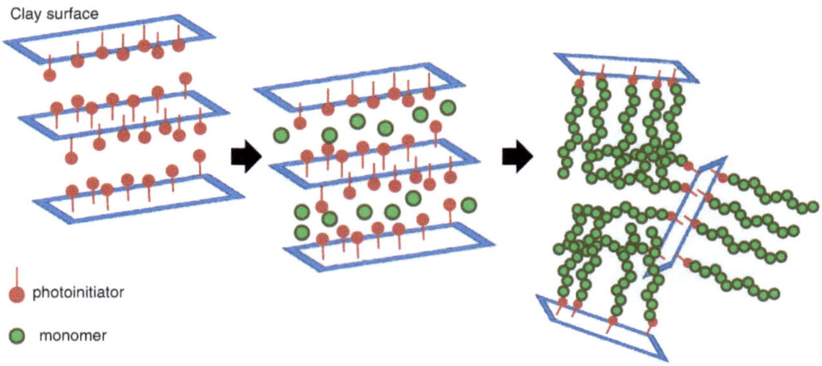

Fig. 2.10 Direct photopolymerization resulting in the generation of exfoliated polymer/clay nanocomposite structures

In another example, ordered organic–inorganic nanocomposites were prepared using lyotropic liquid-crystalline monomers that self-assembled into an inverse hexagonal phase around hydrophilic solutions containing precursors to inorganic (SiO_2) species. In situ photopolymerization was used to "solidify" the ordered matrix structure followed by subsequent chemical reactions to generate the inorganic silica within the cross-linked polymer matrix, yielding highly ordered polymer–SiO_2 nanocomposites. (Gray et al. 1997).

2.2.6 *Intercalation*

In this process, a polymer solution or melt diffuses into the inorganic channels of a layered inorganic structure, such as silicate clay, phosphates, metal oxides, etc., resulting in the exfoliation and nanoscale dispersion of the inorganic layers into the organic polymer matrix and the formation of polymer–inorganic nanocomposites, as schematically presented in Fig. 2.11 (Alexandre and Dubois 2000; Meng et al. 2006). Alternatively, reactive monomers may be polymerized in situ after being diffused inside the inorganic interlayers (as already discussed in the previous subsection) resulting in effective interlayer dissociation/dispersion within the polymer matrix.

2.2.7 *Microwave Irradiation*

The phenomenon of microwave heating is based on the alignment of molecules having permanent dipole moments in respect to an externally applied electromagnetic field, resulting in the generation of heat due to rotation, friction, and collision processes (Hoogenboom and Schubert 2007).

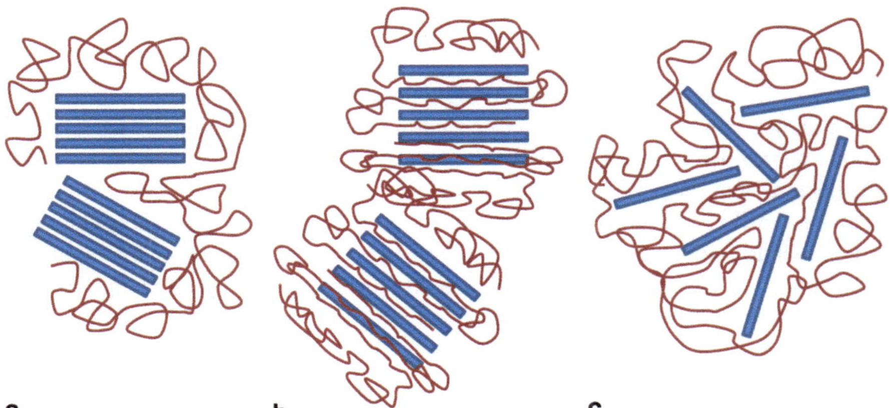

a b c

Fig. 2.11 Inorganic nanolayered-polymer composites obtained by the intercalation method **a** microcomposite structure, **b** intercalated, and **c** exfoliated nanocomposite structures

During the past years, microwave irradiation has been entering the field of polymer chemistry (Wiesbrock et al. 2004) providing the possibility for obtaining polymers in high yields, free of side products, and at fast reaction times.

Microwave irradiation processes have been successfully employed for synthesizing organic–inorganic polymer hybrids. In a recent review, focusing on the synthesis of polymer nanocomposites prepared under microwave irradiation, the authors present nicely the recent developments in this rapidly growing research field providing examples on the fabrication of polymer clays, polymer–MNPs, and polymer–nanowire nanocomposites under microwave conditions (Bogdal et al. 2011).

2.2.8 *Electrochemical Synthesis*

A simple synthetic approach that is used for producing polymer-based nanocomposites in the form of films involves the electrochemical polymerization of electroactive monomers, such as thiophenes, aniline, pyrrole, etc., in the presence of inorganic nanoparticles. The resulting polymer-based nanocomposite is directly deposited onto the electrode surface. As an example, Tombácz et al. (Janaky et al. 2009) have prepared polythiophene–magnetite composite layers via electrodeposition of 3-thiophene-acetic-acid in the presence of Fe_3O_4 nanoparticles to prepare a conducting, magnetoresponsive polymer-based electrode with potential use in magnetic electrocatalysis. In another recent example, polyaniline (PANI) has been electrodeposited into nanoporous WO_3 films (Janaky et al. 2012). Among others, this hybrid polymer–oxide assembly exhibited superior polymer-related cathodic photocurrents in comparison to the pristine polymer film, rendering this particular system highly attractive in photoelectrocatalytic reduction processes.

2.2.9 Surface Grafting

As previously discussed, a key requirement for developing polymer-based hybrids with enhanced physical and chemical properties is the promotion of specific interactions between the polymeric and the inorganic species. For this purpose, three grafting strategies have been developed for tailoring the surface properties of the inorganic nanofillers, aiming to improve the interfacial interactions, and consequently their dispersion within the organic polymer matrix: The "grafting to," the "grafting from," and the "grafting through" methods (Beija et al. 2011). The topic of grafting is also covered in Chapter 3 and grafting techniques in Section 3.2.

2.2.9.1 Surface-Initiated Polymerization (the "grafting from" Approach)

The "grafting from" approach involves the polymerization of monomeric species onto an inorganic surface that has been previously modified to bear active moieties (e.g., initiator molecules, polymerization mediators). In the case where the surface-initiated polymerization is carried out in the presence of polymerization mediators involved in controlled radical polymerization processes, such as nitroxide-mediated polymerization (NMP) (Ghannam et al.), reversible addition-fragmentation chain transfer (RAFT) (Beija et al. 2011), and surface-initiated atom transfer radical (SI-ATRP) polymerization (Hui et al. 2014), that are either anchored or are found free in solution (vice versa to the location of the polymerization initiator), the thickness of the polymer shell covering the inorganic surface can be precisely controlled.

By the "grafting from" method, high-graft densities can be easily achieved. A drawback of this method is the fact that besides the formation of polymer chains onto the inorganic surface, non-anchored polymeric species are also generated in solution during the polymerization process. Consequently, after the completion of the process, purification is required for the removal of free polymer chains and other nonreacting species.

The "grafting from" method has been widely used to synthesize polymer/silica hybrids starting from surface-functionalized SiO_2 nanoparticles or silicon surfaces (Radhakrishnan 2006), whereas fewer reports exist on the synthesis of polymer/MNP hybrids by following this synthetic approach.

2.2.9.2 Post-modification (the "grafting to" Approach)

This technique is based on either physical or chemical anchoring of organic modifiers such as surfactants or polymers onto inorganic surfaces.

Physical grafting (also known as *post-modification*) involves the physical adsorption of macromolecular species onto the inorganic surfaces upon mixing the two components together, whereas in chemical anchoring the formation of a covalent bond takes place between the polymer and the inorganic surface.

Polymers containing end-group functionalities, such as thiols, carboxylic acids, disulfides, phosphine oxides, dithioesters, as well as polymers possessing anchoring moieties along their backbone capable of forming chemical bonds with inorganic species have been used to prepare organic–inorganic polymer hybrids with various inorganic nanoparticles including Au (Du et al. 2008; Gambinossi et al. 2014), Fe_3O_4 (Basuki et al. 2013), TiO_2 (Tao et al. 2011; 2013), semiconductor nanocrystals (Tomczak et al. 2013), etc.

Furthermore, "click chemistry" has also been used in the immobilization of polymer chains onto the surfaces of inorganic nanoparticles. Click chemistry is a general term used to describe a chemical reaction process in which small functional groups are "clicked" together in a rapid and quantitative manner. A well-known example of a click reaction process is the Azide–Alkyne Huisgen Cycloaddition taking place between an azide ($-N_3$) and an alkyne ($\equiv CH$) to form a triazole ring. Polymer–inorganic hybrids have been generated by introducing azide moieties on the surfaces of preformed inorganic nanoparticles, and perform the "click reaction" with polymers bearing alkyne moieties (Chen et al. 2011).

The advantage of following the "grafting to" approach is that the morphological and structural characteristics of both the polymers and the inorganic species can be well-controlled (depending, of course, on the preparation method employed for their production) prior to their intermixing, allowing for the generation of polymer hybrids exhibiting well-defined structural and compositional characteristics.

2.2.9.3 The "grafting through" Approach

This method involves the anchoring of polymerizable groups onto the inorganic materials followed by polymerization that results in the formation of polymer chains directly onto the surface of the inorganic species. Examples include the generation of polystyrene/CdSe hybrids by conventional free-radical polymerization of styrene starting from methacryloxypropyltrimethoxysilane (MPS)-modified CdSe nanoparticles bearing a double bond (Etxeberria et al. 2013) and the synthesis of star-shaped polyhedral oligomeric silsesquioxanes–poly(methyl methacrylate; POSS/PMMA) hybrids via RAFT-controlled radical polymerization by incorporating RAFT mediators (chain transfer agents, CTA) onto the surface of the inorganic core by click chemistry (Ye et al. 2011).

Besides the aforementioned grafting approaches, the "in situ" preparation method has been extensively used for synthesizing nanohybrid materials consisting of polymers and inorganic metal or metal oxide nanoparticles. In this one-pot method, the inorganic nanoparticles are generated in situ starting from a metal ion salt precursor, in the presence of the polymer. Limitations accompanying this synthetic approach, including low graft density, the relatively broad nanoparticle size distribution, and its non-applicability in polymer systems, exhibiting solubility or stability issues in the presence of reducing agents introduced during the reaction, have been overcome with the surface-grafting techniques.

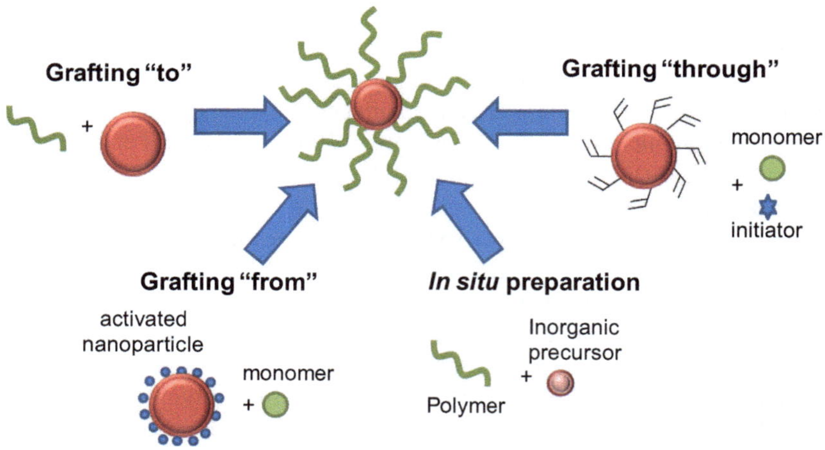

Fig. 2.12 Surface-grafting approaches used for the synthesis of polymer–inorganic hybrid materials

Figure 2.12 summarizes the grafting and the in situ preparation methods for synthesizing polymer–inorganic hybrids.

2.2.10 Self-Assembly

The self-assembly process used for generating polymer–inorganic hybrid materials involves a spontaneous organization of the polymeric species and the inorganic components at the nanoscale (Cong and Yu 2009). This self-assembly process is mainly governed by noncovalent interactions including electrostatic attractive forces, hydrophobic interactions hydrogen and coordination bonding. The most widely used self-assembly strategy involves the preparation of polymer–inorganic nanocomposites based on electrostatic attractive forces developed between oppositely charged organic and inorganic species.

The introduction of the layer-by-layer technology back in 1991 (Decher 1977), created new perspectives for developing polymer-based nanocomposites via the alternative deposition of the polymer and the inorganic compounds possessing opposite electrical charges on selected surfaces (Srivastava and Kotov 2008; Du et al. 2012a) as schematically presented in Fig. 2.13.

In the pioneer work of Caruso et al. (1998a; 1998b), the authors describe the synthesis of polymer–silica nanocomposites, starting from colloidal polystyrene nanospheres, and generating a highly regular multilayer coating around those spheres via sequential layer-by-layer deposition of a positively charged polyelectrolyte and of negatively charged SiO_2 nanoparticles.

As mentioned above, besides the self-assembly processes based on electrostatic interactions, other self-assembly strategies have been reported based on hydrogen

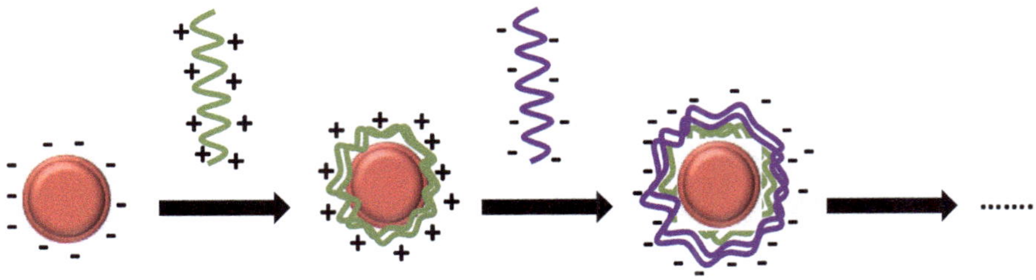

Fig. 2.13 The layer-by-layer self-assembly approach used for the synthesis of polymer-based nanocomposites. This approach is based on the development of strong electrostatic attractive forces between oppositely charged organic (polymer) and inorganic components

bonding interactions. For example, recently, Liu et al. (2012) described the synthesis of polymer–TiO_2 raspberry-like core–corona composites consisting of a poly(4-vinylpyridine; PVPy) core and a poly(ethylene glycol dimethacrylate-co-methacrylic acid)/TiO_2 corona. In this system, the hydrogen bonding takes place between the hydroxyl moieties of titania and the pyridyl groups of PVPy.

2.2.11 Amphiphilic Block Copolymer-Mediated Synthetic Approaches

The ABCs consist of at least two constitutional and/or configurational building blocks that are covalently linked together. Depending on the number of block segments incorporated within their structures, ABCs are classified into diblock, triblock, or multiblock copolymers, whereas by employing different synthetic pathways ABCs of various architectures, including linear, branched, graft, and star copolymers, can be prepared (Hadjichristidis et al. 2003). Due to incompatibility issues existing between the different block segments, ABCs have the tendency for microphase separation in solution or at the solid state. This phenomenon results in the generation of well-defined nanomorphologies in solution (such as micelles of different shapes and vesicles) as well as at the solid-state (ordered continuous morphologies including lamellae, body-centered cubic (BCC) packed spheres, hexagonally ordered cylinders (HEX) or bicontinuous structures) as schematically presented in Fig. 2.14 (Förster and Antonietti 1998).

The ABCs have been extensively employed for the synthesis of nanostructured polymer–inorganic hybrids in which either metal or ceramic nanoparticles have been embedded within nanostructured polymer domains. The use of well-defined ABCs (i.e., characterized by controlled molecular weights and narrow molecular weight distributions) synthesized by means of living/controlled polymerization methods as nanotemplates for the development of such hybrid nanomaterials is very important, since, depending on the nanomorphology adopted by the ABC chains, the morphology, size, and consequently properties of the embedded inorganic nanoparticulates, including optical, electrical, magnetic, catalytic, and others, can be precisely controlled.

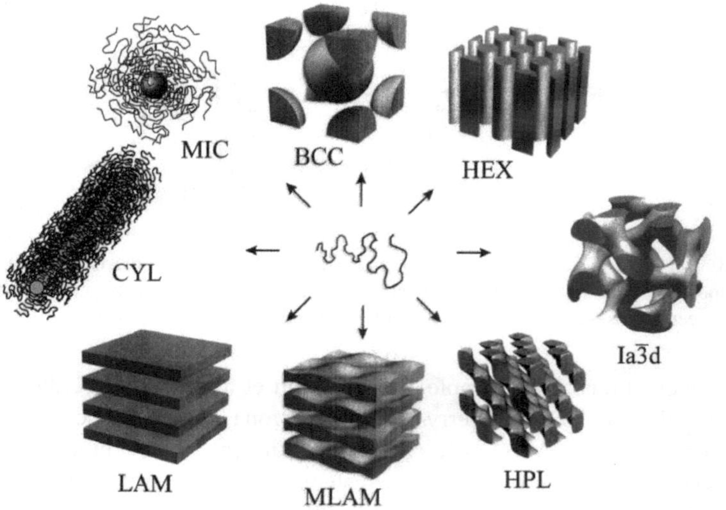

Fig. 2.14 Different nanomorphologies generated by ABC consisting of two different block segments in solution and at the solid state. (Reprinted (adapted) with permission from (Förster and Antonietti 1998) Copyright (1998) WILEY-VCH Verlag GmbH)

Since polymers are characterized by low surface energies rendering them incompatible with inorganic surfaces, the development of highly stable polymer/inorganic hybrid materials requires the presence of functional sites within the block copolymer structure that would enable the development of specific interactions between the polymeric and the nonpolymeric part. Such functionalities include, among others, carboxylic acid groups, bi- and terpyridine moieties, diols, thiols, amines, etc. (Förster and Antonietti 1998). In this long list of ABC-bearing metal-binding functionalities, the *β*-ketoester moiety—known to be a strong bidentate ligand for various metal ions of different geometries and oxidation states—has been introduced for the first time by our team back in 2001 when we have reported the synthesis of well-defined block copolymers based on 2-(acetoacetoxy)ethyl methacrylate (AEMA) (Schlaad et al. 2001).

Figure 2.15 provides the chemical structures of various organic groups incorporated within ABC aiming toward the development of highly stable nanostructured polymer–inorganic hybrids.

A number of synthetic routes have been developed throughout the years for generating block copolymer–inorganic nanohybrids. Typically, complexes of transition metal ions (behaving as soft acids) with hard bases are introduced as metal ion precursors. The use of a block-copolymer-possessing functionalities acting as soft bases results in complex formation between those moieties and the metal ions since, based on the hard soft acids bases (HSAB) principle, soft acids tend to form strong bonds with soft bases, whereas they bind weakly to hard bases, i.e., the hard–soft combination is not stable (Ho 1975). The metal ion precursor loading is then followed by a chemical transformation reaction that results in the generation of either

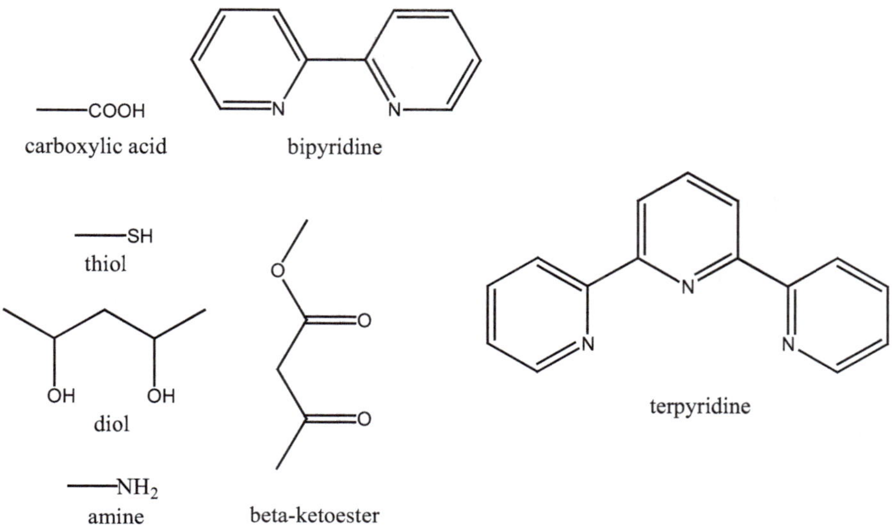

Fig. 2.15 Chemical structures of organic moieties capable of binding onto inorganic surfaces incorporated within ABCs for the development of polymer–inorganic hybrids.

metallic or semiconducting (oxides, sulphides, selenides) nanoparticles, also known as QD. This step is followed by nucleation and growth mechanisms that govern the size, shape, and number of nanoparticles per microdomain. Consequently, inorganic nanoparticles of different sizes and shapes may be obtained, whereas it is possible to generate polymer nanodomains in which either single nanoparticles (Fig. 2.16a) or multiple small nanoparticles may be incorporated, with the latter case resembling the "raspberry" morphology (Fig. 2.16b).

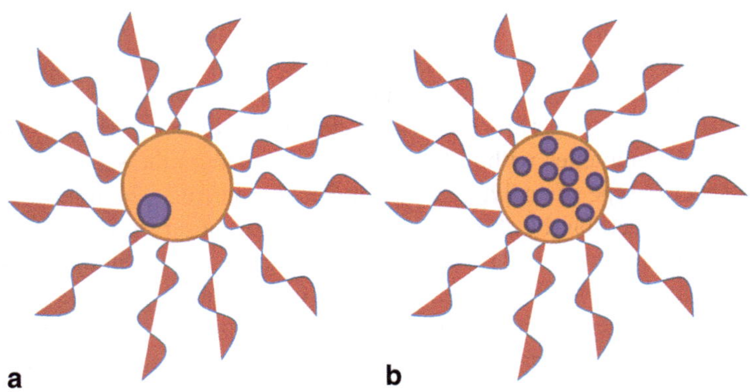

Fig. 2.16 Schematic representation of a block copolymer micellar nanodomain in which **a** a single metallic nanoparticle, and **b** many metallic nanoparticles ("raspberry" morphology) have been generated within its core

More information on the use of ABCs as nanotemplates for the development of structurally controlled nanomaterial hybrids may be found in an excellent review written by M. Antonietti and S. Förster (1998).

2.3 Applications of Organic–Inorganic Polymer Hybrids

Due to their unique and superior properties compared to their individual counterparts, organic–inorganic polymer hybrids have attracted considerable attention in many fields (Sanchez et al. 2011). The broadness of the applicability spectrum restricts the discussion on their exploitation in every area. In this section, a general overview of the most important applications of polymer-based organic–inorganic hybrids, in areas including biomedicine, energy, catalysis, sensing, optoelectronics, and hybrids destined for use in environmental, construction, automotive, and coating applications, is provided.

2.3.1 Sensing

During the past decades, there is has been an exponential increase in the number of publications in the field of sensing. Sensors can be classified into chemical, physical, and biosensors. According to the definitions proposed by the International Union of Pure and Applied Chemistry (IUPAC) "a chemical sensor is a device that transforms chemical information, ranging from the concentration of a specific sample component to total composition analysis, into an analytically useful signal," whereas "a physical sensor is a device that provides information about a physical property of the system". In the case of a biosensor, biological entities, such as enzymes, antibodies, proteins and other biomolecules, are employed as a recognition element (Wang and Uchiyama 2013).

A schematic illustration of a sensor is provided in Fig. 2.17. As seen in the figure, a sensor is composed of an active layer (that can be either organic, inorganic, or hybrid) and a transducer. The entrapment of a targeted molecule into the active layer results in an altering of its physical or chemical properties. The incorporation of target-specific moieties within the active layer enhances the sensing selectivity toward a specific type of analyte, i.e., the targeted molecule. The transducer is responsible for converting the above-mentioned changes observed in the physical or chemical properties into a measurable signal that can be electrical, optical, electrochemical, etc., and proportional to the analyte's concentration.

In a very nice recent review devoted to hybrid sensors, the authors describe the preparation of organic–inorganic hybrids that have been successfully used in gas, humidity, UV, strain, electrochemical, and fluorescent sensing (Wang et al. 2013). Polymer–inorganic hybrid sensors have been fabricated in the form of intercalated structures, films, pellets, mesoporous structures, and core–shell morphologies. The

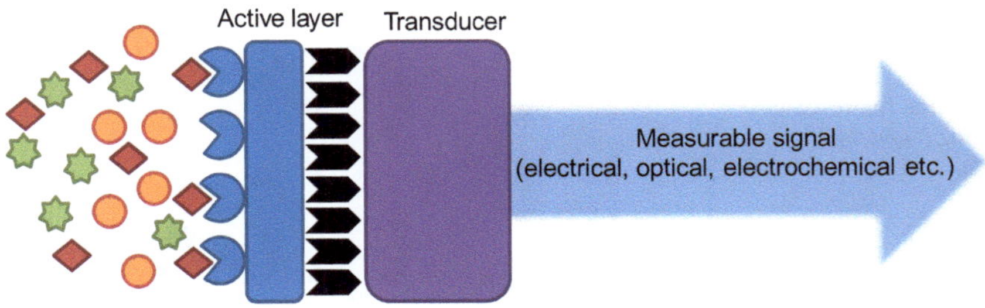

Fig. 2.17 Schematic illustration of a sensor consisting of the active layer serving as the molecular-recognition layer and the transducer, responsible for converting the observed changes into a measurable signal

combination of inorganic elements with polymers leads to novel sensing materials characterized by good chemical, thermal, and mechanical stability as well as enhanced sensing modalities especially in the case where material selection results in synergistic sensing phenomena between the organic and the inorganic elements. Other important characteristics rendering a material a "high-quality" sensor are the linear and high response, fast response/recovery rate time, selectivity, reproducibility, and low cost.

Polymer–inorganic hybrid materials have been developed and evaluated as humidity (Tsigara et al. 2007; Patil et al. 2010) and gas sensors for detecting harmful gases, including ammonia (Zhang et al. 2009; 2013a), H_2S (Shirsat et al. 2009), NO_2 (Xu et al. 2010; Huang et al. 2011a; Zampetti et al. 2013) and alcohols (Wei et al. 2010; Huang et al. 2011b). In these materials, conducting or electroactive polymers, such as PANI, polythiophenes, polypyrrole (PPy), and polyvinylcarbazole (Hatchett and Josowicz 2008; Rajesh et al. 2009; Zanardi et al. 2013), have been combined with metal oxides including TiO_2, SnO_2, In_2O_3, WO_3, ZnO, Fe_xO_y and MoO_3. The use of conducting polymers enables a low-temperature response in contrast to the all-inorganic sensors operating at high temperatures (Rajesh et al. 2009).

Polymer-based hybrids have also been used in the development of optical sensors. Fluorescent hybrid sensors belonging to the general class of optical sensors have been successfully used in the chemosensing of metal ions (Fan and Jones 2006; Niamsa et al. 2013), anions (Zhou et al. 2014), and gases (Ong and Levitsky 2011; Zaharieva et al. 2011). Fluorescence sensing is based on the altering of the fluorescence intensity of organic fluorophores embedded within a polymer matrix via doping or by covalent bonding. When the analyte is found in close proximity to those moieties, either enhancement or suppression of the fluorescent intensity, i.e., a "turn on" and "turn off" response, is observed as schematically presented in Fig. 2.18.

To conclude, literature examples appearing during the past few years emphasize the significance of polymer–inorganic hybrids in the development of novel and highly efficient sensors. The existing possibility of developing multicomponent polymer-based hybrids creates new pathways for the generation of multifunctional materials with enhanced sensor performances.

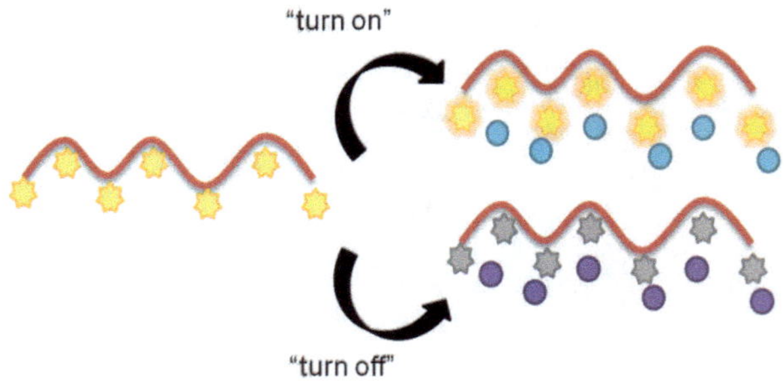

Fig. 2.18 Fluorescent-based sensing: Fluorescence enhancement ("*turn on*" response) and fluorescence quenching ("*turn off*" response)

2.3.2 *Biomedicine*

Polymer-based biomedicine is a broad and highly emerging field with polymers being the key players in the development of new materials destined for use in biosensing, biocatalysis, bioseparation, therapy, diagnostics, as well as in combined therapeutic/diagnostic modalities known as "theranostics." This subsection focuses on the application of organic–inorganic polymer hybrids in biomedicine and biotechnology. Selected literature examples and reviews existing on such topics are referenced aiming to provide a general overview of the applicability of these materials in the biomedical arena.

In the past years, industrial and academic research has been focusing on the design and development of novel and highly efficient materials to be used in the biomedical field as therapeutic, diagnostic, and theranostic (where therapeutic and diagnostic elements are combined within the same system) agents. Organic–inorganic polymer nanocomposites have played a leading role in the abovementioned biomedical areas (Shen and Shi 2010; Wu et al. 2010; Maya et al. 2013; Joglekar and Trewyn 2013; Miyazaki et al. 2013; Sridhar et al. 2013; Liu and An 2014; Gaharwar et al. 2014; Savva and Krasia-Christoforou 2014). An example of the use of organic-inorganic polymer nanocomposites is for drug delivery systems capable of delivering the therapeutic agents at the right time, dosage, and site of interest, thus increasing the therapeutic efficacy and avoiding undesirable side effects (Liong et al. 2008; Forrest and Kwon 2008; Viseras et al. 2008; Soundrapandian et al. 2009; McInnes and Voelcker 2009; Viseras et al. 2010; Bonanno and Segal 2011; de Sousa et al. 2013). Other examples include materials for use in tissue engineering (Sahoo et al. 2013; Gloria et al. 2010; Trachtenberg et al. 2013; Nelson et al. 2013; Mourino et al. 2013; Wahl and Czernuszka 2006), dentistry applications (Choi et al. 2013), in photohyperthermia treatment (Fang and Chen 2013), biocatalysis, bioseparation, and biosensing, including optical, electrochemical, and magnetobiosensing (Gutierrez et al. 2009; Prakash et al. 2013), and diagnostics (Swierczewska et al. 2011). These materials may also combine both diagnostic and therapeutic elements (Krasia-Christoforou and Georgiou 2013).

Silicon-based polymer hybrids have attracted considerable attention for the development of effective *drug-delivery systems*. The combination of polymers with silicon-based nanoparticles exhibiting bioinertness, nontoxicity, and high surface areas, with the latter characteristic enabling high drug-loading capacities, results in materials with tunable and controlled drug-release properties. Moreover, the existing flexibility on the materials' design as far as the polymer chemical structure is concerned, allows for the incorporation of stimuli-responsive moieties within these hybrid drug-delivery systems that may be further exploited in environmentally triggered drug delivery processes (McInnes and Voelcker 2009).

Polymer–inorganic nanohybrids have been evaluated as *diagnostic* (*imaging*) tools in magnetic resonance imaging (MRI) and fluorescence imaging. Magnetic nanoparticles, including Fe, Co, Ni, and corresponding oxides combined with functional polymers, enable the development of nanohybrids employed as MRI contrast enhancement agents (Pyun 2007; Chanana et al. 2009; Zhou et al. 2011; Guo et al. 2013a; Thevenot et al. 2013). The use of an appropriate polymer coating capable of providing stabilization of the inorganic nanoparticles in aqueous media and at the same time exhibiting biocompatibility and/or biodegradability is a key requirement for the applicability of such nanohybrids in the biomedical field. In comparison to low-molecular-weight stabilizing agents, such as surfactants, the use of polymers is highly advantageous since polymers exhibit prolonged blood circulation times, and their structural and compositional characteristics may be tailored allowing for the development of polymer-based multifunctional, next-generation magnetic nanoparticles (Du et al. 2011).

In most examples appearing in the literature, either polymer-stabilized gadolinium(III)-based nanoparticles or superparamagnetic iron oxide nanoparticles, the so co-called SPIONs, have been used as contrast enhancement agents in MRI. The former are employed as T_1 MRI contrast agents, and they are capable of generating a positive image contrast (Aime et al. 2006), whereas the latter are used as T_2 MRI contrast enhancement agents (Wang et al. 2001).

Fluorescence spectroscopy is considered to be one of the most effective imaging modalities in biomedicine (Caldorera-Moore et al. 2011). For in vivo imaging, the use of near-infrared (NIR) light (in the range between 700 and 900 nm) is preferred since it promotes deep-tissue penetration and reduces autofluorescence and tissue-scattering phenomena. Therefore, efforts have been focusing on the synthesis of NIR fluorescent contrast agents (Sevick-Muraca et al. 2002; Luo et al. 2011a; Nolting et al. 2011).

Nanohybrids consisting of various polymers, including poly(vinylpyrrolidone; PVP), poly(ethylene glycol; PEG), chitosan, poly(vinyl alcohol; PVA), etc., combined with semiconducting QD, including CdTe, ZnO, CdSe, or metal and bimetallic nanoparticles (for example, Au and Au/Ag, respectively), with the latter having a surface plasmon resonance (SPR) in the NIR region, exhibit unique optical properties that enable the prolonged fluorescence monitoring in living organisms.

Moreover, the possibility of combining different inorganic imaging agents within the same polymer-based system allows for multifunctional imaging using both fluorescent and MRI modalities (Key and Leary 2014; Yan et al. 2014b).

When therapeutic and diagnostic elements are combined together within the same system, a "theranostic" agent is generated capable of simultaneously acting as a diagnostic and a therapeutic tool. Figure 2.19 provides a schematic presentation of

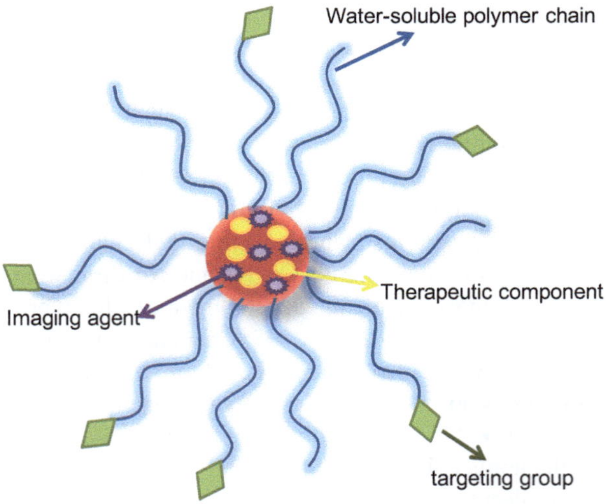

Fig. 2.19 Schematic presentation of a polymer-based theranostic system

a polymer-based theranostic system consisting of (a) a water-soluble and biocompatible polymer chain, (b) an imaging component, and (c) a therapeutic component that can be either a drug or a gene. Moreover, the existing possibility of introducing functional-targeting moieties on the outer surface of the theranostic nanocarrier enables the selective transfer of the theranostic cargo to the area of interest.

Various polymer-based theranostic systems have been developed including magnetoresponsive, fluorescent, as well as dual-imaging drug and gene delivery systems. An overview of this highly interesting and emerging biomedical arena is provided in our recently published review article on polymeric theranostics (Krasia-Christoforou and Georgiou 2013).

2.3.3 Energy

During the past years, many scientific efforts have been targeting toward the generation of new and efficient materials for energy-related applications, including energy conversion and storage. Polymer–inorganic hybrid materials have been used in the development of flexible hybrid solar cells, in fuel cell, and supercapacitor technologies as well as in thermoelectric and hydrogen storage devices (Gomez-Romero et al. 2010).

2.3.3.1 Hybrid Solar Cells

In solar cell technologies, researchers have been working on the discovery of alternative organic-based materials exhibiting high power-conversion efficiencies that could solve the problems of energy costs accompanying, nowadays, the silicon

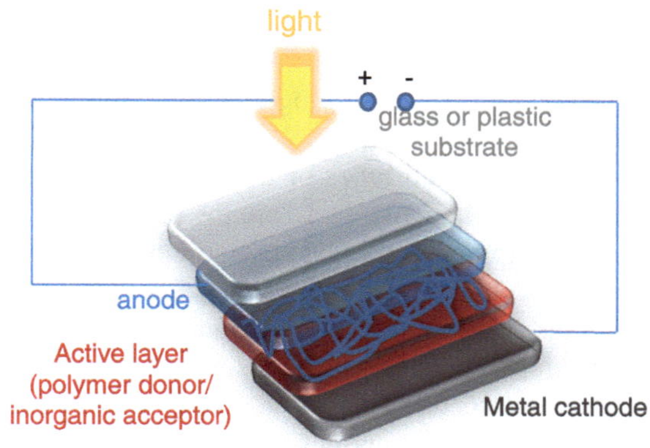

Fig. 2.20 Schematic representation of a polymer–inorganic solar cell device

photovoltaic industry. Those are mainly attributed to the expensive methodologies employed for the production of high-quality silicon and for manufacturing efficient silicon-based solar cells (Wright and Uddin 2012).

On the contrary, the facile solution processing of polymer-based photovoltaic devices allows for their cost-effective production, using the roll-to-roll or ink-jet solution processing techniques (Krebs 2009). Furthermore, conjugated organic polymers employed as the organic electron-donating component in combination with inorganic electron-accepting nanofillers, are characterized by high absorption coefficients, thus allowing the fabrication of flexible thin-film solar cell devices (Ong and Levitsky 2010; Arici et al. 2003). As a consequence, less material will be necessary to reach the desired solar cell efficiencies, whereas, at the same time, the lightweight properties and flexibility of such thin-film photovoltaic devices will lead to the dramatic expansion of the number of organic-based photovoltaic products.

In polymer–inorganic hybrid solar cells, the photoactive layer consisting of a conjugated polymer (electron donor) and inorganic nanocomponents (electron acceptors) is sandwiched between two electrodes of different work functions as shown schematically in Fig. 2.20.

As seen in the figure, the device consists of an anode usually consisting of indium tin oxide (ITO) that lies on glass or on a plastic, flexible substrate (poly(ethylene terephthalate; PET) is usually employed). ITO enables the transmittance of light through the device, and is responsible for the collection of holes. The polymer–inorganic photoactive layer that is placed between the two electrodes is deposited from solution containing both the polymer and the inorganic nanoparticulates via spin coating, to generate a thin nanocomposite layer. The device is completed by depositing a metal on top of the photoactive layer (usually aluminum) that acts as the cathode, and its role is to collect the generated electrons.

The incorporation of inorganic nanocomponents within the photoactive layer provides significant advantages over the use of the typical electron-acceptor layer employed, that combines organic conjugated polymers with (6,6) phenyl C61- butyric acid methyl ester (PCBM) or other organic fullerene derivatives. The use of inorganic acceptors may lead to higher light absorption efficiencies in comparison to that of PCBM (Saunders and Turner 2008; Wright and Uddin 2012; Saunders 2012). In addition, the possibility of altering the electronic properties of the inorganic acceptors by modifying their dimensions and morphology allows for the fine-tuning of their absorption behavior. Moreover, ultrafast charge carrier transfer rates observed in certain types of inorganic nanomaterials, such as semiconductor QD (Knowles et al. 2013), facilitate the effective charge transfer and separation at the donor–acceptor interface suppressing the electron–hole recombination phenomena that lead to low device efficiencies. The development of a wide range of synthetic protocols for the fabrication of inorganic nanomaterials allows in some cases the architectural control at the nanoscale, thus enabling the development of structurally defined architectures such as vertically aligned nanostructures that facilitate exciton (electron-hole pair) dissociation and electron and hole transporting to the "right" electrodes, i.e., electron transporting to the anode through the inorganic acceptor, and hole transporting to the cathode through the polymer donor (Gonzalez-Valls and Lira-Cantu 2009; Jiang et al. 2010; Woo et al. 2012).

Several types or organic semiconducting polymers have been employed as the electron-donating organic matrices in organic solar cells (Guenes and Sariciftci 2008). Figure 2.21 displays the chemical structures of some typical polymers used in such applications, with poly(3-hexylthiophene) (P3HT) being the most preferable one.

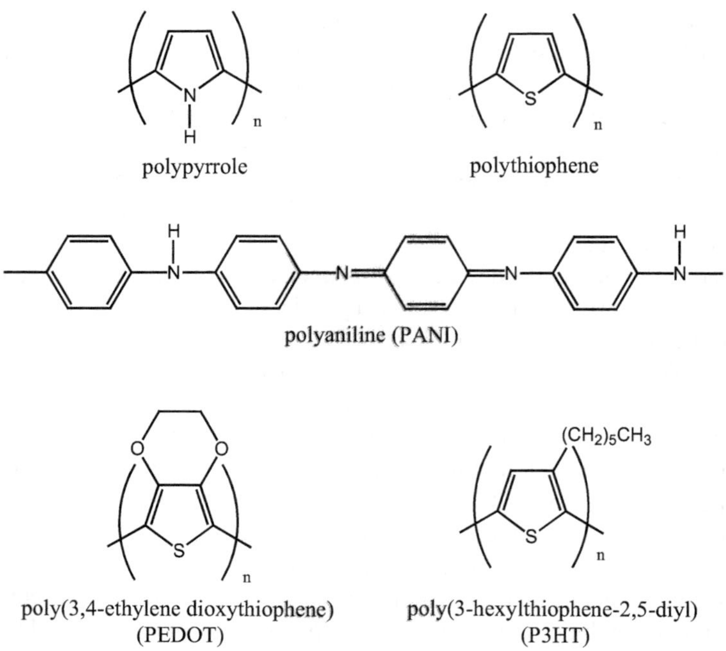

Fig. 2.21 Chemical structures of polymeric materials employed as the electron-donating component in polymer–inorganic hybrid solar cells

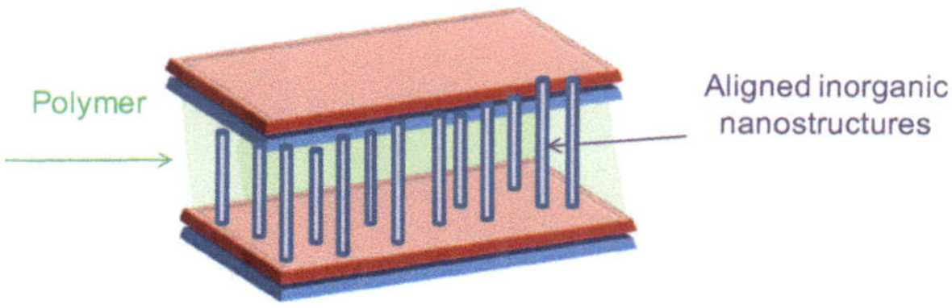

Fig. 2.22 Schematic diagram of a polymer–inorganic hybrid solar cell in which the inorganic electron acceptor nanocomponents are vertically aligned

Various semiconductor inorganic components have been used as the electron acceptor elements in polymer-based hybrid solar cells including CdS and CdSe, TiO_2, ZnO, PbS, etc., in the form of nanoparticles, nanorods, nanowires, and nanotubes (Wright and Uddin 2012; Zhang and Hou 2012; Zhao and Lin 2012; Saunders 2012; Gao et al. 2013; Chen et al. 2013). Anisotropic nanostructures such as nanowires, nanotubes, and nanorods exhibit superior electrical transport properties compared to the corresponding spherical nanostructures, especially when those are found in a well-aligned architecture as seen in Fig. 2.22 (Thien-Phap 2011).

Based on the above, it is expected that polymer–inorganic hybrid solar cells will present higher efficiencies and superior performance compared to the all-organic (polymer–fullerene derivatives) solar cells reaching so far a 10 % power conversion efficiency. However, limiting factors including the attachment of organic insulating ligands onto the nanoparticle surfaces during synthesis for achieving control over the nanoparticle size and shape and for eliminating nanoparticle agglomeration phenomena, as well as nanomorphology control issues must still be overcome (Wright and Uddin 2012).

2.3.3.2 Hybrid Fuel Cells

In addition to photovoltaics, fuel cells also belong to the class of energy conversion devices. In fuel cells, the chemical energy is converted into electrical energy through an electrochemical oxidation process. Typically, a fuel cell consists of an anode, a cathode, and a polymer-based electrolyte membrane that separates the anode from the cathode, and the electrolyte membrane is responsible for the transfer of protons (H^+) from one electrode to the other. This membrane, known as the proton exchange membrane, must be characterized by low cost, high proton conductivity, electrical insulating properties, as well as durable chemical, thermal, and mechanical stability under the operating conditions (Mishra et al. 2012). During this electrochemical process, the produced electrons are transferred from the anode to the cathode through an external circuit resulting in the generation of direct current (DC) . In Fig. 2.23, a schematic presentation of a H_2 fuel cell and the electrochemical reactions taking place in the anode and the cathode are provided.

Organic–inorganic polymer-based hybrids began to play a key role in the development of novel, high performance proton exchange electrolyte membranes used in fuel cell technologies (Peighambardoust et al. 2010; Nagarale et al. 2010; Tripathi

$$Anode: 2H_2 \rightarrow 4H^+ + 4e^-$$

$$Cathode: O_2 + 4H^+ + 4e^- \rightarrow 2H_2O$$

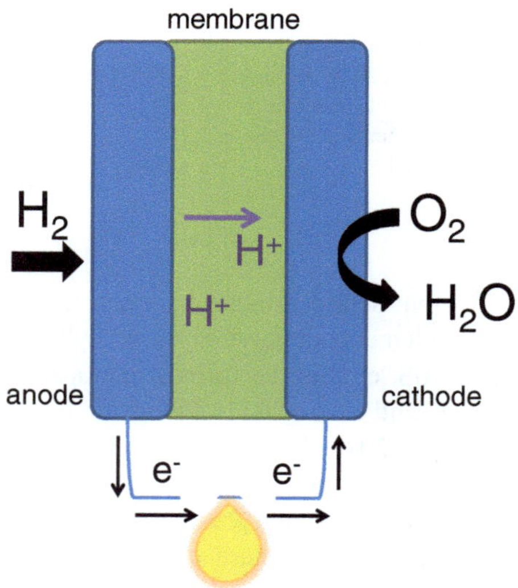

Fig. 2.23 Schematic presentation of a hydrogen fuel cell

and Shahi 2011; Bose et al. 2011; Mishra et al. 2012; Xu and Mu 2014). Nafion®, first introduced by Du Pont during the 1970s, is the most widely used material employed as a proton-exchange electrolyte membrane in fuel cells. This material fulfills many of the abovementioned requirements such as excellent electrochemical properties at low temperatures and under high humidity conditions, high ionic conductivity and good chemical stability, and mechanical properties. However, under reduced humidity conditions and high temperatures, this material suffers from low-proton conductivity. For overcoming this limitation, inorganic nanofillers including silicates, TiO_2, and ZrO_2 have been incorporated within Nafion membranes (Mishra et al. 2012; Cele and Ray 2009) as well as in other polymers such as styrene-, benzimidazole-, chitosan-, polyvinyl alcohol- and polyimide-based ones (Tripathi and Shahi 2011) aiming toward the development of new and highly effective hybrids for proton exchange membrane fuel cell (PEMFC) applications. In organic–inorganic nanocomposite polymer electrolyte membranes, the presence of the polymeric component provides high flexibility, whereas the inorganic counterparts lead to an enhancement in the operating temperature window and in the membrane mechanical performance. Moreover, the possibility of introducing acidic functionalities within these materials results in an improvement in the conductivity and in ion exchange capability. An excellent review on organic–inorganic nanocomposite polymer electrolyte membranes (PEMs) designed for use in fuel cell applications exists covering all related aspects including material preparation methods, characterization, properties, and types of organic–inorganic PEMs (Tripathi and Shahi 2011).

2.3.3.3 Hybrid Nanocomposites: Thermoelectrics

Polymer-based thermoelectric nanocomposites are very promising in the cost-effective and large-scale production of high-performance thermoelectric materials. In such systems, the advantages of a polymer matrix, including low cost, easy processing, and low thermal conductivity, are combined with the superior thermoelectric performance of specific inorganic components (Bubnova and Crispin 2012). To date, conducting polymer matrices including PANI, polycarbazole (PC), and poly(3,4 ethylenedioxythiophene):poly(styrenesulfonate; PEDOT:PSS) have been combined with inorganic thermoelectric nanosystems within composite nanostructures, aiming to enhance the thermoelectric performance of the pristine polymers. The main factor which limits the use of the abovementioned polymers and corresponding nanocomposites in real-life applications is the low-thermoelectric efficiency, attributed to their low *ZT* values ranging between 10^{-4} and 10^{-2} within a temperature range of 150–400 K in most cases (Du et al. 2012b). Those values are significantly lower than the *ZT* of the best materials that are currently used in room temperature thermoelectric devices that is close to 1. *ZT* refers to the dimensionless thermoelectric *figure of merit* that is related to the ability of a material to produce thermoelectric power, and it depends on the Seebeck coefficient, the thermal and electrical conductivity, and temperature.

Different types of inorganic nanocomponents have been combined with conductive polymers aiming to develop new and efficient polymer-based thermoelectric nanocomposites. As demonstrated by P. Mavinaculi and coworkers (Mavinakuli et al. 2010), the electrical conductivity of PPy at room temperature increased when incorporating SiC nanoparticles within the polymer to form PPy/SiC nanocomposites. On the other hand, the inclusion of nano-sized SiC nanoparticles within a polymer matrix, results in a reduction in the thermal conductivity through the enhancement of phonon scattering (Li and Liu 2006). According to the experimental data provided in (Li and Liu 2006), an estimated reduction of 14.5 % is observed in the thermal conductivity of SiC-dispersed Bi_2Te_3 at 385 K containing 0.5 % wt SiC compared to the monolithic Bi_2Te_3 system. Moreover, the inclusion of SiC nanoparticulates within thermoelectric polymer matrices leads to an improvement in the mechanical properties (Young's modulus and tensile strength) of the nanocomposite (Guo et al. 2008). In particular, by introducing SiC nanoparticles (5 % wt) within a polyurethane matrix, an increase of 71 and 166 % is observed in the Young's modulus and the tensile strength, respectively.

In another literature example, core–shell nanorods were fabricated in which the core component was tellurium (a material exhibiting a high thermopower) and the shell counterpart was the organic conductive polymer poly (3, 4-ethylenedioxythiophene) oxidized with poly (4-styrenesulfonate PEDOT: PSS) (See et al. 2010). The presence of the conductive polymer shell around the inorganic nanorods led to an increase in the electrical conductivity, while suppressing the thermal conductivity.

The incorporation of the inorganic nanoparticles within thermoelectric polymers is expected to create a new platform for the development of cost-effective and highly efficient low-temperature thermoelectric materials that can be easily processed into versatile forms. Moreover, the incorporation of selected inorganic nanofillers within a thermoelectric polymer matrix is expected to further increase the *ZT* values, since nanostructuring may lead to an effective decoupling of the electrical and thermal transport mechanisms.

2.3.3.4 Hybrid Energy Storage Devices: Hydrogen Storage, Batteries, and Capacitors

Hydrogen (H_2) is considered to be one of the future environmentally friendly and highly efficient fuel that could potentially substitute the conventional fossil fuels in future mobile and stationary power applications. The interest in using H_2 as an alternative fuel dates back to the late 1960s, whereas during the past two decades there has been a tremendous increase in scientific reports related to the development of new materials for use in hydrogen storage applications (Niemann et al. 2008; Pukazhselvan et al. 2012; Jain et al. 2010).

Depending on the adsorption mechanism involved, hydrogen storage materials can be categorized in physical and in chemical adsorbents. In the first case, physical adsorption of molecular hydrogen onto the storage material takes place via weak noncovalent interactions, whereas in the second case chemical bonding occurs between atomic hydrogen and the storage material.

Polymer–inorganic nanocomposites have been explored in hydrogen storage applications. In a nice recent review by N. F. Attia and K. E. Geckeler (Attia and Geckeler 2013), the authors provide, among others, examples on the use of PANI-based nanocomposites where PANI has been combined with different inorganic nanomaterials such as SnO_2, aluminum, V_2O_5, MNPs, and metal alloys in hydrogen storage applications. Magnesium nanocrystal polymer composites have been investigated as new hydrogen storage materials, demonstrating their high potential in such applications since they exhibited three very important characteristics required: environmental stability, high hydrogen storage density, and rapid hydrogenation kinetics (Bardhan et al. 2011; Ruminski et al. 2013).

Organic–inorganic polymer hybrids are also attractive for energy storage devices such as batteries, capacitors, and supercapacitors (Gomez-Romero et al. 2010; Wang and Zhu 2011; Malinauskas et al. 2005; Nyholm et al. 2011; Orilall and Wiesner 2011; Hanemann and Szabo 2010; Barber et al. 2009; Gomez-Romero 2001). By fine-tuning the structural characteristics of the polymers and of the inorganic nanofillers, and predefining the organic–inorganic interfacial characteristics, materials with enhanced energy densities and fast charge–discharge capabilities can be obtained. One of the main drawbacks appearing in capacitors is the low-energy density that depends on the dielectric polymeric interlayer sandwiched between the two electrodes. The incorporation of inorganic nanofillers within the dielectric polymer matrices appears to be highly promising for developing high-energy-density dielectric films (Wang and Zhu 2011).

Electrochemical capacitors known as supercapacitors belong to the energy-storage devices that receive great attention owing to their distinct advantages over conventional batteries and solid-state conventional capacitors, including fast charge/discharge rates, high-power density, lightweight characteristics, environmental friendliness, and safety (Yan et al. 2014a). Supercapacitors consist of two electrodes, an electrolyte, and a separator that is responsible for electrically isolating the two electrodes. Depending on the type of materials involved, the electrode design, and the charge storage mechanism, supercapacitors are classified as electric double-layer capacitors, pseudocapacitors, and hybrid capacitors such as lithium-ion capacitors (Wang et al. 2012).

Polymer–inorganic nanocomposites have been investigated as electrode materials in supercapacitor structures. For example, polymer composites consisting of either PPy or PANI (both being conductive polymers) and $Li(Mn_{1/3}Ni_{1/3}Fe_{1/3})O_2$ nanoparticles have been used as cathode materials in lithium-ion capacitors functioning in organic electrolytes (Karthikeyan et al. 2013). The PANI-based electrode exhibited excellent supercapacitor properties delivering high specific capacitances as well as superior cyclability compared to the PPy-based systems. In another example, porous conducting PPy–phosphomolybdate hybrid films with tunable pore sizes have been prepared and evaluated in electrochemical pseudosupercapacitor applications. The inorganic component is characterized by fast and reversible redox behavior which is essential for the development of pseudosupercapacitors in which electrochemical storage is based on electron charge-transfer phenomena achieved via redox reactions (Suppes et al. 2008).

In a final example, nanohybrid PPy/Fe_2O_3 materials have been synthesized and characterized in regard to their charge storage capabilities (Mallouki et al. 2007). Among others, the authors demonstrated the improved charge storage ability and charge–discharge kinetics exhibited by the hybrid nanocomposite in comparison to the pristine polymer analogue.

2.3.4 Catalysis

The generation of new materials for catalytic applications is one of the most important goals in the field of nanotechnology. The availability of numerous synthetic strategies and processing tools enables the development of nanostructured materials with tunable properties that would lead to an enhancement in the catalytic efficacy. The combination of catalytic metal or metal oxide nanoparticles with polymers may lead to a significant improvement of the catalytic activity and selectivity in homogeneous and heterogeneous catalytic processes. Organic–inorganic polymer hybrids have been used in photocatalysis (Rammohan and Nadagouda 2013; Khin et al. 2012) and electrocatalysis (Gomez-Romero 2001; Walcarius 2001), as well as in catalytic processes involved in organic synthesis (Magdesieva et al. 2012; Zhang et al. 2011b; Akelah et al. 2006; Mangold et al. 2004).

Photocatalytic processes are mainly used in wastewater treatment, i.e., for the catalytic photodegradation of organic pollutants such as hazardous dyes. Inorganic metal oxides including Fe_xO_y, TiO_2, and ZnO are well known for their ability to generate hydroxyl radicals upon irradiation with light having photon energies equal or higher than the energy band gaps of the inorganic components (Pan et al. 2012). Those radicals are highly oxidative and lead to the degradation of organic pollutants. Polymer–inorganic hybrids have been prepared by different groups and evaluated toward their photocatalytic efficiency for the degradation of organic dyes in aqueous media.

Many researchers have used the electrospinning technique for fabricating polymer-based nanocomposites exhibiting photocatalytic properties. Electrospinning is a simple and cost-effective technique used for the production of fibrous materials with fiber diameters ranging between a few nanometers and a few micrometers (Teo and Ramakrishna 2006). The high versatility of this technique allowing for the fabrication of not only polymer fibers but also of ceramic and composite (nano)fibrous materials has led to a dramatic increase in the number of scientific publications dealing with this method from 2000 onwards. Q. Ding and coworkers have recently described the preparation of flexible polyimide/ZnO nanofibrous electrospun membranes that have been further evaluated as photocatalytic substrates for the degradation of methylene blue (Fig. 2.24; Ding et al. 2013). These composite membranes are highly porous, whereas the nanodimensions of the fiber diameters provide high surface-to-volume ratios which is a critical requirement in catalytical processes. Moreover, the authors have demonstrated that the morphological characteristics of the ZnO nanoparticulates may be altered from nanoplatelets to nanorods, resulting in an enhancement in the photocatalytic activity.

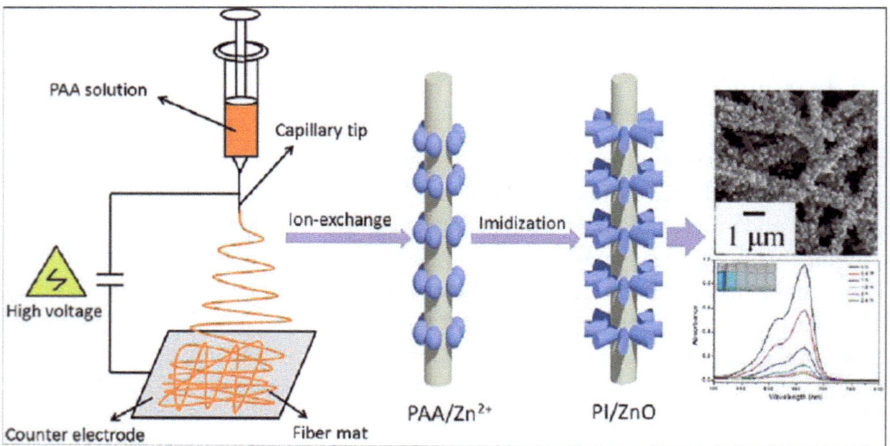

Fig. 2.24 Electrospun polymer–ZnO nanocomposite fibers used as photocatalytic substrates for the degradation of organic dyes in aqueous solutions. (Reprinted (adapted) with permission from (Ding et al. 2013). Copyright (2013) American Chemical Society)

Polymer–inorganic nanocomposite films have been also developed and used in photocatalytic applications (Priya et al. 2011; Xu et al. 2012). For example, TiO_2 nanoparticles have been immobilized in a poly(vinyl alcohol) polymer matrix, and the resulting polymer nanocomposite film has been evaluated in the UV–photocatalytic degradation of methyl orange. In this case, the use of a hydrophilic polymer matrix assisted the accessibility of the TiO_2 catalytic surfaces to the dye molecules resulting in an enhancement in the photocatalytic activity (Lei et al. 2012).

Besides the combination of polymers with metal oxides, functional polymeric materials with metal chelating side-chain moieties have been combined with metal ions exhibiting photosensitizing properties. The resulting polymer hybrids demonstrated high photocatalytic activity upon irradiation with visible light enabling molecular transformation reactions (Shimakoshi et al. 2011).

In electrocatalytic processes, a heterogeneous catalyst is introduced as an electrode surface resulting to the acceleration of an electrochemical reaction. Highly efficient electrocatalytic materials are very interesting in fuel cell technologies. In particular, the development of electrocatalytic materials that can effectively oxidize methanol and reduce oxygen is highly desirable for improving the performance of direct methanol (DMFC) and PEMFC.

Polymer–inorganic hybrid materials have been used as electrocatalysts in fuel cell applications. Examples include hybrid composites consisting of poly(3-methylthiophene)–Pt nanoparticle nanorods (Zhou et al. 2010), polyaniline–Pt composite films (O'Mullane et al. 2006), multilayered poly(ethylene imine)–capped Pt nanoparticles, and PEDOT: PSS conducting films, (Knowles et al. 2012) Pt–PPy–carbon nanocomposites (Zhao et al. 2008), and Pt(or PtRu)–PANI–carbon composite films (Wu et al. 2005).

Besides their photocatalytic and electrocatalytic performance, polymer–inorganic hybrid assemblies have also been studied as effective catalysts in organic synthesis. Various polymer-stabilized MNP colloidal systems have been prepared and evaluated as nanocatalysts for organic reactions in solution (Biffis et al. 2001; Balanta et al. 2011). Furthermore, catalytic MNPs, such as Pd, Au, and Ag, have been incorporated within electrospun polymer fibrous matrices, and have been evaluated as catalytic supports in Heck (Guo et al. 2013b) and hydrogenation (Demir et al. 2004; Gardella et al. 2013) reactions. Moreover, hybrid polymer (chitosan-based)-metal (Au) nanoparticle membranes prepared by the layer-by-layer self-assembly approach have been prepared and evaluated in reduction processes (Dhar and Patil 2012). From their studies, the authors concluded that the number of layers in the resulting nanocomposite films as well as the sizes of the MNPs are critical parameters that affect the catalytic performance of these materials.

2.3.5 Optoelectronics

The development of hybrid nanocomposites with interesting optical and electronic properties has led to a dramatic use of such materials in optoelectronic applications including nonlinear optics, light-emitting diodes (LEDs), transistors, and optical data storage (Mitzi et al. 2001; Tessler et al. 2001; Haberkorn et al. 2009; Luo et al. 2011b; Pardo et al. 2011; Vaynzof et al. 2012).

In most examples existing on polymer-based electronic devices, conjugated polymers are mainly employed due to their good electrical conductivity derived from their conjugated chemical structure (Tessler et al. 2001; Thien-Phap 2011). Conjugated polymers have been readily employed in optoelectronic applications including nonlinear optical devices, transistors, and LEDs. Although commercially available products based on polymer LEDs such as display screens and lighting already exist, such materials often suffer from chemical long-term stability. The incorporation of inorganic nanoparticles, such as TiO_2, ZnO, SiO_2, and semiconducting QD, within conjugated polymers results in materials characterized by improved mechanical, chemical, and thermal stability, whereas the versatility in materials' design allows for the fine-tuning of the energy levels of the organic and the inorganic components that may lead to improved optoelectronic properties translated into enhancement in the device performance (Gangopadhyay and De 2000; Tomczak et al. 2009; Sessolo and Bolink 2011).

It is strongly believed that the future optoelectronic and photonic devices will consist of organic–inorganic hybrid materials demonstrating custom-made nonlinear optical (NLO) response (Innocenzi and Lebeau 2005). Materials with NLO properties exhibit great interest due to their potential use in technological applications such as all-optical switching, signal processing, and fast optical communications.

The ability to create MNPs with definite geometries and sizes has allowed for the systematic investigation of the fundamental physics involved in the optical properties of MNP (Sigalas and Garcia 2000). In that context, it has been shown that large electric field enhancements can be achieved in the surfaces of MNP or in the gaps between MNP near the plasmon frequencies. These high-field enhancements are ideal for improving nonlinear effects, spontaneous emission rates, and sensing of small particles or biological molecules (Subramania et al. 1999).

During the past decades, there has been a remarkable effort in combining MNPs and polymers for the development of organic–inorganic hybrid materials and device fabrication in order to take advantage of the unique physical properties of the nanoparticles and the excellent processing of polymers. A key issue in this area is to develop efficient ways to control the arrangement of nanoparticles within a polymer matrix since the dispersion of nanoparticles significantly impacts the optoelectronic, transport, and mechanical properties of the nanohybrid (Kamps et al. 2010). The electronic and optical properties of metal nanoparticles are of great interest for several optoelectronic and photonic applications and devices. As a result, during the past years, there has been an intense scientific activity concerning the preparation and the detailed investigation of the electro-optical properties of MNPs, mainly focused on noble metals (e.g., gold or silver) (Ganeev 2005; Xenogiannopoulou et al. 2008).

A versatile route followed for the preparation of MNPs makes use of ABCs capable of self-assembling in selective organic media creating well-organized nanomorphologies such as micelles of various shapes and vesicles, as discussed in Sect. 2.11. Block copolymer-mediated self-assembly of MNP leads to advanced hybrid materials, since it promotes the confinement of metallic compounds within well-defined nanostructured domains, providing control of their size and shape (Grubbs 2005; Grubbs 2007). It has been demonstrated by us and other groups that the NLO properties of such materials depend not only on the size and geometrical characteristics of the metal nanoparticles but also on the nature of the polymeric templates used as stabilizers for the formation of the nanocomposites in solution (West et al. 2003; Iliopoulos et al. 2011; Papagiannouli et al. 2014a; Papagiannouli et al. 2014b).

In addition to the polymer hybrids in solution, in recent years, thin films of noble metal nanoparticles (Au, Ag, etc) deposited on or embedded in different dielectric hosts have attracted a lot of interest since, in addition to their inherent large third-order NLO, the localized SPR that they exhibit may be exploited to further enhance their NLO response (Mitsuishi et al. 2007; Adhyapak et al. 2006; Wang et al. 2005). The ability of block copolymers to form a rich variety of nanoscale periodic patterns offers the potential to fabricate ordered, high-density arrays for use in data storage and electronics. As an example, Tsuchiya et al. investigated the enhancement of the optical nonlinearities of gold nanoparticles by using the periodic microphase-separated structure of PS-*b*-PMMA and poly(styrene)-*b*-poly(*tert*-butyl methacrylate) (PS-*b*-PtBuMA) diblock copolymers as templates for their dispersion. The introduction of gold nanoparticles into microphase-separated structures caused the local field effect and the effect of periodic gold nanoparticle arrays and consequent enhancement of the optical nonlinearities of the nanoparticles (Tsuchiya et al. 2008).

2.3.6 Environmental (Water) Remediation

Nowadays, wastewater treatment is a high priority for the protection of the ecosystem. The discharging of hazardous materials of either organic or inorganic nature into the environment from natural and human sources leads to severe environmental problems. Pollutants of inorganic origin include heavy metal ions of high toxicity such as Pd^{2+}, Zn^{2+}, Cd^{2+}, Ni^{2+}, Hg^{2+}, Pb^{2+}, Cu^{2+}, etc., whereas organic pollutants include dyes that are extensively used in manufacturing such as in textile, paper, paint, food, pharmaceutical and cosmetic industries, pharmaceutics and their metabolites, detergents, chemical compounds used in cosmetics, and so on.

The presence of toxic metal ions in water resources may result in their entering in the food chain and consequently their accumulation in humans. Various methodologies have been developed for metal ion removal from wastewater with the most popular being the precipitation, reverse osmosis, ion exchange, and adsorption (Fu and Wang 2011). The latter exhibits significant advantages over the others, due to its high removal efficacy, low cost, and recyclability/reuse.

Different types of materials have been used as adsorbents in heavy metal ion removal from aqueous media (Aditya et al. 2011). These include, among others, carbon-based materials, clays, zeolites, biomass, and polymers (Bailey et al. 1999; Babel and Kurniawan 2003; Chowdhury and Balasubramanian 2014; Mubarak et al. 2014). The abovementioned materials exhibit certain limitations including low adsorption capacity, low selectivity, inferior mechanical properties, and weaknesses in regard to their isolation, regeneration and reuse.

Such limitations may be overcome upon combining polymers with inorganic substances (Pan et al. 2009). In such organic–inorganic polymer hybrids, the organic and inorganic components may be carefully selected so as to result in new, functionalized polymer-based hybrids presenting high adsorption efficiencies, high selectivity, superior mechanical and thermal properties, and regeneration/reuse capability.

Polymer–metal and polymer–metal oxide hybrids in the form of layered structures, core–shell nanocomposites, fibrous membranes, and mesoporous hybrids have been investigated toward their ability to act as adsorbents for heavy metal ions including Hg^{2+}, Pb^{2+}, Cd^{2+}, Fe^{3+}, Zn^{2+}, As^{5+}, and Ni^{2+}(Sarkar et al. 2012). The incorporation of organic metal chelating moieties or ionizable (anionic) groups results in the reinforcement of the interactions between the hybrid adsorbent and the metal ion, whereas depending on the organic chelating moiety introduced, metal ion selectivity may be enhanced (Sarkar et al. 2012).

Besides their use in the removal of toxic metal ions from contaminated water, organic–inorganic polymer hybrids have also been evaluated toward their ability to act as adsorbents for organic molecules. In the case of adsorption of organic dyes and antibiotics, the use of polymer-based hybrids is highly advantageous since besides adsorption, the inorganic component embedded within these materials may serve as a catalyst for the degradation of the organic dye through a catalytic process. In a recent literature example, magnetically recyclable ternary titania–cobalt ferrite–polyaniline adsorbents have been synthesized and successfully employed for the adsorption of a well-known organic dye (methyl orange) from aqueous solutions (Xiong et al. 2013). The presence of TiO_2 within these polymer-based hybrids allowed for the subsequent photodegradation of the dye under visible irradiation. Moreover, due to the incorporation of the magnetoactive $CoFe_2O_4$ nanoparticles within these materials, their facile separation from the aqueous suspension after the completion of the photocatalytic degradation process was possible by using an externally applied magnetic field. In another example, core–shell alginate-Fe^{2+}/Fe^{3+} polymer-coated Fe_3O_4 nanoparticles have been used as heterogeneous nanocatalysts for the degradation of norfloxacin, an antibiotic belonging to one of the most important classes of synthetic antibiotics used worldwide for human and veterinary treatment. The degradation of norfloxacin is realized via a breakdown process of H_2O_2 catalysed by Fe_3O_4 nanoparticles that generates reactive radicals responsible for the degradation of the organic pollutant (Niu et al. 2012).

2.3.7 Construction, Automotive and Coatings

Modern technologies involving construction, automotive, and coating materials have also embraced the field of organic–inorganic polymer hybrids.

The use of polymers and polymer-based composites in construction industries is highly desirable due to their lightweight characteristics, low cost, ease of processing, good mechanical performance (especially in the case of the composite materials), insulating and anticorrosion properties, rendering them highly competitive over traditional construction materials such as steel, concrete, and aluminum.

Reinforced polymer nanocomposites in which polymer matrices including thermoplastic polyesters, epoxy, and vinylester resins are combined with fibrous fillers such as glass, polymer (aramid), and carbon fibers are under investigation for various applications related to the construction sector including rehabilitation and retrofit, concrete reinforcement, and load-bearing structures exclusively comprised of reinforced polymer composites (RPC), though the latter have been rarely used for the construction of all RPC structures (Pendhari et al. 2008; Zaman et al. 2013).

Silica-based aerogels are porous SiO_2 materials that are usually prepared via supercritical drying processes that preserve the highly porous structure of the inorganic matrix upon solvent removal (Dorcheh and Abbasi 2008; Gurav et al. 2010). Aerogels exhibit some highly interesting properties including extremely low density and low thermal conductivity, high porosity, and high specific surface area (Maleki et al. 2014). However, they are characterized by weak mechanical properties such as high fragility and brittleness. The combination of silica-based aerogels with polymers results in materials with improved mechanical performance and flexibility, rendering them appropriate for several applications related to the construction industry. Commercially available polymer/silica hybrid aerogels have been successfully employed in building and pipeline insulation (Baetens et al. 2011).

Polymer–inorganic nanocomposites consisting of polymers and inorganic nanofillers of different types such as metal oxides, clays, metals, semiconducting nanocrystals, etc., have been also used in the tyre and automotive industries (Zeng et al. 2005; Stefanescu et al. 2009). Moreover, polymer–inorganic hybrids in the form of foams, present interesting properties for acoustical and thermal insulation applications in the building and in the transport industries.

The area of hybrid coatings based on polymers and inorganic nanomaterials has faced a tremendous flourishing during the past years, and some of these newly developed materials have been already commercialized (Sanchez et al. 2011).

Functional polymer–inorganic nanocomposite coatings have been used in anticorrosion (Khelifa et al. 2013), scratch-resistant enhancement (Sangermano and Messori 2010), UV-protection (Calvo et al. 2012), flame retardancy (Lu et al. 2011; Chattopadhyay and Raju 2007; Horrocks et al. 2005), as well as protective materials for outdoor applications. Regarding the latter, silicone-containing polymer matrices exhibiting enhanced hydrophobicity have been used as protective coatings for the protection of historic monuments (Sadat-Shojai and Ershad-Langroudi 2009) and as electrical insulation materials (Amin 2013).

2.4 Conclusions and Outlook

This chapter highlights the diversity in existing synthetic methodologies employed for the preparation of organic–inorganic polymer hybrids. The 11 synthetic strategies discussed lead to a great variety of polymer–based nanohybrids/nanocomposites with different compositions, dimensional and morphological characteristics, in the form of solids, solutions, emulsions, or suspensions. It is noteworthy to mention at this point that other (rather than the presented ones) synthetic pathways have been proposed by researchers over the years for the preparation of organic–inorganic polymer-based hybrids; however, within the narrow framework of this chapter, it was not possible to include all of them herein. The continuous synthetic efforts toward the development of new experimental protocols for the preparation of such materials is highly important, enriching the highly emerging field of advanced multifunctional materials, and creating new pathways for their industrial advancement. In the past few years, the worldwide market has been shifting toward this field trying to solve several practical issues on the mass production of inorganic nanoparticles but most importantly trying to generate to the general public a positive mind-set for "Nano," i.e., nanotechnology and nanomaterials. Current and future challenges in regard to nano-containing materials arise from health and safety issues in regard to their production, handling, and use. Therefore, for the development of safe nano-based materials, industrial key players, academia, and governmental authorities have been investing on the evaluation of the risk and safety aspects of such materials, as well as on standardization issues.

The combination of polymers with inorganic nanocomponents within the same material does not simply lead to a new material that exhibits the properties of the individual organic and the inorganic parts. In many cases, synergistic effects arising from organic–inorganic interfacial interaction phenomena give rise to new properties or enhance the materials' performance. Consequently, future research efforts must focus on the in-depth understanding of such polymer/filler interfacial phenomena so as to be able to correlate the materials' structure to their properties. It is expected that the existing and the newly developed synthetic tools will allow for the fine-tuning of the polymer/filler interfacial interactions, thus promoting a new perspective in understanding the structure–property relationships.

The unique properties of organic–inorganic polymer hybrids render them highly interesting and valuable in many diverse fields. Their applicability spectrum is very broad, restricting the discussion on their use in every field in the present chapter. However, an effort to highlight and emphasize on the diversity of applications of these materials has been carried out by providing examples on their use in various fields including biomedicine, energy, catalysis, sensing, optoelectronics, environmental remediation, and in construction, automotive, and coatings technologies. Undoubtedly, with the further future development of the synthetic strategies involved in organic–inorganic polymer hybrid technologies, the applicability window of these materials will become even broader, reinforcing their significance in modern life.

References

Adhyapak PV, Khanna PK, Dadge JW et al (2006) Tuned-optical properties of in-situ synthesized m-nitroaniline doped Ag/PVA nano-composites. J Nanosci Nanotechnol 6:2141–2146

Aditya D, Rohan P, Suresh G (2011) Nano-adsorbents for wastewater treatment: a review. Res J Chem Environ 15:1033–1040
Aime S, Crich SG, Gianolio E et al (2006) High sensitivity lanthanide(III) based probes for MR-medical imaging. Coordination Chem Rev 250:1562–1579
Akelah A, Rehab A, Kenawy ER et al (2006) Catalytic activity of polymer-montmorillonite composites in chemical reactions. J Appl Polym Sci 101:1121–1129
Albdiry M T, Yousif BF, Ku H et al (2013) A critical review on the manufacturing processes in relation to the properties of nanoclay/polymer composites. J Compos Mater 47:1093–1115
Alexandre M, Dubois P (2000) Polymer-layered silicate nanocomposites: preparation, properties and uses of a new class of materials. Mater Sci Eng R-Rep 28:1–63
Amin M (2013) Methods for preparation of nano-composites for outdoor insulation applications. Rev Adv Mater Sci 34:173–184
Antonietti M (2001) Surfactants for novel templating applications. Curr Opin Colloid Interface Sci 6:244–248
Antonietti M, Basten R, Lohmann S (1995) Polymerization in microemulsions—A new approach to ultrafine highly functionalised polymer dispersions. Macromol Chem Phys 196:441–466
Apperley D, Hay JN, Raval HM (2002) Silica-dimethylsiloxane hybrids non-hydrolytic sol-gel synthesis and characterization by NMR spectroscopy. Chem Mater 14:983–988
Arici E, Meissner D, Schaffler F et al (2003) Core/shell nanomaterials in photovoltaics. Int J Photoenergy 5:199–208
Asua JM (2004) Emulsion polymerization: from fundamental mechanisms to process developments. J Polym Sci Part A: Polym Chem 42:1025–1041
Attia NF, Geckeler KE (2013) Polyaniline as a material for hydrogen storage applications. Macromol Rapid Commun 34:1043–1055
Azhdar B, Stenberg B, Kari L (2008) Polymer-nanofiller prepared and high velocity cold by high-energy ball milling compaction. Polym Compos 29:252–261
Babel S, Kurniawan TA (2003) Low-cost adsorbents for heavy metals uptake from contaminated water: a review. J Hazard Mate 97:219–243
Baetens R, Jelle BP, Gustavsen A (2011) Aerogel insulation for building applications: a state-of-the-art review. Energy Build 43:761–769
Bailey SE, Olin TJ, Bricka RM et al (1999) A review of potentially low-cost sorbents for heavy metals. Water Res 33:2469–2479
Balanta A, Godard C, Claver C (2011) Pd nanoparticles for C-C coupling reactions. Chem Soc Rev 40:4973–4985
Bao L, Jiang JS (2005) Evolution of microstructure and phase of Fe_3O_4 in system of Fe_3O_4-polyaniline during high-energy ball milling. Physica B- Condensed Matter 367:182–187
Barber P, Balasubramanian S, Anguchamy Y et al (2009) Polymer composite and nanocomposite dielectric materials for pulse power energy storage. Mater 2:1697–1733
Bardhan R, Ruminski AM, Brand A et al (2011) Magnesium nanocrystal-polymer composites: a new platform for designer hydrogen storage materials. Energy Environ Sci 4: 4882–4895
Basuki JS, Esser L, Zetterlund PB (2013) Grafting of P(OEGA) onto magnetic nanoparticles using Cu(0) mediated polymerization: comparing grafting "from" and "to" approaches in the search for the optimal material, design of nanoparticle MRI contrast agents. Macromol 46:6038–6047
Beck JB, Rowan SJ (2003) Multistimuli, multiresponsive metallo-supramolecular polymers. J Am Chem Soc 125:13922–13923
Beck JS, Vartuli JC, Roth WJ et al (1992) A new family of mesoporous molecular-sieves prepared with liquid-crystal templates. J Am Chem Soc 114:10834–10843
Beija M, Marty J-D, Destarac M (2011) RAFT/MADIX polymers for the preparation of polymer/inorganic hybrids. Prog Polym Sci 36:845–886
Biffis A, Zecca M, Basato M (2001) Palladium metal catalysts in Heck C–C coupling reactions. J Mol Catal A: Chem 173:249–274
Bogdal D, Prociak A, Michalowski S (2011) Synthesis of polymer nanocomposites under microwave irradiation. Curr Organ Chem 15:178–188

Bonanno LM, Segal E (2011) Nanostructured porous silicon-polymer-based hybrids: from biosensing to drug delivery. Nanomed 6:1755–1770
Bose S, Kuila T, Thi XLN et al (2011) Polymer membranes for high temperature proton exchange membrane fuel cell: recent advances and challenges. Prog Polym Sci 36:813–843
Bourgeat-Lami E, Lansalot M (2010) Organic/inorganic composite latexes: the marriage of emulsion polymerization and inorganic chemistry. In: VanHerk AM, Landfester K (eds) Hybrid latex particles: preparation with (mini) emulsion polymerization Book series: Advances in Polymer Science, vol 233. Springer, pp 53–123
Brinker CJ (2004) Evaporation-induced self-assembly: functional nanostructures made easy. MRS Bulletin 29:631–640
Bubnova O, Crispin X (2012) Towards polymer-based organic thermoelectric generators. Energy Environ Sci 5:9345–9362
Caldorera-Moore ME, Liechty WB, Peppas NA (2011) Responsive theranostic systems: integration of diagnostic imaging agents and responsive controlled release drug delivery carriers. Acc Chem Res 44:1061–1070
Calvo ME, Castro Smirnov JR, Miguez H (2012) Novel approaches to flexible visible transparent hybrid films for ultraviolet protection. J Polym Sci Part B-Polym Phys 50:945–956
Capek I, Chern CS (2001) Radical polymerization in direct mini-emulsion systems. New polymerization techniques and synthetic methodologies book series: Advances in Polymer Science, vol 155. Springer, pp 101–165
Caruso F, Caruso RA, Möhwald H (1998a) Nanoengineering of inorganic and hybrid hollow spheres by colloidal templating. Science 282:1111–1114
Caruso F, Lichtenfeld H, Giersig M, Möhwald H (1998b) Electrostatic self-assembly of silica nanoparticle–polyelectrolyte multilayers on polystyrene latex particles. J Am Chem Soc 120:8523–8524
Castrillo PD, Olmos D, Amador DR et al (2007) Real dispersion of isolated fumed silica nanoparticles in highly filled PMMA prepared by high energy ball milling. J Colloid Interface Sci 308:318–324
Ceccia S, Turcato EA, Maffettone PL, Bongiovanni R (2008) Nanocomposite UV-cured coatings: organoclay intercalation by an epoxy resin. Prog Organ Coat 63:110–115
Cele N, Ray SS (2009) Recent progress on nafion-based nanocomposite membranes for fuel cell applications. Macromol Mater Eng 294:719–738
Chanana M, Mao Z, Wang D (2009) Using polymers to make up magnetic nanoparticles for biomedicine. J Biomed Nanotechnol 5:652–668
Chattopadhyay DK, Raju KVSN (2007) Structural engineering of polyurethane coatings for high performance applications. Prog Polym Sci 32:352–418
Chen J, Liu M, Chen C et al (2011) Synthesis and characterization of silica nanoparticles with well-defined thermoresponsive PNIPAM via a combination of RAFT and click chemistry. ACS Appl Mater Interfaces 3:3215–3223
Chen G, Seo J, Yang C et al (2013) Nanochemistry and nanomaterials for photovoltaics. Chem Soc Rev 42: 8304–8338
Chern CS (2006) Emulsion polymerization mechanisms and kinetics. Prog Polym Sci 31:443–486
Chibac A, Melinte V, Buruiana T et al (2012) One-pot synthesis of photocrosslinked sol-gel hybrid composites containing silver nanoparticles in urethane-acrylic matrixes. Chem Eng J 200:577–588
Choi AH, Ben-Nissan B, Matinlinna JP et al (2013) Current perspectives: calcium phosphate nanocoatings and nanocomposite coatings in dentistry. J Dent Res 92:853–859
Chowdhury S, Balasubramanian R (2014) Recent advances in the use of graphene-family nanoadsorbents for removal of toxic pollutants from wastewater. Adv Coll Interface Sci 204:35–56
Chu W-C, Chiang S-F, Li J-G et al (2014) Mesoporous silicas templated by symmetrical multiblock copolymers through evaporation-induced self-assembly. RSC Adv 4:784–793
Chujo Y, Saegusa T (1992) Organic polymer hybrids with silica-gel formed by means of the sol-gel method. Adv Polym Sci 100:11–29
Chujo Y, Ihara E, Kure S et al (1993) Synthesis of triethoxysilyl-terminated polyoxazolines and their cohydrolysis polymerisation with tetraethoxysilane. Macromolecules 26:5681–5686

Cong H-P, Yu S-H (2009) Self-assembly of functionalized inorganic-organic hybrids. Curr Opin Colloid Interface Sci 14:71–80
Coradin T, Allouche J, Boissiere M et al (2006) Sol-gel biopolymer/silica nanocomposites in biotechnology. Curr Nanosci 2:219–230
De Juan F, Ruiz-Hitzky E (2000) Selective functionalization of mesoporous silica. Adv Mater 12:430–432
de Sousa R, Luis A, Figueiras A, Veiga F et al (2013) The systems containing clays and clay minerals from modified drug release: a review. Colloids Surf B-Biointerfaces 103:642–651
Decher G (1977) Fuzzy nanoassemblies: toward layered polymeric multicomposites. Science 277:1232–1237
Decker C (1998) The use of UV irradiation in polymerisation. Polym Int 45:133–141
Demir MM, Gulgun MA, Mencelogu YZ et al (2004) Palladium nanoparticles by electrospinning from poly(acrylonitrile-co-acrylic acid)-$PdCl_2$ solutions. Relations between preparation conditions, particle size, and catalytic activity. Macromolecules 37:1787–1792
Deng Y, Wei J, Sun Z et al (2013) Large-pore ordered mesoporous materials templated from non-Pluronic amphiphilic block copolymers. Chem Soc Rev 42:4054–4070
Dhar J, Patil S (2012) Self-assembly and catalytic activity of metal nanoparticles immobilized in polymer membrane prepared via layer-by-layer approach. ACS Appl Mater Interfaces 4:1803–1812
Ding Q, Miao Y-E, Liu T (2013) Morphology and photocatalytic property of hierarchical polyimide/ZnO fibers prepared via a direct ion-exchange process. ACS Appl Mater Interfaces 5:5617–5622
Dorcheh AS, Abbasi MH (2008) Silica aerogel; synthesis, properties and characterization. J Mater Process Technol 199:10–26
Du B, Zhao B, Tao P et al (2008) Amphiphilic multiblock copolymer stabilized Au nanoparticles. Colloids Surf A-Physicochem Eng Aspects 317:194–205
Du K, Zhu Y, Xu H et al (2011) Multifunctional magnetic nanoparticles: synthesis, modification and biomedical applications. Prog Chem 23: 2287–2298
Du N, Zhang H, Yang D (2012a) One-dimensional hybrid nanostructures: synthesis via layer-by-layer assembly and applications. Nanoscale 4:5517–5526
Du Y, Shen SZ, Cai K et al (2012b) Research progress on polymer-inorganic thermoelectric nanocomposite materials. Prog Polym Sci 37:820–841
El-Safty SA (2011) Instant synthesis of mesoporous monolithic materials with controllable geometry, dimension and stability: a review. J Porous Mater 18:259–287
Etxeberria H, Zalakain I, Tercjak A et al (2013) Functionalisation of CdSe semiconductor nanoparticles with polystyrene brushes by radical polymerization. J Nanosci Nanotechnol 13:643–648
Fan LJ, Jones WE (2006) A Highly selective and sensitive inorganic/organic hybrid polymer fluorescence “turn-on” chemosensory system for iron cations. J Am Chem Soc 128:6784–6785
Fang J, Chen Y-C (2013) Nanomaterials for photohyperthermia: a review. Curr Pharm Des 19:6622–6634
Faucheu J, Gauthier C, Chazeau L et al (2010) Miniemulsion polymerization for synthesis of structured clay/polymer nanocomposites: short review and recent advances. Polym 51:6–17
Forrest ML, Kwon GS (2008) Clinical developments in drug delivery nanotechnology. Adv Drug Deliv Rev 60:861–862
Förster S, Antonietti M (1998) Amphiphilic block copolymers in structure-controlled nanomaterial hybrids. Adv Mater 10:195–217
Fu F, Wang Q (2011) Removal of heavy metal ions from wastewaters: a review. J Environ Manag 92:407–418
Gaharwar AK, Peppas NA, Khademhosseini A (2014) Nanocomposite hydrogels for biomedical applications. Biotechnol Bioeng 111:441–453
Gambinossi F, Chanana M, Mylon SE et al (2014) Stimulus-responsive Au@(MeO(2)MA(x)-co-OEGMA(y)) nanoparticles stabilized by non-DLVO interactions: implications of ionic strength and copolymer (x:y) fraction on aggregation kinetics. Langmuir 30:1748–1757
Ganeev RA (2005) Nonlinear refraction and nonlinear absorption of various media. J Opt A-Pure Appl Opt 7:717–733

Gangopadhyay R, De A (2000) Conducting polymer nanocomposites: a brief overview. Chem Mater 12:608–622
Gao F, Ren S, Wang J (2013) The renaissance of hybrid solar cells: progresses, challenges, and perspectives. Energy Environ Sci 6:2020–2040
Gardella L, Basso A, Prato M et al (2013) PLA/POSS nanofibers: a novel system for the immobilization of metal nanoparticles. ACS Appl Mater Interfaces 5:7688–7692
Ghannam L, Parvole J, Laruelle G et al (2006) Surface-initiated nitroxide-mediated polymerization: a tool for hybrid inorganic/organic nanocomposites 'in situ' synthesis. Polym Int 55:1199–1207
Giri AK (1997) Magnetic properties of iron-polyethylene nanocomposites prepared by high energy ball milling. J Appl Phys 81:1348–1350
Gloria A, De Santis R, Ambrosio L (2010) Polymer-based composite scaffolds for tissue engineering. J Appl Biomater Biomech 8: 57–67
Gohy J-F, Lohmeijer BGG, Schubert US (2003) From supramolecular block copolymers to advanced nano-objects. Chem—A Eur J 9:3472–3479
Gomez-Romero P (2001) Hybrid organic–inorganic materials—In search of synergic activity. Adv Mater 13:163–174
Gomez-Romero P, Ayyad O, Suarez-Guevara J et al (2010) Hybrid organic-inorganic materials: from child's play to energy applications. J Sol State Electrochem 14:1939–1945
Gonzalez-Benito J, Gonzalez-Gaitano G (2008) Interfacial conformations and molecular structure of PMMA in PMMA/silica nanocomposites. Effect of high-energy ball milling. Macromolecules 41:4777–4785
Gonzalez-Valls I, Lira-Cantu M (2009) Vertically-aligned nanostructures of ZnO for excitonic solar cells: a review. Energy Environ Sci 2:19–34
Gray DH, Hu SL, Juang E et al (1997) Highly ordered polymer-inorganic nanocomposites via monomer self-assembly: in situ condensation approach. Adv Mater 9:731–736
Grubbs RB (2005) Hybrid metal-polymer composites from functional block copolymers. J Polym Sci Part A-Polym Chem 43:4323–4336
Grubbs RB (2007) Roles of polymer ligands in nanoparticle stabilization. Polym Rev 47:197–215
Gu D, Schueth F (2014) Synthesis of non-siliceous mesoporous oxides. Chem Soc Rev 43:313–344
Guenes S, Sariciftci NS (2008) Hybrid solar cells inorganic chimica acta. Inorganica Chimica Acta 361:581–588
Guo Z, Kim TY, Lei K et al (2008) Strengthening and thermal stabilization of polyurethane nanocomposites with silicon carbide nanoparticles by a surface-initiated-polymerization approach. Compos Sci Technol 68: 164–170
Guo J, Yang W, Wang C (2013a) Magnetic colloidal supraparticles: design, fabrication and biomedical applications. Adv Mater 25:5196–5214
Guo L, Bai J, Liang H et al (2013b) A facile approach to preparing palladium nanoparticles-embedded polyvinylpyrrolidone (PVP) heterogeneous hybrid nanofibers mats by electrospinning. Korean J Chem Eng 30:2142–2150
Gurav JL, Jung I-K, Park H-H et al (2010) Silica aerogel: synthesis and applications. J Nanomater 409310
Gutierrez MC, Ferrer ML, Tartaj P, del Monte F (2009) Biomedical applications of organic-inorganic hybrid nanoparticles In: Merhari L (ed) Hybrid nanocomposites for nanotechnology. Springer, pp 707–768
Haberkorn N, Lechmann MC, Sohn BH et al (2009) Templated organic and hybrid materials for optoelectronic applications. Macromol Rapid Commun 30:1146–1166
Hadjichristidis N, Pispas S, Floudas GA (2003) Block copolymers synthetic strategies, physical properties and applications. John Wiley & Sons, Inc., New Jersey
Hanemann T, Szabo DV (2010) Polymer-nanoparticle composites: from synthesis to modern applications. Materials 3:3468–3517
Hatchett DW, Josowicz M (2008) Composites of intrinsically conducting polymers as sensing nanomaterials. Chem Rev 108, 746–769
Hay JN, Raval HM (2001) Synthesis of organic-inorganic hybrids via the non-hydrolytic sol-gel process. Chem Mater 13:3396–3403

Hellweg T (2011) Block copolymer surfactant mixtures in aqueous solution: can we achieve size and shape control by co-micellization? In: Muller AHE, Borisov O (eds) Self-organized nanostructures of amphiphilic block copolymers II book series: advances in polymer science, vol 242. Springer, pp 1–27

Henderson MJ, Gibaud A, Bardeau JF et al (2006) An X-ray reflectivity study of evaporation-induced self-assembled titania-based films. J Mater Chem 16:2478–2484

Herregods SJF, Mertens M, Van Havenbergh K et al (2013) Controlling pore size and uniformity of mesoporous titania by early stage low temperature stabilization. J Colloid Interface Sci 391:36–44

Ho TL (1975) Hard soft acids bases (HSAB) principle and organic chemistry. Chem Rev 75:1–20

Holmberg K, Berggren A, Palmqvist AEC (2005) Surfactant-templated mesostructured materials from inorganic silica. Soft Matter 1:219–226

Hoogenboom R, Schubert US (2007) Microwave-assisted polymer synthesis: recent developments in a rapidly expanding field of research. Macromol Rapid Commun 28:368–386

Horrocks AR, Kandola BK, Davies PJ et al (2005) Developments in flame retardant textiles—A review. Polym Degradation Stab 88:3–12

Hu J, Chen M, Wu L (2011) Organic-inorganic nanocomposites synthesized via miniemulsion polymerization. Polym Chem 2:760–772

Huang SQ, Fan DQ, Lei YQ et al (2004) Alkoxysilane-functionalized acrylic copolymer latexes. I. Particle size, morphology, and film-forming properties. J Appl Polym Sci 94:954–960

Huang J, Kang Y, Yang T et al (2011a) Preparation of polythiophene/WO_3 organic-inorganic hybrids and their gas sensing properties for NO_2 detection at low temperature. J Nat Gas Chem 20:403–407

Huang J, Yang T, Kang Y et al (2011b) Gas sensing performance of polyaniline/ZnO organic-inorganic hybrids for detecting VOCs at low temperature. J Nat Gas Chem 20:515–519

Hui CM, Pietrasik J, Schmitt M et al (2014) Surface-initiated polymerization as an enabling tool for multifunctional (nano-)engineered hybrid materials. Chem Mater 26:745–762

Ijeri VS, Nair JR, Gerbaldi C et al (2010) Metallopolymer capacitor in "one pot" by self-directed UV-assisted process. ACS Appl Mater Interfaces 2:3192–3200

Iliopoulos K, Chatzikyriakos G, Demetriou M et al (2011) Preparation and nonlinear optical response of novel palladium-containing micellar nanohybrids. Opt Mater 33:1342–1349

Imai Y, Itoh H, Naka K et al (2000) Thermally reversible IPN organic-inorganic polymer hybrids utilizing the Diels-Alder reaction. Macromolecules 33:4343–4346

Innocenzi P, Lebeau B (2005) Organic-inorganic hybrid materials for non-linear optics. J Mater Chem 15:3821–3831

Jain IP, Jain P, Jain A (2010) Novel hydrogen storage materials: a review of lightweight complex hydrides. J Alloys Compd 503:303–339

Janaky C, Visy C, Berkesi O et al (2009) Conducting polymer-based electrode with magnetic behavior: electrochemical synthesis of poly(3-thiophene-acetic-acid)/magnetite nanocomposite thin layers. J Phys Chem C 113:1352–1358

Janaky C, de Tacconi NR, Chanmanee W et al (2012) Electrodeposited polyaniline in a nanoporous WO_3 matrix: an organic/inorganic hybrid exhibiting both p- and n-type photoelectrochemical activity. J Phys Chem C 116:4234–4242

Jeon I-Y, Baek J-B (2010) Nanocomposites derived from polymers and inorganic nanoparticles. Mater 3:3654–3674

Jiang X, Chen F, Xu H et al (2010) CdS nanorods and its application in hybrid solar cells. Sol Energy Mater Sol Cells 94:338–344

Joglekar M, Trewyn BG (2013) Polymer-based stimuli-responsive nanosystems for biomedical applications. Biotechnol J 8:931–945

Julián B, Gervais C, Rager M-N et al (2004) Solid-State ^{17}O NMR characterization of PDMS–M_xO_y (M = Ge(IV), Ti(IV), Zr(IV), Nb(V), and Ta(V)) organic–inorganic nanocomposites. Chem Mater 16:521–529

Kahraman MV, Bayramoğlu G, Boztoprak Y, Güngör A, Kayaman-Apohan N (2009) Synthesis of fluorinated/methacrylated epoxy based oligomers and investigation of its performance in the UV curable hybrid coatings. Prog Organ Coat 66:52–58

Kamps AC, Sanchez-Gaytan BL, Hickey RJ et al (2010) Nanoparticle-directed self-assembly of amphiphilic block copolymers. Langmuir 26:14345–14350
Karthikeyan K, Amaresh S, Aravindan V (2013) Unveiling organic–inorganic hybrids as a cathode material for high performance lithium-ion capacitors. J Mater Chem A 1:707–714
Key J, Leary JF (2014) Nanoparticles for multimodal in vivo imaging in nanomedicine. Int J Nanomed 9:711–726
Khelifa F, Druart M-E, Habibi Y et al (2013) Sol-gel incorporation of silica nanofillers for tuning the anti-corrosion protection of acrylate-based coatings. Prog Organ Coat 76:900–911
Khin MM, Nair AS, Babu VJ et al (2012) A review on nanomaterials for environmental remediation. Energy Environ Sci 5:8075–8109
Knowles KR, Hanson CC, Fogel AL et al (2012) Layer-by-layer assembled multilayers of polyethylenimine-stabilized platinum nanoparticles and PEDOT:PSS as anodes for the methanol oxidation reaction. ACS Appl Mater Interfaces 4:3575–3583
Knowles KE, Peterson MD, McPhail MR et al (2013) Exciton dissociation within quantum dot-organic complexes: mechanisms, use as a probe of interfacial structure, and applications. J Phys Chem C 117:10229–10243
Krasia-Christoforou T, Georgiou TK (2013) Polymeric theranostics: using polymer-based systems for simultaneous imaging and therapy. J Mater Chem B 1:3002–3025
Krebs FC (2009) Fabrication and processing of polymer solar cells: a review of printing and coating techniques. Sol Energy Mater Sol Cells 93:394–412
Lai S-M, Chen J-R, Han J-L et al (2013) Preparation and properties of melt-blended polylactid acid/polyethylene glycol-modified silica nanocomposites. J Appl Polym Sci 130:496–503
Landfester K (2003) Miniemulsions for nanoparticle synthesis In: colloid chemistry II book series: topics in current chemistry, vol 227. Springer, Heidelberg, pp 75–123
Landfester K (2009) Miniemulsion polymerization and the structure of polymer and hybrid nanoparticles. Angew Chem Int Ed Engl. 48:4488–4507
Lei P, Wang F, Gao X et al (2012) Immobilization of TiO_2 nanoparticles in polymeric substrates by chemical bonding for multi-cycle photodegradation of organic pollutants. J Hazard Mater 227:185–194
Li J-F, Liu J (2006) Effect of nano-SiC dispersion on thermoelectric properties of Bi_2Te_3 polycrystals. Phys Status Sol A-Appl Mater Sci 203:3768–3773
Liong M, Lu J, Kovochich M et al (2008) Multifunctional inorganic nanoparticles for imaging, targeting and drug delivery. ACS Nano 2:889–896
Lionti K, Toury B, Boissiere C et al (2013) Hybrid silica coatings on polycarbonate: enhanced properties. J Sol-Gel Sci Technol 65:52–60
Liu YR (2009) One-pot synthesis of ordered mesoporous carbon-silica nanocomposites templated by mixed amphiphilic block copolymers. J Mater Sci 44:3600–3607
Liu G, An Z (2014) Frontiers in the design and synthesis of advanced nanogels for nanomedicine. Polym Chem 5:1559–1565
Liu H, Wang D, Yang X (2012) Preparation of polymer@titania raspberry-like core-corona composite via heterocoagulated self-assembly based on hydrogen-bonding interaction. Colloids Surf A-Physicochem Eng Aspects 397:48–58
Lu H, Song L, Hu Y (2011) A review on flame retardant technology in China. Part II: flame retardant polymeric nanocomposites and coatings. Polym Adv Technologies 22:379–394
Luan J, Wang S, Hu Z et al (2012) Synthesis, techniques, properties and applications of polymer nanocomposites. Curr Organ Synth 9:114–136
Luo J, Huang S, Shi Z et al (2011a) Tailored organic electro-optic materials and their hybrid systems for device applications. Chem Mater 23:544–553
Luo S, Zhang E, Su Y et al (2011b) A review of NIR dyes in cancer targeting and imaging. Biomater 32:7127–7138
Ma YR, Qi LM, Ma JM (2003) Supramolecular-templating synthesis of mesoporous silica: control of structure and morphology. Prog Chem 15:477–486
Ma T-Y, Lin X-Z, Yuan Z-Y (2010) Periodic mesoporous titanium phosphonate hybrid materials. J Mater Chem 20:7406–7415

Magdesieva TV, Nikitin OM, Levitsky OA et al (2012) Polypyrrole-palladium nanoparticles composite as efficient catalyst for Suzuki-Miyaura coupling. J Mol Catal A-Chem 353:50–57
Maleki H, Duraes L, Portugal A (2014) An overview on silica aerogels synthesis and different mechanical reinforcing strategies. J Non-Crystalline Solids 385:55–74
Malinauskas A, Malinauskiene J, Ramanavicius A (2005) Conducting polymer-based nanostructurized materials: electrochemical aspects. Nanotechnology 16:R51–R62
Mallouki M, Tran-Van F, Sarrazin C et al (2007) Polypyrrole-Fe_2O_3 nanohybrid materials for electrochemical storage. J Solid State Electrochem 11:398–406
Mangiacapra P, Gorrasi G, Sorrentino A et al (2006) Biodegradable nanocomposites obtained by ball milling of pectin and montmorillonites. Carbohydr Polym 64:516–523
Mangold KM, Meik F, Juttner K (2004) Polypyrrole/palladium composites for the electrocatalyzed Heck reaction. Synth Metals 144:221–227
Manners I (1996) Polymers and the periodic table: recent developments in inorganic polymer science. Angew Chemie-Int Ed Engl 35:1603–1621
Mark JE (1996) Ceramic-reinforced polymers and polymer-modified ceramics. Polym Eng Sci 36:2905–2920
Mavinakuli P, Wei S, Wang Q et al (2010) Polypyrrole/silicon carbide nanocomposites with tunable electrical conductivity. J Phys Chem C 114:3874–3882
Maya S, Sarmento B, Nair A et al (2013) Smart stimuli sensitive nanogels in cancer drug delivery and imaging: a review. Curr Pharm Des 19:7203–7218
McDowell JJ, Zacharia NS, Puzzo D et al (2010) Electroactuation of alkoxysilane-functionalized polyferrocenylsilane microfibers. J Am Chem Soc 132:3236–3237
McInnes SJP, Voelcker NH (2009) Silicon-polymer hybrid materials for drug delivery. Future Medicinal Chem 1:1051–1074
Meng XY, Wang Z, Tang T (2006) Controlling dispersed state and exfoliation process of clay in polymer matrix. Mater Sci Technol 22:780–786
Minhao Wong M, Guenther J, Sun L et al (2012) Synthesis and fabrication of multifunctional nanocomposites: stable dispersions of nanoparticles tethered with short, dense and polydisperse polymer brushes in poly(methyl methacrylate). Adv Funct Mater 22:3614–3624
Mishra AK, Bose S, Kuila T et al (2012) Silicate-based polymer-nanocomposite membranes for polymer electrolyte membrane fuel cells. Prog Polym Sci 37:842–869
Mitsuishi M, Ishifuji M, Endo H et al (2007) Hybrid polymer nanoassemblies: polymer nanosheets organized with metal nanoparticle arrays for surface plasmon photonics. Polym J 39:411–422
Mitzi DB, Chondroudis K, Kagan CR (2001) Organic-inorganic electronics. IBM J Res Develop 45:29–45
Miyazaki T, Ishikawa K, Shirosaki Y et al (2013) Organic-inorganic composites designed for biomedical applications. Biol Pharm Bull 36:1670–1675
Morselli D, Messori M, Bondioli F (2011) Poly(methyl methacrylate)-TiO_2 nanocomposite obtained by non-hydrolytic sol–gel synthesis. J Mater Sci 46:6609–6617
Mourino V, Cattalini JP, Roether JA et al (2013) Composite polymer-bioceramic scaffolds with drug delivery capability for bone tissue engineering. Expert Opin Drug Deliv 10:1353–1365
Mubarak NM, Sahu JN, Abdullah EC et al (2014) Removal of heavy metals from wastewater using carbon nanotubes. Sep Purif Rev 43:311–338
Nagarale RK, Shin W, Singh PK (2010) Progress in ionic organic-inorganic composite membranes for fuel cell applications. Polym Chem 1:388–408
Nelson C, Magge A, St Bernard T et al (2013) Nanostructured composites for bone repair. J Biomater Tissue Eng 3:426–439
Nese A, Sen S, Tasdelen MA et al (2006) Clay-PMMA nanocomposites by photoinitiated radical polymerization using intercalated phenacyl pyridinium salt initiators. Macromol Chem Phys 207:820–826
Niamsa N, Kaewtong C, Srinonmuang W et al (2013) Hybrid organic–inorganic nanomaterial sensors for selective detection of Au^{3+} using rhodamine-based modified polyacrylic acid (PAA)-coated FeNPs. Polym Chem 4:3039–3046

Niemann MU, Srinivasan SS, Phani AR et al (2008) Nanomaterials for hydrogen storage applications: a review. J Nanomater 950967

Niu H, Zhang Di, Meng Z et al (2012) Fast defluorination and removal of norfloxacin by alginate/Fe@Fe_3O_4 core/shell structured nanoparticles. J Hazard Mater 227:195–203

Nolting DD, Gore JC, Pham W (2011) Near-infrared dyes: probe development and applications in optical molecular imaging. Curr Organ Synth 8:521–534

Nyholm L, Nystrom G, Mihranyan A et al (2011) Toward flexible polymer and paper-based energy storage devices. Adv Mater 23:3751–3769

Ogoshi T, Chujo Y (2005) Organic-inorganic polymer hybrids prepared by the sol-gel method. Compos Interfaces 11:539–566

Oh JK (2008) Recent advances in emulsion and in controlled/living radical polymerization dispersion. J Polym Sci Part A-Polym Chem 46:6983–7001

Olmos D, Rodriguez-Gutierrez E, Gonzalez-Benito J (2012) Polymer structure and morphology of low density polyethylene filled with silica nanoparticles. Polym Compos 33:2009–2021

Olmos D, Gonzalez-Gaitano G, Kholkin AL et al (2013) Flexible PVDF-$BaTiO_3$ nanocomposites as potential materials for pressure sensors. Ferroelectrics 447:9–18

O'Mullane AP, Dale SE, Day TM et al (2006) Formation of polyaniline/Pt nanoparticle composite films and their electrocatalytic properties. J Sol State Electrochem 10:792–807

Ong P-L, Levitsky IA (2010) Organic/IV, III-V semiconductor hybrid solar cells. Energies 3:313–334

Ong P-L, Levitsky IA (2011) Fluorescent gas sensors based on nanoporous optical resonators (microcavities) infiltrated with sensory emissive polymers. IEEE Sens J 11:2947–2951

Orilall MC, Wiesner U (2011) Block copolymer based composition and morphology control in nanostructured hybrid materials for energy conversion and storage: solar cells, batteries, and fuel cells. Chem Soc Rev 40:520–535

Ouyang G, Wang K, Chen XY (2012) TiO_2 nanoparticles modified polydimethylsiloxane with fast response time and increased dielectric constant. J Micromechanics Microengineering 22:074002

Palmqvist AEC (2003) Synthesis of ordered mesoporous materials using surfactant liquid crystals or micellar solutions. Curr Opin Colloid Interface Sci 8:145–155

Pan B, Pan B, Zhang W et al (2009) Development of polymeric and polymer-based hybrid adsorbents for pollutants removal from waters. Chem Eng J 151:19–29

Pan Y, Deng S, Polavarapu L et al (2012) Plasmon-enhanced photocatalytic properties of Cu_2O nanowire-Au nanoparticle assemblies. Langmuir 28:12304–12310

Pandey S, Mishra SB (2011) Sol-gel derived organic-inorganic hybrid materials: synthesis, characterizations and applications. J Sol-Gel Sci Technol 59:73–94

Pantaleon R, Gonzalez-Benito J (2010) Structure and thermostability of PMMA in PMMA/silica nanocomposites: effect of high-energy ball milling and the amount of the nanofiller. Polym Compos 31:1585–1592

Papagiannouli I, Demetriou M, Chatzikyriakos G et al (2014a) Palladium micellar nanohybrids with tunable nonlinear optical response. Opt Mater 36:123–129

Papagiannouli I, Demetriou M, Krasia-Christoforou T et al (2014b) Palladium-based micellar nanohybrids: preparation and nonlinear optical response. RSC Adv 4:8779–8788

Pardo R, Zayat M, Levy D (2011) Photochromic organic-inorganic hybrid materials. Chem Soc Rev 40:672–687

Patil D, Patil P, Seo Y-K et al (2010) Poly(o-anisidine)-tin oxide nanocomposite: synthesis, characterization and application to humidity sensing. Sens Actuators B-Chem 148:41–48

Pavel FM (2004) Microemulsion polymerization. J Dispers Sci Technol 25:1–16

Peighambardoust SJ, Rowshanzamir S, Amjadi M (2010) Review of the proton exchange membranes for fuel cell applications. Int J Hydrog Energy 35:9349–9384

Pendhari SS, Kant T, Desai YM (2008) Application of polymer composites in civil construction: a general review. Composite Struct 84:114–124

Prakash S, Chakrabarty T, Singh AK et al (2013) Polymer thin films embedded with metal nanoparticles for electrochemical biosensor applications. Biosens Bioelec 41:43–53

Priya DN, Modak JM, Trebse P et al (2011) Photocatalytic degradation of dimethoate using LbL fabricated TiO2/polymer hybrid films. J Hazard Mater 195:214–222
Psarras GC, Siengchin S, Karahaliou PK et al (2011) Dielectric relaxation phenomena and dynamics in polyoxymethylene/polyurethane/alumina hybrid nanocomposites. Polym Int 60:1715–1721
Pukazhselvan D, Kumar V, Singh SK (2012) High capacity hydrogen storage: basic aspects, new developments and milestones. Nano Energy 1:566–589
Pyun J (2007) Nanocomposite materials from functional polymers and magnetic colloids. Polym Rev 47:231–263
Radhakrishnan B, Ranjan R, Brittain WJ (2006) Surface initiated polymerizations from silica nanoparticles. Soft Matter 2:386–396
Rajesh C, Tarushee A, Devendra K (2009) Recent progress in the development of nanostructured conducting polymers/nanocomposites for sensor applications. Sens Actuators B-Chem 136:275–286
Rammohan G, Nadagouda MN (2013) Green photocatalysis for degradation of organic contaminants: a review. Curr Organ Chem 17:2338–2348
Reddy CS, Das CK (2005) HLDPE/organic functionalized SiO_2 nanocomposites with improved thermal stability and mechanical properties. Compos Interfaces 11:687–699
Ruminski AM, Bardhan R, Brand A et al (2013) Synergistic enhancement of hydrogen storage and air stability via Mg nanocrystal-polymer interfacial interactions. Energy Environ Sci 6:3267–3271
Sadat-Shojai M, Ershad-Langroudi A (2009) Polymeric coatings for protection of historic monuments: opportunities and challenges. J Appl Polym Sci 112:2535–2551
Sahoo NG, Pan YZ, Li L et al (2013) Nanocomposites for bone tissue regeneration. Nanomedicine 8:639–653
Sanchez C, Belleville P, Popall M et al (2011) Applications of advanced hybrid organic-inorganic nanomaterials: from laboratory to market. Chem Soc Rev 40:696–753
Sangermano M, Messori M (2010) Scratch resistance enhancement of polymer coatings. Macromol Mater Eng 295:603–612
Sangermano M, Amerio E, Epicoco P et al (2007) Preparation and characterization of hybrid nanocomposite coatings by cationic UV-curing and the sol-gel process of a vinyl ether based system. Macromol Mater Eng 292:634–640
Sangermano M, Bongiovanni R, Longhin M et al (2009) Hybrid organic/inorganic UV-cured acrylic films with hydrophobic surface properties. Macromol Mater Eng 294:525–531
Sangermano M, Gaspari E, Vescovo L et al (2011) Enhancement of scratch resistance properties of methacrylate UV-cured coatings. Prog Organ Coat 72:287–291
Sardon H, Irusta L, Fernandez-Berridi MJ et al (2010) Synthesis of room temperature self-curable waterborne hybrid polyurethanes functionalized with (3-aminopropyl)triethoxysilane (APTES). Polymer 51:5051–5057
Sarkar S, Guibal E, Quignard F et al (2012) Polymer-supported metals and metal oxide nanoparticles: synthesis, characterization and applications. J Nanoparticle Res 14:715
Sarwar MI, Zulfiqar S, Ahmad Z (2009) Investigating the property profile of polyamide-alumina nanocomposite materials. Scripta Materialia 60:988–991
Saunders BR (2012) Hybrid polymer/nanoparticle solar cells: preparation, principles and challenges. J Colloid Interface Sci 369:1–15
Saunders BR, Turner ML (2008) Nanoparticle-polymer photovoltaic cells. Adv Colloid Interface Sci 138:1–23
Savva I, Krasia-Christoforou T (2014) Electrospun magnetoactive fibrous nanocomposites: fabrication and applications in biomedicine. In: Sabbas NP (ed) Magnetic nanoparticles synthesis, physicochemical properties and role in biomedicine. Nova Science Publishers, New York, pp 163–199
Schlaad H, Krasia T, Patrickios CS (2001) Controlled synthesis of coordination block copolymers with beta-dicarbonyl ligating segments. Macromolecules 34:7585–7588
Schmatloch S, van den Berg AMJ, Alexeev AS et al (2003) Soluble high-molecular-mass poly(ethylene oxide)s via self-organization. Macromolecules 36:9943–9949

Schork FJ, Luo YW, Smulders W et al (2005) Miniemulsion polymerization. In: Okubo M (ed) Polymer particles book series: advances in polymer science, vol 175. Springer, pp 129–255
Schubert US, Eschbaumer C (2002) Macromolecules containing bipyridine and terpyridine metal complexes: towards metallosupramolecular polymers. Angew Chemie-Int Ed 41:2893–2926
See KC, Feser JP, Chen CE et al (2010) Water-processable polymer-nanocrystal hybrids for thermoelectrics. Nano Lett 10:4664–4667
Serra-Gomez R, Gonzalez-Gaitano G, Gonzalez-Benito J (2012) Composites based on EVA and barium titanate submicrometric particles: preparation by high-energy ball milling and characterization. Polym Compos 33:1549–1556
Sessolo M, Bolink HJ (2011) Hybrid organic-inorganic light-emitting diodes. Adv Mater 23:1829–1845
Sevick-Muraca EM, Houston JP, Gurnkel M (2002) Fluorescence-enhanced, near infrared diagnostic imaging with contrast agents. Curr Opin Chem Biol 6:642–650
Shen M, Shi X (2010) Dendrimer-based organic/inorganic hybrid nanoparticles in biomedical applications. Nanoscale 2:1596–1610
Shimakoshi H, Nishi M, Tanaka A et al (2011) Photocatalytic function of a polymer-supported B12 complex with a ruthenium trisbipyridine photosensitizer. Chem Commun 47:6548–6550
Shirsat MD, Bangar MA, Deshusses MA et al (2009) Polyaniline nanowires-gold nanoparticles hybrid network based chemiresistive hydrogen sulfide sensor. Appl Phys Lett 94:083502
Sigalas M, Garcia N (2000) Theoretical study of three dimensional elastic band gaps with the finite difference time domain method. J Appl Phys 87:3122–3125
Solans C, Izquierdo P, Nolla J et al (2005) Nano-emulsions. Curr Opin Colloid Interface Sci 10:102–110
Soler-Illia GJAA, Azzaroni O (2011) Multifunctional hybrids by combining ordered mesoporous materials and macromolecular building blocks. Chem Soc Rev 40:1107–1150
Soler-Illia GJDA, Crepaldi EL, Grosso D et al (2003) Block copolymer-templated mesoporous oxides. Curr Opin Colloid Interface Sci 8:109–126
Song X, Wang X, Wang H et al (2008) PMMA–silica hybrid thin films with enhanced thermal properties prepared via a non-hydrolytic sol–gel process. Mater Chem Phys 109:143–147
Soundrapandian C, Sa B, Datta S (2009) Organic-inorganic composites for bone drug delivery. AAPS Pharm Sci Tech 10:1158–1171
Sridhar R, Sundarrajan S, Venugopal JR et al (2013) Electrospun inorganic and polymer composite nanofibers for biomedical applications. J Biomater Sci Polym Ed 24:365–385
Srivastava S, Kotov NA (2008) Composite Layer-by-Layer (LBL) assembly with inorganic nanoparticles and nanowires. Acc Chem Res 41:1831–1841
Stefanescu EA, Daranga C, Stefanescu C (2009) Insight into the broad field of polymer nanocomposites: from carbon nanotubes to clay nanoplatelets, via metal nanoparticles. Materials 2:2095–2153
Subramania G, Constant K, Biswas R, Sigalas MM, Ho KM (1999) Optical photonic crystals fabricated from colloidal systems. Appl Phys Lett 74:3933–3935
Suppes GM, Deore BA, Freund MS (2008) Porous conducting polymer/heteropolyoxometalate hybrid material for electrochemical supercapacitor applications. Langmuir 24:1064–1069
Swierczewska M, Lee S, Chen X (2011) Inorganic nanoparticles for multimodal molecular imaging. Mol Imaging 10:3–16
Tamaki R, Samura K, Chujo Y (1998) Synthesis of polystyrene and silica gel polymer hybrids via pi-pi interactions. Chem Commun 10:1131–1132
Tao P, Li Y, Rungta A et al (2011) TiO_2 nanocomposites with high refractive index and transparency. J Mater Chem 21:18623–18629
Tao P, Viswanath A, Li Y et al (2013) Bulk transparent epoxy nanocomposites filled with poly(glycidyl methacrylate) brush-grafted TiO2 nanoparticles. Polymer 54:1639–1646
Teo WE, Ramakrishna SA (2006) A review on electrospinning design and nanofibre assemblies. Nanotechnology 17:R89–R106
Tessler N, Pinner DJ, Ho PKH (2001) Optoelectronic devices based on hybrid organic-inorganic structures. Opt Mater 17:155–160

Tew GN, Aamer KA, Shunmugam R (2006) Novel block copolymers with terpyridine pendant groups. In: Schubert US, Newkome GR, Manners I (eds) Metal-containing and metallosupramolecular polymers and materials book series: ACS symposium series. Am Chem Soc 928:126–140

Thevenot J, Oliveira H, Sandre O et al (2013) Magnetic responsive polymer composite materials. Chem Soc Rev 42:7099–7116

Thien-Phap N (2011) Polymer-based nanocomposites for organic optoelectronic devices. A review. Surf Coat Technol 206:742–752

Tomczak N, Janczewski D, Han M et al (2009) Designer polymer-quantum dot architectures. Prog Polym Sci 34:393–430

Tomczak N, Liu R, Vancso JG (2013) Polymer-coated quantum dots. Nanoscale 5:12018–12032

Tomoki O, Hideaki I, Kyung-Min K et al (2001) Synthesis of organic–inorganic polymer hybrids having interpenetrating polymer network structure by formation of ruthenium–bipyridyl complex. Macromolecules 35:334–338

Trachtenberg JE, Mountziaris PM, Kasper FK et al (2013) Fiber-based composite tissue engineering scaffolds for drug delivery. Israel J Chem 53:646–654

Tripathi BP, Shahi VK (2011) Organic–inorganic nanocomposite polymer electrolyte membranes for fuel cell applications. Prog Polym Sci 36:945–979

Tsigara A, Mountrichas G, Gatsouli K et al (2007) Hybrid polymer/cobalt chloride humidity sensors based on optical diffraction. Sens Actuators B-Chem 120:481–486

Tsiourvas D, Tsetsekou A, Papavasiliou A et al (2013) A novel hybrid sol-gel method for the synthesis of highly porous silica employing hyperbranched poly(ethyleneimine) as a reactive template. Microporous Mesoporous Mater 175:59–66

Tsuchiya K, Nagayasu S, Okamoto S et al (2008) Nonlinear optical properties of gold nanoparticles selectively introduced into the periodic microdomains of block copolymers. Opt Express 16:5362–5371

Uilk JM, Mera AE, Fox RB et al (2003) Hydrosilation-cured poly(dimethylsiloxane) networks: Intrinsic contact angles via dynamic contact angle analysis. Macromolecules 36:3689–3694

Vaynzof Y, Kabra D, Brenner TJK et al (2012) Recent advances in hybrid optoelectronics. Israel J Chem 52:496–517

Vertuccio L, Gorrasi G, Sorrentino A et al (2009) Nano clay reinforced PCL/starch blends obtained by high energy ball milling. Carbohydr Polym 75:172–179

Viseras C, Aguzzi C, Cerezo P et al (2008) Biopolymer-clay nanocomposites for controlled drug delivery. Mater Sci Technol 24:1020–1026

Viseras C, Cerezo P, Sanchez R et al (2010) Current challenges in clay minerals for drug delivery. Appl Clay Sci 48:291–295

Wahl DA, Czernuszka JT (2006) Collagen-hydroxyapatite composites for hard tissue repair. Eur Cells Mater 11:43–56

Walcarius A (2001) Electrochemical applications of silica-based organic-inorganic hybrid materials. Chem Mater 3:3351–3372

Wang X, Uchiyama S (2013) Polymers for biosensor construction. In: Rinken T (ed) State of the art in biosensors—general aspects. InTech, pp 67–86

Wang Q, Zhu L (2011) Polymer nanocomposites for electrical energy storage. J Polym Sci Part B-Polym Phys 49:1421–1429

Wang YXJ, Hussain SM, Krestin GP (2001) Superparamagnetic iron oxide contrast agents: physicochemical characteristics and applications in MR imaging. Eur Radiol 11:2319–2331

Wang Y, Xie XB, Goodson T (2005) Enhanced third-order nonlinear optical properties in dendrimer-metal nanocomposites. Nano Lett 5:2379–2384

Wang G, Zhang L, Zhang J (2012) A review of electrode materials for electrochemical supercapacitors Chem Soc Rev 41:797–828

Wang S, Kang Y, Wang L et al (2013) Organic-inorganic hybrid sensors: a review. Sens Actuators B-Chem 182:467–481

Wei X, Jiao L, Sun J et al (2010) Synthesis, electrochemical, and gas sensitivity performance of polyaniline/MoO_3 hybrid materials. J Sol State Electrochem 14:197–202

Weiss CK, Landfester K (2010) Miniemulsion polymerization as a means to encapsulate organic and inorganic materials. In: VanHerk AM, Landfester K (eds) Hybrid latex particles: preparation with (mini) emulsion polymerization book series: advances in polymer science, vol 233. Springer, pp 185–236

Wen JY, Wilkes GL (1996) Organic/inorganic hybrid network materials by the sol-gel approach. Chem Mater 8:1667–1681

West R, Wang Y, Goodson T (2003) Nonlinear absorption properties in novel gold nanostructured topologies. J Phys Chem B 107:3419–3426

Whittell GR, Hager MD, Schubert US, Manners I (2011) Functional soft materials from metallopolymers and metallosupramolecular polymers. Nat Mater 10:176–188

Wiesbrock F, Hoogenboom R, Schubert US (2004) Microwave-assisted polymer synthesis: state-of-the-art and future perspectives. Maromol Rapid Commun 25:1739–1764

Woo S, Jeong JH, Lyu HK et al (2012) Hybrid solar cells with conducting polymers and vertically aligned silicon nanowire arrays: The effect of silicon conductivity. Physica B-Condensed Matter 407:3059–3062

Wright M, Uddin A (2012) Organic-inorganic hybrid solar cells: a comparative review. Sol Energy Mater Sol Cells 107:87–111

Wu G, Li L, Li JH et al (2005) Polyaniline-carbon composite films as supports of Pt and PtRu particles for methanol electrooxidation. Carbon 43:2579–2587

Wu C-J, Gaharwar AK, Schexnailder PJ et al (2010) Development of biomedical polymer-silicate nanocomposites: a materials science perspective. Materials 3:2986–3005

Xenogiannopoulou E, Iliopoulos K, Couris S et al (2008) Third-order nonlinear optical response of gold-island films. Adv Funct Mater 18:1281–1289

Xia XP, Cai SZ, Xie CS (2006) Preparation, structure and thermal stability of Cu/LDPE nanocomposites. Mater Chem Phys 95:122–129

Xiong P, Wang L, Sun X et al (2013) Ternary titania-cobalt-ferrite-polyaniline nanocomposite: a magnetically recyclable hybrid for adsorption and photodegradation of dyes under visible irradiation. Industrial Eng Chem Res 52:10105–10113

Xu F, Mu S (2014) Nanoceramic oxide hybrid electrolyte membranes for proton exchange membrane fuel cells. J Nanosci Nanotechnol 14:1169–1180

Xu M, Zhang J, Wang S et al (2010) Gas sensing properties of SnO_2 hollow spheres/polythiophene inorganic-organic hybrids. Sens Actuators B-Chem 146:8–13

Xu S, Gu L, Wu K et al (2012) The influence of the oxidation degree of poly(3-hexylthiophene) on the photocatalytic activity of poly(3-hexylthiophene)/TiO_2 composites. Sol Energy Mater Sol Cells 96:286–291

Yan J, Wang Q, Wei T et al (2014a) Recent advances in design and fabrication of electrochemical supercapacitors with high energy densities. Adv Energy Mater 4:xx–xx

Yan K, Li H, Wang X et al (2014b) Self-assembled magnetic luminescent hybrid micelles containing rare earth Eu for dual-modality MR and optical imaging. J Mater Chem B 2:546–555

Ye Y-S, Shen W-C, Tseng C-Y et al (2011) Versatile grafting approaches to star-shaped POSS-containing hybrid polymers using RAFT polymerization and click chemistry. Chem Commun 47:10656–10658

Yu B, Guo S, He L et al (2013) Synthesis and characterization of a luminescence metallosupramolecular hyperbranched polymer. Chem Commun 49:3333–3335

Yusuf Yagci Y (2012) New photoinitiating systems designed for polymer/inorganic hybrid nano-coatings. J Coat Technol Res 9:125–134

Zaharieva J, Milanova M, Todorovsky D (2011) SiO_2/polyester hybrid for immobilization of Ru(II) complex as optical gas-phase oxygen sensor. J Mater Chem 21:4893–4903

Zaman A, Gutub SA, Wafa MA (2013) A review on FRP composites applications and durability concerns in the construction sector. J Reinf Plast composites 32:1966–1988

Zampetti E, Pantalei S, Muzyczuk A et al (2013) A high sensitive NO_2 gas sensor based on PEDOT-PSS/TiO_2 nanofibres. Sens Actuators B-Chem 176:390–398

Zanardi C, Terzi F, Seeber R (2013) Polythiophene and polythiophene-based composites in amperometric sensing. Anal Bioanal Chem 405:509–531

Zeng QH, Yu AB, Lu GQ et al (2005) Clay-based polymer nanocomposites: research and commercial development. J Nanosci Nanotechnol 5:1574–1592

Zhang H, Hou X (2012) Development of organic polymer/inorganic semiconductor hybrid solar cells. Prog Chem 24:2106–2115

Zhang J, Wang S, Xu M et al (2009) Polypyrrole-coated SnO_2 hollow spheres and their application for ammonia sensor. J Phys Chem C 113: 1662–1665

Zhang P, Weng Z, Guo J et al (2011a) Solution-dispersible, colloidal, conjugated porous polymer networks with entrapped palladium nanocrystals for heterogeneous catalysis of the Suzuki-Miyaura coupling reaction. Chem Mater 23:5243–5249

Zhang SW, Ren L, Jiang JQ et al (2011b) Facile synthesis of waterborne UV-curable polyurethane/silica nanocomposites and morphology, physical properties of its nanostructured films. Prog Organ Coat 70:1–8

Zhang J, Liu X, Wu S et al (2013a) One-pot fabrication of uniform polypyrrole/Au nanocomposites and investigation for gas sensing. Sens Actuators B-Chem 186:695–700

Zhang K, Zha Y, Bo Peng B et al (2013b) Metallo-supramolecular cyclic polymers. J Am Chem Soc 135:15994–15997

Zhao L, Lin Z (2012) Crafting semiconductor organic-inorganic nanocomposites via placing conjugated polymers in intimate contact with nanocrystals for hybrid solar cells. Adv Mater 24: 4353–4368

Zhao H, Li L, Yang J et al (2008) Nanostructured polypyrrole/carbon composite as Pt catalyst support for fuel cell applications. J Power Sour 184:375–380

Zhou Y, Xian H, Li F et al (2010) Construction of hybrid nanocomposites containing Pt nanoparticles and poly(3-methylthiophene) nanorods at a glassy carbon electrode: characterization, electrochemistry, and electrocatalysis. Electrochem Acta 55:5905–5910

Zhou L, Yuan J, Wei Y (2011) Core-shell structural iron oxide hybrid nanoparticles: from controlled synthesis to biomedical applications. J Mater Chem 21:2823–2840

Zhou Z, Wang Q, Tan C (2014) Soft matter anion sensing based on lanthanide (Eu^{3+} and Tb^{3+}) luminescent hydrogels. Soft Mater 12:98–102

Zhu YG, Li ZQ, Zhang D et al (2006) ABS/iron nanocomposites prepared by cryomilling. J Appl Polym Sci 99:501–505

Zhu Y, Li Z, Zhang D (2008) Electromagnetic nanocomposites prepared by cryomilling of polyaniline and Fe nanoparticles. J Polym Sci Part B: Polymer Physics 46:1571–1576

Zou H, Wu S, Shen J (2008) Polymer/silica nanocomposites: preparation, characterization, properties, and applications. Chem Rev 108:3893–3957

Chapter 3
Polymer-Tethered Nanoparticle Materials—An Emerging Platform for Multifunctional Hybrid Materials

Golda L. Chakkalakal, Subramanian Ramakrishnan and Michael R. Bockstaller

Abstract The grafting of polymeric chains to inorganic (as well as organic) particle interfaces has become an indispensable tool to engineer the physicochemical and/or biochemical properties of material interfaces. For example, polymer grafting is ubiquitously being used to compatibilize particles to polymer matrices to augment the properties of polymers in applications such as biomedical devices, lightweight aircraft wings, energy generation and storage, and for separation and environmental remediation to name a few. The recent emergence of surface-initiated controlled radical polymerization has further expanded the scope of polymer-grafted particulate materials, as the precise control of the structure of the polymer grafts offers new opportunities to tailor the properties of polymer-grafted particle systems. This chapter summarizes recent developments in synthesis of polymer-tethered nanoparticle interfaces that have afforded this fine control in the structure and properties of the resultant composite. Particular emphasis is given to the concept of "one-component hybrid materials"—that is the ability to synthesize multifunctional nanocomposite materials by the self-assembly of polymer-tethered particle systems. The role of polymer-graft modification on the interaction, dynamics, and assembly of particle brush materials is discussed to provide the context to showcase studies that have demonstrated the opportunity to harness the precision-engineered polymer-grafted particle systems for the fabrication of innovative nanocomposite material technologies.

S. Ramakrishnan (✉) · G. L. Chakkalakal
Department of Chemical and Biomedical Engineering, FAMU-FSU College of Engineering, 2525 Pottsdamer Street, Tallahassee, FL 32310, USA
e-mail: sramakrishnan@fsu.edu

M. R. Bockstaller
Department of Materials Science and Engineering, Carnegie Mellon University, 5000 Forbes Ave, Pittsburgh, PA 15213, USA
e-mail: Bockstaller@cmu.edu

C.-S. Kim et al. (eds.), *Hybrid and Hierarchical Composite Materials*,
DOI 10.1007/978-3-319-12868-9_3

3.1 Introduction

The grafting of polymeric chains to inorganic (as well as organic) interfaces has become an indispensable tool to engineer the physicochemical and/or biochemical properties of material interfaces (Advincula 2006; Barbey et al. 2009; Milner 1991; Tsujii et al. 2006). Polymer brushes—a term coined by de Gennes to describe the molecular brush-like conformation of densely grafted polymer chains—have gained technological relevance in the context of a wide range of material technologies. Classic examples for applications of polymer-tethered interfaces encompass the design of surfaces with controlled wetting, corrosive, or tribological characteristics (Advincula 2006). Polymer brushes have also emerged as an enabling technology in areas ranging from biomedical devices and energy generation and storage to separation and environmental remediation (Advincula 2006). In general, the properties of polymer brushes intimately depend on the conformation of grafted chains—spatial distribution of chains as one moves away from the tethered surface. In the absence of chain–surface interactions, the conformation of surface-tethered chains is predominantly determined by excluded volume interactions between adjacent chains that, in turn, depend on the distance between the grafted chains (Milner 1991). There are two distinguishable regimes (Fig. 3.1) that depend on the grafting density (or the distance between grafted chains, D)—sparse and densely grafted regimes: first, the sparse grafting regime with $D > 2R_G$ in which chains assume a relaxed conformation (often called "mushroom" structure); second, the dense grafting regime for $D < 2R_G$ in which excluded volume interactions give rise to stretched chain conformations (similar to a "molecular brush"). Here, the conformation relates to the particular set of bond and torsional angles that identify the three-dimensional structure of polymer chains, D denotes the distance between adjacent tethered chains and R_G denotes the chains' radius of gyration, that is—in the case of ideal chains—related to the number n and length l of segments via $R_G^2=(1/6)nl^2$. It should be noted that the classification of polymer brushes into unique and distinct conformational regimes based on grafting density is applicable only to macroscopic surfaces. This is because for nanostructured surfaces—i.e., surfaces with topographical features comparable to the polymer chain dimension—the curvature of the surface implies a systematic variation of the polymer segment density and thus of the chain conformation within the brush.

Particle brushes (or polymer-tethered inorganic (nano) material interfaces) hence represent a special case of the general "polymer brush" (Fig. 3.1). Here, the surface is presented by the outer surface of a nanoparticle. The convex curvature of the particle surface gives rise to structural transitions of the grafted chains (or confirmations) such that chains are more stretched in the vicinity of the surface and assume more relaxed conformations at greater distance from the surface (Binder and Milchev 2012). These conformational transitions are particularly relevant in particle brushes in which the particle core radius equals the dimension of tethered chains and have been shown to influence the structure formation, interaction, as well as dynamic properties of particle brush materials (Choi et al. 2010, 2012, 2013). The sensitive dependence of the physicochemical properties on the architecture of particle brush materials has fueled research into synthetic methodologies to control

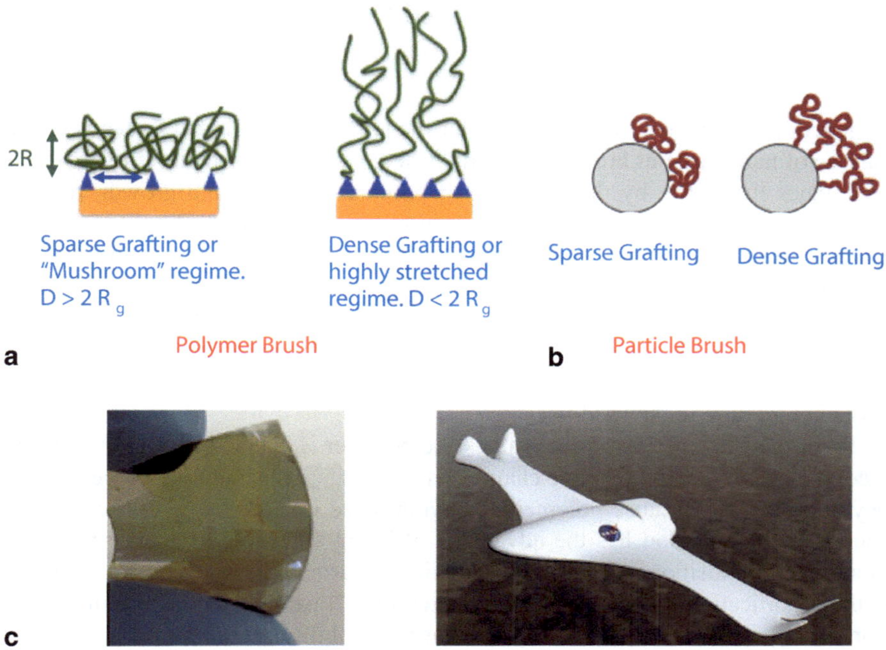

Fig. 3.1 **a** Schematic of a polymer brush. There are two distinguishable regimes that depend on the grafting density (or the distance between grafted chains—D). First, the sparse grafting regime with $D>2R_G$ in which chains assume a relaxed conformation (often called "mushroom" structure); second, the dense grafting regime for $D<2R_G$ in which excluded volume interactions give rise to stretched chain conformations. **b** Schematic of a particle brush. The convex curvature of the particle surface gives rise to structural transitions of the grafted chains (or confirmations) such that chains are more stretched in the vicinity of the surface and assume more relaxed conformations at a greater distance from the surface. **c** Examples to illustrate the use of particle brushes in different applications—resonant structures for transparent conductors and lightweight aircraft coatings

the architecture of grafted polymer chains and the implications of polymer-graft modification on the material properties and processability of particle brush materials (Fernandes et al. 2013; Hui et al. 2014).

The objective of this chapter is to review the recent advancements in the molecular engineering of polymer-tethered inorganic (nano) material interfaces as well as to highlight experimental studies that illustrate some of the opportunities to facilitate novel functional materials based on polymer-tethered particulates. Examples have been chosen with the aim to illustrate "core concepts" and potentially "transformative ideas and directions" in order to provide a context for ongoing research and to stimulate the reader to further pursue this fascinating emerging field. First, we discuss developments in synthesis of polymer-tethered nanoparticle interfaces (Sect. 3.2). This is followed by the role of polymer-graft modification on the interaction, dynamics, and assembly of particle brush materials (Sect. 3.3). Finally, we conclude by highlighting studies that hold promise and might result in novel technologies which utilize the particle brush materials (Sect. 3.4).

3.2 Synthesis of Polymer-Tethered (Nano) Particle Interfaces

In general, two distinct methodologies for the tethering of polymeric chains to the surface of nanoparticles are being distinguished, i.e., *grafting-to* and *grafting-from* techniques that differ by the sequence of polymer coupling and growth, respectively. (The topic of grafting techniques is also covered in section 2.2.9.) In the following, the key distinguishing features of the two techniques will be contrasted and the particular relevance of *grafting-from* techniques to the synthesis of well-defined particle brush materials will be illustrated. We then present a detailed discussion of the most relevant form of *grafting-from* techniques—surface-initiated atom transfer radical polymerization (SI-ATRP)—to enable the reader to benefit from the recent surge in literature in the field.

In *grafting-to* reactions, a presynthesized polymer is coupled to the particle surface, typically by using reactive chain ends to facilitate the coupling process (Barbey et al. 2009). Because polymers are synthesized independently of the particle chemistry, this method has the advantage of being amenable to a wide range of monomer compositions as well as scalability. A particularly well-known example is the modification of gold (and other heavy metal) nanocrystals with thiol-functionalized polymers. Due to their experimental viability, *grafting-to* modification reactions have been among the most widely used polymer-modification techniques, for example, to achieve compatibilization of inorganic particles to organic embedding media (Ojha et al. 2013). Major advantages of the *grafting-to* functionalization process involve the wide range of possible polymer chemistries as well as the ability to control chain architectures such as linear, comb, or star (as long as a proper end-functionality allows for tethering of the chain to the surface), as well as the scalability and possibility of continuous processing. A major disadvantage of the *grafting-to* process is the limitation to low graft densities and degree of polymerization of surface-grafted chains. While the first is a consequence of the "screening effect" that is exerted by surface-bonded chains, the latter results from the reduced number of chain ends in high molecular weight (MW) polymeric systems.

In contrast to *grafting-to* techniques, the polymerization of polymers off the particle surface—i.e., the so-called *grafting-from* the family of synthetic techniques—facilitates a high level of control of all structural characteristics of tethered polymer chains. In particular, *grafting-from* techniques allow superior control of the MW and density of grafted chains. The enhanced control is due to the sequential nature of the *grafting-from* process that involves two steps: first, the binding of low-molecular initiator molecules to the surface of particles, and second, the polymerization of monomers off the initiator-modified surface. Because the coupling step involves low molecular initiator species, *grafting-from* reactions generally provide a higher level of control of the density of tethered chains. An important driving force for the surge of interest in *grafting-from* modification reactions has been the advancement of surface-initiated polymerization techniques that provide a range of versatile and efficient methodologies to precisely grow polymer chains off an initiator-modified surface (Barbey et al. 2009). Surface-initiated controlled radical polymerization (SI-

CRP) has been particularly successfully applied to the synthesis of polymer brushes due to the combination of versatility and the high level of control over molecular parameters such as composition, dispersity, and chain connectivity (Hui et al. 2014). Among the most relevant forms of SI-CRP are SI-ATRP, surface-initiated nitroxide-mediated polymerization (SI-NMP), and surface-initiated reversible-addition fragmentation chain transfer polymerization (SI-RAFT)—the reader will find comprehensive review articles describing the application of these techniques to the synthesis of (planar) polymer brushes in the literature (Edmondson et al. 2004; Hui et al. 2014; Tsujii et al. 2006). More recently, SI-CRP techniques have been applied to facilitate precise control of the architecture of particle brush materials. While the relevant parameters mimic those of planar surfaces, additional challenges arise, for example, due to the need to retain stability of particle dispersions during polymerization reactions or due to the different reactivity of nanoscale surfaces as compared to their macroscopic counterpart (Matyjaszewski and Tsarevsky 2009). The objective of this section is to provide a status report on synthetic strategies focusing on well-defined particle brush materials and to showcase examples of recent research that illustrate the potential for innovative material technologies based on particle brush materials. The focus of the discussion will be on the most relevant concepts, for a more comprehensive discussion of technical details of the surface polymerization process the reader is referred to the reference (Hui et al. 2014) and citations therein.

3.2.1 Surface Modification by Surface-Initiated Atom Transfer Controlled Radical Polymerization

Atom transfer radical polymerization is among the most efficient and robust controlled radical polymerization techniques (Braunecker and Matyjaszewski 2007). Conceptually, the idea of controlled radical polymerization processes (such as ATRP) is that by cycling growing chains through a sequence of dormant and active states (where only in the active state the chain is able to add repeat units), polymerization occurs via a "living-type" mechanism that minimizes irreversible termination reactions and gives rise to the first-order reaction kinetics (Matyjaszewski et al. 1997; Wang and Matyjaszewski 1995). Generation of active and dormant states is accomplished by the reversible dissociation of halide end groups in the presence of a Cu(I/II) catalyst that generates active radicals. The ATRP process is hence often represented by the illustration of the "ATRP equilibrium" that is shown in Fig. 3.2. For a comprehensive review on the mechanism of ATRP, the reader is referred to the reference (Matyjaszewski and Xia 2001).

Surface-initiated atom transfer radical polymerization refers to a subset of ATRP reactions in which the initiator (i.e., the halogenated species) is bound to the surface (planar or curved) as illustrated in Fig. 3.3. The upper right insert of Fig. 3.3 shows a typical relationship between polymerization rate (R_p) and dispersity (M_w/M_n) on kinetic parameters as well as reagent concentrations. R_p depends on the propagation rate constant (k_p), and the ratio of the rate constants of activation (k_a) and deactivation (k_{da}) and on the concentrations of the reagents involved in the process.

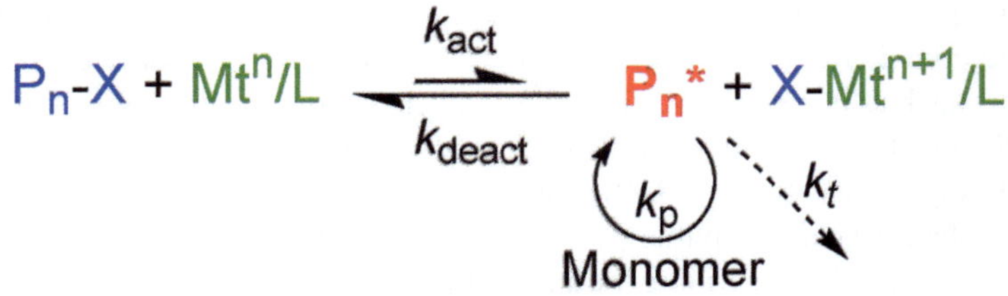

Fig. 3.2 Principal reaction scheme of an atom-transfer controlled radical polymerization process. P_n–X denotes the halide functionalized growing chain (dormant state) that upon reaction with the metal/ligand catalyst forms the active radical species P_n^*

Molecular weight disparity decreases with increasing deactivation rate—this can be accomplished, for example, by catalysts with a higher value of k_{da} and at increased deactivator concentrations. The dispersity decreases with the monomer conversion and generally decreases with increasing MW (smaller $[P_n–X]_0$; Matyjaszewski 2012).

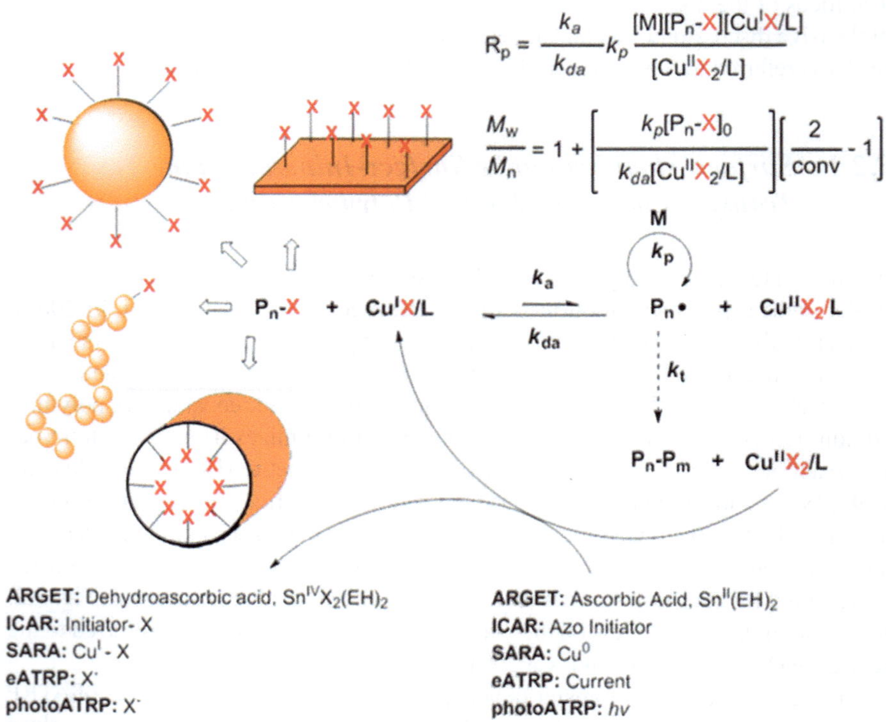

Fig. 3.3 Illustration of the reaction scheme of SI-ATRP as well as the distinct low-catalyst variants of ATRP. Indicated as "X" is the surface-bound initiator species. Equation shows the governing parameters determining the rate of polymerization R_p as well as molecular weight distribution M. (Figure reproduced from Hui et al. (2014) with permission from the American Chemical Society). *SI-ATRP* surface-initiated atom transfer radical polymerization

The need of Cu catalyst in ATRP can present a challenge since it introduces impurities that might limit the application of products and require extensive purification. However, during the past several years, it has become possible to reduce the amount of catalysts to a few parts per million (ppm) in the presence of various reducing agents (Tsarevsky and Matyjaszewski 2007). The bottom part of Fig. 3.3 provides a summary of the different ATRP variants allowing for a reduced catalyst concentration. The distinct variants share the objective to regenerate "lost" catalyst by the addition of a reducing agent—the different ATRP variants are then distinguished based on the particular type of reducing mechanism. For example, in activators regenerated by electron transfer ATRP (ARGET ATRP) reducing agents such as ascorbic acid, sugars, tin(II) octanoate, sulfites, or amines are being applied (Jakubowski and Matyjaszewski 2006). In the initiators for continuous activator regeneration ATRP process (ICAR ATRP), conventional radical initiators are being used (Matyjaszewski et al. 2006). The "supplemental activators and reducing agents ATRP" process (SARA ATRP) is based on zerovalent metals (Abreu et al. 2012). Electrochemistry, that can be run under either potentiostatic or galvanostatic conditions, is used in electrochemical ATRP (eATRP; Magenau et al. 2011). In all these systems, the regeneration of lost catalyst allows the reduction of the amount of Cu(I/II) to the ppm level and hence substantially reduces the need for purification of the final product. However, it should be noted that the decrease of catalyst concentration beyond a threshold value can also cause problems. For example, it may be accompanied by the increase of MW disparity (see equation in inset of Fig. 3.3). This relation of catalyst concentration and MW disparity offers new opportunities to synthesize polymers with designed dispersity that could exhibit favorable physicochemical properties for some applications. For example, disperse chain ligands have been proposed to enhance the miscibility of particle brushes in polymer embedding media (Ojha et al. 2013). Note that if the contribution of radical termination is negligible, the catalyst concentration can be reduced even without reducing agents. Since k_p/k_t (with k_t denoting the rate constant for termination reactions) values are 30 times larger in polymerization of acrylates than those of methacrylates, chain termination is much less likely to occur and hence a lower amount of Cu catalyst can be used (Zhong and Matyjaszewski 2011a). Whether special care needs to be taken to regenerate lost catalyst will hence depend on the particular system under study. In general, acrylate-based monomers can be polymerized with native ATRP (or SI-ATRP) while the reduced Cu variants are particularly helpful for the polymerization of styrenic monomers. In the following section, the basic features of SI-ATRP are summarized and the control parameters for tailoring the composition, architecture, and properties of particle brush materials are discussed.

3.2.2 Fundamentals of SI-ATRP

SI-ATRP denotes the application of ARTP to functionalize organic or inorganic interfaces. SI-ATRP utilizes the same mechanism as a conventional ATRP from an untethered initiator, which involves initiation, propagation, activation/deactivation, and termination processes. There are, however, several differences due to the confinement of the initiating group at the (inorganic) surface. In particular, the rate of propagation could be limited due to constraints on the diffusion of monomer to the chain ends, thus affecting chain growth. Second, the slow generation of deactivators mandates the addition of sacrificial initiator or deactivator in SI-ATRP. Note that in the limit of small deactivator concentration, irreversible radical termination can dominate. In typical particle brush systems—due to the large number of tethered chains—even very small amounts of interparticle termination will result in macroscopic gelation. SI-ATRP has been applied to various substrates, including alumina, iron oxides, germanium, gold, titania, silica, quantum dots, as well as a range of organic and biological materials (Barbey et al. 2009; Hui et al. 2014). The following sections describe specific aspects of the application of SI-ATRP to the modification of inorganic surfaces.

3.2.2.1 Initiation

SI-ATRP has been applied to modify the surface of many different inorganic (as well as organic or biological) materials. Of the different inorganic materials, silica, and iron oxide have received the most attention, due to their availability, physical properties, and potential technological applications. Coupling to silica can readily be achieved by reacting the silanol (Si–OH) surface functionalities with reactive groups such as chloro- or alkoxy silane (e.g., Cl–Si–, MeO–Si–, or EtO–Si– are frequently used) that are capable of forming stable siloxane bonds. Organic acids have been found to act as effective ligands for metal oxide systems. The modification of metal (oxide) surfaces with organic acids has been widely observed to follow Pearson's concept of hard/soft acid/base. Thus, coupling chemistries should be selected to facilitate hard–hard (ionic character) and soft–soft (covalent character) core-ligand interactions. For example, phosphoric acid, carboxylic acid, catechol, or acyl halides are suitable functional groups to modify the surface of Fe by forming stable Fe–O bonds. Figure 3.4 summarizes different coupling chemistries that are suitable for coupling initiator groups to a variety of organic or inorganic interfaces.

The properties of particle brush materials (as well as other composite materials) can be finely tuned by controlling the grafting density at the organic/inorganic interface (Fernandes et al. 2013). One effective strategy to control the initiator density is by intermixing the initiator with a nonreactive "dummy" coupling agent as illustrated in Fig. 3.4b. In this case, the distribution of active and inactive centers across the particle surface is generally thought to be stochastic (Bombalski et al. 2007).

Just as in the case of conventional ATRP initiators, the appropriate ATRP initiating functionality on a surface-bound initiator should be selected to assure that the rate of activation is equal to or greater than the rate of propagation as this is a prerequisite for high initiation efficiency and narrow MW distribution (MWD). In general, the activation rate constant is expected to increase with initiator substitution from primary to tertiary, as well as with better radical stabilizing groups; CN > Ph > C(O)OR (Matyjaszewski and Xia 2001; radical stabilization is generally increased with increasing delocalization of the radical).

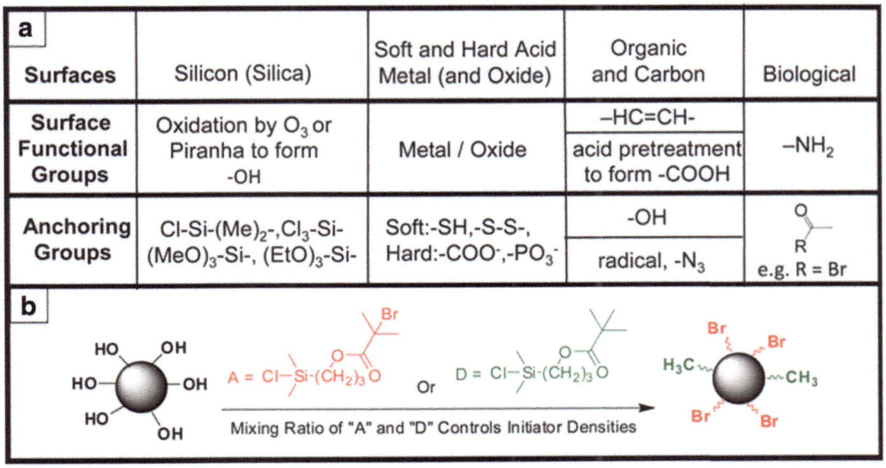

Fig. 3.4 Examples of anchoring groups for surface modification (**a**), and a strategy to control grafting density via inactive ("dummy") initiators (**b**). (Figure reproduced from Hui et al. 2014 with permission from the American Chemical Society)

3.2.2.2 Propagation

The MW and MWD during SI-ATRP reactions are typically evaluated from untethered polymer chains that are formed by small amounts of purposefully added sacrificial initiators. An analysis of MW and MWD has shown that the reaction kinetics in SI-ATRP follows similar trends as conventional ATRP; however, deviations can be observed due to steric confinement that can interfere with the chain growth process. The "confinement effect" relates to the slowdown of monomer and catalyst diffusion in situations of steric confinement that can occur, for example, if surface polymerization is performed on the inside of nanotubes or framework structures. Furthermore, differences between surface-initiated and bulk ATRP reactions have been reported depending on the choice of initiators, catalysts, and reaction conditions, such as solvent, targeted MW, temperature, concentration, etc.

3.2.2.3 Exchange Reactions

Controlling the concentration of deactivator in SI-ATRP is a prerequisite to facilitate a controlled/living mechanism that enables well-defined molecular characteristics of tethered chains. The impact of deactivator concentration on the MW and MWD can be deduced from the equation shown in the inset of Fig. 3.3. While for regular ATRP performed in the bulk or homogeneous state the control of deactivator concentration is not a problem, challenges arise in the case of SI-ATRP due to the low concentration of tethered initiators (that imply a small concentration of deactivator). The challenge can be resolved by addition of sacrificial initiator before or during the polymerization process thus resulting in more uniform growth, higher grafting density, and higher MW.

3.2.2.4 Termination

Termination in SI-ATRP is of central importance since it reduces the number of chain end functionalities and—in the case of interparticle termination—can result in macroscopic gelation of the particle brush system. The mechanism of termination reactions during SI-ATRP is complex and remains the subject of research; however, several important general aspects have been highlighted in the literature that will be discussed in the following.

Because the rate of intraparticle termination (i.e., the recombination of adjacent radicals on a surface) is dependent on the average distance between radicals, the termination rate coefficient is proportional to the catalyst concentration. Thus, intraparticle termination is expected to increase with catalyst concentration (Gao et al. 2010). A second termination mechanism might be through interparticle radical recombination. This process is generally considered as undesirable since small fractions of interparticle termination can induce macroscopic gelation of the system. The latter is a particular challenge in the case of small particle systems due to the increase of particle number density (at a constant mass concentration) as well as surface area. To prevent interparticle termination, the concentration of terminated chains should be small, for example, by conducting the polymerization at a slower rate, by limitation of the monomer conversion or by the reduction of the targeted MW (Carrot et al. 2001; Zhong and Matyjaszewski 2011b). However, achieving these conditions still presents a challenge for monomers with propagation-to-termination rate ratio such as styrene or methyl methacrylate. Here, miniemulsion methods can provide a solution since the "trapping" of particles within miniemulsion droplets or vesicles prevents interparticle termination even at a high concentration (Pyun et al. 2003). However, it should be considered that taking full advantage of miniemulsion techniques requires empirical determination of the phase behavior of particle/monomer/surfactant/solvent systems, and ideal process conditions are typically particular to a given material system. If the required data is not available then the performance of SI-ATRP at elevated pressures can be an alternative, since under high-pressure conditions the termination rate constants decrease and the propagation rate constants increase (Pietrasik et al. 2011). For example, the rapid synthesis of hybrid particles with polymethyl methacrylate (PMMA) or polybutyl methacrylate (PBMA) grafts with $M_n > 10^6$ g/mol at room temperature was recently demonstrated at 6 kbar pressure.

3.2.3 Toward More Complex Polymer-Graft Architectures with SI-ATRP

While the above sections discussed the basic SI-ATRP process, it is important to note that the ATRP mechanism also allows the fine control of a range of the "low-level" characteristics of tethered polymer chains that could have important implications on the properties and applications of particle brush materials. For example, the MWD of tethered chains can be controlled by variation of the concentration of catalyst

[$Cu^{II}LnX$], according to the equation shown in the inset of Fig. 3.3. This can be achieved in a straightforward way by means of ARGET, ICAR, or eATRP processes. Broad MWD of tethered chains have attracted interest as a means to aid the dispersion of particle fillers within high molecular polymer matrices (Jayaraman 2013; Li et al. 2013). SI-ATRP also enables the synthesis of a wide range of intricate polymer architectures that further expand the property spectrum of particle brush materials. For example, SI-ATRP has been applied to the synthesis of tethered block, gradient, and statistical copolymer grafts on a variety of inorganic material substrates (Ignatova et al. 2006; Pyun et al. 2003; Xu et al. 2005). Random copolymer grafts have attracted particular attention since tailoring of the composition of the random copolymer allows fine control of the interaction of the brush with its environment. Miktoarm polymer grafts (where, the term "miktoarm" refers to a Y-link-type molecular structure) were shown to facilitate particle systems with superior solubility (Luzinov et al. 2004). To realize miktoarm architectures, a two-step reverse ATRP method was applied, in which azo-initiators are coupled to the surface, and served to initiate the polymerization of brushes with different compositions. Chemically cross-linked brush architectures have been prepared via homo- or copolymerization of a difunctional cross-linker by SI-ATRP. For example, grafting of butyl acrylate in the presence of a dimethacrylate-based cross-linker was shown to be an effective route toward gold nanoparticle brush materials with cross-linked polymer shell. Interestingly, cross-linking of the polymer-graft layer was found to inhibit coarsening of nanoparticles at elevated temperature and to stabilize embedded gold nanoparticles even at high temperature (Dong et al. 2008).

While the examples presented above illustrate the opportunities facilitated by regular SI-ATRP, recent research has shown that the balance of polymer-graft interactions and the surface-bonding strength of polymer grafts allows further expansion of the opportunities to design particle brush architectures. For example, the phase separation of chemically distinct polymeric ligands (Polystyrene-PS and PMMA) in mixed particle brush systems in which polymeric tethers are weakly coupled to inorganic substrates (such as gold-thiol bonds) was shown to provide a path toward Janus-type particle systems (Li et al. 2011). Figure 3.5 summarizes the parameter space that is accessible via SI-ATRP and related methodologies to tailor the structure and ultimately properties of particle brush materials.

The excellent control over the architecture, composition, and distribution of polymer grafts, facilitated by SI-ATRP (and other SI-CRP techniques), provides new opportunities to tailor interactions, and to control the assembly and physical properties of particle brush-based materials. This opens up a wealth of opportunities for novel scientific insights into the physics of nanostructured hybrid materials as well as the development of novel functional materials. Two major threads can be distinguished in current research on particle brush-based materials, depending on whether the particle brush takes the role as an exclusive building block or as an additive in the material system of interest. Research in the first thread primarily investigates the role of architecture and composition of polymer tethers on the interaction and dynamical properties of particle brush materials (in the solution or solid state) as well as the assembly of particle brushes into novel functional materials. Research within the second thread takes advantage of the high level of control of particle/polymer

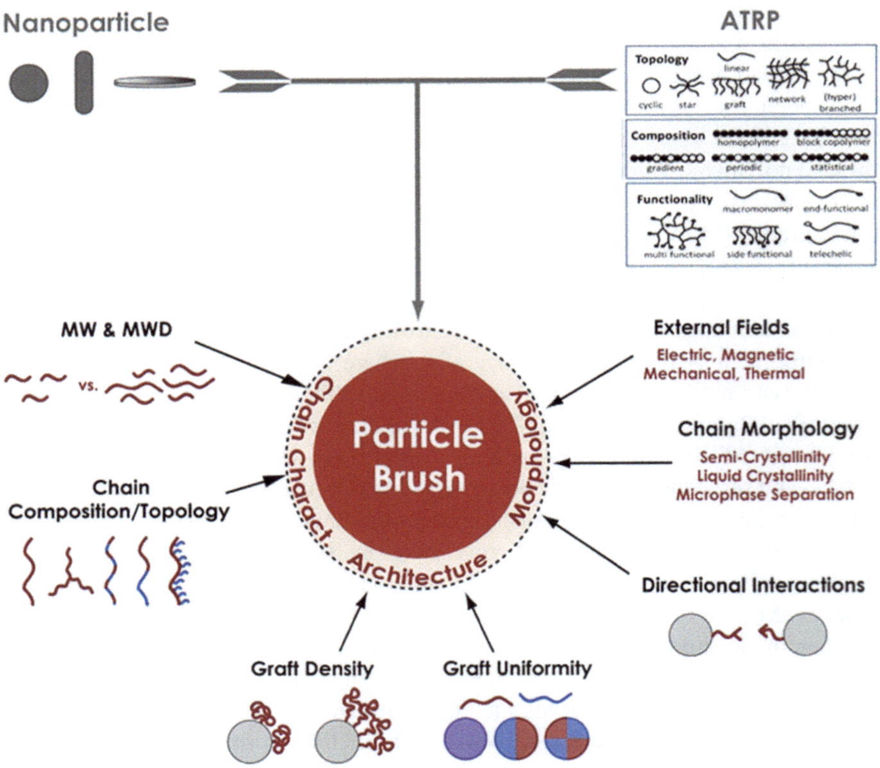

Fig. 3.5 Particle brush structure map illustrating the parameter space for controlling the composition and architecture of particle brush materials that is afforded by SI-ATRP. (Figure reproduced from Hui et al. (2014) with permission from the American Chemical Society). *SI-ATRP* surface-initiated atom transfer radical polymerization

interactions that is enabled by precision graft modification, to understand the interaction and organization of particle brush additives in polymeric media. The following sections discuss the recent progress in understanding the static and dynamic properties of particle brush materials and showcase examples that illustrate the potential for transformative material technologies based on particle brush materials.

3.3 Role of Polymer-Graft Modification on the Interaction, Dynamics, and Assembly of Particle Brush Materials

Single-component hybrid materials based on particle brush systems hold the promise for the development of novel functional materials due to a wide parameter space that enables control of the microstructures and tailoring of properties (such as mechanical, optical). The control parameters for structure–property relations are (i) nature, size, and shape of the core, (ii) synthetic route of grafting chains, (iii) the nature and composition of grafted shell, (iv) graft density (σ), MW of grafted chains, degree of polymerization (N), (v) the spatial distribution of tethered chains, (vi) morphology of grafted chains in the solid state, (vii) the composition (and shape) of the inorganic core, and (viii) the presence of external fields capable of coupling to the particle brush. By controlling the abovementioned parameters, one can tailor the interaction between particle brush systems, their self-assembly, and final properties.

Because physical properties of particle brush materials arise due to the superposition of interactions and properties of the inorganic particle core and the tethered polymer shell, particle brush materials have been considered to be an intermediate state between hard particulate systems and soft polymer materials. Analogies have been drawn to that of star polymers to explain the polymer conformation on the surface of the particles (Daoud and Cotton 1982; Ohno et al. 2006, 2007). Understanding of the structure–property relations in particle brush materials is a prerequisite for the development of materials with deliberately engineered properties. Critical parameters involve: the conformation of tethered polymer chains, their density profile from the particle surface, and how these characteristics change as one changes the graft density and MW. These factors play a key role in determining the overall interaction potential between two particles which in turn determine the structure and resulting dynamics and hence the overall macroscopic properties. In this section, we first focus on theories dealing with conformation of polymers on spherical particle surfaces before summarizing recent studies that have shed light on the relationship between structure and dynamics of particle brush materials.

The scaling model by Daoud and Cotton (1982) for star polymers provides a basis to evaluate the important aspects of structure and various interactions in polymer-tethered nanoparticles. Here we note that star polymer architecture corresponds to a polymer consisting of n-number of chains that are bonded to a central point (hence a "star polymer" could be viewed as a "particle brush" in the limit of very small particle size). The Daoud–Cotton approach presents a generalized treatment for curved brushes and extends the scaling arguments of Alexander and de Gennes (Alexander 1977; de Gennes 1980) that were originally devised for planar brush geometries to star polymers. According to the Daoud–Cotton model (DC model), a star polymer with f number of branches, degree of polymerization, N, and Kuhn length of b can be categorized into two distinct regimes differing in the conformation of polymeric chains: (i) the concentrated regime in which excluded volume interactions give rise to stretched chain conformations and (ii) the semidilute regime in which chains assume a relaxed chain conformation; the brush height is predicted to scale approximately as $h \sim N^{3/5} v^{1/5} f^{1/5} b$ (v is the excluded volume parameter). Later, Ohno et al. (Ohno et al. 2006; Ohno et al. 2007) extended this model to treat chains grafted to a nanoparticle of core radius, r_c. The effective graft density $\sigma_{eff} = \sigma_0 (r_c/r)^2$ decreases with increasing radial distance where σ_0 is the graft density on the core surface and r is the distance from the core center. They postulate the existence of a critical radius r_{DC} ($r_{DC} = r_c (\sigma^*)^{1/2} (v^*)^{-1}$) to determine the transition between two brush regimes with the reduced grafting density, $\sigma^* = \sigma_0 b^2$ and the reduced excluded volume, $v^* = v/(4\pi)^{1/2}$. If the total particle radius ($r = [r_c + h] < r_{DC}$, where brush height, $h = [h_{CPB} + h_{SDPB}]$) is less than the critical radius, the polymer-grafted particles are in concentrated particle brush (CPB) regime while for $r > r_{DC}$ the particle brush is considered to be in the semidilute particle brush (SDPB) regime (Fig. 3.6). The conformational regimes predicted by the DC model are expected to show a significant change of the scaling exponent x as it relates the brush thickness h with the degree of polymerization N ($h = N^x$).

Initially, Savin and coworkers (Savin et al. 2002) synthesized and investigated experimentally the brush height of PS-grafted chains on silica nanoparticles (core diameter, D_0 =20 nm). The graft molecular masses were varied up to ~34 kDa with high graft densities ~0.6 chains/nm^2 and polydispersity varying between 1.2 and 1.3. The dynamic light scattering (DLS) experiments showed a linear variation of brush height with N and they deduced that the brushes were in planar scaling regime. Ohno et al. (2007) demonstrated experimentally and theoretically the influence of graft chain length L_c on the hydrodynamic diameter D_h of PMMA grafted silica nanoparticles ($D_{0=}$130 nm) having high graft density, σ_0=0.7 chains/nm^2. They conducted DLS experiments on a series of grafted particles with nearly same graft density and varied the length of PMMA chains. The value of hydrodynamic thickness h (where, h=[D_h/2–r_c]) of the polymer brush layer was found to increase with molecular mass M_w with a scaling exponent 0.83. When interpreted with a modified DC scaling model for short graft chains, h obeyed the relation, $h[1 + (h/2r_c)] \sim L_c\sigma_0^{1/2}$. This led them to the interpretation that for $r_c < r_{DC} > r$, grafted chains at a sufficiently high density exhibits CPB behavior at lower MWs. If the chains are longer than the critical length r_{DC}, the brush layer crosses over into the SDPB regime. They predicted this crossover transition from CPB to SDPB at an MW ~ 120 kDa for their system.

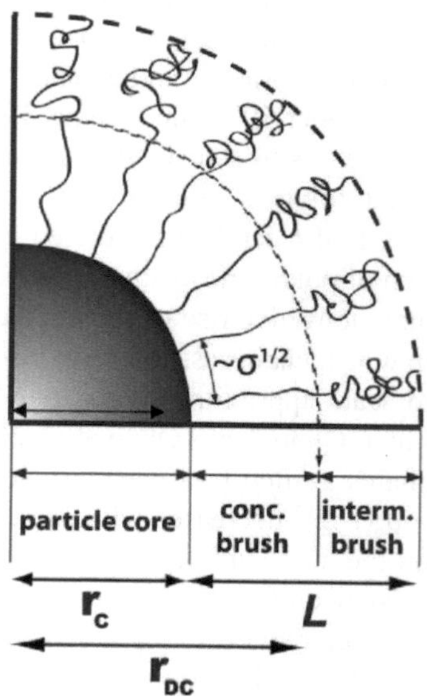

Fig. 3.6 Illustration of the brush regimes of particles systems: concentrated brush regime characterized by thickness of the grafted layer $L \sim N$ and the intermediate brush regime characterized by $L \sim N^{0.5}$ (assuming ideal chain of N monomers). (Reproduced with permission from Voudouris et al. (2009))

Self-consistent field (SCF) models have been used as an efficient tool to quantify the polymer brush behavior and the interactions between them (Kim and Matsen 2008; Dan and Tirrell 1992). Milner–Witten–Cates (MWC; Milner et al. 1988) made use of SCF calculations to predict the density profile and free energy of a grafted polymer brush considering binary repulsions between chain segments. MWC model found that the equilibrium parabolic profile of grafted brush layer corresponds to a unique global minimum of the free energy. The investigation of non-equilibrium effects like deformational and dynamical behavior of grafted polymer layer gave a different picture to the layer structure as compared to scaling approach. This revealed the fact that the polymer concentration decreases monotonically on going away from the grafting surface, and free chain ends are distributed throughout the whole brush layer. Ball et al. (1991) extended the SCF approach for the first time to chains grafted onto a convex (cylindrical) surface. For the case of densely grafted chains at a cylindrical interface immersed in a melt, increasing the radius of curvature resulted in an exclusion zone next to the surface from which free chain ends were excluded. The height of the zone increased with increasing radius of curvature. Later, Dolan–Edwards SCF equations were numerically solved by Dan and Tirrell (1992) in the vicinity of a spherical grafting surface and found that the scaling of brush height with N varied from $h \; \alpha \; bN^{3/5}v^{1/5}f^{1/5}$ to $h \; \alpha \; bN\sigma^{*1/3}$ as the particle radius was systematically increased. They also found that the ungrafted chain ends were uniformly distributed in the brush except for an exclusion zone near the particle surface. Wijmans and Zhulina (WZ model; Wijmans and Zhulina 1993) extended the model given by Scheutjens and Fleer (1979, 1980) to polymer brushes at spherical interfaces immersed in a low MW solvent and made it possible to study the characteristics of grafted polymer chains with varying chain lengths. According to the extended WZ model, for polymer brushes grafted on spherical surfaces, the brush height h depends upon the relative curvature, ω^{-1} and is given by the following relationship for particles in a good solvent,

$$(h/h_0)^3\left[1+3/_4(h/\omega h_0)+1/5(h/\omega h_0)^2\right]=1 \tag{3.1}$$

where $\omega = r_c/h_0$ and h_0 ($h_0=(8/\pi)^{1/3}bNv^{1/3}\sigma^{*1/3}$) is the planar brush height with an identical graft density and the degree of polymerization as on a spherical surface. The limit of small h/r_c yields brush heights that scale linearly with N and a continuous transition to $h \sim N^{3/5}$ with increasing h/r_c. From their equation, changing the particle surface curvature merely alters the onset of this transition with the transition at higher curvatures occurring at shorter chain lengths. Thus, similar to the DC model, at intermediate curvatures, a critical radius r_{DC} is found to separate two regions with distinct chain conformations—for $r<r_{DC}$, surface chains adopt a stretched conformation while for $r>r_{DC}$, more relaxed conformations are observed. The advantage of the extended WZ model is that it is more sophisticated than the DC model and is able to predict the dependence of the critical radius on the degree of polymerization of grafted chains. Investigations (theoretical and experimental) by Dukes et al. (2010) on the structural transitions of polymer-grafted particles in solution showed that the extended WZ model was more accurate in predicting the scaling relationships in both the CPB and the SDPB regime. Even though the two models describe the structural transitions in a solution state and are based on pairwise interactions, their agreement with self-consistent theories coupled with the ability to capture structural transitions in solution (as seen from the agreement with experimental data) has extended their use to predict the transitions that occur even in the one-component systems in the solid state. The DC model has been a more attractive model for experimentalists as it easily allows one to categorize the particle brush systems into CPB and SDPB regimes just based on the compositional data of the particle brushes.

Recently, Voudouris et al. (2009) quantitatively presented for the first time the effect of architecture of tethered polymer chains on the static and dynamic properties of PS-grafted silica particle brushes in good solvents using DLS. For particles in the intermediate brush regime, significant interpenetration of the grafted polymer chains was observed at concentrations near the overlap concentration resulting in slowing down of the dynamics and occurrence of fast cooperative modes related to the relaxation of the polymer brushes. The changes in structure and dynamics for particles in SDPB also occur at lower concentrations than the corresponding hard-sphere analogs. However, for the CPB, increased osmotic pressure due to the stretched chains results in a higher threshold concentration (by about one order of magnitude) before interpenetration of grafts could be detected. In summary, this work was the first to demonstrate that structure of the brush has a profound influence on how the particles interact and on the static and dynamic properties in solution. The latter is an important prerequisite to understand the role of polymer-graft modification on the interaction and structure formation characteristics of particle brushes in solid film assemblies.

Choi et al. (2010, 2013) synthesized and characterized a number of polymer-grafted nanoparticles (silica coated with polystyrene and poly methyl methacrylate) using the criteria given by Ohno et al. It was found that the ability of these particles to form long-range ordered structures depended sensitively on the polymer brush architecture. In the limit of high grafting density and stretched chain conformations, particle array structures, which resembled close-packed hard-sphere arrays, were observed. As the MW of the grafted chains was increased, the reduction of structural regularity was seen which was attributed to the increase of volume occupied by the relaxed polymer segments. The order–disorder transition was well captured by the model of Fukuda and coworkers that interprets the structure of the polymer brush as being constituted of two distinct (stretched and relaxed) conformational regimes that exert an opposing influence on the order formation process. They observed with increasing MW of surface-grafted polymer ligands, a transition from hexagonal close packing (HCP) or face-centered cubic (FCC) to predominantly FCC ordering. Thus, it was proposed to use this model as design criteria for the synthesis of systems for a variety of interesting applications.

Goel and coworkers (Goel et al. 2011) studied the structure of polymer-tethered highly grafted nanoparticles. The nanoparticles crystallized into highly ordered crystal structures (FCC) as measured by small angle X-ray scattering (Fig. 3.7a, b).

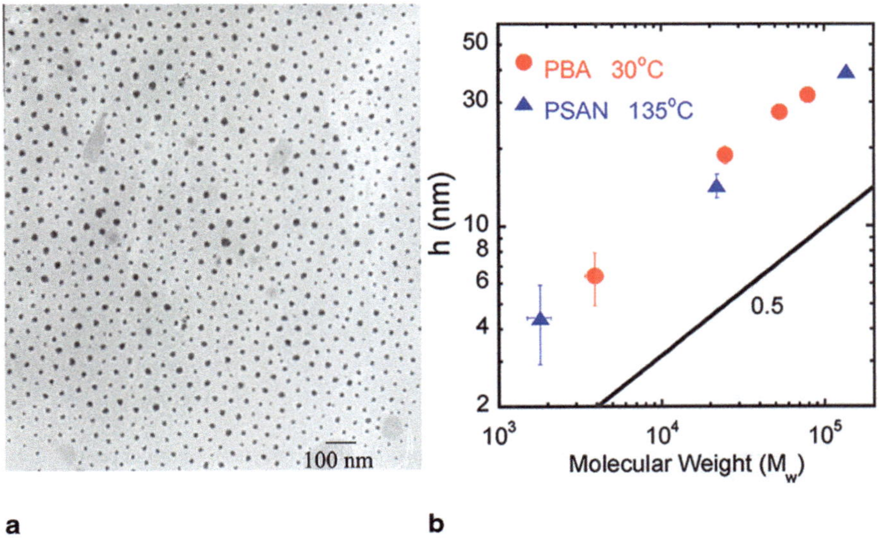

Fig. 3.7 **a** TEM of an unstained poly(*n*-butyl acrylate)—silica hybrid thin film that demonstrates a well-ordered structure. Particle diameter is 16.5 nm, volume fraction of silica is 0.027, and the molecular weight (M_n) of the polymer hairs is 79,400 with a graft density of 0.8 chains/nm^2. **b** The minimum distance between the spherical nanoparticles, *h,* which is a good measure of the brush height is shown for the poly(*n*-butyl acrylate)-silica and poly(styrene-co-acrylonitrile)-silica hybrids (with 16.5 nm diameter SiO_2 nanoparticles and containing ~1000 chains/nanoparticle) as a function of molecular weight of the nearly monodisperse polymer chain attached to the silica nanoparticles. The distance between the particles scales as $N^{0.5}$ as expected for a star polymer-like system (reproduced with permission from Goel et al. (2011). *TEM* transmission electron microscopy

The particle spacing in the crystal structure was controlled by adjusting the MW of the polymer hairs (M_w) and it scaled as $M_w^{1/2}$ as expected for a star polymer system (based on a DC model). Thus, one could achieve good control of the spacing by adjusting the degree of polymerization of the chains. Even though the solid loading was as low as 0.7 vol% silica, the bulk relaxation times were $>10^5$ s indicative of solid-like behavior. Similarities were drawn with colloidal crystallization and an interaction potential $U(r) \propto r^{-(12.5\pm2.0)}$ was hypothesized to predict the elastic properties of the nanocomposite. Such a potential confirmed the similarities to colloidal crystallization and was unlike the exponential potential functions seen for star polymers.

Choi and coworkers (Choi et al. 2012) systematically studied the mechanical characteristics of the one-component particle–polymer assemblies in which the characteristics of the brush were varied from CPB to SDPB. First, they clearly demonstrated that by tuning the properties of the polymer brush, the nanoparticles could assemble into mechanically robust yet ordered solid structures (predominantly FCC structures). In the CPB regime and for low MWs in the SDPB regime, the elastic modulus of the composite was "hard sphere like" and increased 200 % from that of the pure polymer. In this regime, the modulus was dominated by the interactions

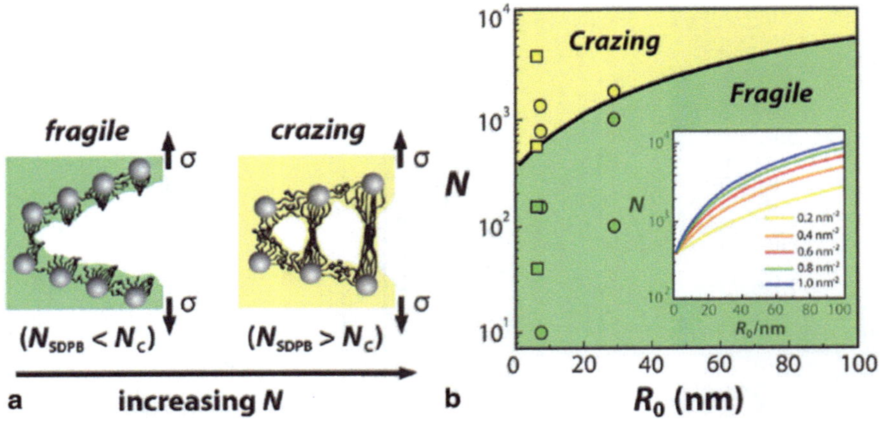

Fig. 3.8 **a** Illustration of the transition in the deformation characteristics of particle brush solids. Craze formation (and thus increased resistance to fracture of the particle solid) is expected if the segment length in the SDPB regime exceeds the threshold segment length for entanglement formation. **b** Predicted dependence of the threshold degree of polymerization to facilitate entanglement formation of grafted chains on the particle core size calculated assuming a grafting density of 0.5 chains/nm^2 (the inset shows theoretical prediction for various grafting densities). The curves are expected to approximately delineate the transition from fragile particle-like to polymer-like fracture. *Circles* and *squares* represent poly(styrene) and poly(methyl methacrylate) grafted particle brush systems, respectively. Colored regions correspond to predicted mechanical response of the particle brush system. (Reproduced with permission from Choi et al. (2012). *SDPB* semidilute particle brush)

between the particles and upon application of a stress, the fracture was fragile and dominated by the particles. Upon increasing the MW of the grafted polymer, the fracture toughness increased by an order of magnitude and the deformation of the composite was more "polymer like" (Fig. 3.8a, b). The increase in toughness was interpreted as a consequence of entanglements between surface grafted chains that resulted in formation of crazes during fracture of particle brush solids. The transition from "particle like" to "polymer like" was well captured by the DC-type scaling model for determining the minimum degree of polymerization, N_{min} ($N_{min} = 2N_e + (a^{-1}[r_{DC} - r_c])1/^x$) required for grafted chains needed to create toughened arrays. In the latter relation, N_e denotes the entanglement segment length (a material constant) and a is the length of a repeat. It is envisioned that the expression could be used as a guiding tool to develop mechanically robust and toughened arrays of nanoparticles (similar to thermoplastic elastomers) by tuning the available parameters.

Koerner et al. (2013) recently studied the structure and elongational deformation of polymer-grafted nanoparticles as the architecture of the polymer hairs transitions from CPB to SDPB. The MW of the grafted chains was kept constant at 120 kDa while the graft density was kept low and changed from 0.01 to 0.05 and 0.1 chains/nm^2. The packing led to non-FCC structures at all three graft densities and the morphology evolution during deformation was unique and resembled those

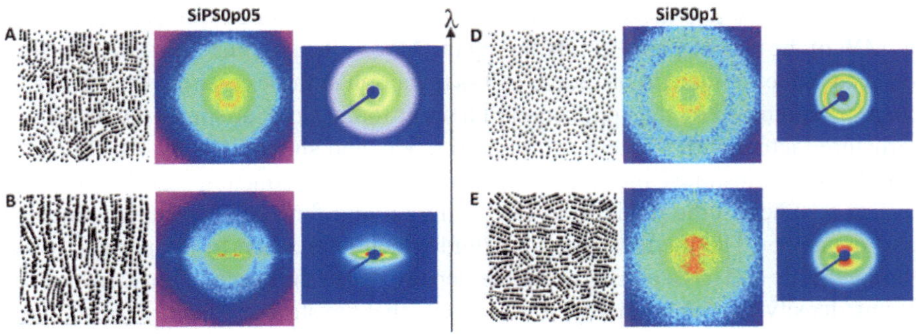

Fig. 3.9 One-component composite morphology evolution during deformation. Deformation direction is vertical (*arrow*, λ). *Left*: silica–polystyrene composites (SiPS0p05—graft density of 0.05 chains/nm^2). *Right*: SiPS0p1 (0.1 chains/nm^2). Three columns are presented for each sample. Column 1 contains a schematic representation of possible morphology; column 2 contains a digital Fourier transform of the schematic in column 1; and column 3 contains the experimental 2D SAX pattern. *Rows* A (SiPS0p05) and D (SiPS0p1) show unstretched data; rows B and E show data for $\lambda = 3$. (Reproduced with permission from Koerner et al. (2013))

typically observed in thermoplastic polymers. The graft density of 0.01 chains/nm^2 gave rise to brittle films at room temperature while the nanocomposites at 0.05–0.1 chains/nm^2 could be stretched >200 %. However, structures of the composites under elongation were very different for 0.05 and 0.1 chains/nm^2. While the particles at 0.05 chains/nm^2 are aligned parallel to the stretching direction, the particles at higher graft density aligned perpendicular to the stretching direction giving rise to anisotropic structures (Fig. 3.9). The differences in elongation characteristics were attributed to the differences in local particle structure and entanglements of the polymer chains which results in nonuniform local strain distributions thus resulting in different alignments. Thus, it was shown that with the right corona architecture, one-component composites afford opportunities to design high inorganic fraction hybrids that retain processability to create thin films or fibers for the next-generation dielectrics or gradient refraction index materials for optoelectronic applications.

The structure of the one-component composites can be changed from crystalline to fluid like by tuning the chemistry and MW of the attached polymer. The composites have been called "nanoscale ionic liquids" and "nanoscale organic hybrid materials" and have been the subject of recent studies. Kim and coworkers (Kim et al. 2012) recently studied the structure and dynamics of self-suspended poly(ethylene glycol) (PEG)-tethered silica. The MW of the attached PEG was low such that the formed composites exhibited Newtonian flow behavior on long time scales and chain relaxation at intermediate time scales. An interesting feature of this system is that it exhibits liquid-like order for the nanoparticles and the viscosity decreases as the polymer MW is increased at a fixed grafting density. This is unlike star polymers where the viscosity increases. A novel tool—X-ray photon correlation spectroscopy (XPCS)—was used to investigate the dynamics of the nanoparticles and the measured relaxation times were the slowest for the lowest MW chains. This was again in contrast to star polymers where the relaxation times become exponentially slower as MW of the chains increases. By tuning the MW of the polymers and graft density, one can induce a liquid to disordered glass transition in these composites, which can then be unjammed by the application of a shear force. Comparison of the measured structure factors with a recent soft-colloid model and simulations by Yu and Koch (2010) have illustrated the importance of space-filling constraint of the polymer chains attached to the particles as one of the driving forces for observed structure and dynamics. The mediate oligomer fluid which is attached to the particle is incompressible and has a constant number density due to the space-filling constraints. Using density functional theory, it was shown that the structure factor at zero wave vector goes to zero under all conditions of core volume fraction and oligomer radius of gyration. Thus, the polymer-grafted particles behave as a single-component fluid rather than a suspension of spheres suspended in a solvent. Another advantage of the method of Yu and Koch was that an insight could be gained into the nonadditive interparticle potential between the particles by direct calculation of the potential of mean force which depends upon the variables such as the core volume fraction, graft density, and the size ratio of chains and the core. Recently, they have extended this approach to predict the effect of interaction potential on the self diffusion and linear viscoelasticity of the single-component particles (Yu and Koch 2014).

Archer and coworkers (Srivastava et al. 2014) envisioned using the silica–PEG composites as electrolytes for lithium ion batteries. The advantages of these materials are that the particles provide stability to the polymer, inhibit its crystallization, and enhance the conductivity at room temperature without compromising colloidal or thermal stability. The ability to manipulate flow properties by inducing jamming transition also gives one the ability to decouple mechanical properties and ionic conductivities in these systems. Recent work by Giannelis and coworkers (Bourlinos et al. 2005, 2006; Rodriguez et al. 2008) has demonstrated the ability to graft ionic polymers to silica and iron oxide nanoparticles. When the MW of the corona polymer was varied, the solution transformed from a free flowing liquid to a waxy solid. Nuclear magnetic resonance (NMR) measurements of the corona dynamics illustrated the fast hopping effect of the corona chains from one particle surface to another while the particle diffusion time scales were a couple of orders of magnitude higher. The authors envision the use of these nanoscale "fluids" in a variety of applications such as thermal management materials, magnetic fluids and as conductive lubricants.

Jayaraman and Schweizer (2008a, b, 2009) applied microscopic polymer reference interaction site model (PRISM) theory to study the structure factors and intermolecular pair correlation functions and the tendency for microphase separation of dense solutions and melt of polymer-tethered particles. Their primary goal was to explore the role of tethers and their variable placement on the nanoparticle spatial organization and aggregation. The role of interparticle attraction strength (ε_{cc}) and total packing fraction (η) was studied for a system having the degree of polymerization $N=8$ with one, two, or four tethered polymer chains and particle-to-monomer diameter ratios (D/d) of 2 and 3. They found a complex interplay of entropy (translational, conformational, and packing) and enthalpy (particle–particle attraction) results in different structural arrangements. Of particular interest was the microphase separation that was observed as the total packing fraction and the particle–particle attraction increased. The above calculations were carried out under different packing fractions (η) and showed that, under athermal conditions, tendency to microphase separation is weak and decreases with increasing number of tethers due to reduced anisotropy of the particle. As the number of tethers increases, the particles are effectively pushed apart and locally order at larger interparticle separations, and the microphase separation boundary is shifted to higher particle–particle attraction strengths or lower temperatures. The tendency for microphase separation is also enhanced when the tethers were asymmetrically placed on one side of the particle due to the high anisotropy of steric hindrance. The microphase spinodal calculations of Jayaraman et al. were found to be in good agreement with the computer simulation studies of the phase behavior of single-tethered spherical nanoparticles by Glotzer and coworkers (Phillips et al. 2010). It was observed that upon increasing the nanoparticle diameter, the microphase spinodal temperature decreases and its packing fraction dependence qualitatively changes due to the competition between an increased tendency for macrophase separation and a decreased driving force for tether-induced microphase ordering. Their recent simulation studies predicted, under suitable conditions, polymer-tethered nanospheres can self-assemble to form microphases similar to block copolymer systems including the complex double gyroid phase for polydispersities up to 10 %. Higher polydispersities kinetically inhibits the formation of the phase while the inclusion of smaller or larger nanospheres encourages low-energy icosahedral clusters and increases gyroid stability while facilitating its formation. These studies illustrate the ability to tune the macroscopic structure of the composite by modulating the interparticle potential by careful placement of tethers. It is anticipated that the observed structures find use in photonic and biosensing applications.

3.4 Conclusions and Future Directions

Significant progress has been made in understanding the physics behind the structure, assembly, and macroscopic properties of single-component composites. A large part of the progress has been due to the novel synthesis techniques that have been developed for polymers and to graft these polymers to the particle surfaces. In particular, the recent advancements in surface-initiated controlled radical polymerization have enabled new opportunities to tailor the composition, topology, and molecular characteristics of polymer-tethered particulate materials. The superposition of the various modes of interactions that are contributed by the inorganic and polymeric constituents and their sensitive dependence on the composition and architecture of particle brush materials presents far-reaching opportunities for the design of novel functional materials. To illustrate these opportunities we highlight some examples of the recent literature below. As this is an emerging area, the chosen examples should only be considered as the "tip of the iceberg" and are meant to stimulate the reader to explore even more creative applications of the concepts around particle brush materials.

Bockstaller and coworkers (Ojha et al. 2013) have recently demonstrated that the enhanced control of the chemical composition of polymer grafts that is afforded by SI-ATRP facilitates the synthesis of thermoplastic polymer nanocomposite materials with high inorganic filling fraction. In particular, it was found that favorable interactions between grafted polymer and matrix polymer (here, poly(styrene-*r*-acrylonitrile)/poly(methyl methacrylate)) facilitate thermodynamically stable and uniform particle dispersion with inorganic filling fractions exceeding 50 % by volume (the term "polymer-filled inorganics" was introduced to highlight the majority content inorganic). The approach was shown to be applicable across a wide range of particle sizes, short grafted chains, and high molecular matrix chains. For example, one can envision harnessing this concept to single-component composites. By mixing particles with different surface chemistries, one can achieve a higher loading and a stronger composite than just a composite with particles of a single type. The structure can be further tuned by changing the chemistry of the polymer grafted to the second particle. Such high loading thermoplastic composite materials can find use in a number of applications such as optical filters and lightweight coating for aircraft wings. Schadler and coworkers (Li et al. 2013, 2014) have recently highlighted the advantages of surface ligand engineering on nanoparticles for the use in nanocomposites (Fig. 3.10). A single population of polymer chains with tunable graft density and graft MW has a limited set of conditions under which a stable dispersion can be produced in a melt. In order to overcome this difficulty, they have come up with a novel idea of grafting a bimodal polymer brush onto the surface of a particle—a brush with densely grafted short chains and a sparsely grafted long brush. Grafting a bimodal brush increases the parameter space to control interparticle interactions and also provides the ability to decouple enthalpic and entropic interactions. The densely grafted short brushes enthalpically screen the core–core interactions while the longer grafts promote entanglements between neighboring particles and suppress entropic dewetting. As an example, they grafted a higher density of electrically active short molecules (oligothiophene or ferrocene) and a longer but sparse matrix compatible poly(glycidyl methacrylate, PGMA) to silica nanoparticles (Virtanen et al. 2014). The resulting bimodal brush particle was dispersible in a commercial epoxy resin and at the same time the dielectric breakdown strength and permittivity increased considerably with only 2 wt.% solids loading. In a recent article, they review ligand engineering and its potential to create nanocomposites for optoelectronic and dielectric applications. This technique holds a lot of promise for single-component nanocomposites as well—where multimodal brush particles can be combined to produce a composite for different applications.

Single-component polymer-grafted particles are ideally suited for dielectric applications as recently demonstrated by the work of Vaia and coworkers (Tchoul et al. 2010). Thermoplastic polymers possess high breakdown strength and are amenable to be processed into thin films but they possess a low dielectric constant which limits the maximum energy storage density. Ceramics such as titanium dioxide and barium titanate have a high dielectric constant but have low breakdown strength

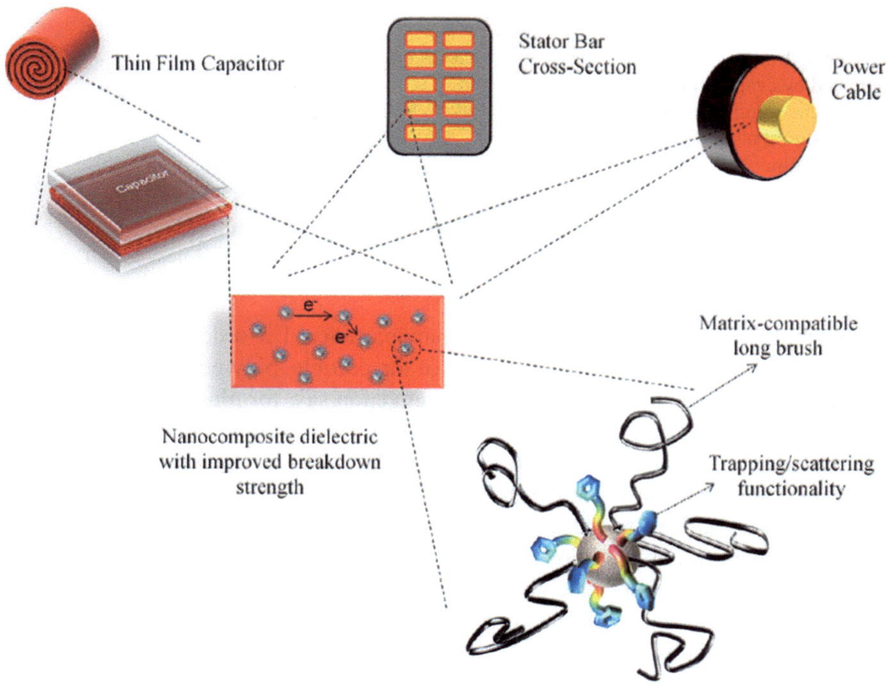

Fig. 3.10 Schematic view of example dielectric applications critically dependent on breakdown strength. Composites utilizing mixed bimodal brush grafted filler particles comprised of long chains to control dispersion and short electronically functional chains for breakdown strength enhancement, may revolutionize materials design for dielectrics. (Reproduced with permission from (Li et al. 2014))

and are brittle which prevents them from being used. Polymer–ceramic composites which combine the properties of the individual polymer and ceramic are thus being actively investigated for use in dielectric applications. A key in this application is to get to a high loading of the inorganic component (30–50 % by volume) to take advantage of the high dielectric permittivity. Just dispersing the particle in the matrix leads to agglomeration and reduction in effective surface area. Vaia and coworkers grafted polystyrene onto titanium dioxide (TiO_2) nanoparticles and studied the structure and dielectric properties of the resulting composite. It was found that a uniform distribution of particles was achieved even at high loadings of 30–50 % by volume, the films were transparent, and the resulting composite had enhanced dielectric properties. For TiO_2 grafted with 10^5 g/mol polystyrene, the films could also be extended by 100 % which illustrates the extensibility and enhanced processability of the formed composite (Fig. 3.11). In addition to the increase in toughness and dielectric permittivity, the refractive index (n) of the film could be controlled by the mass fraction of TiO_2 particles resulting in a transparent film. Refractive indices greater than three could be achieved by controlling the mass fraction. This example illustrates the ability to form ordered arrays of nanoparticles with tailored properties

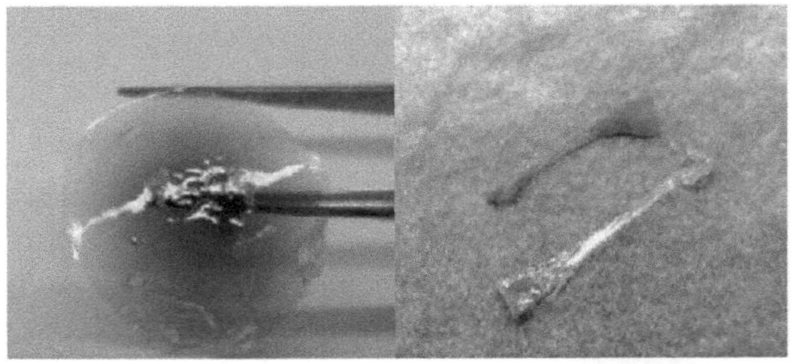

Fig. 3.11 *Left*: A 2-mm thick film of the assembly of TiO_2-PS100k hybrid nanoparticles obtained by evaporation of the toluene solution in the round-bottom tube. *Right*: the same film stretched after being heated above T_g. (Reproduced with permission from Tchoul et al. (2010))

using single-component polymer-grafted nanoparticles. The order and functionality can be tuned by controlling the chemistry of the particle, polymer, the graft density of the polymer, and its MW. One can now envision say, forming photonic crystals using single-component particles. Recent work by Bockstaller and coworkers (Narayanan et al. 2013) has illustrated the ability of gold–polystyrene composites to form photonic crystals which can be used in plastic electronics. Such composites are transparent and retain the flexural stability of the polymer matrix thus finding use in a number of applications.

Archer and coworkers recently demonstrated the ability of single-component nanoparticles for use as shape memory polymers (Agarwal et al. 2011). First, they grafted functional oligomer chains (PEG) to the particle (silica) surfaces. These single-component materials display fluid like properties which can be tuned by changing the core volume fraction, graft density, and graft MW. The dual functionalized PEG was then cross-linked to result in a network in which the silica particles served as network junctions. The resulting material manifested shape memory properties in that they can be cooled to remember their shape and when heated again, they quickly recover their original fixed shape. The rubbery modulus (~ 100 MPa) for a particle volume fraction of 0.16 was also substantially higher than that for traditional shape memory polymer composites which have moduli in the range of 1–10 MPa. It is anticipated that by utilizing the different shapes, sizes, and chemistries of the nanoparticles and that of the polymer, versatile shape memory polymers can be produced for a diverse range of applications such as remote actuation, biodegradability and for drug delivery. Balazs and coworkers (Iyer et al. 2013a, b; Hamer et al. 2014) developed a computational multiscale model to predict the behavior of the nanocomposites under tensile deformation. The composites had polymers grafted to the particles and the grafted polymers contain reactive end groups which could form weak labile bonds, which could reform after breakage or permanent bonds which rupture irreversibly. The system was referred to as a "dual cross-linked" nanoparticle network. The advantage of such a system was that the permanent bonds provides enhanced toughness which prevents the material from failure—it leads to the formation of long threads under tensile stress and these threads can extend over large gaps and hold high loads without rupturing, thus, resulting in dramatic improvements to mechanical properties. These results provide valuable guidelines for developing novel self-healing materials, which display enhanced ductility and strength.

The possibility to balance interactions common to inorganic and polymeric materials afforded by precision-engineered particle brush materials also opens opportunities to mimic nature's material paradigms to material design. For example, structural biological materials exhibiting property combinations such as superior performance and resilience in changing environments as well as low-energy fabrication and processing conditions have always inspired the development of innovative material technologies. One pervasive paradigm of material design that is often observed in nature is the presence of various structural (or kinetic) frustrations in hierarchical organized systems that give rise to a range of nonequilibrium and metastable states that are associated with distinct property characteristics and that can be accessed reversibly by interaction with the environment. To harness this design principle for the development of interactive and multifunctional advanced materials is a grand challenge in materials science. One potential strategy toward realizing this goal is the design of multicomponent materials in which competing driving forces give rise to a range of (well-defined) metastable states that can reversibly be populated by coupling of the material with an external field. The superposition of different interaction modes along with the constraints that are imparted by the chemical confinement of surface grafted chains render polymer-tethered hybrid particles ideal model systems to mimic nature's design principles.

For example, recent work by Glotzer and coworkers (Shah et al. 2013; Phillips et al. 2010; Glotzer and Solomon 2007) has shown that precise arrangements of patches combined with patch "recognition" or selectivity may be used to control the relative position of particles and the overall structure of particle assemblies. Using molecular simulations, it was shown that a variety of structures—chains, sheets, rings, icosahedra, square pyramids, tetrahedra, twisted, and staircase can be achieved by the self-assembly of patchy nanoparticles. They demonstrated how the additional constraints introduced by the nanoparticle geometry and the nano building block topology combined with tether and nanoparticle immiscibility lead to various self-assembled structures. Such self-assembled structures can be used in a number of applications—the 2D sheets of self-assembled patchy spheres with tunable lattice structures have interesting optical and mechanical properties and can be used in photonic applications. The more complex structures such as rings and staircase structures can be used to prepare tubes, helices, and other assemblies which can find use as scaffolds for biological applications and as channels (membranes) for transport of liquids and molecules. Patchy nanoparticle interactions can be achieved in a number of ways. A most common method is to tether oligomeric, polymeric, and biomolecular polymers at selective locations on the surface (a synthetic challenge) and also to selectively tune the chemistry of the tethers. Inspired from the self-assembly of biological systems, an emerging approach is to confer upon nanoparticles, the predetermined "instructions" for assembly by anisotropically decorating the surface with biological molecules. Mirkin and coworkers (Park et al. 2008; Mirkin et al. 1996) have demonstrated that by grafting DNA oligonucleotides to gold nanoparticle surfaces, a variety of assemblies can be achieved depending on the DNA sequences grafted to the particle surface. The gold nanoparticles could be organized into FCC or body centered cubic structures by utilizing

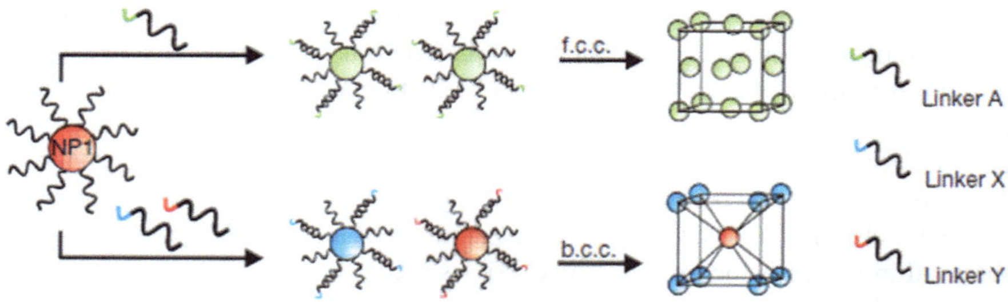

Fig. 3.12 Scheme of gold nanoparticle assembly method: Gold nanoparticle–DNA conjugates can be programmed to assemble into different crystallographic arrangements by changing the sequence of the DNA linkers. (Reproduced with permission from Park et al. (2008)

different DNA strands (Fig. 3.12). This work clearly demonstrated that synthetically programmable colloidal crystallization is possible and that a single-component system can be directed to form different structures. It is anticipated that such assemblies find use in biosensing and photonic applications due to the programmable crystal structures.

Acknowledgments S. Ramakrishnan and G. L. Chakkalakal were supported by AMSRD-ARL-RO-SI proposal number: 62885-MS-REP, agreement number: W911NF-13-1-0132 from the Department of Defense (Army Research Office).

References

Abreu CMR, Mendonca PV, Serra AC, Popov AV, Matyjaszewski K, Guliashvili T, Coelho JFJ (2012) Inorganic sulfites: efficient reducing agents and supplemental activators for atom transfer radical polymerization. ACS Macro Lett 1(11):1308–1311. doi:10.1021/mz300458x

Advincula R (2006) Polymer brushes by anionic and cationic surface-initated polymerization (SIP). In: Jordan R (ed) Advances in polymer science: surface initiated polymerization, vol 197. Springer, Berlin, pp 107–136

Agarwal P, Chopra M, Archer LA (2011) Nanoparticle netpoints for shape- memory polymers. Angew Chem-Int Ed 50(37):8670–8673. doi:10.1002/anie.201103908

Alexander S (1977) Adsorption of chain molecules with polar head: a scaling approach. J De Phys 38(8):983–987. doi:10.1051/jphys:01977003808098300

Ball RC, Marko JF, Milner ST, Witten TA (1991) Polymers grafted to a convex surface. Macromolecules 24(3):693–703. doi:10.1021/ma00003a011

Barbey R, Lavanant L, Paripovic D, Schuwer N, Sugnaux C, Tugulu S, Klok HA (2009) Polymer brushes via surface-initiated controlled radical polymerization: synthesis, characterization, properties, and applications. Chem Rev 109(11):5437–5527. doi:10.1021/cr900045a

Binder K, Milchev A (2012) Polymer brushes on flat and curved surfaces: how computer simulations can help to test theories and to interpret experiments. J Polym Sci Part B-Polym Phys 50(22):1515–1555. doi:10.1002/polb.23168

Bombalski L, Dong HC, Listak J, Matyjaszewski K, Bockstaller MR (2007) Null-scattering hybrid particles using controlled radical polymerization. Adv Mater 19(24):4486–4490. doi:10.1002/adma.200700928

Bourlinos AB, Herrera R, Chalkias N, Jiang DD, Zhang Q, Archer LA, Giannelis EP (2005) Surface-functionalized nanoparticles with liquid-like behavior. Adv Mater 17(2):234–237. doi:10.1002/adma.2004001060

Bourlinos AB, Giannelis EP, Zhang Q, Archer LA, Floudas G, Fytas G (2006) Surface-functionalized nanoparticles with liquid-like behavior: the role of the constituent components. Eur Phys J E 20(1):109–117. doi:10.1140/epje/i2006-10007-3

Braunecker WA, Matyjaszewski K (2007) Controlled/living radical polymerization: features, developments, and perspectives. Prog Polym Sci 32(1):93–146. doi:10.1016/j.progpolymsci.2006.11.002

Carrot G, Diamanti S, Manuszak M, Charleux B, Vairon IP (2001) Atom transfer radical polymerization of n-butyl acrylate from silica nanoparticles. J Polym Sci Part A-Polym Chem 39(24):4294–4301. doi:10.1002/pola.10085

Choi J, Dong H, Matyjaszewski K, Bockstaller MR (2010) Flexible particle array structures by controlling polymer graft architecture. J Am Chem Soc 132(36):12537–12539. doi:10.1021/ja105189s

Choi J, Hui CM, Pietrasik J, Dong HC, Matyjaszewski K, Bockstaller MR (2012) Toughening fragile matter: mechanical properties of particle solids assembled from polymer-grafted hybrid particles synthesized by ATRP. Soft Matter 8(15):4072–4082. doi:10.1039/c2sm06915f

Choi J, Hui CM, Schmitt M, Pietrasik J, Margel S, Matyjazsewski K, Bockstaller MR (2013) Effect of polymer-graft modification on the order formation in particle assembly structures. Langmuir 29(21):6452–6459. doi:10.1021/la4004406

Dan N, Tirrell M (1992) Polymers tethered to curved interfaces-a self-consistent-field analysis. Macromolecules 25(11):2890–2895. doi:10.1021/ma00037a016

Daoud M, Cotton JP (1982) Star shaped polymers—a model for the conformation and its concentration dependence. J De Phys 43(3):531–538. doi:10.1051/jphys:01982004303053100

de Gennes PG (1980) Conformations of polymers attached to an interface. Macromolecules 13(5):1069–1075. doi:10.1021/ma60077a009

Dong HC, Zhu MZ, Yoon JA, Gao HF, Jin RC, Matyjaszewski K (2008) One-pot synthesis of robust core/shell gold nanoparticles. J Am Chem Soc 130(39):12852–12853. doi:10.1021/ja8038097

Dukes D, Li Y, Lewis S, Benicewicz B, Schadler L, Kumar SK (2010) Conformational transitions of spherical polymer brushes: synthesis, characterization, and theory. Macromolecules 43(3):1564–1570. doi:10.1021/ma901228t

Edmondson S, Osborne VL, Huck WTS (2004) Polymer brushes via surface-initiated polymerizations. Chem Soc Rev 33(1):14–22. doi:10.1039/b210143m

Fernandes NJ, Koerner H, Giannelis EP, Vaia RA (2013) Hairy nanoparticle assemblies as one-component functional polymer nanocomposites: opportunities and challenges. MRS Commun 3(1):13–29. doi:10.1557/mrc.2013.9

Gao X, Feng W, Zhu SP, Sheardown H, Brash JL (2010) Kinetic modeling of surface-initiated atom transfer radical polymerization. Macromol React Eng 4(3–4):235–250. doi:10.1002/mren.200900063

Glotzer SC, Solomon MJ (2007) Anisotropy of building blocks and their assembly into complex structures. Nat Mater 6(8):557–562. doi:10.1038/nmat1949

Goel V, Pietrasik J, Dong H, Sharma J, Matyjaszewski K, Krishnamoorti R (2011) Structure of polymer tethered highly grafted nanoparticles. Macromolecules 44(20):8129–8135. doi:10.1021/ma200621r

Hamer MJ, Iyer BVS, Yashin VV, Kowalewski T, Matyjaszewski K, Balazs AC (2014) Modeling polymer grafted nanoparticle networks reinforced by high-strength chains. Soft Matter 10(9):1374–1383. doi:10.1039/c3sm52300d

Hui CM, Pietrasik J, Schmitt M, Mahoney C, Choi J, Bockstaller MR, Matyjaszewski K (2014) Surface-Initiated polymerization as an enabling tool for multifunctional (nano-)engineered hybrid materials. Chem Mater 26(1):745–762. doi:10.1021/cm4023634

Ignatova M, Voccia S, Gilbert B, Markova N, Cossement D, Gouttebaron R, Jerome R, Jerome C (2006) Combination of electrografting and atom-transfer radical polymerization for mak-

ing the stainless steel surface antibacterial and protein antiadhesive. Langmuir 22(1):255–262. doi:10.1021/la051954b
Iyer BVS, Salib IG, Yashin VV, Kowalewski T, Matyjaszewski K, Balazs AC (2013a) Modeling the response of dual cross-linked nanoparticle networks to mechanical deformation. Soft Matter 9(1):109–121. doi:10.1039/c2sm27121d
Iyer BVS, Yashin VV, Kowalewski T, Matyjaszewski K, Balazs AC (2013b) Strain recovery and self-healing in dual cross-linked nanoparticle networks. Polym Chem 4(18):4927–4939. doi:10.1039/c3py00075c
Jakubowski W, Matyjaszewski K (2006) Activators regenerated by electron transfer for atom-transfer radical polymerization of (meth)acrylates and related block copolymers. Angew Chem-Int Ed 45(27):4482–4486. doi:10.1002/anie.200600272
Jayaraman A (2013) Polymer grafted nanoparticles: effect of chemical and physical heterogeneity in polymer grafts on particle assembly and dispersion. J Polym Sci Part B-Polym Phys 51(7):524–534. doi:10.1002/polb.23260
Jayaraman A, Schweizer KS (2008a) Effect of the number and placement of polymer tethers on the structure of concentrated solutions and melts of hybrid nanoparticles. Langmuir 24(19):11119–11130. doi:10.1021/la801432b
Jayaraman A, Schweizer KS (2008b) Structure and assembly of dense solutions and melts of single tethered nanoparticles. J Chem Phys 128(16):164904. doi:10.1063/1.2907717
Jayaraman A, Schweizer KS (2009) Effective interactions and self-assembly of hybrid polymer grafted nanoparticles in a homopolymer matrix. Macromolecules 42(21):8423–8434. doi:10.1021/ma901631x
Kim D, Srivastava S, Narayanan S, Archer LA (2012) Polymer nanocomposites: polymer and particle dynamics. Soft Matter 8(42):10813–10818. doi:10.1039/c2sm26325d
Kim JU, Matsen MW (2008) Interaction between polymer-grafted particles. Macromolecules 41(12):4435–4443. doi:10.1021/ma8002856
Koerner H, Drummy LF, Benicewicz B, Li Y, Vaia RA (2013) Nonisotropic self-organization of single-component hairy nanoparticle assemblies. ACS Macro Lett 2(8):670–676. doi:10.1021/mz4001805
Li F, Josephson DP, Stein A (2011) Colloidal assembly: The road from particles to colloidal molecules and crystals. Angew Chem-Int Ed 50(2):360–388. doi:10.1002/anie.201001451
Li Y, Tao P, Viswanath A, Benicewicz BC, Schadler LS (2013) Bimodal surface ligand engineering: the key to tunable nanocomposites. Langmuir 29(4):1211–1220. doi:10.1021/la3036192
Li Y, Krentz TM, Wang L, Benicewicz BC, Schadler LS (2014) Ligand engineering of polymer nanocomposites: from the simple to the complex. ACS Appl Mater Interfaces 6(9):6005–6021. doi:10.1021/am405332a
Luzinov I, Minko S, Tsukruk VV (2004) Adaptive and responsive surfaces through controlled reorganization of interfacial polymer layers. Prog Polym Sci 29(7):635–698. doi:10.1016/j.progpolymsci.2004.03.001
Magenau AJD, Strandwitz NC, Gennaro A, Matyjaszewski K (2011) Electrochemically mediated atom transfer radical polymerization. Science 332(6025):81–84. doi:10.1126/science.1202357
Matyjaszewski K (2012) Atom transfer radical polymerization (ATRP): current status and future perspectives. Macromolecules 45(10):4015–4039. doi:10.1021/ma3001719
Matyjaszewski K, Tsarevsky NV (2009) Nanostructured functional materials prepared by atom transfer radical polymerization. Nat Chem 1(4):276–288. doi:10.1038/nchem.257
Matyjaszewski K, Xia JH (2001) Atom transfer radical polymerization. Chem Rev 101(9):2921–2990. doi:10.1021/cr940534g
Matyjaszewski K, Coca S, Gaynor SG, Wei ML, Woodworth BE (1997) Zerovalent metals in controlled "living" radical polymerization. Macromolecules 30(23):7348–7350. doi:10.1021/ma971258l
Matyjaszewski K, Jakubowski W, Min K, Tang W, Huang JY, Braunecker WA, Tsarevsky NV (2006) Diminishing catalyst concentration in atom transfer radical polymerization with reducing agents. In: Proceedings of the National Academy of Sciences of the United States of America 103(42):15309–15314. doi:10.1073/pnas.0602675103

Milner ST (1991) Polymer brushes. Science 251(4996):905–914. doi:10.1126/science.251.4996.905
Milner ST, Witten TA, Cates ME (1988) A parabolic density profile for garfted polymers. Europhys Lett 5(5):413–418. doi:10.1209/0295-5075/5/5/006
Mirkin CA, Letsinger RL, Mucic RC, Storhoff JJ (1996) A DNA-based method for rationally assembling nanoparticles into macroscopic materials. Nature 382(6592):607–609. doi:10.1038/382607a0
Narayanan S, Choi J, Porter L, Bockstaller MR (2013) Flexible transparent metal/polymer composite materials based on optical resonant laminate structures. ACS Appl Mater Interfaces 5(10):4093–4099. doi:10.1021/am303211g
Ohno K, Morinaga T, Takeno S, Tsujii Y, Fukuda T (2006) Suspensions of silica particles grafted with concentrated polymer brush: a new family of colloidal crystals. Macromolecules 39(3):1245–1249. doi:10.1021/ma0521708
Ohno K, Morinaga T, Takeno S, Tsujii Y, Fukuda T (2007) Suspensions of silica particles grafted with concentrated polymer brush: effects of graft chain length on brush layer thickness and colloidal crystallization. Macromolecules 40(25):9143–9150. doi:10.1021/ma071770z
Ojha S, Dang A, Hui CM, Mahoney C, Matyjaszewski K, Bockstaller MR (2013) Strategies for the synthesis of thermoplastic polymer nanocomposite materials with high inorganic filling fraction. Langmuir 29(28):8989–8996. doi:10.1021/la401522v
Park SY, Lytton-Jean AKR, Lee B, Weigand S, Schatz GC, Mirkin CA (2008) DNA-programmable nanoparticle crystallization. Nature 451(7178):553–556. doi:10.1038/nature06508
Phillips CL, Iacovella CR, Glotzer SC (2010) Stability of the double gyroid phase to nanoparticle polydispersity in polymer-tethered nanosphere systems. Soft Matter 6(8):1693–1703. doi:10.1039/b911140a
Pietrasik J, Hui CM, Chaladaj W, Dong HC, Choi J, Jurczak J, Bockstaller MR, Matyjaszewski K (2011) Silica-Polymethacrylate hybrid particles synthesized using high-pressure atom transfer radical polymerization. Macromol Rapid Commun 32(3):295–301. doi:10.1002/marc.201000531
Pyun J, Jia SJ, Kowalewski T, Patterson GD, Matyjaszewski K (2003) Synthesis and characterization of organic/inorganic hybrid nanoparticles: Kinetics of surface-initiated atom transfer radical polymerization and morphology of hybrid nanoparticle ultrathin films. Macromolecules 36(14):5094–5104. doi:10.1021/ma034188t
Rodriguez R, Herrera R, Archer LA, Giannelis EP (2008) Nanoscale ionic materials. Adv Mater 20(22):4353–4358. doi:10.1002/adma.200801975
Savin DA, Pyun J, Patterson GD, Kowalewski T, Matyjaszewski K (2002) Synthesis and characterization of silica-graft-polystyrene hybrid nanoparticles: effect of constraint on the glass-transition temperature of spherical polymer brushes. J Polym Sci Part B-Polym Phys 40(23):2667–2676. doi:10.1002/polb.10329
Scheutjens J, Fleer GJ (1979) Statistical-theory of the adsorption of interacting chain molecules.1. Partition-function, segment density distribution and adsorption isotherms. J Phys Chem 83(12):1619–1635. doi:10.1021/j100475a012
Scheutjens J, Fleer GJ (1980) Statistical-theory of the adsorption of interacting chain molecules.2. Train, loop and tail size distribution. J Phys Chem 84(2):178–190. doi:10.1021/j100439a011
Shah AA, Schultz B, Kohlstedt KL, Glotzer SC, Solomon MJ (2013) Synthesis, assembly, and image analysis of spheroidal patchy particles. Langmuir 29(15):4688–4696. doi:10.1021/la460317t
Srivastava S, Schaefer JL, Yang Z, Tu Z, Archer LA (2014) Polymer-particle composites: phase stability and applications in electrochemical energy storage. Adv Mater 26(2):201–233. doi:10.1002/adma.201303070
Tchoul MN, Fillery SP, Koerner H, Drummy LF, Oyerokun FT, Mirau PA, Durstock MF, Vaia RA (2010) Assemblies of titanium dioxide-polystyrene hybrid nanoparticles for dielectric applications. Chem Mater 22(5):1749–1759. doi:10.1021/cm903182n
Tsarevsky NV, Matyjaszewski K (2007) "Green" atom transfer radical polymerization: from process design to preparation of well-defined environmentally friendly polymeric materials. Chem Rev 107(6):2270–2299. doi:10.1021/cr050947p

Tsujii Y, Ohno K, Yamamoto S, Goto A, Fukuda T (2006) Structure and properties of high density polymer brushes prepared by surface-initaited living radical polymerization. In: Jordan R (ed) Advances in polymer science: surface initiated polymerization, vol 197. Springer, Berlin, pp 1–47

Virtanen S, Krentz TM, Nelson JK, Schadler LS, Bell M, Benicewicz B, Hillborg H, Zhao S (2014) Dielectric breakdown strength of epoxy bimodal-polymer-brush-grafted core functionalized silica nanocomposites. IEEE Trans Dielectr Electr Insulation 21(2):563–570. doi:10.1109/tdei.2014.004415

Voudouris P, Choi J, Dong H, Bockstaller MR, Matyjaszewski K, Fytas G (2009) Effect of shell architecture on the static and dynamic properties of polymer-coated particles in solution. Macromolecules 42(7):2721–2728. doi:10.1021/ma802878r

Wang JS, Matyjaszewski K (1995) Controlled living radical polymerization—atom-transfer radical polymerization in the presence of transition-metal complexes. J Am Chem Soc 117(20):5614–5615. doi:10.1021/ja00125a035

Wijmans CM, Zhulina EB (1993) Polymer brushes at curved surfaces. Macromolecules 26(26):7214–7224. doi:10.1021/ma00078a016

Xu C, Wu T, Mei Y, Drain CM, Batteas JD, Beers KL (2005) Synthesis and characterization of tapered copolymer brushes via surface-initiated atom transfer radical copolymerization. Langmuir 21(24):11136–11140. doi:10.1021/la051853d

Yu H-Y, Koch DL (2010) Structure of solvent-free nanoparticle-organic hybrid materials. Langmuir 26(22):16801–16811. doi:10.1021/la102815r

Yu HY, Koch DL (2014) Self-diffusion and linear viscoelasticity of solvent-free nanoparticle-organic hybrid materials. J Rheol 58(2):369–395. doi:10.1122/1.4862316

Zhong M, Matyjaszewski K (2011a) How fast can a CRP be conducted with preserved chain end functionality? Macromolecules 44(8):2668–2677

Zhong MJ, Matyjaszewski K (2011b) How fast can a CRP be conducted with preserved chain end functionality? Macromolecules 44(8):2668–2677. doi:10.1021/ma102834s

Chapter 4
Multiferroic Magnetoelectric Composites/Hybrids

Yongke Yan and Shashank Priya

Abstract The multiferroic magnetoelectric (ME) effect describes the coupling between the electric and magnetic fields, and is defined as a generated electric polarization *P* in response to an externally applied magnetic field *H* (direct ME effect), or an induced magnetization *M* with an applied electric field *E* (converse ME effect). Unfortunately, the ME coupling of all the known single-phase materials is usually small at room temperature to be practically applicable. Alternatively, multiferroic composites (ferroelectric and ferri/ferromagnetic phases) typically yield a giant ME coupling response above room temperature, which makes them attractive for technological applications. In the composites, the ME effect is generated as a product property of the magnetostrictive effect (magnetic/mechanical effect) and piezoelectric effect (mechanical/electric effect). To achieve a large ME response, piezoelectric constituent with a high piezoelectric coefficient, magnetostrictive constituent with a high piezomagnetic coefficient, and good coupling between the piezoelectric and magnetostrictive constituent are required. In this chapter, we begin with a brief overview of the development of each material's constituent (piezoelectrics and magnetostriction) providing a list of state-of-the-art piezoelectric and magnetostrictive materials in multiferroic ME hybrid. Next, a discussion is provided on the composite structure and interface elastic coupling between the piezoelectric and magnetostrictive phases. After that we describe the fabrication process of several important ME hybrids with different phase connectivity, interface, and configuration. Considering the importance of nanostructure and 2–2-type ME composite, the scaling effect and theoretical modeling for these architectures are presented in some detail. Following these sections, some of the potential applications for ME hybrids are reviewed and illustrated by examples. Lastly, the chapter is concluded with a brief summary and future perspective.

Y. Yan (✉) · S. Priya
Bio-inspired Materials and Devices Laboratory (BMDL),
Center for Energy Harvesting Materials and Systems (CEHMS),
Virginia Tech, Blacksburg, VA 24061, USA
e-mail: yanthu@vt.edu

S. Priya
e-mail: spriya@vt.edu

C.-S. Kim et al. (eds.), *Hybrid and Hierarchical Composite Materials*,
DOI 10.1007/978-3-319-12868-9_4

Keywords Magnetoelectric · Multilferroic · Piezoelectric · Magnetostrictive · Composite

Abbreviations

α_{ME}	Magnetoelectric coefficient
BT	Barium titanate, $BaTiO_3$
d	Piezoelectric charge coefficient
E	Electric field
EBSD	Electron backscatter diffraction
E_c	Coercive electric field
EDS	Energy-dispersive X-ray spectroscopy
EMR	Electromechanical resonance
ε	Dielectric permittivity
f_c	Resonance frequency
g	Piezoelectric voltage coefficient
H	Magnetic field
IDE	Interdigitated electrode
k	Electromechanical coupling coefficient
λ	Magnetostriction coefficient
LTCC	Low-temperature co-fired ceramics
M	Magnetization
ME	Magnetoelectric
MFC	Macro-fiber composites
MPB	Morphotropic phase boundary
μ	Magnetic permeability
P	Polarization
PFM	Piezoresponse force microscopy
PLD	Pulsed laser deposition
PMN–PT	Lead magnesium niobate–lead titanate, $Pb(Mg_{1/3}Nb_{2/3})O_3–PbTiO_3$
PPT	Polymorphic phase boundary
P_r	Remnant polarization
PVDF	Polyvinylidene difluoride
PZT	Lead zirconate titanate, $Pb(Zr, Ti)O_3$
q	Piezomagnetic coefficient
Q_m	Mechanical quality factor
SEM	Scanning electron microscopy
tanδ	Dielectric loss
T_c	Curie temperature
$T_{O\text{-}T}$	Orthorhombic-tetragonal phase transition temperature
XRD	X-ray diffraction
Z	Acoustic impedance

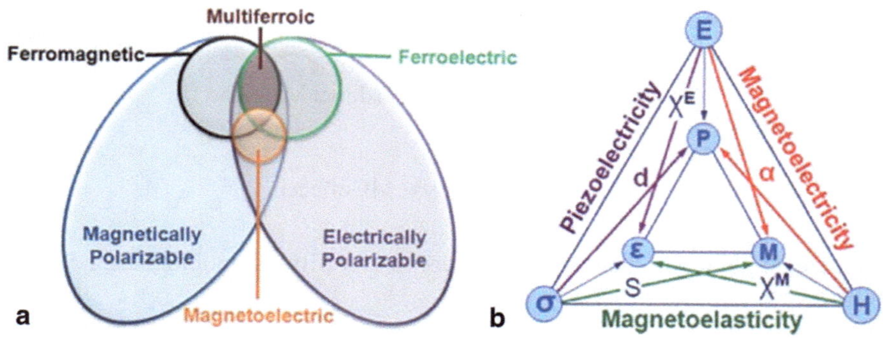

Fig. 4.1 **a** Relationship between *multiferroic* and *magnetoelectric* materials. **b** Schematically illustrating different types of coupling present in the materials. (Reprinted with permission from Martin (2010). Copyright, 2010, The Royal Society of Chemistry)

4.1 Introduction

Magnetic and ferroelectric materials are found in many aspects of modern science and technology. Ferromagnetic materials with switchable spontaneous magnetization M driven by an externally applied magnetic field H have been widely used in data-storage industries. Similarly, dynamic random access memory has been built upon the ferroelectric materials with spontaneous polarization P reversible upon the application of an external electric field E. All the ferroelectrics, especially perovskite oxides, are high-performance ferroelastics or piezoelectrics exhibiting spontaneous strain. The coexistence of field-induced strain and polarization allows these ferroelectric materials to be used for various applications where elastic energy can be converted into electric energy or vice-versa (Spaldin and Fiebig 2005; Wang et al. 2009).

With the ongoing trend towards device miniaturization, integration of multifunctionality in a single material system has become highly desirable. Multiferroic magnetoelectric (ME) materials with the coexistence of ferroelectric and ferromagnetic orders are emerging and particularly appealing not only because they have the properties of their parent compounds but also because interactions between the magnetic and electric order provide additional functionalities. The relationship between multiferroic and ME materials is shown in Fig. 4.1a. In multiferroic ME materials, the ME effect could provide the possibility to control magnetization or/and polarization by an electric field or/and magnetic field (Fig. 4.1b; Spaldin and Fiebig 2005; Wang et al. 2009). The realization of a material with the simultaneous presence of strong coupling between electric and magnetic order would be a milestone for modern electronics. It will open the gateway for very high-density memory storage media using both magnetic and electric polarization, and the possibility of electrically reading or writing magnetic memory devices (and vice-versa).

The ME effect describes the coupling between the electric and magnetic fields. It is defined as the generated electric polarization P in response to an externally applied magnetic field H (direct ME effect), or an induced magnetization M with an applied electric field E (converse ME effect) (Wang et al. 2009; Ma et al. 2011):

$$\Delta P = \alpha \Delta H \quad \text{or} \quad \Delta E = \alpha_E \Delta H \, (\text{direct ME effect}) \tag{1}$$

$$\Delta M = \alpha \Delta E \, (\text{converse ME effect}) \tag{2}$$

In the single-phase ME materials, the upper limit for ME susceptibility can be given as (Brown et al. 1968; O'Dell 1963):

$$\alpha_{ij} < (\varepsilon_{ii} \mu_{jj})^{1/2} \tag{3}$$

where ε and μ are the electric permittivity and magnetic permeability, respectively.

Unfortunately, most of the single-phase materials possess either low permittivity (ε) or low permeability (μ) or both, as a consequence of which the ME coupling of all the known single-phase materials is usually small at room temperature to be practically applicable. Several researchers have investigated single-phase ME materials and have presented alternative ways to overcome the fundamental problem related to the filling of electronic states (Wang et al. 2009; Hill 2000). Hill summarized that of all the conventional ferroelectric perovskites, the transition metal (TM) ions, such as Ti^{4+}, Ta^{5+}, at the B-sites have an empty *d* shell to allow the formation of a partially covalent bond with the surrounding oxygen anions, whereas magnetism arises from unpaired electrons in the *d* or *f* shell of TM ions at the B-site, such as Cr^{3+}, Mn^{3+}, and Fe^{3+}. The difference in filling of the TM ion *d*-shells at the B-site makes these two ordered states mutually incompatible. To solve this paradox, efforts have been made to search for ferroelectricity intrinsically generated by special spin orders. This enables not only an effective combination of the two orders but also the spontaneous mutual control of them.

Alternatively, a class of artificial multiferroic ME materials in the form of piezoelectric and magnetostrictive composite has drawn significant interest in recent years (Nan et al. 2008; Srinivasan 2010). In the composites, the ME effect is generated as a property product of the magnetostrictive effect (magnetic/mechanical effect) and piezoelectric effect (mechanical/electric effect). As a result, this two-phase composite material can be represented as (Nan et al. 2008):

$$\frac{\partial P}{\partial H} = \alpha_{ME} = k_c e^m e \tag{4}$$

where

$$e^m = \frac{\partial S}{\partial H} \text{ and } e = \frac{\partial P}{\partial S} \tag{5}$$

k_c is the coupling factor ($0 \leq k_c \leq 1$) between the two phases (Nan 1993), α_{ME} is the ME coefficient of the composite, S is the strain, and e^m and e are piezomagnetic and piezoelectric coefficients, respectively. Therefore, a high piezoelectric coefficient, e, a high piezomagnetic coefficient, e^m, and good coupling (large k_c) will result in large ME coefficient α_{ME}.

Figure 4.2 illustrates the parameters that should be taken into account for developing high-performance ME hybrids. In this chapter, we begin with a brief overview of the development of each material's constituent (piezoelectrics and magnetostriction) providing a list of state-of-the-art piezoelectric and magnetostrictive materials in multiferroic ME hybrids. Next, a discussion is provided on the composite structure and interface elastic coupling between the piezoelectric and magnetostrictive phases. After that we describe the fabrication process of several important ME hybrids with different phase connectivity, interface, and configuration. Considering the importance of nanostructure and 2–2-type ME composites, the scaling effect and theoretical modeling for these architectures are presented in some detail. Following these sections, some of the potential applications for ME hybrids are reviewed and illustrated by examples. Lastly, the chapter is concluded with a brief summary and future perspective.

4.2 Materials Constituent: Piezoelectrics

4.2.1 Piezoelectrics and Ferroelectrics

a. *Piezoelectrics and relation with ferroelectrics*

Piezoelectricity is the ability of certain crystalline materials that possess noncentrosymmetry (point group lacks an inversion center) to develop an electric charge proportional to the applied mechanical stress (direct piezoelectric effect) or develop a mechanical strain (deformation) proportional to an applied electric field (converse piezoelectric effect (Damjanovic 1998). Piezoelectricity is utilized extensively in the fabrication of various devices such as transducers, sensors, actuators, surface acoustic wave devices, frequency control, etc. (Jaffe 1971; Uchino 2010).

According to Neumann's principle, the symmetry elements of any physical property of a crystal must include the symmetry elements of the point group of the crystal (Nye 1985; Newnham 2005). Piezoelectric property can exist only in 20 noncentrosymmetric point groups. Of the 20 piezoelectric point groups, 10 point groups show the presence of spontaneous polarization (electric polarization that a substance possesses in the absence of an external electric field). These materials are termed as pyroelectric materials, in which the magnitude of the polarization varies with change in temperature. All ferroelectric crystals belong to one of these ten polar point groups. Ferroelectrics are a subgroup of the pyroelectric family that exhibit at least two equilibrium orientations of the spontaneous polarization in the absence of an external electric field. The spontaneous polarization can be switched between

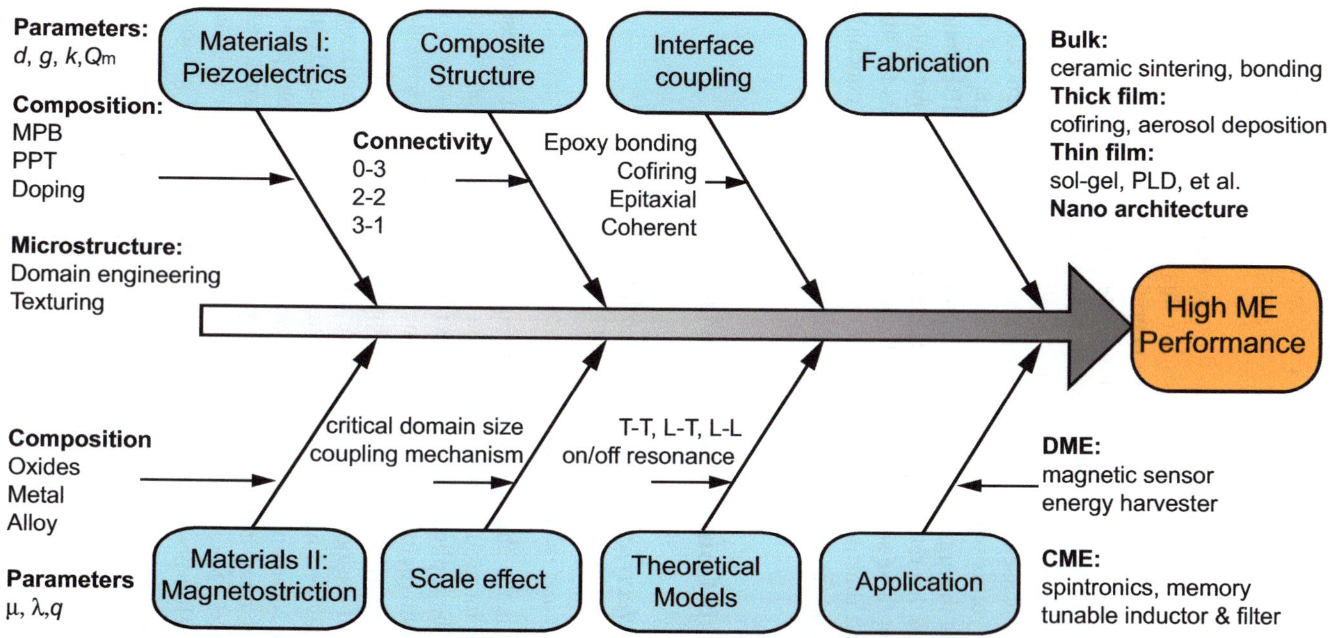

Fig. 4.2 Fish-bone diagram describing the parameters required for developing high-performance ME hybrids. *ME* magnetoelectric

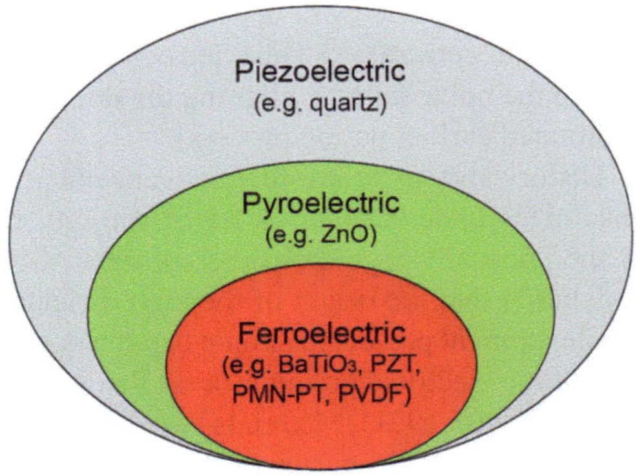

Fig. 4.3 The relationship among *piezoelectric*, *pyroelectric*, and *ferroelectric* materials. (*ZnO* zinc oxide, *$BaTiO_3$* barium titanate, *PZT* lead zirconate titanate, PMN–PT lead magnesium niobate–lead titanate, *PVDF* polyvinylidene difluoride)

the two orientations by applying an electric field (Damjanovic 1998). Figure 4.3 illustrates the relationship among the piezoelectric, pyroelectric, and ferroelectric materials.

The Neumann principle can be applied to all classes of materials with varying crystalline states and dimensions (Newnham 2005). Piezoelectric ceramic materials with random orientation of grains possess spherical symmetry, ∞∞m, which is centrosymmetric and cannot exhibit a piezoelectric effect even if the symmetry of each grain belongs to one of the piezoelectric or polar point groups. If the ceramic is ferroelectric, the spontaneous polarization direction in each grain may be reoriented by an external electric field in the direction of the applied field. Such poled or polarized ceramic possesses symmetry of a cone, ∞m, and exhibits both piezoelectric and pyroelectric effects. The poling process (a process of polarizing a ferroelectric material by applying a high electric field for an extended period of time at a specific temperature) is possible only in ferroelectric materials. Polycrystalline thin films of zinc oxide (ZnO), which is a polar but nonferroelectric material shows a decent magnitude of piezoelectric and pyroelectric response. This is possible because grains of ZnO films usually grow with the crystal polar axis perpendicular to the substrate surface, and in most if not in all cases, the grains and the polarization are oriented in the same direction. [0001]/c axis textured ZnO films show the largest piezoelectric response among all the crystallographic orientation. ZnO ceramics with randomly distributed grains possess spherical symmetry ∞∞m and do not exhibit the piezoelectric or pyroelectric effect. ZnO ceramics cannot become piezoelectric and pyroelectric by poling with an external electric field because ZnO is a nonferroelectric material (Damjanovic 1998).

It should be noted here that as-synthesized ferroelectric materials, either in the form of grain-oriented polycrystalline thin films and bulk ceramics or single crystals, do not exhibit piezoelectric and pyroelectric response. The reason is that as-sintered ferroelectric ceramic or as grown thin film and single crystals usually contain randomly oriented domains. The pyroelectric and piezoelectric effects of individual domains will cancel each other resulting in net zero polarization. Ferroelectric domains are formed below the Curie temperature in order to minimize the electrostatic energy of depolarizing fields and the elastic energy associated with mechanical constraints. Thus, ferroelectric materials need a strong electric field to induce the polar state by aligning the domains along the applied field direction (as mentioned earlier, poling process).

Historically, there are three advancements that have made a significant difference in the understanding of ferroelectricity and piezoelectricity: (1) the discovery of the unusually high dielectric constant of barium titanate ($BaTiO_3$, BT), (2) the realization that the origin of the high dielectric constant was a permanent internal dipole moment present in simple oxygen octahedral ABO_3 materials, and (3) the design of the electrical poling process that aligns the domains within the ferroelectric material (Jaffe 1971). Materials based on barium titanate and lead zirconate titanate ($Pb(Zr, Ti)O_3$, PZT) currently dominate the piezoelectric applications. Recent developments include PZT-based ternary solid solutions, relaxor ferroelectrics such as $Pb(Mg_{1/3}Nb_{2/3})O_3$–$PbTiO_3$ (PMN–PT) and lead-free complex perovskites.

b. *Piezoelectric parameters*

There are five important material parameters for the design of piezoelectric material based applications: the piezoelectric strain constant *d*, the piezoelectric voltage constant *g*, the electromechanical coupling factor *k*, the mechanical quality factor Q_m, and the acoustic impedance *Z* (Uchino 2010; Zhang and Li 2012).Each of these parameters is defined below.

Piezoelectric Strain Constant *d* The magnitude of the induced strain *x* by an external electric field *E* is represented as: $x = dE$. The magnitude of *d* is an important consideration for actuator applications.

Piezoelectric Voltage Constant *g* The induced electric field *E* is related to applied external stress *X* through the piezoelectric voltage constant *g* (an important figure of merit for sensor applications): $E = gX$.

Electromechanical Coupling Factor *k* The electromechanical coupling factor k_{ij} of a piezoelectric material refers to the efficiency of conversion between electric and mechanical energy and vice-versa. This parameter is closely related to the bandwidth of resonant devices.

Mechanical Quality Factor Q_m For a piezoelectric resonator, the mechanical quality factor Q_m characterizes the resonator's bandwidth relative to its center frequency: $Q_m = f_r/\Delta f$, where f_r is resonance frequency, and Δf is the frequency difference at −3dB of the maximum admittance. A high Q_m is important for resonant applications while high Q_e (inverse of dielectric loss) is crucial for off-resonance devices.

Acoustic Impedance *Z* The acoustic impedance *Z* of a material is also related to the elastic constant and/or phase velocity through the relationship: $Z = \sqrt{\rho c}$, where ρ is the density and *c* is the elastic stiffness of material. Analogous to electrical impedance, the transfer of acoustic energy from one medium to another will be maximized when the two media have the same acoustic impedance.

Besides piezoelectric properties, dielectric properties are also very important in designing the piezoelectric device. Dielectric constant (ε_r), dielectric loss ($\tan\delta$), and Curie temperature (T_c) are the most commonly characterized parameters. Dielectric constant (ε_r) is important for the device performance, since the capacitance is closely related to electrical impedance. Dielectric loss ($\tan\delta$) is the dissipation of energy in the form of heat. Low $\tan\delta$ is very important for high-power device and sensor application. Curie temperature (T_c) represents the point above which the crystal structure becomes symmetrical and the material becomes nonpiezoelectric. This point represents the maximum operating temperature for piezoelectric materials. Ferroelectric materials can be characterized by measuring their polarization as a function of the applied AC electric field (*P–E* hysteresis loop). Remnant polarization (P_r) is the polarization that remains when the electric field is removed, and coercive field (E_c) is the electric field required to reduce the polarization to zero. High E_c implies the need for high electric field to pole the sample, while low E_c materials may have low electrical stability as they can be easily depolarized.

There are mainly two approaches to achieve enhanced electric, ferroelectric, and piezoelectric properties: (1) composition selection and modification, and (2) microstructure design. The composition selection approach focuses on identifying binary and ternary phase diagrams that offer unique phase boundaries. The composition modification focuses on doping and substitutions in order to tailor the defect chemistry and phase transitions. The microstructure design focuses on the optimization of grain size (GS) and distribution, grain orientation or texture, domain size, and domain engineering.

4.2.2 Composition Selection Methodology

a. *Morphotropic phase boundary*

The common approach toward achieving a high piezoelectric composition is to find a boundary between two ferroelectric phases (Liu and Ren 2009). Such a boundary known as the "morphotropic phase boundary (MPB)" reflects the instability of the polarization state such that the polarization vector can be easily rotated by applying external stress or electric field. The ferroelectric PZT system has been extensively studied because of its interesting physical properties close to the MPB which is nearly vertical between the tetragonal and rhombohedral regions of the phase diagram (Jaffe 1971; Uchino 2010). The high magnitude of piezoelectric coefficient exhibited in the vicinity of MPB in polycrystalline materials has been explained in terms of the ease of domains' reorientation (Guo et al. 2000).

b. *Polymorphic Phase Transition*

Polymorphic phase transition (PPT) such as tetragonal to orthorhombic ($T_{O\text{–}T}$) shown in Fig. 4.4 differs from the MPB in two significant ways: (i) the PPT boundary separates two ferroelectric phases over a much wider range of composition as compared to narrow range for MPB and (ii) the PPT boundary occurs at much lower temperatures than Curie temperature over the whole range of composition while the MPB stretches from low temperature until Curie temperature. It should

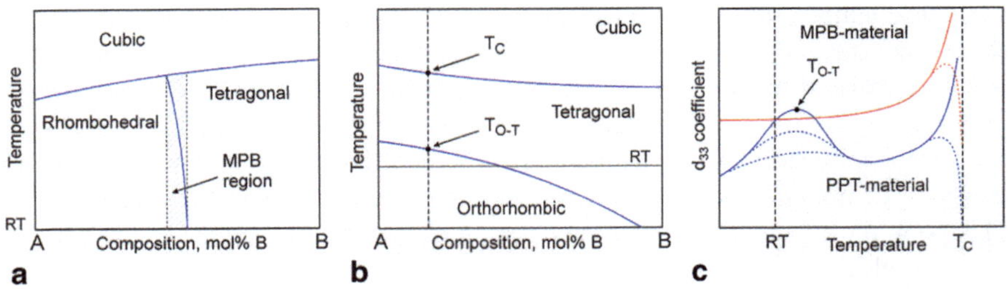

Fig. 4.4 Characteristic (hypothetical) phase diagrams for perovskite solid solutions exhibiting **a** MPB or **b** polymorphic phase transformation (*PPT*). The resulting temperature dependence of the piezoelectric coefficient of **c** hypothetical MPB and PPT materials with the same room-temperature (*RT*) piezoelectric coefficient and same Curie temperature (T_c), where *dotted trend lines* indicate degradation of piezoelectric properties due to depolarization at temperatures near T_{O-T} and T_C. (Reprinted with permission from eontsev and Eitel (2010). Copyright, 2010, IOP PUBLISHING)

be pointed out here that triclinic point shown on the phase diagram in Fig. 4.4a has been recently the subject of intense investigation. The dielectric and piezoelectric properties show local maxima when the temperature is in the vicinity of polymorphic phase transition. It is possible to shift the local maxima toward room temperature through doping and substitutional reactions. This technique has been used for improving the piezoelectric response of the (K, Na)NbO_3 (KNN) ceramics where T_{O-T} was modulated by modifying with additives such as $LiTaO_3$, $LiNbO_3$, $LiSbO_3$, and $SrTiO_3$ (Zhang et al. 2007).

It should be noted that both PPT and MPB lead to improved dielectric and piezoelectric properties in ferroelectric systems. However, the property enhancement at PPT is strongly temperature dependent. As illustrated in Fig. 4.4, MPB-based piezoceramics typically exhibit relatively flat temperature dependence of the piezoelectric properties and broad usage temperatures approaching ~ ½ T_C. Conversely, PPT-based piezoceramics generally exhibit large temperature dependence of the piezoelectric coefficient and rapid degradation of the poled piezoelectric coefficient (due to partial depoling) as the temperature approaches close to the PPT temperature (Leontsev and Eitel 2010).

c. *Doping effect*

Dielectric, ferroelectric, and piezoelectric properties can be tailored through composition modification. As mentioned earlier, there are different figures of merit (FOM) in piezoelectrics depending upon the application. In the case of PZT ceramics, the use of dopants in the range of 0.05–5.0 at.% is quite common in order to modify the properties for specific applications. Table 4.1 lists the common dopants used for modifying the PZT ceramics.

4.2.3 Microstructure Design Methodology

In contrast to the composition selection, the microstructure design focuses on controlling the micro- and/or nanoscale structure of the piezoelectric material. A significant

Table 4.1 Common dopants for modifying the PZT ceramics

Type	Site	Dopants	Defects	Effect
Donor	A	Pb^{2+} by La^{3+}	Pb^{2+} vacancies, enhanced domain motion	High ε, d, k, P_r, tanδ; low Q_m, E_c
	B	Zr^{4+} by Nb^{5+}		
Acceptor	A	Pb^{2+} by K^{+}	Oxygen vacancies, domain pinning	Low ε, d, k, P_r, tanδ; high Q_m, E_c
	B	Zr^{4+} by Fe^{3+}, Mn^{3+}		
Isovalent	A	Pb^{2+} by Ba^{2+}, Sr^{2+}		Tuning elastic compliance, aging rate, and transitions
	B	Zr^{4+} by Sn^{4+}		

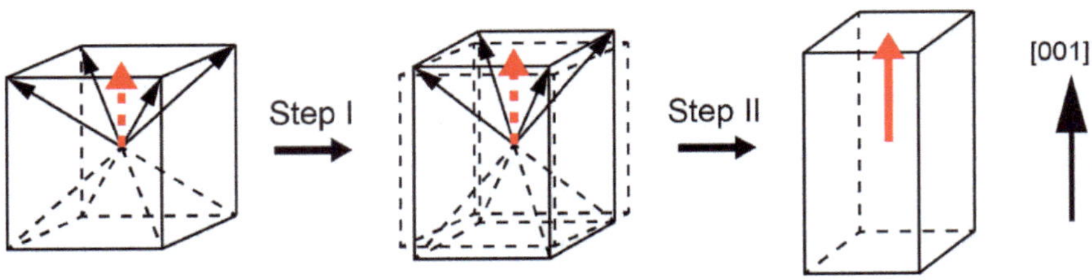

Fig. 4.5 Schematic diagram of domain configurations in <001> oriented rhombohedral crystals (step I: piezoelectricity, step II: electric-field-induced phase transition). (Adapted from Park and Shrout 1997)

enhancement in electromechanical properties has been achieved by utilizing various microstructure-design techniques including domain engineering, domain size/wall engineering (GS optimization, poling optimization), and texturing.

a. *Domain engineering*

The number and possible orientation of domains (domain configuration) are determined by the crystal symmetry, which are closely related to their dielectric and piezoelectric properties (Zhang and Li 2012). The technique to control the desirable domain configurations is called domain engineering.

In an engineered domain configuration, relaxor-PT crystals have been shown to possess an ultrahigh piezoelectric coefficient and electromechanical coupling factor k_{33} on the order of >2000 pC/N and ~0.9, respectively, far outperforming the state-of-art PZT polycrystalline ceramics (Park and Shrout 1997). This excellent piezoelectric property is due to the easy tilt of the spontaneous polarization by applying electric field along the nonpolar direction. As depicted in Fig. 4.5, the rhombohedral relaxor-PT single crystals exhibit eight degenerate domain variants with the spontaneous polarization (polar vector) along one of the <111> directions. When poled along the nonpolar axis [001], four of the eight domain variants will be energetically favored by the poling field with polar vectors along [111], $[\bar{1}\bar{1}\bar{1}]$, $[\bar{1}\bar{1}\bar{1}]$, and $[\bar{1}\bar{1}\bar{1}]$.

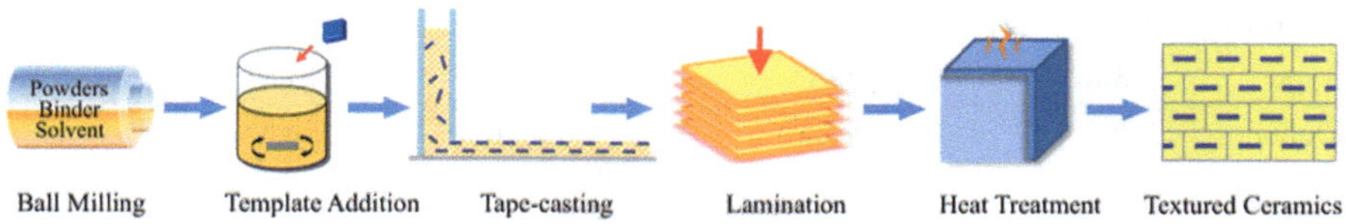

Fig. 4.6 Schematic diagram of template grain growth (TGG) process

Thus, four domain variants exist after the poling process with polar vectors equally inclined along the poling direction (~54.7°). The resultant domain-engineered configuration will be stable since there is no driving force for the domain-wall motion (only polarization rotation or inclination, step I). This accounts for the high piezoelectric response, the hysteresis-free strain characteristics, and low dielectric loss observed in the relaxor-PT crystals. With a further increase in the applied electric field, polarization inclination toward [001] finally results in the collapse of all the polarization along the [001] direction inducing a structural transition to the tetragonal phase (step II) (Zhang and Li 2012; Park and Shrout 1997).

b. *Texture Engineering*

As described above, domain-engineered relaxor-PT ferroelectric single crystals possess an ultrahigh piezoelectric coefficient d_{33} and electromechanical coupling factor k_{33} on the order of >2000 pC/N and ~0.9, respectively, far outperforming the state-of-art PZT polycrystalline ceramics (Zhang and Li 2012; Park and Shrout 1997). These excellent piezoelectric properties are due to the stable engineered domain configuration as described above. However, the application of single crystals has been limited due to the high production cost relating to the use of platinum crucibles and small size of useful samples due to the compositional heterogeneity across the ingot. To overcome the high production cost and limited size of single crystal, <001> textured relaxor-PT/PZT piezoelectric ceramics via templated grain growth (TGG) method such as PMN–PT, $Pb(Mg_{1/3}Nb_{2/3})O_3$-$PbZrO_3$-$PbTiO_3$ (PMN–PZT) have been developed. Textured ceramics have shown a significant improvement in the piezoelectric property comparable to that of single crystals, but the cost is similar to that of traditional ceramics (Messing et al. 2004; Yan et al. 2012a, b; Yan et al. 2013a; Yan et al. 2014).

Figure 4.6 shows the schematic diagram of the TGG texturing process. In order to obtain the textured ceramics, micro-sized template crystals are aligned in the ceramic matrix powders using a tape-casting method. During high-temperature sintering, epitaxial growth occurs on the template seeds giving rise to textured grains (Messing et al. 2004). The anisotropic-shaped (whisker or platelet) template seeds should have a smaller (<15 %) lattice mismatch with the matrix phase for getting high degree of texturing (Messing et al. 2004). The anisotropic-shaped BT templates were synthesized using the topochemical microcrystal conversion process (Yan et al. 2012a; Liu et al. 2007). These templates were used to fabricate various highly textured piezoelectric materials with excellent properties as described below.

[001] Textured PMN–PT Ceramics Due to ultrahigh piezoelectric properties from domain-engineered PMN–PT single crystals, [001] textured PMN–PT ceramics

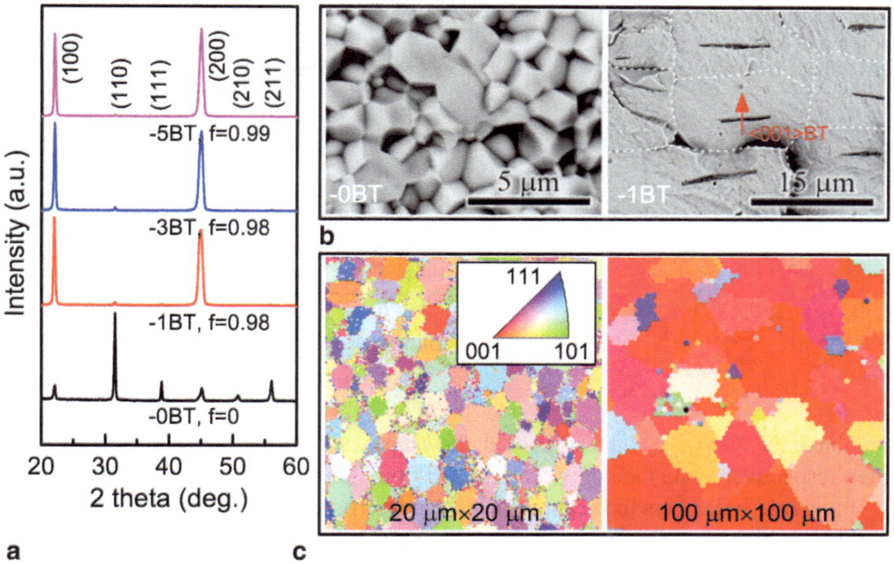

Fig. 4.7 **a** X-ray diffraction (*XRD*) patterns of PMN–PT–*x*BT samples; **b** Cross-sectional scanning electron microscope (*SEM*) images of PMN–PT–0BT (*left*) and PMN–PT–1BT (*right*); **c** Electron backscatter diffraction (*EBSD*) images of PMN–PT–0BT (*left*) and PMN–PT–1BT (*right*) surfaces. (Yan et al. 2013a)

were synthesized by using the TGG method to achieve low cost with high performance (Yan et al. 2012b).

Figure 4.7a shows the X-ray diffraction (XRD) patterns of PMN–PT–*x*BT samples. Here, PMN–PT–0BT represents the random ceramic without BT seeds ($x=0$), and PMN–PT–*x*BT ($x=1, 3, 5$) represents the textured ceramics with different BT template content. Domination of (00 *l*) peaks in PMN–PT–*x*BT ($x=1, 3, 5$) samples indicated high <001> orientation. The Lotgering factor for these textured samples was found to be almost the same ($f<98\,\%$). The Lotgering factor (f_{00l}) is defined by the following equations (Lotgering 1959):

$$f_{(00l)} = \frac{P_{(00l)} - P_0}{1 - P_0} \tag{6}$$

$$P_{(00l)} = \frac{\sum I_{(00l)}}{\sum I(hkl)} \tag{7}$$

$$P_0 = \frac{\sum I_{0(00l)}}{\sum I_0(hkl)} \tag{8}$$

Where $\sum I_{(00l)}$ and $\sum I_{0(00l)}$ are the summation of the XRD peak intensities of all the 00*l* peaks in the textured and random samples, and $\sum I_{(hkl)}$and$\sum I_{0(hkl)}$ are the sum-

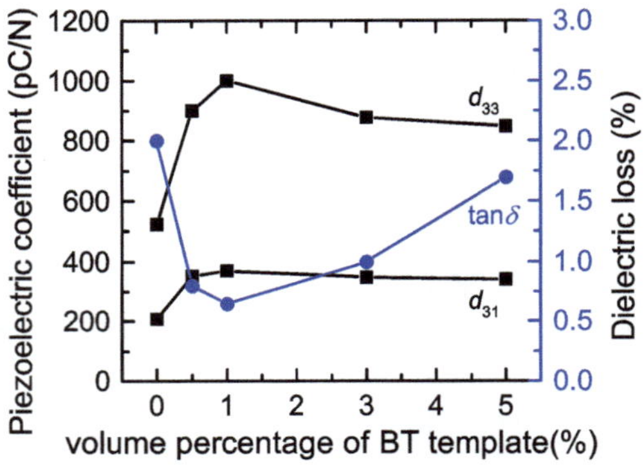

Fig. 4.8 Piezoelectric properties and dielectric loss of PMN–PT–xBT ceramics

mation of the XRD peak intensities of all the *hkl* peaks in the textured and random samples, respectively. The *f* factor was calculated from 2θ scan in the range of 20 to 60°.

Figure 4.7b displays the scanning electron microscopy (SEM) images of random and textured cross-sectional samples. Compared to the equiaxed grains in random ceramics (left), all the matrix grains in the textured sample (right) were well aligned with a brick-wall microstructure. Another feature of textured ceramics is the existence of an aligned plate-like BT template inside the oriented PMN–PT matrix. The BT template microcrystals had a length of 5–10 μm and thickness of 0.5–1 μm. Therefore, the textured ceramics had two unique characteristics: <001> grain orientation and the existence of the heterogeneous BT "core." Figure 4.7c shows the electron backscatter diffraction (EBSD) inverse pole figures of the random ceramic and textured ceramic clearly illustrating that the textured grains exhibit strong <001> preferred orientation.

The piezoelectric response and dielectric loss (tanδ) for PMN–PT–xBT is shown in Fig. 4.8. The d_{33} value was found to achieve a maximum value of 1000 pC/N at $x=1$. The enhanced piezoelectric response was attributed to the high degree of texturing and domain structure similar to that of single crystals. The trend in variation of the value of d_{31} were similar to that of d_{33}. Interestingly, the textured PMN–PT–1BT ceramics were found to exhibit a loss tangent value as low as 0.6 %, which is about one third of the magnitude obtained for most of the soft piezoelectric ceramics (>2.0 %). Therefore, grain-oriented PMN–PT–1BT ceramics, with high piezoelectric response (*d*, *g*, and *k*) and low dielectric loss (tanδ), are an ideal substitute for the currently used soft piezoelectric ceramics.

[001] Textured Pb($Mg_{1/3}Nb_{2/3}$)O_3-$PbZrO_3$-$PbTiO_3$ (PMN–PZT) Ceramics The domain-engineered PMN–PT relaxor piezoelectric single crystals have been characterized by a high piezoelectric response. However, low ferroelectric rhombohedral to tetragonal phase transition temperature ($T_{R\text{-}T}$ ~60–90 °C) and low coercive fields (E_C ~2–3 kV cm^{-1}) were found to limit the stability of these relaxor-based single crystals. Many attempts have been made to grow PZT-based single crystals near MPB to achieve high and stable piezoelectric properties, but so far there has been only limited success in growing PZT single crystals with MPB composition due to the inherent technical challenges (incongruent melting behavior; Clarke 1976).

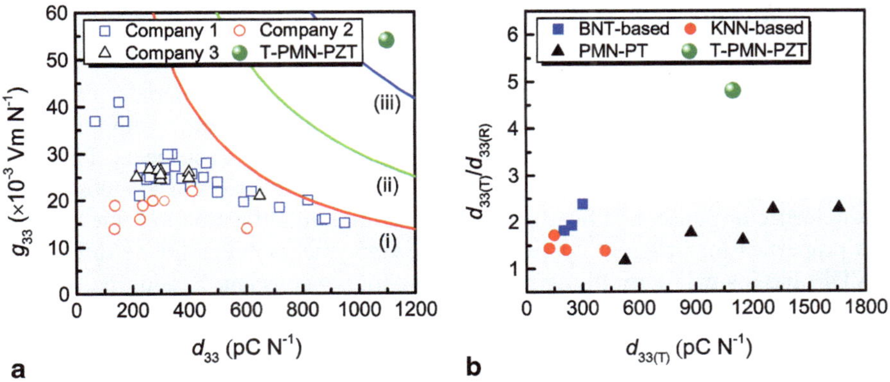

Fig. 4.9 **a** Comparison of g_{33} and d_{33} values of various piezoelectric ceramics. Different colored lines (i), (ii), and (iii) indicate the plots of $g_{33} \cdot d_{33} = c$ function; (i) $c = 16{,}500 \times 10^{-15}$ m^2 N^{-1}, (ii) $c = 30{,}000 \times 10^{-15}$ m^{2}N^{-1}, and (iii) $c = 50{,}000 \times 10^{-15}$ m^{2}N^{-1}; **b** Comparison of $d_{33(T)}/d_{33(R)}$ ratio versus $d_{33(T)}$ of various textured piezoelectric ceramics. The $d_{33(T)}$ and $d_{33(R)}$ represent d_{33} of textured ceramic and that of randomly oriented counterpart, respectively. BNT and KNN represent $(Bi_{1-x}Na_x)TiO_3$ and $(K_{1-x}Na_x)NbO_3$ compositions, respectively

Table 4.2 Piezoelectric and dielectric properties of PMN–PZT piezoelectrics: randomly oriented ceramic (R-ceramic), <001> textured ceramic (T-ceramic), and <001> single crystal (S-crystal)

Properties	R-ceramic	T-ceramic	S-crystal
Piezoelectric charge constant, d_{33} (pC/N)	230	1100	1530
Electromechanical coupling constant, k	0.4 (k_p)	0.84 (k_p)	0.93 (k_{33})
Relative dielectric permittivity, $\varepsilon_{33}/\varepsilon_0$	915	2310	4850
Piezoelectric voltage constant, g_{33} (× 10^{-3} Vm/N)	28.4	53.8	35.6
$d_{33} \cdot g_{33}$ (× 10^{-15} m^2/N)	6532	59,180	54,468
Remnant polarization, P_r (μC/cm^2)	30	36	29
Coercive field, E_c (kV/cm)	8.2	7.4	4.5
Curie temperature, T_c (°C)	233	204	211

An alternative is to synthesize textured PMN–PZT ceramics. We have synthesized >90% [001] textured PMN–PZT ceramics with enhanced piezoelectric properties ($d_{33} > 1000$ pC/N, $g_{33} > 50 \times 10^{-3}$ Vm/N) (Yan et al. 2013b). Generally, high d_{33} piezoelectric materials possess low g_{33} value and vice-versa as shown in Fig. 4.9a because the d_{33} is mostly proportional to the square root of dielectric constant of the piezoelectric materials. Interestingly, the textured PMN–PZT ceramics exhibited both high d_{33} and high g_{33} values due to the template-controlled dielectric characteristics. Table 4.2 summarizes the piezoelectric and dielectric properties of randomly oriented ceramic (R-ceramic), <001> textured ceramic (T-ceramic), and <001> single crystal (S-crystal) of PMN–PZT composition. The textured ceramic exhibited giant d_{33} of 1100 pC N^{-1} which was 4.8 times higher than that of its randomly oriented counterpart having d_{33} ~230 pC N^{-1}. This increased ratio of d_{33} value between R- and T-ceramic (4.8) was much higher than that of the other textured piezoelectric ceramics (usually less than 2.5) as shown in Fig. 4.9b.

4.2.4 Piezoelectric Materials

This section summarizes the current status of piezoelectric materials: ceramic, single crystals, polymers, thin films, and textured ceramics. Quartz with the highest mechanical quality factor is used for low-loss resonant transducers. PZT ceramics are the most widely used piezoelectric ceramics due to their low cost, high piezoelectric response, and flexibility of composition modification to achieve various properties for different applications. Piezopolymer polyvinylidene difluoride (PVDF) has a small permittivity resulting in a high *g* constant. In addition, PVDF has high mechanical flexibility which makes it suitable for pressure/stress sensor applications. Table 4.3 summarizes the properties of these piezoelectric materials.

4.3 Materials Constituent: Magnetostriction

4.3.1 Magnetostriction

Magnetostriction is a property where ferromagnetic materials change their shape or dimensions under an applied magnetic field or change magnetization under an applied mechanical stress (Engdahl 2000).

The origin of magnetostriction at the atomic level is related to the spin–orbit coupling. The spin refers to the angular momentum of an electron spinning around its axis giving rise to a magnetic moment. In atoms with partially filled electron shells where the spins do not exactly cancel each other, there arises a net magnetic moment. This net magnetic moment has to be coupled strongly with the orbit in spin–orbit coupling. This coupling gets larger as the nucleus increases and is thus greatest in the rare earths. Magnetostriction is a result of an applied magnetic field rotating the magnetic moment and its rigidly attached anisotropic electron charge density. The rotating electron charge density moves the neighboring ions inducing a strain. This happens due to the magnetoelastic coupling whereby the lattice is distorted in order to minimize the energy of the system. Figure 4.10 schematically illustrates the $4f$ charge density in rare earths rotating in a magnetic field. Conversely, the magnetoelastic coupling can be given by the change in magnetic anisotropy energy due to a specified strain. In the magnetic $3d$ transition metals, the spin–orbit coupling is weaker than in rare earths by approximately an order of magnitude.

The origin of magnetostriction at a macroscopic level is related to the rotation of magnetic domains under applied magnetic field as shown in Fig. 4.11. This rotation and reorientation causes internal strain. The strain results in the stretching of the material in the direction of the applied magnetic field. As the magnitude of magnetic field increases, more and more domains get aligned along the direction of magnetic field. When all the magnetic domains become aligned, magnetostriction reaches its saturation value, λ.

Table 4.3 Dielectric and piezoelectric properties of the various classes of piezoelectric materials

Sample	$\varepsilon_{33}{}^{T}/\varepsilon_0$	$\tan\delta$	T_c	k_{33}	k_{31}	d_{33}	d_{31}	g_{33}	g_{31}	Q_m	$d_{33} \times g_{33}$	$d_{31} \times g_{31}$
			(°C)			(pC/N)		(10^{-3} m/N)			(10^{-15} m²/N)	
(A) Ceramic												
BT	1700	0.01	120	0.5		190		12.6			2394	
PZT-4	1200	0.003	320	0.70	−0.33	285	−122	24.8	−10.6	500	7068	1354
PZT-5A	1700	0.020	374	0.71	−0.34	374	−171	24.8	−11.4	75	9275	1949
PZT-5H	3400	0.025	195	0.75	−0.39	593	−274	19.7	−9.1	65	11,682	2496
PZT-8	1000	0.004	300	0.64	−0.30	225	−97	25.4	−10.9	1000	5715	1057
(B) Polymer												
PVDF	10	0.02	129	0.19	0.12	−33	16.5			4		
(C) [001] Single crystal												
PMN-29PT	5400	0.006	137	0.91	−0.43	1540	−699	31.3	−14.2	150	48,202	9926
PMN–PZT	4850	0.005	211	0.93	−0.44	1530	−718	35.6	−16.7	100	54,468	11,991
Mn–PMN–PZT	3410	0.002	203	0.92	−0.45	1140	−513	37.8	−17.0	1050	43,092	8721
(D) <001> Textured ceramics												
PMN–PT	2591	0.006	162		−0.57	1000		44			44,000	
PMN–PT–C	2820	0.006	162		−0.60		−411		−16.5			6766
PMN–PZT	2310		211		−0.60	1100				70	59,180	
Mn–PMN–PZT	1865	0.003	210		−0.53	720				400		

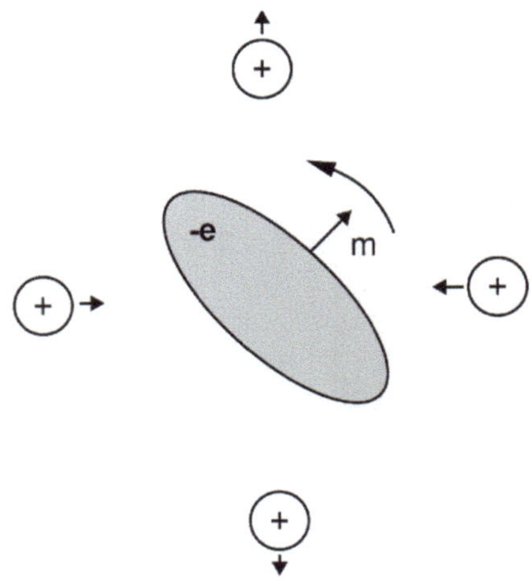

Fig. 4.10 Schematic of the 4*f* charge density rotating in a magnetic field. (Reprinted with permission from Engdahl (2000). Copyright, 2000, Elsevier.)

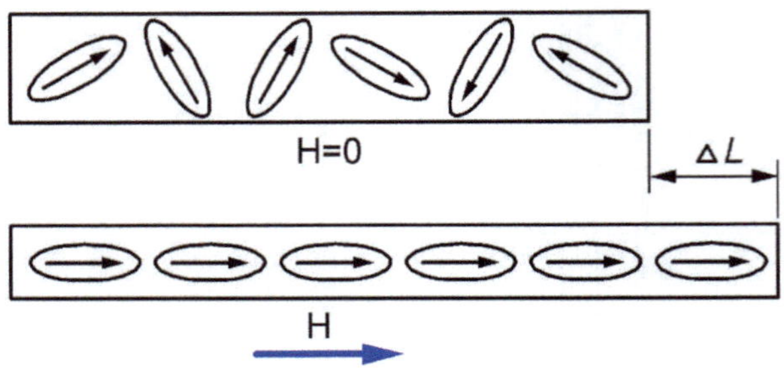

Fig. 4.11 Ferromagnetic domain rotate under an applied magnetic field

4.3.2 Magnetostriction Materials

The development of magnetic materials in large part consists of efforts to combine the two groups of magnetic elements (rare earths and transition metals) with other elements to achieve desired coupling. To achieve useful magnetostriction, there are several important requirements (Engdahl 2000):

1. The intrinsic magnetoelastic coupling must be large.
2. The magnetic ordering temperature (Curie temperature) must be above the desired operating temperatures.
3. Low driving magnetic field, small coercive field, and small magnetic anisotropy.

The transition metals have larger moments at room temperature and above but small magnetoelastic coupling. Rare earth elements and alloys exhibit strong magnetoelastic coupling but have low Curie temperatures. For room temperature applications, an alloy or compound of rare earth and transition metals may be the best choice. Table 4.4 provides a summary of the properties of magnetostrictive materials that have previously been used in ME laminate composites. Here, we briefly describe each of these materials listed in Table 4.4. (1) Ferrites: Ferrites have large

Table 4.4 List of magnetostrictive materials used for fabricating magnetoelectric (ME) laminate

Materials	λ	μ	T_c	R (Ω m)
$NiFe_2O_4$	27	20	535	1×10^6
Terfenol-D	1400	6–10	377	5.8×10^{-7}
Fe–Ga	200	20		6×10^{-7}
Metglas 2605	40	>40,000	395	1.3×10^{-6}

resistances (relative to the other magnetic materials) but small magnetostriction. Nickel ferrite (NFO) is one of the best candidates for ME laminates due to the large magnetostriction and good sinterability. The advantage of using ferrite is that it can be co-fired with ferroelectric oxides. (2) Terfenol-D: The alloy has the largest magnetostriction of any known material, more than 50 times than that of nickel ferrite. But it cannot be co-processed with ferroelectrics oxide, as sintering will dramatically reduce the alloys superior magnetostriction due to oxidization (Sandlund et al. 1994). (3) Fe–Ga: The magnetostriction of Fe–Ga is less than that of Terfenol-D, but it is mechanically ductile whereas Terfenol-D is brittle. (4) Metglas: the amorphous alloy $Fe_{81}Si_{3.5}B_{13.5}C_2$ with its trade name Metglas 2605SC is a promising magnetostrictive alloy. A favorable property of this material is its enormous relative magnetic permeability of > 10,000 (accordingly, its magnetization saturates at very low DC magnetic biases). Although its magnetostriction ($\lambda\sim40$ ppm) is far smaller than Terfenol-D, the maximum piezomagnetic coefficient is 4× larger than that of Terfenol-D.

4.4 Composite Design and Interface Coupling

4.4.1 Connectivity Schemes

The ME response of an ME composite is highly dependent on the connectivity (Nan et al. 2008). Connectivity determines the magnetic flux (related to demagnetizing fields) and the strain coupling in the active phase in the ME composite. Newnham introduced the nomenclature for "connectivity" of the piezoelectric composites (Newnham et al. 1978). When considering a two-phase composite, each phase in the composite is self-connected in 0, 1, 2, or 3 dimensions. Therefore, a diphasic composite can be identified with two numbers "*m*–*n*," where *m* stands for the connectivity of an active phase and *n* for an inactive phase.

Similar to piezoelectric composites, the ME composites could have various connectivity schemes but the commonly used configurations are 3–0-type particulate composites, 2–2-type laminate composite, and 1–3-type fiber composites as shown in Fig. 4.12 (Nan et al. 2008). The 3-0-type particulate ME composites were constructed by embedding one-phase particles in a matrix of another phase and found to exhibit a coupling coefficient of several 10–100 mV/cm·Oe at room temperature.

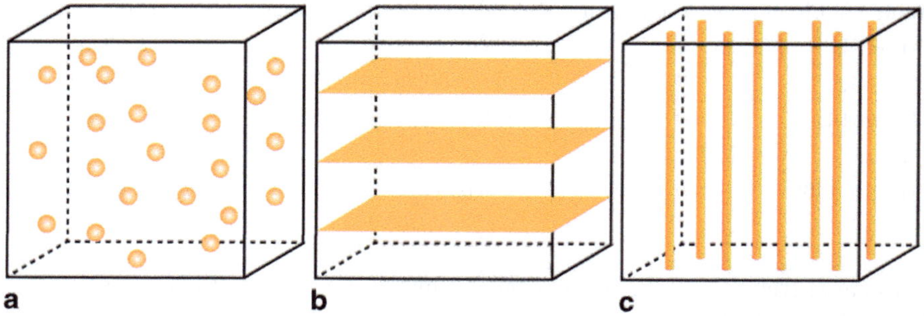

Fig. 4.12 Schematic diagram for connectivity of multiphase ME composites: **a** 3–0-type particulate structure, **b** 2–2-type laminate structure, and **c** 1–3-type fiber/rod–matrix structure

However, this magnitude is much below the theoretical magnitude due to the problems related to the interdiffusion between piezoelectric and magnetostrictive phases during high-temperature sintering, thermal expansion mismatch between the two phases, and low resistivity. In order to improve the resistivity in ME composites, 2–2-type ME laminates have been synthesized by using piezoelectric and magnetostrictive layers and shown to exhibit ME coefficient of several thousand mV/cm·Oe at room temperature due to reduction of the leakage problem. The 1–3-type ME composites were investigated for high ME coefficient by exploiting contribution from both piezoelectric coefficient d_{33} and piezomagnetic coefficient q_{11}. The PZT/Terfenol-D composites have been reported with an ME coefficient of 500 mV/cm·Oe at off-resonance condition and 18,200 mV/cm·Oe at a resonance condition by Ma et al. (2007). Laminate composites are generally fabricated by bonding magnetostrictive and piezoelectric layers using silver epoxy followed by annealing at lower temperature of ~100 °C. The most common geometry for laminate composites is a "sandwich" structure where the piezoelectric layer is arranged between two magnetostrictive ones. Under applied magnetic field, the strain in magnetostrictive layer is transferred to the piezoelectric layer through the bonding layer resulting in the generation of electrical charge.

4.4.2 Working Modes of ME Composites

In 2–2 ME laminate composites, there can be four possible operating modes as shown in Fig. 4.13. In the *L–L* mode, both the applied *H* field and the measured *E* field are parallel to the principle vibration mode; in the *T–L* mode, the applied *H* field is perpendicular to the induced *E* field which is measured parallel to the principle mode; in the *L–T* mode, the applied *H* field is parallel and the induced *E* field is measured perpendicular to the principle mode; and in the *T–T* mode, the applied *H* field and induced *E* are perpendicular to the principle mode. Here, *L* stands for longitudinal and *T* for transverse (Priya et al. 2007).

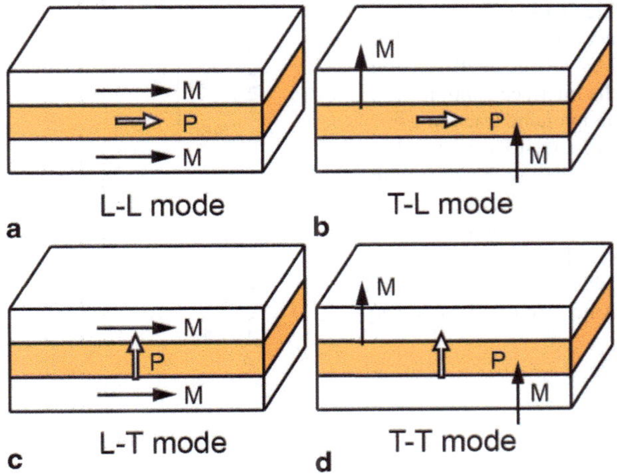

Fig. 4.13 Laminate ME composites operating in different vibration modes. Here, *L* stands for longitudinal and *T* for transverse

Table 4.5 lists the typical values of the ME voltage coefficient (α_{ME}) for these four basic types of operational modes. In this table, it can be seen that the L–L mode has a larger value of α_{ME} than the L–T one. This is because the longitudinal piezoelectric coefficients d_{33} of PZT, PMN–PT, and PZN–PT are larger than the transverse one d_{31}. In addition, the value of α_{ME} was notably lower for the T–T mode. This undoubtedly is due to the effect of demagnetization fields on the magnetostrictive layers, which results in much larger required DC magnetic biases to reach the maximum value of the effective piezomagnetic coefficient. Theoretically, the L–L mode should have the largest ME voltage coefficient out of all the four basic modes. However, in actuality, this was not found to be the case, simply because the capacitance of the piezoelectric layer in the laminate was very small.

4.4.3 Interface Coupling

Bichurin et al. developed a generalized effective medium method for calculating the composite ME effect by introducing an interface coupling parameter k for characterizing the actual bonding conditions at the interface, i.e., $k = ({}^{p}S_i - {}^{p}S_{i0}) / {}^{m}S_i - {}^{p}S_{i0}$ (i=1,2), where ${}^{p}S_{i0}$ are strain tensor components with no friction between layers. k depends on the interface quality and is a measure of the differential deformation

Table 4.5 Magnetoelectric (ME) voltage coefficient of a ME laminate under different operating modes (Priya et al. 2007). All laminates were made by using two pieces of Terfenol-D (14 mm × 6 mm × 1.2 mm) and one piece of PZT (16 mm × 6 mm × 2 mm)

	L–L	L–T	T–L	T–T
H_{dc} (Oe)	500	500	2200	2200
α_{ME} (V cm^{-1} Oe^{-1})	2.4	0.73	0.78	0.5

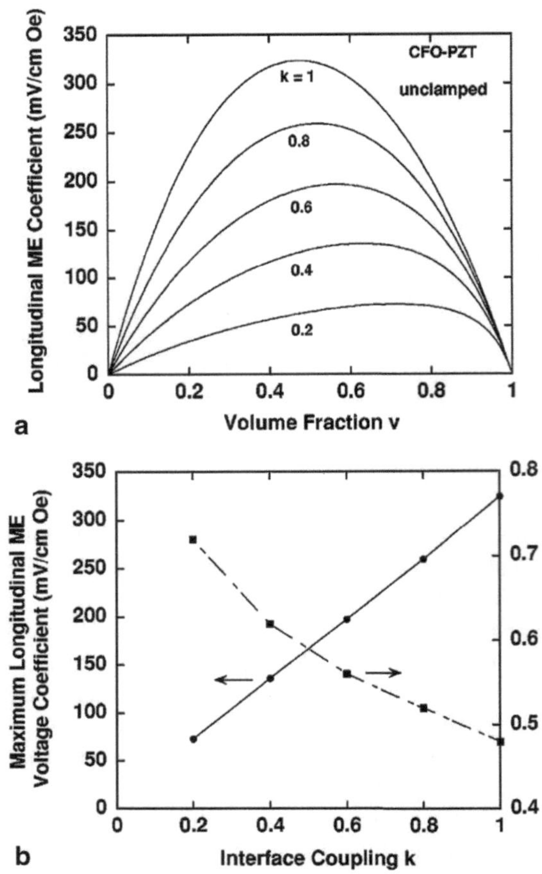

Fig. 4.14 **a** Estimated dependence of longitudinal ME voltage coefficient on the interface coupling k and volume fraction v for $CoFe_2O_4$-PZT bilayer. **b** Variation with k of maximum α_{E33} and the corresponding v_{max}. (Reprinted with permission from (Bichurin et al. 2003). Copyright, 2003, The American Physical Society)

between the piezoelectric and magnetostrictive layers. The coupling factor $k=1$ for an ideal interface and $k=0$ for the case without friction. In practical cases, the ME coefficients of the laminate composites can be given as (Bichurin et al. 2003):

$$\alpha_{E33} = \frac{E_3}{H_3} = \frac{2kv(1-v)d_{31}q_{31}\mu_0\,\underline{s}}{(2vd_{31}^2 - {}^p\varepsilon_{33}\,\underline{s})[\bar{\mu}\,\underline{s} + 2kq_{31}^2(1-v)^2]} \tag{9}$$

$$\alpha_{E31} = \frac{E_3}{H_1} = \frac{-kv(1-v)(q_{11}+q_{21})d_{31}}{{}^p\varepsilon_{33}\,\underline{s} - 2kvd_{31}^2} \tag{10}$$

$$\text{with } \underline{s} = v({}^p s_{11} + {}^p s_{12}) + k(1-v)({}^m s_{11} + {}^m s_{12})$$

Figure 4.14 shows a theoretical example for the 2–2 laminate composite fabricated from PZT and $CoFe_2O_4$ with different k values. It can be seen that the ME coefficient decreases linearly with the decrease of interface coupling factor, *k*. Most of the 2–2 laminates reported in the literature have been synthesized by epoxy bonding. Therefore, there is an additional polymer layer between the piezoelectric and magnetostrictive layer which weakens the interface coupling.

4.5 Important Hybrid Material Systems

4.5.1 *Overview of Fabrication Process*

As discussed above, the ME performance is not only related to properties of individual material constituents (piezoelectric and magnetostrictive) but also strongly dependent on the connectivity of the composite structure and its interface coupling. Several fabrication processes have been developed to achieve different connectivity schemes. For 0–3-type particulate composites, the widely used method consists of sintering the mixture of ceramic oxide powders at high temperature. The ceramic sintering process has the advantage of being relatively simple, scalable, and low cost. However, the leakage problem due to the low resistivity in the magnetic phase makes the electric poling of the ME composites difficult, resulting in low ME properties.

For 2–2-type composites, the leakage problem in the particulate composite ceramics can be eliminated. A giant ME effect has been reported in a variety of 2–2-type magnetostrictive–piezoelectric composites consisting of ferrite, Terfenol-D, and Metglas as the magnetostrictive phase, and PZT and PMN–PT as the piezoelectric phase. There are several processes to fabricate 2–2-type composites including epoxy bonding, co-firing, and thin film deposition. Epoxy bonding is widely used because this method is easy to implement in the laboratory. However, the epoxy layer is much softer than both the magnetostrictive alloy and the ferroelectric ceramic, and thus it will dampen the generated strain resulting in loss of efficiency. To achieve the effective bonding, co-firing process for the 2–2-structured composite is employed. Co-firing of the layered ME composite is compatible with the current multilayer capacitor (MLC) process, which provides the possibility of cost-effective mass production. However, co-firing is challenging due to the different shrinkage rates, thermal expansion mismatch, and atomic interdiffusion and/or chemical reactions between the two ceramic layers during the sintering process at high temperature. Recently, a number of multiferroic ME films have been prepared by the thin film deposition method. This method has the ability of yielding excellent epitaxial growth of thin films with atomic scale thickness control and coherent interfaces. However, most ME composite films reported in the literature have very low ME properties due to the large constraint from the stiff substrate. Using the thin film deposition method, a 1–3 ME composite via self-assembly was reported to have a reduced clamping effect by the substrate and efficient strain coupling, resulting from a large interfacial surface area. However, the design and control of such a structure remains a challenge, and ME coefficients could not be directly measured because of the leakage problem resulting from low resistance of the magnetic pillars penetrating through the films.

Table 4.6 Summary of the advantages and disadvantages of different fabrication process

Connectivity	Process method	Interface coupling	Advantage	Disadvantage
0–3	Ceramic sintering	Good	Simple processing	Low resistivity (difficult to pole piezoelectric phase) Low ME property
2–2	Epoxy bonding	Weak	Easy and flexible processing Good ME property Wide range of magnetic materials (oxide, metal or alloy)	Difficult scaling and mass production
	Co-firing	Good	Simpler processing Excellent ME property Integrated with thick film device	Interdiffusion Limited range of magnetic materials Thermal expansion mismatch
	Thin film deposition	Excellent	Integrated with MEMS devices or chip	Weak ME property Strong clamping effect
1–3	Thin film deposition	Excellent	Reduced clamping effect Efficient strain coupling due to large interfacial area	Low resistivity Hard to fabricate Strong interface residual stress

ME magnetoelectric, *MEMS* microelectromechanical systems

Table 4.6 summarizes the advantage and disadvantage of different fabrication processes. Table 4.7 summarizes the ME properties with different combination of piezoelectric and magnetostrictive materials, and different interface conditions. The ME properties of thin film composite/hybrids are not listed here due to the weak ME response or unavailability of direct measurement.

4.5.2 ME Composite/Hybrid with 0–3 Connectivity

a. *Ceramic sintering*

0–3-type particulate ME composites have generally been obtained by sintering the mixture of piezoelectric and ferromagnetic phases. The sintered composites have several advantages such as freedom in the selection of constituent phases, starting particle size, and processing parameters. Various particulate composites of piezoelectric (such as BT, PZT) and magnetostrictive (such as nickel zinc ferrite (NZF), cobalt ferrite (CFO)) oxides have been reported.

Table 4.7 Characteristics of different types of layered magnetoelectric (ME) composites

M	P	d*E*/d*H* [mV/cm Oe, kHz]	Reference	Characteristics
Epoxy bonding				
Ni	PZT	450	(Laletin et al. 2005)	
NZF	PZT	240	(Mandal et al. 2010)	
Terfenol-D	PMN–PT crystal	4300	(Dong et al. 2011)	High-cost Terfenol-D,
Metglas	PZN–PZT	150	(Cho et al. 2010)	High cost and limited thickness of Metglas
Metglas	Mn–PZN–PZT	220	(Cho et al. 2010)	
Epoxy bonding, IDE				
Metglas	PZT	22,000	(Dong et al. 2006)	High cost, high properties
Metglas	PMN–PT crystal	52,000	(Wang et al. 2011)	Complex structure, limited production
Multilayer co-firing ceramics (MLCC)				
Ni	BT	7.1	(Israel et al. 208)	Low cost, low properties
Co-firing, tape casting				
CFO	BT	0	(Harshe 1991)	Pt inner electrode
CFO	PZT	75	(Harshe Harshe 1991)	Pt inner electrode
NZF	PZT (20 layer)	400	(Srinivasan et al. Harshe 2001)	No inner electrode
LSM	PZT	32	(Srinivasan et al. Harshe 2002)	No inner electrode
NCZF	PZT–PZN	82	(Park and Priya 2011)	Without inner electrode, 930 °C
NCZF	PZT–PZN	107	(Park and Priya Park and Priya 2011)	Ag inner electrode, 930 °C
NCZF	Textured PMN–PT	1346	(Yan et al. 2013c)	Ag inner electrode, 930 °C

CFO cobalt ferrite, *LSM* lanthanum strontium manganite; *NZF,* nickel zinc ferrite; *NCZF,* nickel copper zinc ferrite, *PZN,* zinc lead niobate, *IDE* interdigitated electrode

Yang et al. investigated the effect of the sintering temperature and ferrite concentration on the sintering behavior, microstructure evolution, and ME properties of 0.948 $K_{0.5}Na_{0.5}NbO_3$–0.052 $LiSbO_3$ (KNNLS) lead-free 0–3 particulate composites. Figure 4.15a, b, c, d shows the SEM images for KNNLS–NZF composites sintered at 1060 °C. All the compositions exhibited dense microstructure.

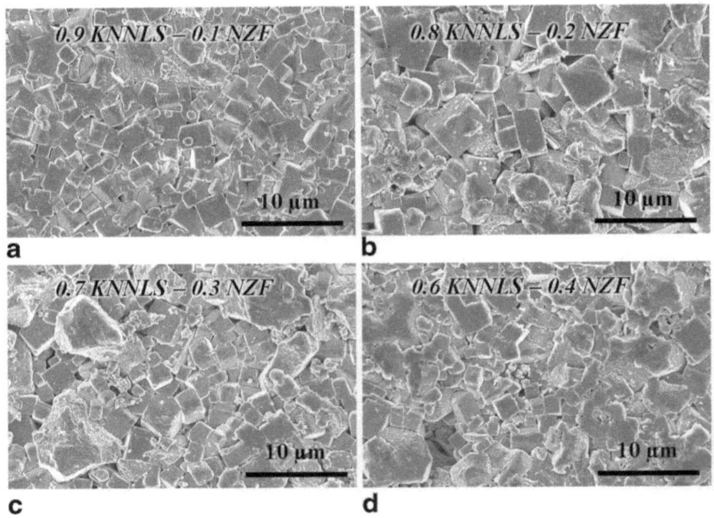

Fig. 4.15 SEM images for (1– x) KNNLS–x NZF composites sintered at 1060 °C. **a** $x=0.1$, **b** $x=0.2$, **c** $x=0.3$, and **d** $x=0.4$

The energy-dispersive spectroscopy (EDS) elemental mapping images shown in Fig. 4.16a, b, c, d, e, f, g and h confirm that the NZF islands were randomly dispersed in the KNNLS matrix. The element Nb (white color in Fig. 4.16a, b, c and d) represents the KNNLS matrix, and Fe (white color in Fig. 4.16e, f, g and h) represents the NZF islands. It can be seen in this figure that the NZF islands were isolated from each other for the composition $x=0.1$–0.3 but the connectivity pattern changes at high mole fraction of $x=0.4$.

Figure 4.17a, b shows the variation of the ME coefficient (α_E) and the optimum DC bias (H_{bias}) for the (1–x) KNNLS–x NZF ($x=0.1$, 0.2, 0.3, and 0.4) composites with increasing sintering temperatures. With increasing NZF concentration, the magnitude of α_E was found to increase to 20.14 mV/cm Oe under $H_{\text{bias}}=380$ Oe at $x=0.3$ and then slightly drop to 18.17 mV/cm·Oe under $H_{\text{bias}}=330$ Oe at $x=0.4$. For all the compositions, the composites sintered at 1050 and 1060 °C were found to exhibit maximum α_E due to their optimized piezoelectric and dielectric values. The magnitude of optimum H_{bias} decreased from 545–326 Oe with increasing NZF concentration. The reason for the shift in H_{bias} is related to the difference in magnetic susceptibilities of the composites and variation in the peak piezomagnetic coefficient.

In particulate ME composites, by controlling the sintering process and composition, ME coefficients of about 10–100 mV/cm Oe have been obtained. These measured values of the ME coefficients are much lower than the theoretically predicted coefficients. One of the main reasons is unavoidable interdiffusion and/or chemical reactions between the piezoelectric and ferrite phases during high-temperature sintering and the structural defects generated due to the thermal expansion mismatch between two phases. Another reason is low resistivity, which makes the electrical poling of the composites difficult and the ME properties are reduced because of the

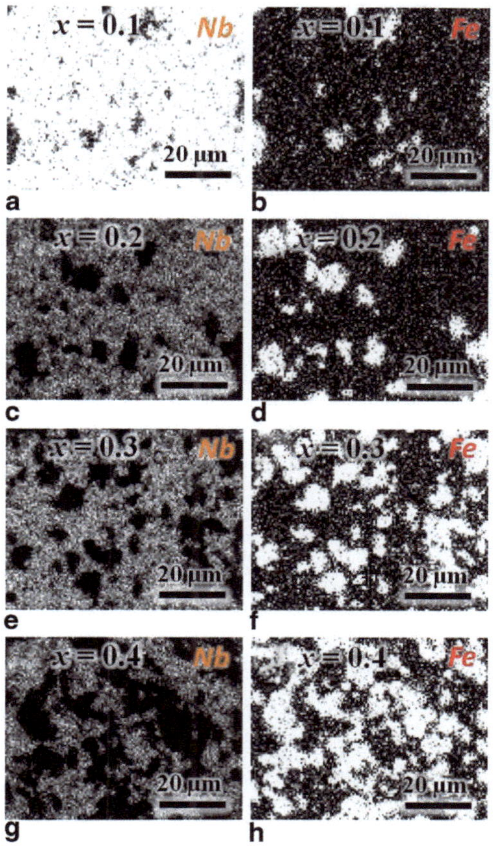

Fig. 4.16 The energy-dispersive spectroscopy (*EDS*) elemental mapping images for $(1-x)$ KNNLS–x NZF composites sintered at 1060 °C. **a–d** SEM images for $x=0.1$–0.4, **e–h** Nb mapping for $x=0.1$–0.4

leakage problem. To achieve high ME response in 0–3 particulate composites, the concentration of magnetic ferrite particles and the resistivity should be increased simultaneously. Good dispersion of the low-resistance ferrite particles in the piezoelectric matrix is required in order to suppress the percolation.

b. *Core–shell Structure*

For the 0–3-type particulate ceramic composites synthesized by mechanical mixing of the two-phase powders, the randomly distributed ferrite particles have a low percolation threshold. In order to achieve a good dispersion of a high concentration of ferrite particles in the composite ceramics, an effective approach is to utilize core–shell structured particles with ferrite-core and piezoelectric-shell, thereby restricting the contact of the ferrite particles during sintering.

Figure 4.18 schematically illustrates the sol–gel or hydrothermal method to prepare the core–shell nanoparticles. In most cases, the magnetic ferrite is obtained in the form of nanoparticles by precipitation of metal chlorides or nitrates in basic aqueous solutions. These particles are then coated with a shell of piezoelectric material such as BT or PZT either by dispersing the particles in a sol, which consists of

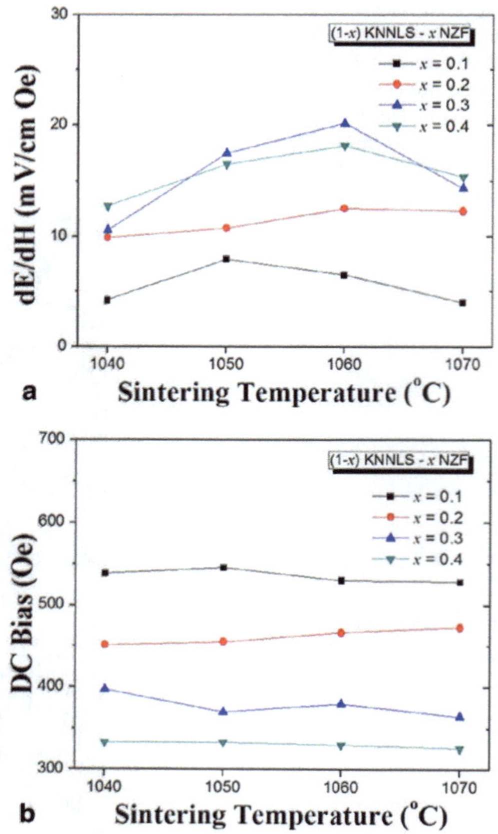

Fig. 4.17 Variation of **a** ME coefficients and **b** H_{bias} for (1–x) KNNLS–x NZF composites with x=0.1, 0.2, 0.3, and 0.4 as a function of sintering temperature. Measured at 1 kHz under H_{ac}=1 Oe

the metallorganic precursors, or by hydrothermal treatment, in which case carbonates or hydroxides can be used. The resulting gel or particle is dried and annealed at various temperatures for crystallization (Liu et al. 2010; Schileo 2013).

Unlike the majority of cases, Islam et al. (2008) reported a PZT core/nickel ferrite (NFO) shell ME composite. To avoid the obvious problem of very high conductivity through the ferrite phase, a 100-mm layer of PZT was used to cover both faces

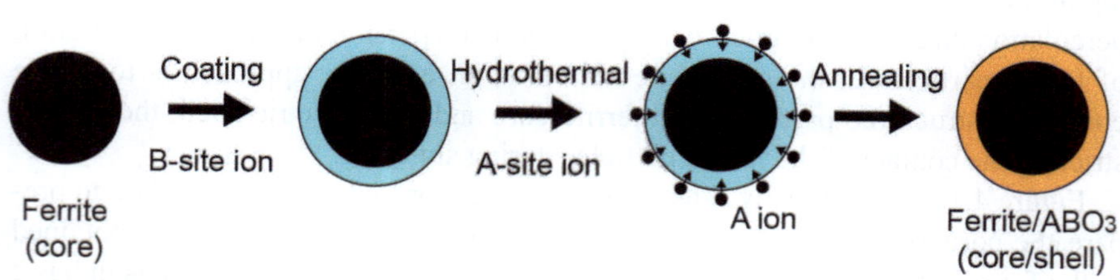

Fig. 4.18 Illustration of the sol–gel or hydrothermal method to prepare ferrite core and piezoelectric shell nanoparticles

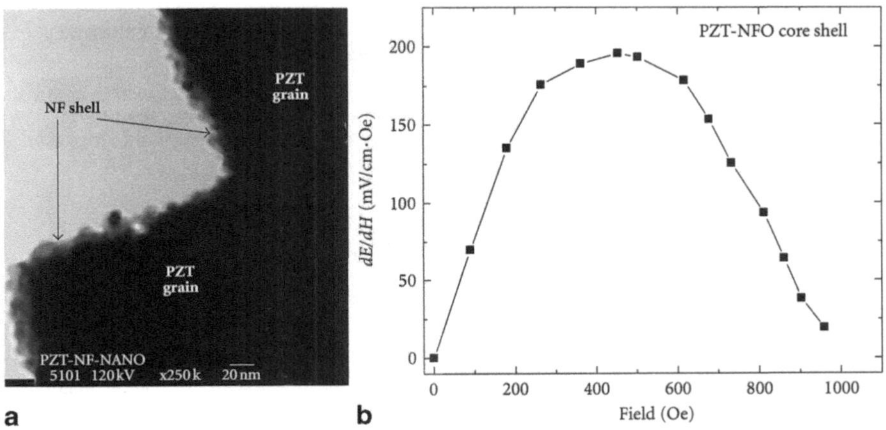

Fig. 4.19 TEM image of sintered core–shell structure, (**b**) magnetoelectric coefficient as a function of DC bias field for sintered PZT-NF core shell structure. *TEM* transmission electron microscopy

of the pellet. Figure 19a shows the TEM image of the grain structure for a sintered PZT–NF composite. NF particles forming the shell had sizes in the range of 20–30 nm. Figure 19b shows the variation of the ME coefficient as a function of DC bias for the sintered composite. A maximum ME coefficient of $\alpha_E = 195$ mV cm^{-1} Oe^{-1} was obtained, which is about 130 times higher than that reported for the BT–CFO core–shell composites. This large ME response is attributed to good interface bonding and minimum misfit strain between the two phases, which effectively transfers the mechanical strain from one another. But so far, such ceramics with ferrite core and piezoelectric shell particles have not yet been ideally obtained due to the difficulty of synthesizing good core–shell structured particles. The demonstrated ME effect is much inferior to that of 2–2-type composites (Schileo 2013).

4.5.3 ME Composite/Hybrid with 2–2 Connectivity

Compared to the particulate composite ceramics, the laminate composite ceramics exhibit high ME coefficients due to the elimination of the leakage problem. There are mainly three types of processes for fabricating 2–2-type ME composites, namely, high-temperature co-firing, epoxy bonding, and thin film deposition.

a. *Epoxy bonding*

A simple method to prepare a 2–2-type ME composite is by simply bonding the piezoelectric ceramic layer (such as PZT, PMN–PT) and magnetostrictive layer (such as ferrite ceramic, Terfenol-D, Metglas) together with epoxy. This process avoids the challenge of different shrinkages, thermal expansion mismatch, atomic interdiffusion, and/or chemical reactions between different phases during the sintering process at high temperature. Using this process, a giant ME effect has been reported in a number of laminate composites.

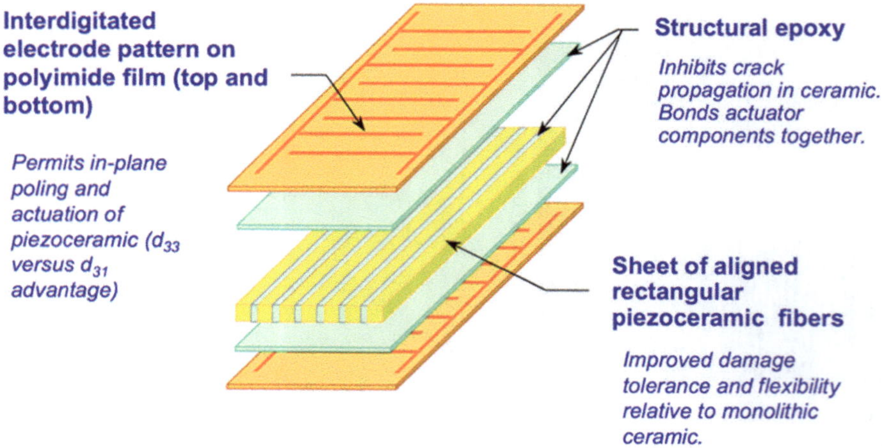

Fig. 4.20 Langley Macro-Fiber Composite™ actuator components. (Reprinted with permission from (Wilkie et al. 2000). Copyright, 2000, SPIE)

Several laminate composites have been fabricated by bonding magnetostrictive (Table 4.4) and piezoelectric layers (Table 4.3) using an epoxy resin, followed by curing at a modest temperature of 80–100 °C. Based on this process, various configurations (disk, rectangular, ring shape) and different working modes (Fig. 4.13) have been studied. Due to the fact that piezoelectric properties and electromechanical coupling in the 33 mode (d_{33} and k_{33}) are generally larger than those in the 31 mode (d_{31} and k_{31}), ME composites with L–L (longitudinally magnetized and longitudinally poled) configuration have been shown to exhibit higher electric field output than that of the L–T (longitudinally magnetized and transversely poled) configuration. In this kind of structure, the configuration of piezoelectric material plays a very important role in achieving high ME performance. Below, several examples with high ME coefficients are described in order to illustrate the structure–property relationships.

a) Macro-Fiber Composites

The general method to achieve d_{33} mode is by utilizing the interdigitated electrode (IDE). Macro-fiber composites (MFC) were developed at the National Aeronautics and Space Administration (NASA) Langley Research Center and commercialized by the licensed manufacturer Smart Material. The primary components of the MFC and their arrangement in the actuator package are illustrated in Fig. 4.20. The MFC actuator consists of three primary components: (1) a sheet of uniaxially aligned active piezoceramic fibers, (2) a pair of thin polymer films with a conductive IDE pattern, and (3) an adhesive epoxy matrix material. Here, IDE represents a pattern of interlocking electrode fingers with positive and negative electric potential that allow the induction of alternating polarization direction in the plane of the film. Piezoelectric MFC has shown better flexibility, higher performances (allows the use of d_{33} mode), and durability, compared to the traditional monolithic piezoelectric ceramics. MFCs have been used for high displacement actuators and transducers for energy conversion.

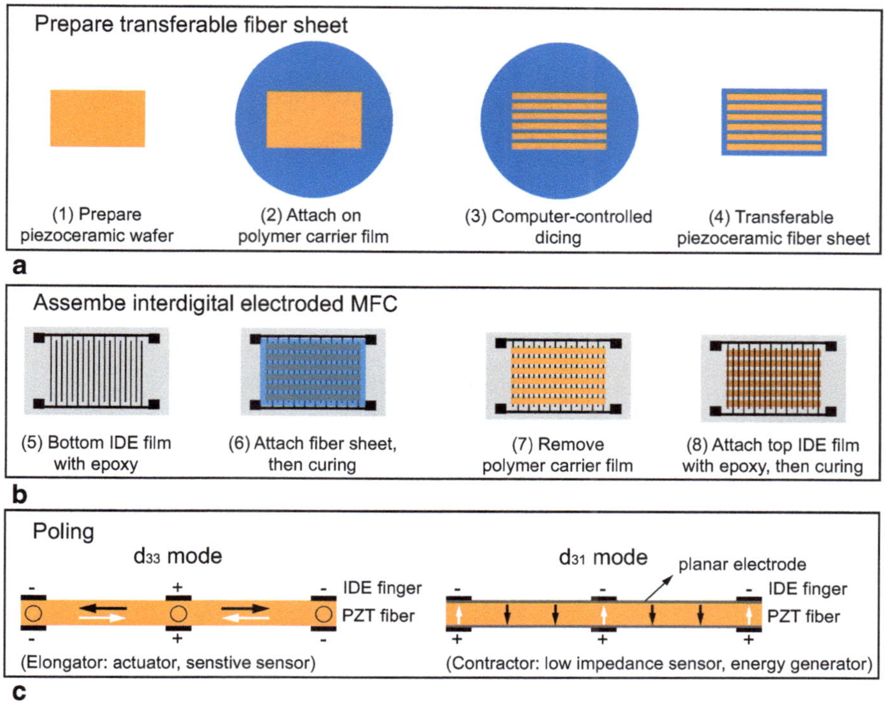

Fig. 4.21 **a** Fabrication of transferable piezoceramic fiber sheet. **b** Assemble IDE MFC. **c** Poling. IDE MFC interdigitated electrode macro-fiber composites

Figure 4.21a depicts the fabrication of MFC piezoceramic fiber sheets. The piezoceramic wafer is attached on a polymer carrier film. The polymer carrier film holds the fibers in place during and after machining. The fibers are machined from this wafer using a computer-controlled dicing saw. These sheets allow the piezoceramic fibers to be precisely aligned within the actuator package. The transfer of the piezoceramic fibers from the carrier film to the IDE and further assembly are illustrated in Fig. 4.21b. This assembly consists of the application of an additional coat of adhesive to the fibers and curing the complete package under heat and pressure. As shown in Fig. 4.21c, there are two types of MFC depending upon the poling direction. The poling direction of the d_{33} type is parallel along the fiber length and each segment has opposite poling direction through the IDE. For the d_{31} type, the poling direction is from the top to the bottom along the fiber thickness. The MFC is extremely flexible, durable, and has the advantage of higher electromechanical coupling coefficients granted through the IDE.

Based on this MFC structure, Metglas/piezofiber ME composites have been developed. An example structure shown in Fig. 4.22 is composed of six layers of magnetostrictive Metglas and a piezoelectric core consisting of five PMN–PT fibers interrogated by a pair of Kapton IDE (Wang et al. 2011). Metglas is the amorphous magnetostrictive alloy $Fe_{81}Si_{3.5}B_{13.5}C_2$ with its trade name Metglas 2605SC.

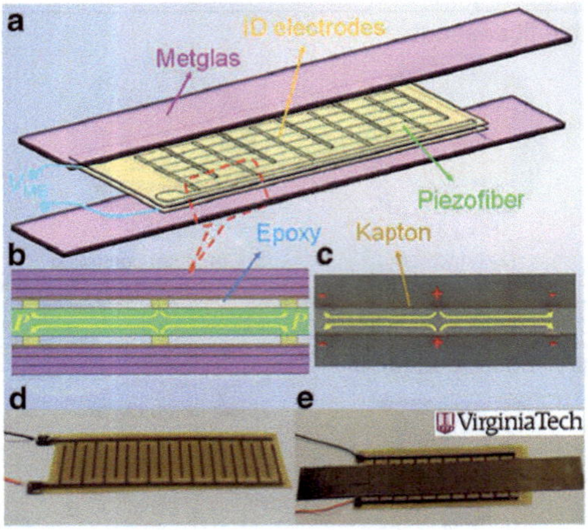

Fig. 4.22 **a** Schematic diagram of the Metglas/piezofiber configuration consisting of an ID electrodes/PMN–PT fibers core composite and symmetric three-layer Metglas actuators on the bottom and top of the core composite. **b** Illustration of the numerous alternating push–pull mode units. **c** Optical microscopy image of a longitudinally poled push–pull element in the core composite. **d**, **e** Photographs of the ID electrode/piezofiber core composite and the complete Metglas/piezofiber ME sensor. (Reprinted with permission from (Wang et al. 2011). Copyright, 2011, Wiley)

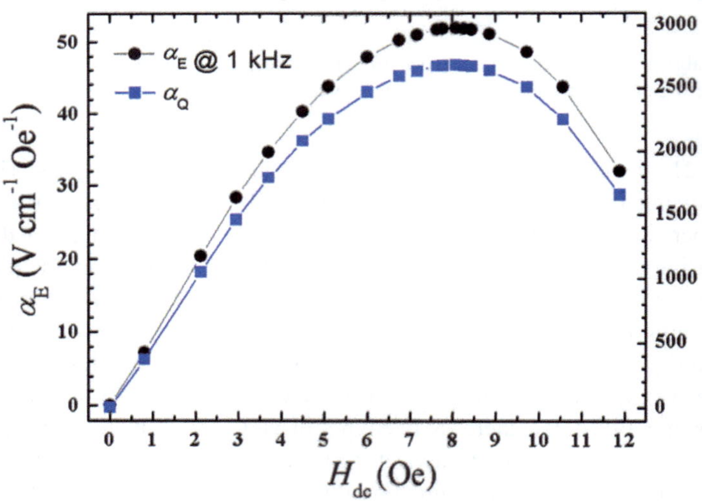

Fig. 4.23 The magnetic field dependence of the ME voltage coefficient and ME charge coefficient of the Metglas/piezofiber sensor. (Reprinted with permission from Wang et al. (2011). Copyright, 2011, Wiley)

Figure 4.23 shows the magnetic field dependence of the ME voltage coefficient and ME charge coefficient of the Metglas/piezofiber sensor. The ME voltage coefficient (α_E) increases approximately linearly with increasing DC magnetic fields for low DC magnetic biases (H_{dc} < 4 Oe). Under optimal DC magnetic bias ($H_{dc} \approx 8$ Oe), α_E reaches a maximum value of 52 V cm^{-1} Oe^{-1}. This ME coefficient is significantly higher than that previously reported for two-phase ME composites, regardless of constituent material types. The high ME coefficient is due to several factors: (i) optimum stress transfer in the multi-push–pull mode, (ii) large $d_{33,\,m}$ and high permeability of magnetostrictive Metglas, (iii) high piezoelectric properties ($d_{33,p}$) of the PMN–PT single crystal piezofiber, (iv) optimum thickness ratio of Metglas to piezofiber, and (v) optimal ID electrodes distribution on Kapton (Wang et al. 2011).

b) Zigzag shaped

The ME laminates composed of the fiber composite and Metglas M-layers showed a strong L–L mode ME interaction. However, as described above, in the fiber composite, IDE-printed polymer films were glued on the top and bottom surfaces of piezoelectric fiber arrays by using epoxy to form an electrical contact between the IDE and piezoelectric fibers. Recently, Wang et al. pointed that the interfacial bonding condition between the polymer film and the P-layer is critical for the polarization of the piezoelectric fibers and hence the ME signal detection (Wang et al. 2012). The presence of residual epoxy between electrodes and piezoelectric fibers had an adverse effect on the polarized state of the P-layer and the ME composite. ME composites fabricated by using smaller amount of epoxy showed a better ME performance. Cho et al. reported a new 33-mode design of P-layer with zigzag shape in which the epoxy adhesive is excluded. This enables a direct contact between the IDE and the piezoelectrics so that the polarization state of the P-layer can be improved, resulting in the enhancement of the piezoelectric response. Moreover, due to the absence of epoxy bonding step, reproducibility, and reliability of the ME characteristics were significantly increased.

Figure 4.24a shows the schematic diagram of the proposed P-layer. The P-layer is composed of a zigzag-shaped piezoelectric ceramic (dark portion) and conducting IDE (bright portion). By applying a DC bias between the left and right electrodes (poling process), a polarized state with push–pull mode is formed in the P-layer. The electrode type and operation mode of this structure are similar to those of the MFC. However, unlike the MFC, the IDE of this structure is formed on the piezoelectric ceramic by filling the interdigitated channels patterned through the piezoelectric body with conducting material. So, direct contact between the electrodes and piezoelectric ceramic can be achieved. Figure 4.24b shows the fabrication process of the zigzag-shaped P-layer. First, a bulk piezoelectric ceramic is prepared (process (i)). On one side of the ceramic, groove patterns are created with proper depth by using a thin dicing saw (process (ii)). Then, the ceramic is turned over and the same groove patterns are formed on the other side of the ceramic (process (iii)). During this step, the position of the groove patterns is shifted from that of the patterns on the opposite side by half of the inter-groove spacing. Next, an electrode material is deposited on both sides of the patterned ceramic (process (iv)), and the poling process is followed (process (v)). Finally, the zigzag-shaped P-layers are obtained by dicing the ceramic body into thin plates as shown in process (vi). This novel process provides several advantages over that of the MFC. First, mass pro-

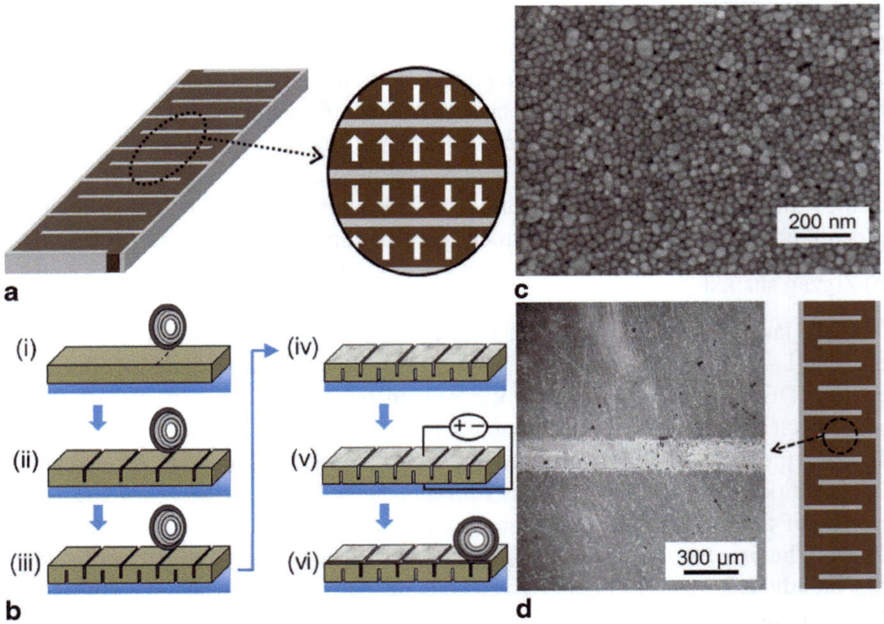

Fig. 4.24 **a** Schematic diagram of the zigzag-shaped piezoelectric layer. **b** Fabrication process of the zigzag-shaped piezoelectric layer. **c** SEM image of nano-Ag particles dispersed in Ag ink. **d** SEM image of Ag electrode sintered in between piezoelectric ceramics. *Arrows* in Fig. 4.1a indicate a direction of polarization in the piezoelectric layer

duction is possible by simply repeating the process (vi). Second, the thickness of the P-layer can be easily controlled by programming the blade position in the process (vi). The fabrication of thick layer samples is also possible using this process, while the thickness of the piezoelectric layer in the MFC is limited to around 200 μm or less for effective poling. Third, the adverse effects of the bonding epoxy layer, such as the decrease in polarizability and reproducibility problem can be avoided.

In the process (iv), nano-Ag ink is deposited as the electrode material. Nano-Ag ink can easily penetrate into the groove patterns due to the capillary force. After filling the whole surfaces of the groove-patterned sides with nano-Ag ink, the sample is heated up to 70 °C for initial drying of solvents. Then, the sintering process of the Ag electrode is conducted at 200 °C for 30 min. Figure 4.24c shows the SEM image of the nano-Ag particles dispersed in the ink. Since the average diameter of the nanoparticle is around 30 nm, the temperature of 200 °C is enough to sinter the Ag grains. As seen in Fig. 4.24d, the Ag grains were well grown on the surface of the piezoelectric ceramic with a good contact state. Based on this process, a P-layer with dimensions of $70 \times 9 \times 0.56$ mm^3, overlapped digit length of 7 mm, and interdigitated spacing of 1.3 mm was fabricated. Instead of PZT, a <001> textured PMN–PT ceramic was employed to prepare this sample since its electromechanical characteristics are superior to those of PZT. Then, magnetostrictive Metglas layers each with thickness of 180 μm were attached on the top and bottom surfaces of this P-layer to form an ME laminate composite.

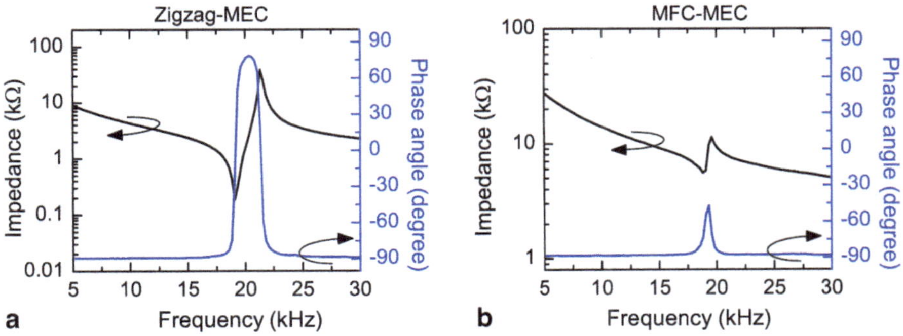

Fig. 4.25 Frequency dependence of impedance and phase angle of **a** magnetoelectric composite with zigzag-shaped textured PMN–PT (Zigzag-MEC) and **b** magnetoelectric composite with macro-fiber composite (*MFC–MEC*)

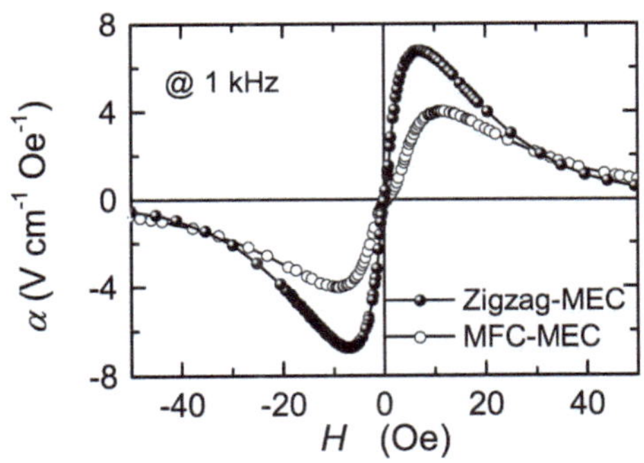

Fig. 4.26 Magnetoelectric coefficient (α) of the Zigzag-MEC and MFC–MEC as a function of DC magnetic field measured at 1 kHz. *MFC–MEC* macro-fiber composite magnetoelectric composite

Figure. 4.25a shows the frequency dependence of impedance and phase angle of the ME composite with zigzag-shaped textured PMN–PT (Zigzag-MEC). It exhibited a 170° shift in phase angle at 20 kHz indicating an excellent polarized state of the P-layer. In the example shown in Fig. 4.25, the shift of phase angle from −90° at resonance frequency is a measure of the polarization degree of the ME transducer and it should be 180°. For the sake of comparison, the impedance characteristics of the ME device composed of commercial MFC (Smart Material Corp., M-8507-P1), which has dimensions similar to those of our P-layer, are depicted in Fig. 4.25b. The phase angle change of MFC is only −45°. Due to this insufficient poling condition, the electromechanical coupling factor, $k = \sqrt{\left(f_a^2 - f_r^2 / f_a^2\right)}$, in the MFC structure is only 0.27, which is much less than 0.45 in the zigzag structure. The superior polarized state of our device is believed to be due to the direct electrical contact between the electrode and piezoelectric material.

Figure 4.26a shows the variation of ME coefficient ($\alpha = \mathrm{d}E/\mathrm{d}H_{ac}$) of the Zigzag-MEC as a function of the DC magnetic field (H_{dc}) measured at 1 kHz. The α of

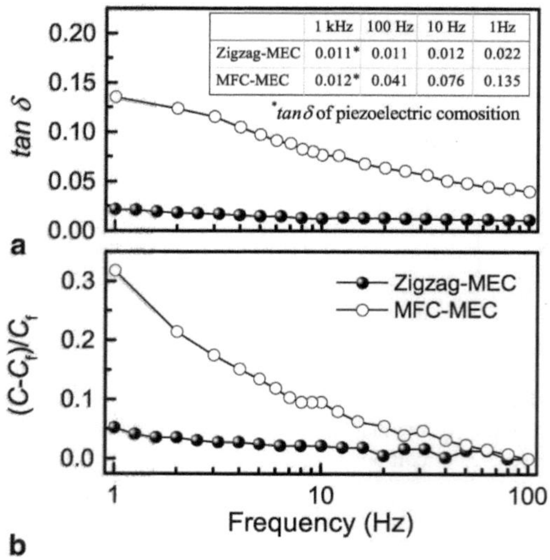

Fig. 4.27 **a** Dielectric loss (*tan δ*). **b** Rate of capacitance variation ($(C\text{-}C_f)/C_f$) as a function of frequency for the Zigzag-MEC and MFC–MEC measured in the low-frequency region. Inset table shows *tan δ* values of the MEC devices at different frequencies. *MFC–MEC* magnetoelectric composite macro-fiber composite

the MFC–MEC is also displayed in this figure for comparison. The Zigzag-MEC showed a typical α versus H_{dc} characteristic and exhibited a giant maximum α of 6.75 $Vcm^{-1}Oe^{-1}$, higher than that of MFC–MEC.

Besides a higher ME coefficient, Zigzag-MEC shows much stable loss characteristics in the low-frequency region, which is very important for magnetic sensor application. As shown in Fig. 4.27a, the *tan δ* of the Zigzag-MEC was 0.011 at 100 Hz and slightly increased to 0.012 at 10 Hz, and reached 0.022 at 1 Hz. While the *tan δ* of the MFC–MEC was 0.041 at 100 Hz, which is 3.4 times larger than the piezoelectric composition's value, and increased rapidly as the frequency decreased, reaching the value of 0.135 at 1 Hz. The trend of capacitance for the two composites with respect to the frequency variation is also similar to that of the *tan δ*, as shown in Fig. 4.27b. It indicates that dielectric behavior of MFC P-layer has a low-frequency dispersion, which is observed in epoxy samples exposed to humidity (Park and Shrout 1997). Since perfect exclusion of humidity from the epoxy resin is difficult during fabrication of MFCs, water molecules can exist in the epoxy, contributing to space charge or ionic polarization at low frequencies. Therefore, it is believed that the residual epoxy between electrode and piezoelectric fiber, as well as between adjacent piezoelectric fibers dominantly contributes to the loss increase. The absence of the interfacial epoxy in the Zigzag-MEC leads to a very stable loss behavior that may result in stable ME response at low frequencies. The proposed zigzag-shaped piezoelectric structure is promising for many other piezoelectric applications as well as ME devices, due to its superior piezoelectric properties along with simple and reliable fabrication process.

c) Co-fired IDE

From the ME properties and loss comparison between Zigzag-MEC and MFC–MEC above, it was found that the ME response and resultant ultralow noise level are highly dependent on the interfacial bonding condition between the IDE/Kap-

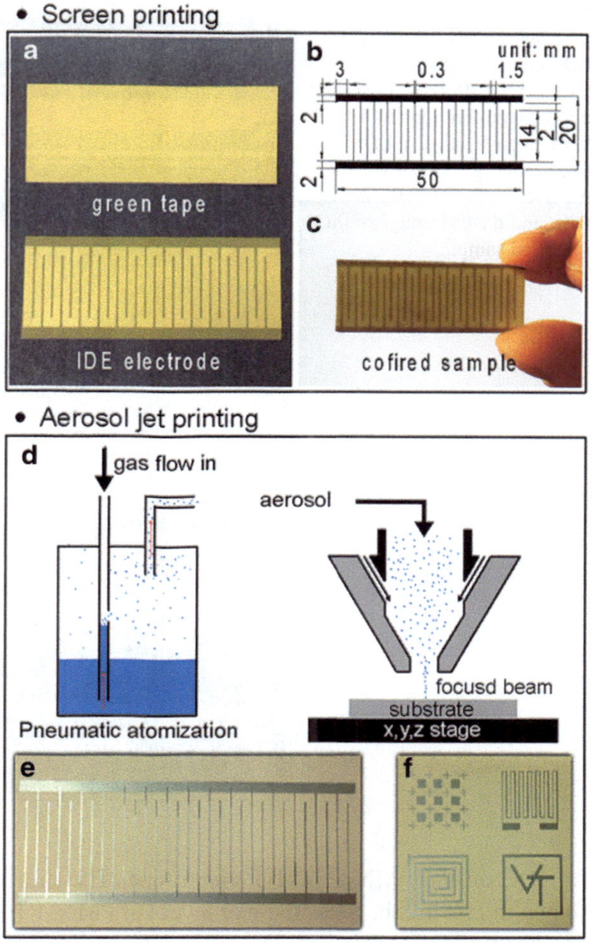

Fig. 4.28 Screen printing: **a** PMN–PT–1BT *green tape* and printed IDE, **b** configurations of IDE pattern, **c** co-fired PMN–PT–1BT/IDE sample; aerosol jet printing: **d** working principle, **e** printed IDE, and **f** other printed patterns on PMN–PT–1BT *green tape*

ton and the piezofibers. The zigzag-shaped piezoelectric structure is a promising structure to achieve high ME response and low loss due to the absence of epoxy between the IDE/Kapton and the piezofibers. However, the fabrication of MFC–MEC and Zigzag-MEC involve the dicing process, resulting in the limitation of the miniaturization. A low-temperature co-firing technique was used to embed the IDE inside the piezoelectric ceramics to achieve a direct bonding that completely excludes the epoxy in the composite. This *L–L*/IDE ME composite was found to show a 3.4 times higher ME voltage coefficient compared to the *L–T* mode (*L–T*/Ag). By reducing the loss originating from the epoxy, the ME sensitivity at low frequency was shown to significantly improve. This co-fired IDE ME composite is a significant advancement in the transition of the ME composites toward practical fabrication processes, as the co-firing process is compatible with the industrial thick film fabrication methods.

Figure 4.28 describes two methods for fabricating the IDE pattern on the piezoelectric ceramic green tape/thick film. Figure 4.28a shows the screen-printed silver/

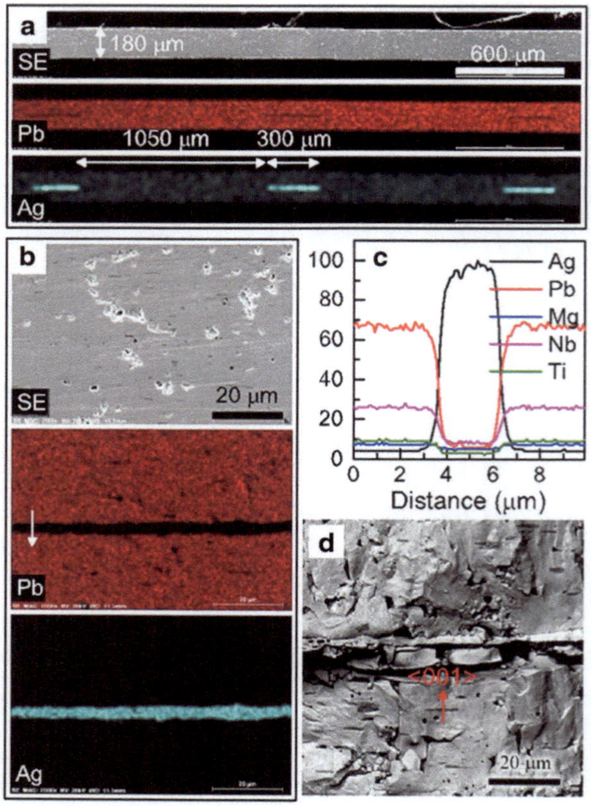

Fig. 4.29 Cross-sectional (polished) SEM images and EDS element mapping of co-fired IDE/PMN–PT–1BT, **a** low magnification and **b** high magnification, **c** EDS line scanning analysis across the IDE/PMN–PT–1BT interface, and **d** SEM image of the fractured sample

Ag IDE on the PMN–PT–1BT green tape. The dimension of the IDE pattern is shown in Fig. 4.28b. Next, the five layers of PMN–PT–1BT green tapes along with one layer of IDE patterned PMN–PT–1BT green tape were stacked (one layer of IDE in the middle and three layers of PMN–PT green tapes on either side), and then laminated. The co-fired samples had high density as shown in Fig. 4.28c. Another approach for patterning the IDE electrode is using aerosol jet printing. Figure 4.28d shows a schematic of the Aerosol Jet® process. The Ag ink was first aerosolized in the pneumatic atomizer. After that, the aerosol was carried by nitrogen gas to the deposition nozzle, focused into fine lines by sheath gas (a secondary nitrogen stream), and patterned into the specific features on PMN–PT green tape as shown in Fig. 4.28e. The advantage of aerosol jet printing is to produce ultrafine features with precise control as shown in Fig. 4.28f. Features as small as 10 μm with a 20-μm pitch can be written using this method.

Figure 4.29 shows the microstructure of co-fired IDE/PMN–PT–1BT samples. It can be seen from Fig. 4.29a that the thickness of the sample was 180 μm, the width of each IDE finger was 300 μm, and the distance between two IDE fingers was 1050 μm. There was no delamination, crack, or any obvious elemental diffusion across the IDE/PMN–PT–1BT interfaces from Figs. 4.29b, c. The thickness of IDE

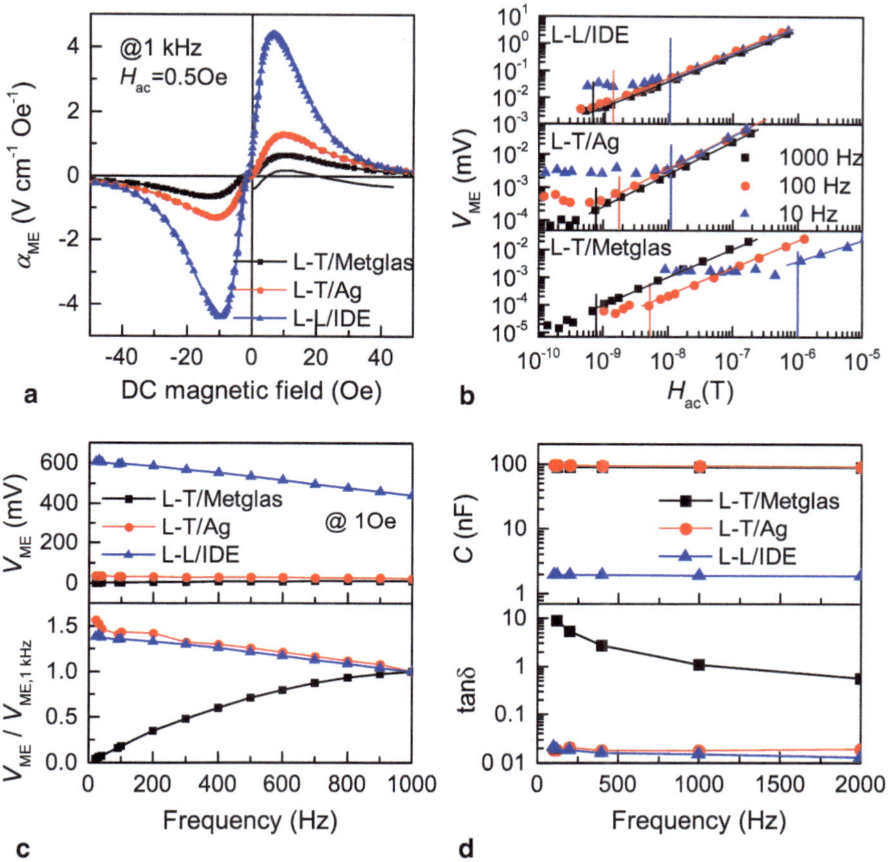

Fig. 4.30 **a** Magnetoelectric voltage coefficients (α_{ME}) under different modes (*L–L, L–T*) and different connection with external circuit (IDE, Ag electrode and Metglas), **b** ME voltage versus AC magnetic field (H_{ac}) under different frequency (the vertical lines indicate the limits of detection) under laboratory condition without any electromagnetic shielding, **c** ME voltage and normalized ratio as a function of frequency, normalized by the ME voltage at 1 kHz, **d** Capacitance and dielectric loss as a function of frequency

was about 3 μm (Fig. 4.29c). From Fig. 4.29d, it can be seen that the matrix grains grew from the aligned BT templates and displayed strong <001> preferred orientation/texture normal to the casting plane. The piezoelectric properties are known to increase 2–5 times by this texturing technique resulting in the piezoelectric charge coefficient (d_{33}) of textured ceramics (such as PMN–PT, PMN–PZT) in the range of 1000–1100 pC/N (Yan et al. 2012b; Yan et al. 2013a). The corresponding ME voltage coefficients in the *L–T* mode increased by 2–3 times (Yan et al. 2013b, c).

Figure 4.30a shows the ME voltage coefficients (α_{ME}) of the *L–L* mode PMN–PT–1BT with IDE sample (*L–L*/IDE) and *L–T* mode textured PMN–PT–1BT (*L–T*/Ag). The α_{ME} of *L–L*/IDE was 4.41 Vcm^{-1}Oe^{-1}, which is ~3.4 times that of *L–T*/Ag (1.29 Vcm^{-1}Oe^{-1}). This improvement was mainly due to a larger longitudinal elec-

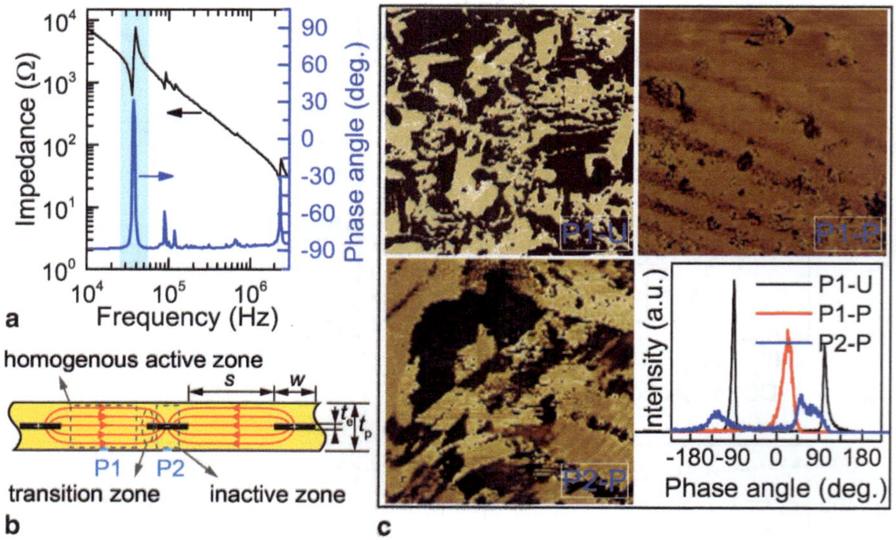

Fig. 4.31 **a** Impedance spectroscopy of co-fired IDE/PMN–PT–1BT sample (piezoelectric part in *L–L*/IDE ME composite), **b** Illustration of electric field line distribution, **c** Piezoelectric force microscopy (*PFM*) phase image and histogram at P1 and P2 position in (**b**), -U and -P represent the unpoled and poled state

tromechanical coupling coefficient k_{33} and piezoelectric charge constant d_{33} than the corresponding transverse coefficients, k_{31} and d_{31}, from the piezoelectric layer. The α_{ME} (0.64 V cm^{-1}Oe^{-1}) of *L–T*/Metglas was much lower than that of *L–T*/Ag (1.29 V cm^{-1}Oe^{-1}) and *L–L*/IDE 4.41 V cm^{-1}Oe^{-1} due to existence of epoxy in the circuit. Figure 4.30b shows the ME voltage as a function of AC magnetic field. Under laboratory conditions, the *L–L*/IDE has the best AC magnetic field sensitivity and largest ME voltage signal. *L–T*/Ag had a similar AC magnetic field sensitivity but ME voltage was 20 times smaller than *L–L*/IDE sample. *L–T*/Metglas not only had the smallest ME voltage but also had much lower AC magnetic field sensitivity, especially at low frequency. Further, the ME voltages of *L–L*/IDE and *L–T*/Ag were very stable, while the ME voltage of *L–T*/Metglas decreased rapidly with decreasing frequency as shown in Fig. 4.30c. From Fig. 4.30d, it can be seen that the loss from the epoxy layer plays a dominating role in the ME composite. The epoxy loss can be avoided in the *L–T* mode and *L–L* (push–pull) mode by directly connecting the external circuit with the electrodes printed on the piezoelectric layer similar to *L–T*/Ag and co-fired *L–L*/IDE.

Compared to the reported 22 V/cm Oe in the push–pull piezoelectric fiber/IDE structure, the ME voltage coefficient (4.41 V/cm Oe) of the co-fired *L–L*/IDE ME was low. The main reason for this low magnitude is that the co-fired PMN–PT–1BT/IDE structure was not well poled as indicated by the low-impedance phase angle change in Fig. 4.31a. Figure 4.31b illustrates the electric field distribution in the co-fired IDE/PMN–PT–1BT samples. It can be seen that the electric field distribution is distinctly inhomogeneous. The effect of inhomogeneous electric field distribution on the poling condition of IDE/PMN–PT–1BT piezoelectric layer was experimentally elucidated by piezoelectric force microscopy (PFM). As shown in Figs. 4.31b, c, the phase and domain-wall density changes between P1-P (homogenous active zone) and P2-P (inactive zone) indicate the difference of poling degree, which is consistent with the electric field distribution shown in Fig. 4.31b. To achieve a high percentage of poling condition, the electrode separation to substrate thickness (s/t_p) ratio and the electrode width (w) should be optimized. The aerosol jet printing as described in Figs. 4.28d, e and f will become highly relevant to create this fine IDE pattern.

b. *Co-fired thick film ME composite*

In the ME composites, the measured ME coefficient was found to be directly related to the effectiveness of elastic coupling at the interface of two phases. The widely used method for the synthesis of laminate composites is bonding of magnetostrictive layer using epoxy resin. This method limits the misfit strain at the interface arising due to thermal expansion mismatch between the layers and atomic interdiffusion and/or chemical reaction between the layers. However, the epoxy layer is much softer than both magnetostrictive alloy and ferroelectric ceramic and thus it will dampen the generated strain resulting in loss of efficiency. Further, the epoxy layer also limits the ME response due to low mechanical strength, and contributes toward the noise floor (noise floor is the sum of all the noise sources and unwanted signals within a measurement system, where noise is defined as any signal other than ME signal) and aging (Srinivasan 2010).

In contrast to bonding, co-firing of the layered ME composite provides compatibility with the current industrial production process commonly used for fabrication of MLCs. Recently, the ME sensor based upon Ni-BT MLCs was reported which laid the foundation for future commercialization of this technology (Israel et al. 2008). The magnitude of α_E was found to be 7.1 mV cm^{-1} Oe^{-1} which is extremely low but the projected overall cost was also extremely low. In order to increase the ME effect while keeping the cost down, materials with large magnetostrictive and piezoelectric coefficient are needed in the co-fired configuration. Srinivasan et al. (2001) have reported the performance of PZT/NFO layered composites synthesized by tape casting and co-firing process. A high α_E of 400 mV cm^{-1} Oe^{-1} was found for the PZT/NFO multilayer stacks (2–21 layers). However, high co-firing temperature (1400–1500 K) and no inner electrode limited the poling of piezoelectric materials and the process was not compatible with the low-temperature co-fired ceramics (LTCC) process required for capacitors, inductors, resistors, transformers, and hybrid circuits.

We investigated a co-fired magnetostrictive/piezoelectric/magnetostrictive (M/P/M) laminate structure with a Ag inner electrode. Compositions corresponding to $Pb(Mg_{1/3}Nb_{2/3})O_3$–32.5$PbTiO_3$ (PMN–PT) and $(Ni_{0.6}Cu_{0.2}Zn_{0.2})Fe_2O_3$ (NCZF) were used as piezoelectric and magnetostrictive materials, respectively. Figure 4.32a shows the optical image of co-fired NCZF/textured–PMN–PT/NCZF (abbreviated as C-N/T/N) laminate composite synthesized at 930 °C for 4 h with dimensions 16 × 4 × 0.4 mm^3. Figure 4.32b shows the cross-sectional SEM images

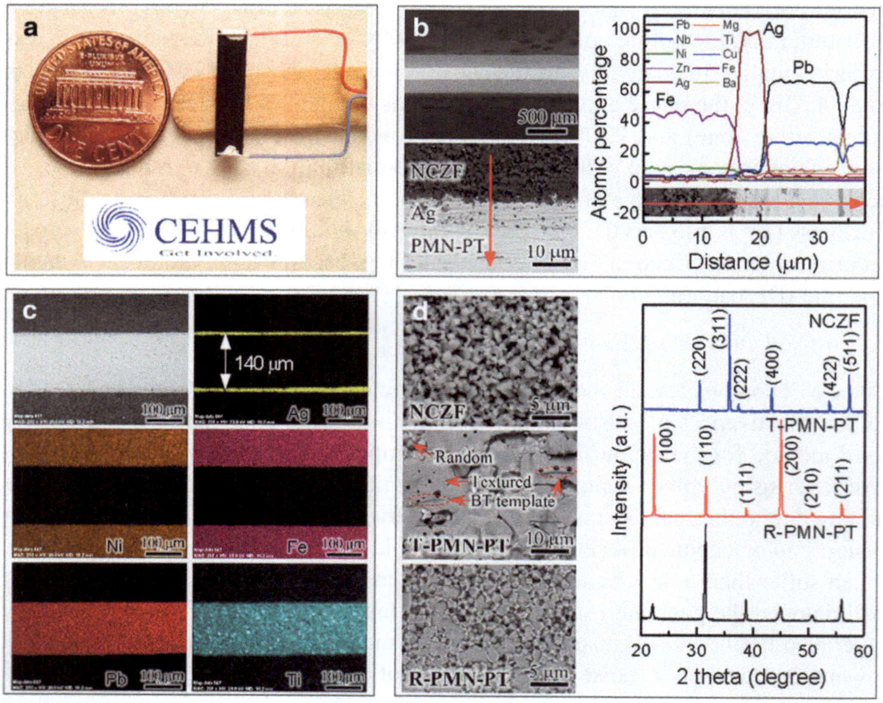

Fig. 4.32 **a** Optical image of co-fired NCZF/T–PMN–PT/NCZF laminate, **b** polished cross-sectional SEM images and EDS/SEM element line scanning analysis, **c** element mapping of co-fired NCZF/T–PMN–PT/NCZF laminate, **d** fracture cross-sectional SEM images and planar XRD patterns of different layers

of the C-N/T/N composite. There was no trace of delamination or micro-crack and the thickness of the textured PMN–PT piezoelectric layer and NCZF ME layer was similar. The interfacial atomic diffusion and/or chemical reactions occurring during the sintering process at high temperature were found to be negligible. EDS analysis of the C-N/T/N cross-sectional area (Fig. 4.32c) shows the dense microstructure and sharp interface across different layers. The thickness of the Ag electrode and PMN–PT piezoelectric layer was 10 and 140 μm, respectively. Figure 4.32d shows the fracture surface SEM images and corresponding planar XRD patterns of different layers. For a textured PMN–PT layer synthesized by the TGG method, a brick wall-like microstructure with well-aligned BT templates (black lines) in the PMN–PT matrix was obtained. An increase of {001} XRD peak intensity and decrease of other peaks clearly indicate the formation of texture. At the co-firing temperature of 930 °C, 66 % degree of texture was obtained.

Figure 4.33a shows the change in ME voltage coefficient (α_E) as a function of DC magnetic field at an off-resonance frequency of 1 kHz. Although the magnetostriction of Metglas ($\lambda = 40$ ppm) is twice that of NCZF ($\lambda \approx 20$ ppm), the maxi-

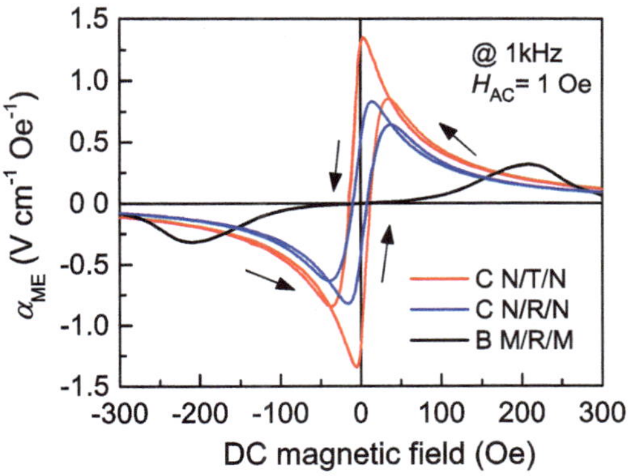

Fig. 4.33 ME voltage coefficients (α_{ME}) of co-fired NCZF/T–PMN–PT/NCZF (C-N/T/N), NCZF/R–PMN–PT/NCZF (C-N/R/N), and epoxy bonded Metglas/R–PMN–PT/Metglas (B-M/R/M) laminate

mum α_E of co-fired NCZF/random-PMN–PT/NCZF (abbreviated as C-N/R/N) is two times that of epoxy bonded Metglas/random-PMN–PT/Metglas (abbreviated as B-M/R/M) laminate. Further, improvement was achieved for the co-fired textured sample (C-N/T/N). The large enhancement in <001> textured ceramics was related to the formation of "engineered domain state" which facilitates the rotation of <111> polarization toward the <001> direction (Messing et al. 2004; Yan et al. 2012a). In Fig. 4.33a, it can also be observed that there is a strong hysteretic ME response in the co-fired composite. A giant ME voltage coefficient (> 1200 mV cm^{-1} Oe^{-1}) at zero bias was achieved in the C-N/T/N composite. The ME behavior can be correlated to magnetostriction (λ) and magnetization (M), and the self-bias effect was mainly related to the hysteretic response of ferrite layer.

The significant enhancement of ME properties and giant self-bias phenomenon in this cost-effective co-fired NCZF/textured PMN–PT/NCZF layered composites opens the possibility of mass production of numerous ME applications.

Figure 4.34 summarizes the development of sintered ME composites. Creation of 0–3 particulate composites can be achieved with ferrite particles well dispersed in the matrix of PZT (Ryu et al. 2001; Islam 2008). The α_{ME} of this composite exhibited the magnitude of 0.1 V/cm.Oe (Islam 2008). By increasing the interphase contact area, this value can be further improved to ~0.2 V/cm.Oe via a core–shell structure (Islam 2008; Islam et al. 2008). To minimize the leakage problem from the low resistivity ferrite phase, 2–2 ferrite/PMN–PT co-fired laminates were developed, with a significant improvement of the α_{ME} (0.8 V/cm Oe) in comparison with the 0–3 particulate ME composites (Yan et al. 2013c). By optimizing the piezoelectric material via texturing/grain orientation method, the α_{ME} of the co-fired 2–2 laminates can be further improved to 1.3 V/cm Oe (Yan et al. 2013c). By utilizing longitudinal poling via the co-fired IDE/piezoelectric laminate, a large α_{ME} with the magnitude of 4.4 V/cm. Oe from *L–L* mode can be obtained.

c. *Thin Film ME Composites*

The investigation of multiferroic ME nanostructured composite films has recently been accelerated by the advances in thin film growth techniques. A number of multiferroic ME films with a combination of ferroelectric (e.g., BT, PT, PZT) and mag-

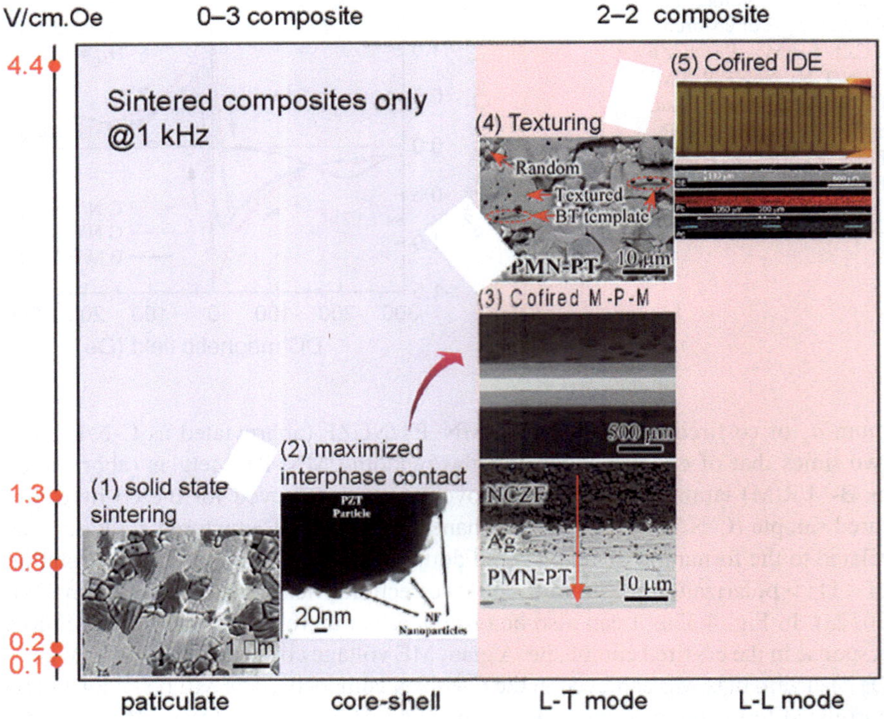

Fig. 4.34 ME voltage coefficient (α_{ME}) for different types of sintered ME composites

netic (e.g., CFO, NFO) materials have been prepared via physical deposition techniques (e.g., pulsed laser deposition (PLD), molecular beam epitaxy (MBE), and sputtering) and chemical processing (e.g., spin coating, and metal-organic chemical vapor deposition (MOCVD)). Among them, spin coating is a convenient and inexpensive chemical method to fabricate polycrystalline films or textured films with preferential orientation. However, the traditional chemical methods are not applicable to grow thin films with high structural perfection at the atomic level. Alternatively, the physical deposition techniques have the ability of yielding excellent epitaxial growth of thin films with coherent interfaces and thickness control at the atomic level.

In thin films, the substrates play an important role in controlling the orientation and strain state of the epitaxial films. For the 2–2-type layered heterostructures, CFO and BTO thin films epitaxially grown on the (001) or (111) strontium titanate (ST) substrates show different morphologies (Deng et al. 2008). Besides orientation dependence, the heterostructures are also restricted by the strain due to the differences in lattice parameters and thermal expansions between the films and substrates. Figure 4.35 shows the lattice constants of typical single-crystal substrates and ferroelectric or magnetic oxides. If the lattice parameters of the film and substrate materials are slightly mismatched, strained-layer epitaxy is formed. Due to

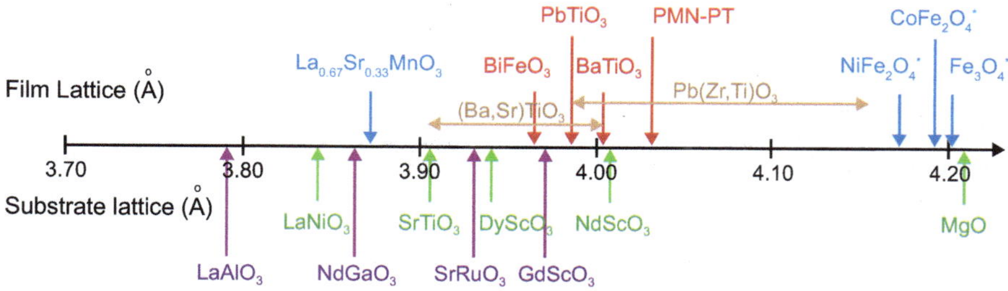

Fig. 4.35 Comparison of the lattice constants of some single crystal substrates (*below* the number axes), and typical ferroelectric and magnetic oxides of interest (*above* the number axes). (Reprinted with permission from Ma et al. (2011). Copyright, 2011, Wiley)

the large constraint from a stiff substrate, most ME composite films reported in the literature have very low magnitude of the ME coefficient. For example, some lead zirconate titanate–nickel ferrite structures show ME coupling values (16 mV cm^{-1} Oe^{-1}) which are much lower than that of their bulk counterpart (Ryu et al. 2006).

Despite the large body of literature on the synthesis of ME films of different compositions, comparatively very few studies have published direct measurements of the ME coupling coefficient, as it can be difficult to make these measurements at the nanoscale. Indirect measurement is used to verify the multiferroic properties such as the change in magnetization or dielectric properties at the ferroelectric, ferromagnetic, or antiferromagnetic Curie/Neel temperatures (Neel temperature is the temperature above which an antiferromagnetic or ferrimagnetic material becomes paramagnetic), the variation of polarization or magnetization hysteresis loops under the presence of an applied magnetic or electric field.

4.5.4 ME Composite/Hybrid with 1–3 Connectivity

Besides 0–3- and 2–2-type structures, the 1–3 structure is another important structure investigated in the films. The prototypical 1–3 ME nanostructure composites were first reported by Zheng et al. in self-assembled BT–CFO complex oxide nanostructures, where CFO nanopillar arrays with diameters of 20–30 nm were embedded in a ferroelectric BT matrix (Fig. 4.36). The formation of these self-organized epitaxial arrays of CFO columns in a BT matrix via spontaneous phase separation occurred due to the thermodynamic instability of the solutions at the growth temperature. This approach to the formation of self-assembled nanostructures is generic, and has been applied to other perovskite-spinel systems. Although this 1–3 nanostructure ME composite is expected to exhibit a reduced clamping effect by the substrate and efficient strain coupling resulting from the large interfacial surface area, the design and control of such heterostructured remains a challenge. The ME coefficients could not be directly measured because of the leakage problem resulting from the low resistance of the magnetic pillars penetrating through the films.

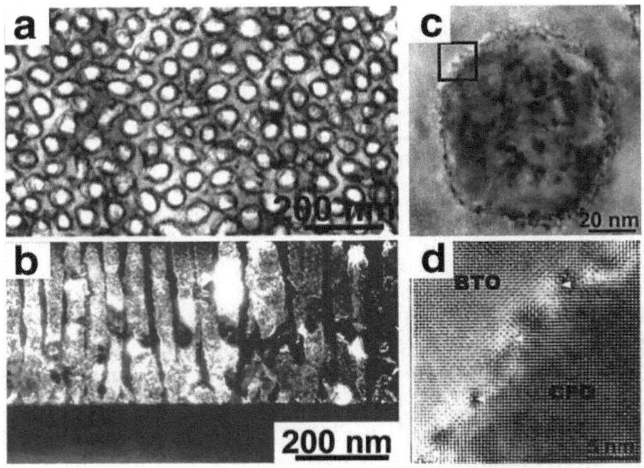

Fig. 4.36 **a** Plan view TEM image. **b** Cross-section dark field TEM image $CoFe_2O_4$ (CFO) pillar embedded in the BT. **c** In-plane high-resolution TEM image of one CFO pillar embedded in the BT matrix. **d** Magnified segment of the interface between the CFO pillar and the BT matrix from (c). *TEM* transmission electron microscopy. (Reprinted with permission from Zheng et al. (2004). Copyright, 2004, American Institute of Physics)

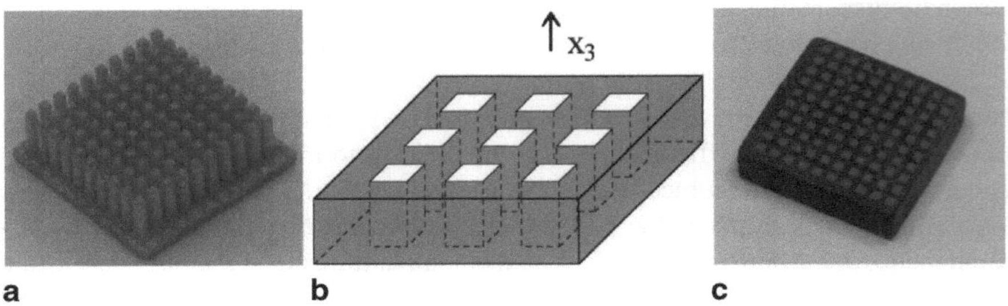

Fig. 4.37 **a** A diced PZT ceramic sample, **b** schematic illustration, **c** a typical micrograph of the PZT rod arrays in the TDE medium. (Reprinted with permission from Shi et al. (2005). Copyright, 2005, American Institute of Physics)

There is another quasi 1–3 structure ME composite (actually it is a three-phase (PZT, epoxy and Terfenol-D) composite, but the mixture of epoxy and magnetic is matrix and functionally can be considered as one phase) structure which is similar to that of well-known 1–3-type piezoelectric composites with PZT rod arrays embedded in an epoxy matrix. This structure was originally proposed by Newnham and coworkers. Instead of using an inert epoxy for the matrix, a ferromagnetic Terfenol-D/epoxy (TDE) mixture is used as the matrix in the PZT rod arrays. The PZT rod arrays in the TDE medium were produced by the dice-and-fill method. Two sets of closely spaced parallel cuts were made in a poled PZT ceramic sample, the two sets being perpendicular to each other (Fig. 4.37a). Then, the diced ceramic samples were filled with an epoxy resin solution containing Terfenol-D particles. After hardening at room temperature, the samples were polished to achieve 1–3-type composites with PZT rod (square rods with width of 0.5 mm) arrays embedded in the TDE medium, as shown in Fig. 4.34. The samples with 2.5 mm in thickness and 10 mm in width were electroded on the top and bottom surfaces by silver paint. The volume fraction *f* of the PZT rods in the TDE medium was controlled by changing

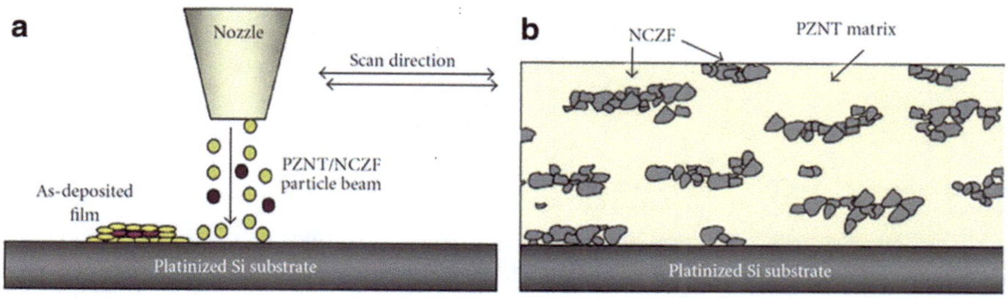

Fig. 4.38 Schematic illustration of **a** ME composite film fabrication and **b** microstructure of nanocomposite ME films by aerosol deposition (*AD*). *ME* magnetoelectric. (Reprinted with permission from Park et al. (2009). Copyright, 2009, IOP Publishing)

the periodicity of the square PZT rod array. The maximum α_{E33} was about 280 mV/cm Oe at the $f=0.12$, which is comparable with that of pseudo-2–2 laminate ME composites and larger than that for the 0–3 particulate ME composites.

4.5.5 ME Composite/Hybrid with 2–3 Connectivity

Due to the difference in thermal mismatch between films and substrates as well as slow deposition rate, most nanocomposite/heterostructured films were limited in their thickness. Therefore, the real output voltage from the ME thin film is very small (~μV level). In order to obtain enough signal, the thickness of film should be in the several micron range.

Recently, the aerosol deposition (AD) process has been demonstrated to fabricate the ME composite thick films. Figure 4.38 shows the schematic of fabrication and microstructure of the 3–2 nanocomposite ME films. For AD, the mixture powders of $Pb(Zr, Ti)O_3$–$Pb(Zn, Nb)O_3$ (PZT–PZN) and NCZF powders (4:1 weight ratio) were sprayed into an evacuated deposition chamber through nozzle. The 3–2 nanocomposite ME films were formed onto a platinized silicone substrate at room temperature. By controlling the number of repetitions of the nozzle scan, highly dense 3–2 nanocomposite thick films of PZT–PZN and NCZF with thickness > 10 μm were synthesized on platinized silicon substrate at room temperature. The maximum ME coefficient was measured to be 150 mV cm^{-1} Oe^{-1}. This magnitude is about three times higher than the previously reported nanocomposite films synthesized by other thin film fabrication processes. This high ME response was associated with reduced clamping effect, effective poling of piezoelectric matrix, and enhanced elastic coupling due to 3–2 composite structure. The XRD and TEM microstructure is shown in Fig. 4.39 along with elemental mapping which illustrates well-defined 3–2 connectivity. In addition to improved ME characteristics, the deposition rate of ME films was exceptionally higher (more than 1 μm/min) than any other conventional thin film processes (Priya et al. 2009).

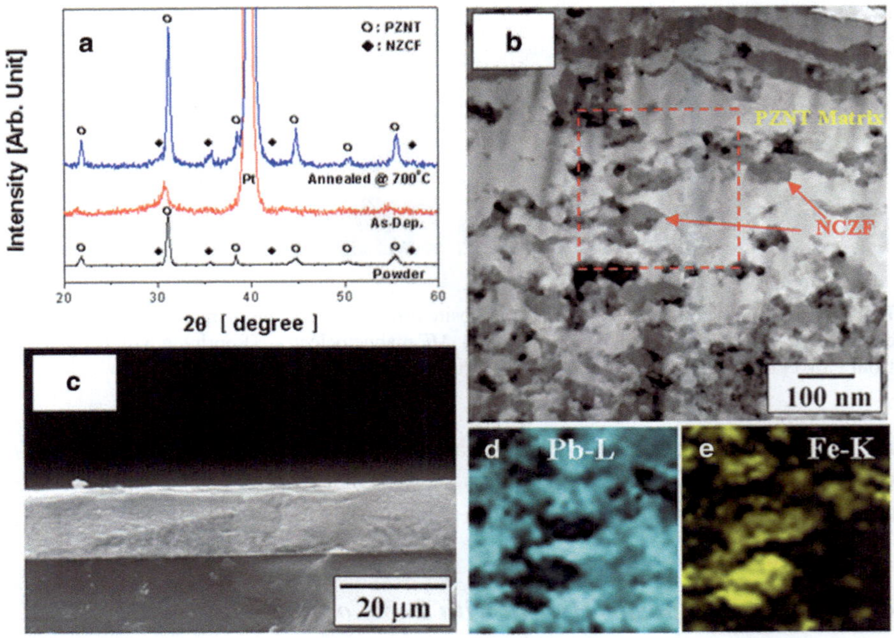

Fig. 4.39 **a** XRD patterns of the powders and composite thick films. **b** Cross-sectional SEM image. **c** STEM. **d** and **e** EDX mapping. *XRD* X-ray diffraction, *SEM* scanning electron microscope, STEM scanning transmission electron microscope, *EDX*, energy-dispersive X-ray spectroscopy

4.6 Scaling Effect

4.6.1 Critical Size in Ferroelectric Materials

Several studies have been conducted to investigate the effect of grain size (GS) on the dielectric, ferroelectric, and piezoelectric properties of ceramics. Figure 4.40 shows the ε', d_{33}, and k_p dependence on the GS of BT ceramics. Remarkably, d_{33} and k_p exhibited a trend similar to that of ε'. The values of d_{33}, k_p, and ε' increased significantly with decreasing GS, reaching maximum values ($\varepsilon'=6079$, $d_{33}=519$ pC/N, and $k_p=39.5\,\%$) at approximately 1 μm, and then decreased rapidly with further decreasing GS (Huan et al. 2014). Uchino et al. investigated the degree of tetragonality (i.e., c/a ratio) as a function of particle size in pure BT at room temperature (Fig. 4.41). It was found that the c/a value decreases drastically below 0.2 μm and became 1 (i.e., cubic) at 0.12 μm, defined as a critical particle size (Uchino et al. 1989).

To understand the effect of GS on the dielectric, ferroelectric, and piezoelectric properties, the domain structures have been theoretically and experimentally investigated to provide new insights into the intrinsic and extrinsic effects and relevant size effects in ferroelectric polycrystalline materials. Stable domain configurations in ferroic materials reflect the balance of anisotropy energy, domain-wall energy,

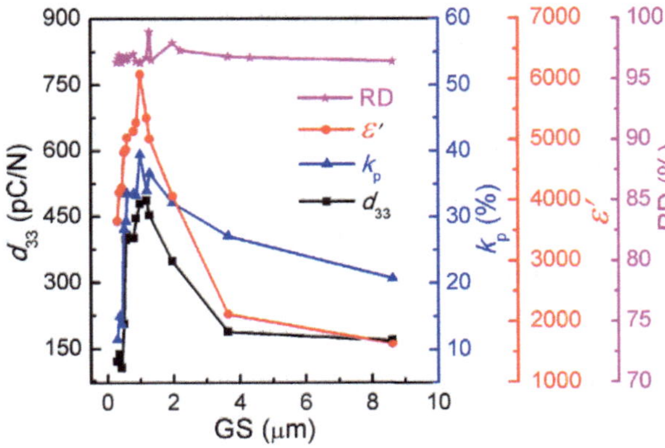

Fig. 4.40 ε', d_{33}, and k_p dependence on grain size (*GS*) of BT ceramics. (Reprinted with permission from Huan et al. (2014). Copyright, 2014, Elsevier)

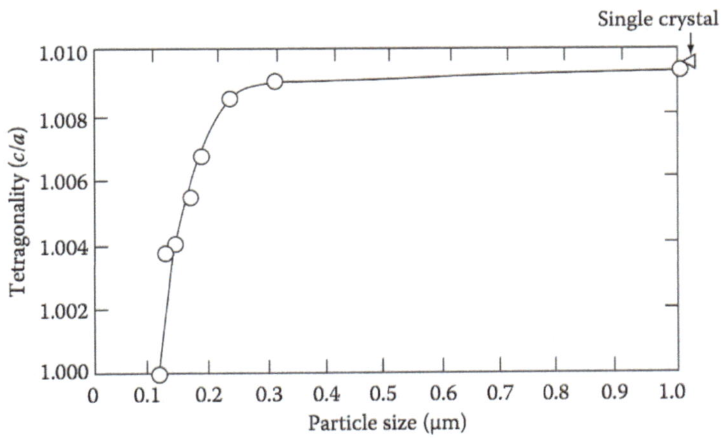

Fig. 4.41 Change in tetragonality with grain size in BT at room temperature. Note that the tetragonal crystal distortion disappears below a critical particle size around 0.1 μm. (Reprinted with permission from Uchino et al. (1989). Copyright, 1989, Wiley)

and electric and elastic energies. The basic variation of crystal size influences the domain size. From the balance of the depolarization energy and the domain-wall energy, theoretically there is a parabolic scaling relationship between the domain size and the GS given as (Randall et al. 1998):

$$d = \left[\left(\frac{\sigma}{\varepsilon * P_0^2} \right) t \right]^{1/2} \quad (11)$$

where σ is the energy density of the domain wall, P_0 is the spontaneous polarization, and ε^* is the effective dielectric constant. Experimentally, in the case of BT ceramics, the variation of domain size displays three distinct ranges and is well consistent with the change in dielectric and piezoelectric properties: (1) Above 10 μm, the domain size does not change and the permittivity does not depend on size in this size range; (2) Between 0.3 and 10 μm, domain size varies with the square root of the GS, following the theoretical prediction given by Eq. (11), and thereby the permittivity increases when the size decreases. However, permittivity does not continuously increase but presents a maximum around 0.8–1 μm. The reason is that the strong coupling between the grain boundaries and the domain walls makes the domain reorientation more difficult and severely constrains the domain-wall motion. This clamping stress reduces the domain-wall mobility, resulting in the decrease of permittivity below 0.8–1 μm. (3) Below 0.3 μm, the stresses reach their highest levels with the disappearance of the 90° domain walls at ~0.3 μm.

In thin films, the surfaces begin to dominate the behavior with the decrease in the thickness. The structure of the thin film surface may be different from that of the core due to the uncompensated atomic bonds at the surface, which can result in the depolarization field that tends to cancel the ferroelectricity. The critical size where the ferroelectric state becomes unstable was estimated to be of the order of 7–10 nm for the well-known ferroelectrics such as BT and PT. It should be noted that the experimentally obtained critical thickness for a ferroelectric thin film is strongly related to the sample quality, such as dead layers, grain boundaries, and defects. Charge screening of the depolarization field is a very important method to develop ferroelectric devices based on the ultra-thin dimensions.

4.6.2 Critical Size in Ferromagnetic Materials

Domains are formed to minimize the total energy of the material. Although a single ferromagnetic domain in which all the magnetic moments are aligned would minimize the exchange energy, it also maximizes the magnetostatic energy which is the main driving force for domain formation. The size of an actual domain is influenced by many factors, including the size/dimension, surface energy, and the defect structure. Considering size/dimension, the magnetostatic energy is proportional to the volume of the particle (i.e., r^3, where r is the particle radius), and the domain-wall energy is proportional to the area of the wall, r^2. Below some certain radius, r_c, it is energetically unfavorable to form domain walls and a single-domain particle is formed.

As the particle size is reduced into the single-domain region, the coercivity increases. This is related to fact that the magnetization-reversal process cannot take place via domain-wall motion. However, below a critical particle size, the coercive field goes to zero, and it becomes superparamagnetic. This occurs when thermal fluctuations are able to exceed the magnetic anisotropy energy that aligns the magnetization in a specific direction. The critical size of the particle at which superparamagnetism occurs is typically around 100 Å in diameter at room temperature.

In contrast, ferroelectricity can be observed at very small sizes, even on the scale of a few lattice constants. The origin of both the different size dependencies and different domain-wall widths in ferroelectrics and ferromagnets can be understood

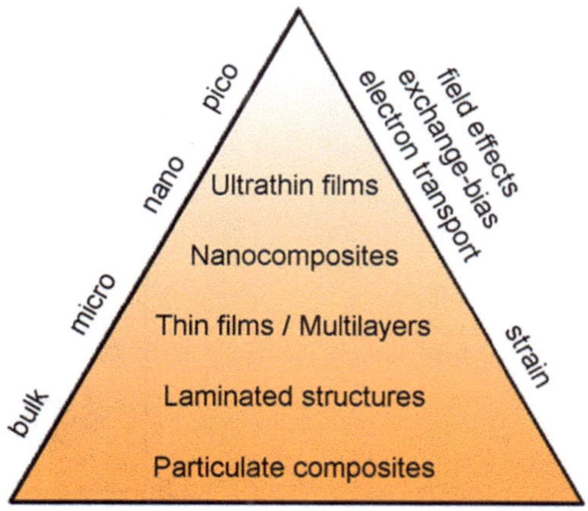

Fig. 4.42 Schematic of the different types of multiferroic structures showing hierarchy of size and type of coupling mechanism. (Reprinted with permission from Vaz et al. (2010). Copyright, 1989, Wiley)

to result from the different atomic-level mechanisms driving ferromagnetism and ferroelectricity. Ferromagnets prefer wide domain walls in which adjacent magnetic moments are oriented close to each other because the exchange coupling between the magnetic moments is strong and the magnetocrystalline anisotropy energy which couples the spin to the lattice is driven by spin–orbit coupling that is comparatively weak. In ferroelectrics, the situation is quite different as the energies of the "anisotropy" and "exchange" effects are comparable.

4.6.3 Coupling Mechanism

In composite structures, an effective coupling between the magnetic and electric order parameters across the composite is achieved by exploring the sensitivity of the ferroelectric and/or magnetic state to strain, charge, or magnetic interactions. From these basic components, a hierarchy of the different types of multiferroics based on the size and type of coupling developed thus far can be established, as illustrated in Fig. 4.42 (Vaz et al. 2010). The mechanisms exploited thus far rely (i) on the elastic coupling at the interface which couples the magnetic and electric order parameters via strain and piezo-, electro-, and magnetostrictive effects; (ii) on the exchange bias between antiferromagnetic ferroelectric and ferromagnetic components of the composite; or (iii) on the modulation of the charge carrier concentration in the ferromagnetic component using a field effect approach.

Strain Mediated It has been widely accepted that the ME coupling in bulk ME composites is strain mediated. In nanostructured films, residual stress/strain resulting from the lattice and/or thermal expansion misfit between the film and substrate and spontaneous polarization have to be taken into account.

Exchange Bias Mediated This coupling offers the possibility of controlling the magnetic properties of an adjacent ferromagnetic layer through the exchange anisotropy effect, which can potentially result in a much larger ME coupling effect.

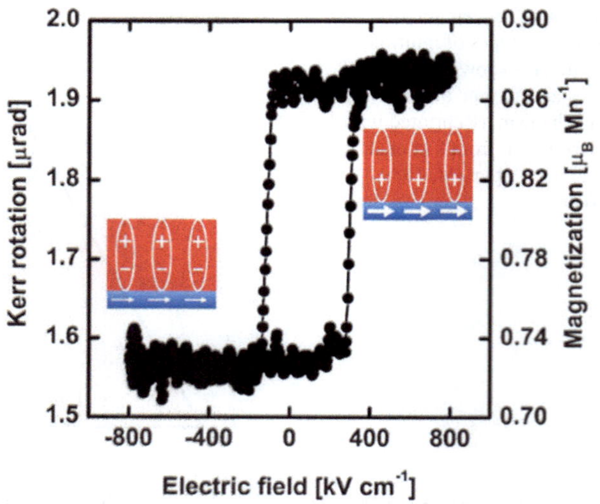

Fig. 4.43 Magnetoelectric hysteresis curve showing the magnetic response of the PZT/LSMO system at 100 K as a function of the applied electric field. Insets represent the magnetic and electric states of the LSMO and PZT layers, respectively. (Reprinted with permission from Molegraaf et al. (2009). Copyright, 2009, Wiley)

The exchange coupling at the ferromagnet/antiferromagnet interface gives rise to a directional biasing of the ferromagnetic magnetization that is manifested by a horizontal shift in the *M–H* hysteresis loop by an amount called the exchange bias field.

Charge Mediated This coupling is via modulating the charge carrier density through the electric field effect to induce changes in the magnetic state. This sensitivity of the magnetic state to the charge was explored in PZT/LSMO heterostructures, where a large change in the magnetic moment of the LSMO layer was found to be a function of the PZT polarization state (Molegraaf et al. 2009). This effect is shown in Fig. 4.43, where the magnetic response of the system is plotted against the applied electric field, showing an abrupt switching between a high and low magnetic state for the PZT polarization states that deplete and accumulate hole carriers from the LSMO layer, respectively.

4.7 Theoretical Models for ME Composites

The constitutive equation for describing the coupling of mechanical–electric–magnetic response in the ME composites using linear approximation can be written as

$$\begin{aligned} S_i &= s_{ij}T_j + d_{ki}E_k + q_{ki}H_k, \\ D_k &= d_{ki}T_j + \varepsilon_{kn}E_n + \alpha_{kn}H_n, \\ B_k &= q_{ki}T_i + \alpha_{kn}E_n + \mu_{kn}H_n, \end{aligned} \tag{12}$$

where S_i and T_j are strain and stress tensor components; E_k, D_k, H_k, and B_k are the vector components of the electric field, electric displacement, magnetic field,

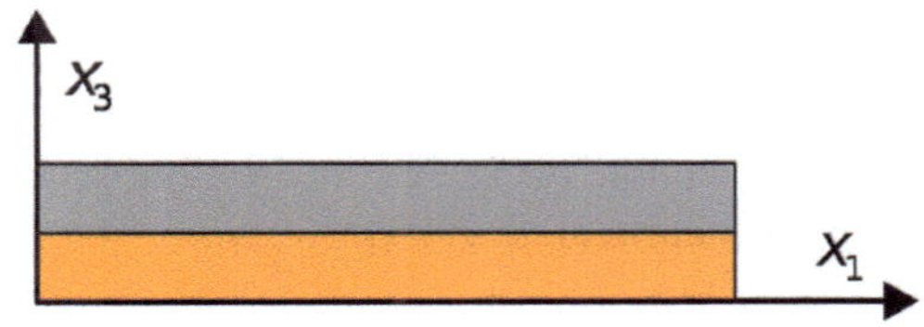

Fig. 4.44 A simple bilayer ME composite consisting of piezoelectric and magnetostrictive layers. *ME* magnetoelectric

and magnetic induction; s_{ij}, d_{ki}, and q_{ki} are effective compliance, piezoelectric, and piezomagnetic coefficients; and ε_{kn}, μ_{kn}, and α_{kn} are effective permittivity, permeability, and ME coefficient.

Theoretically, several methods have been employed to solve the constitutive Eq. (12) above and model the ME coupling at both low frequency and near electromechanical resonance (EMR), including (i) an average-field method; (2) Green's function (Nan et al. 2001); and (3) equivalent-circuit approach (Dong et al. 2003). The ME theoretical models have predicted that the 2–2 connectivity in laminate composites results in the highest ME coupling amongst the various composite architectures. This prediction has been validated by experimental data. Here, we concentrate on describing an average-field method to predict the ME response in composites under both quasi-static and resonance conditions based on magnetostrictive and piezoelectric constitutive equations. Consider a simple bilayer composite of piezoelectric and magnetic ceramic phases with free boundary condition (Fig. 4.44). The constitutive equation Eq. (12) can be directly solved by an averaging method to estimate the effective material parameters.

4.7.1 Low-Frequency ME Effect in Free Standing Bilayers

a. *Longitudinal ME Effect*

The bilayer is poled with an electric field E along direction-$x3$. The bias field H and the AC field δH are along the same direction as E. Using continuity conditions for magnetic and electric fields and open and closed circuit conditions, one obtains the following expressions for longitudinal ME voltage coefficient (Bichurin et al. 2003):

$$\alpha_{E,33} = \frac{E_3}{H_3} = 2\frac{\mu_0 v(1-v)^p d_{31}^m q_{31}}{\left\{2^p d_{31}^2(1-v) + {}^p\varepsilon_{33}[({}^p s_{11} + {}^p s_{12})(v-1) - v({}^m s_{11} + {}^m s_{12})]\right\}}$$
$$\times \frac{[({}^p s_{11} + {}^p s_{12})(v-1) - v({}^m s_{11} + {}^m s_{12})]}{\left\{[\mu_0(v-1) - {}^m\mu_{33}v][v({}^m s_{11} + {}^m \varepsilon_{12}) - ({}^p s_{11} + {}^p s_{12})(v-1) + 2^m q_{31}^2 v^2]\right\}} \quad (13)$$

where $v = {}^p v / ({}^p v + {}^m v)$ and ${}^p v$ and ${}^m v$ denote the volume of piezoelectric and magnetostrictive phase, respectively.

b. *Transverse ME Effect*

This case corresponds to E and δE along direction-$x3$ and H and δH along direction-$x1$ (in the sample plane). Expression for transverse ME voltage coefficient is given below (Bichurin et al. 2003):

$$\alpha_{E,31} = \frac{E_3}{H_1} = \frac{-v(1-v)(^m q_{11} +^m q_{21})^p d_{31}}{^p \varepsilon_{33}[(^m s_{12} +^m s_{11})v +^p \varepsilon_{33}(^p s_{11} +^p s_{12})(1-v) - 2^p d_{31}^2 (1-v)]} \tag{14}$$

c. *In-plane longitudinal ME effect*

Lastly, consider a bilayer poled with an electric field E in the plane of the sample; the in-plane fields H and δH are parallel and the induced electric field δE is measured in the same direction (x−1). Expression for in plane longitudinal ME voltage coefficient was derived to be (Bichurin et al. 2003):

$$\begin{aligned}\alpha_{E,11} = &\Big(\big(^m q_{11}\left(^p s_{33}{}^p d_{11} -^p s_{12}{}^p d_{12}\right) +^m q_{12}\left(^p s_{11}{}^p d_{12} -^p s_{12}{}^p d_{11}\right)\big)(1-v) \\ &+\left(^m q_{11}\left(^m s_{11}{}^p d_{11} -^m s_{12}{}^p d_{12}\right) +^m q_{11}\left(^m s_{11}{}^p d_{12} -^m s_{12}{}^p d_{11}\right)\right)v\Big)v(1-v) \\ &\Big/\Big(\left((1-v)^m \varepsilon_{11} + v^p \varepsilon_{11}\right)\Big[(1-v)^2\left(^p s_{11}{}^p s_{33} -^p s_{12}^2\right) + (1-v)v\left(^m s_{11}{}^p s_{11}\right. \\ &\left. +^p s_{33}{}^p s_{11} - 2^p s_{12}{}^m s_{12}\right)\Big] + V^2\left(^m s_{11}^2 -^m s_{12}^2\right) - v(1-v)^2\Big[2^p s_{12}{}^p d_{11}{}^p d_{12} \\ &-^p s_{33}{}^p d_{11}^2 -^p s_{11}{}^p d_{12}^2\Big] + v^2(1-v)\left(^m s_{11}{}^p d_{12}^2 +^m s_{11}{}^p d_{11}^2 - 2^m s_{12}{}^p d_{12}{}^p d_{11}\right)\Big)\end{aligned} \tag{15}$$

The in-plane ME coefficient is the strongest among the cases discussed in this section due to the high piezomagnetic q-values, piezoelectric d-values, and the low magnitude of demagnetizing fields.

4.7.2 ME Effect at EMR

a. *ME Effect at Longitudinal Modes of EMR*

Since the ME coupling in the composites is mediated by the mechanical stress, one would expect that the ME effect would be greatly enhanced when the piezoelectric or magnetic phase undergoes resonance. For the orientation when the poling direction of piezoelectric phase, DC and AC magnetic fields are parallel to the x_1-axis, the ME voltage coefficient under the open circuit condition is given as (Bichurin et al. 2010a):

$$\alpha_{E,31} = \frac{2^p d_{31}{}^m g_{31}\mu_{eff}{}^p s_{11} v(1-v)\tan(kL/2)}{s_2(pd_{31}^2 -^p s_{11}{}^p \varepsilon_{33})kL - 2^p d_{31}^2 v^m s_{11}^2 \tan(kL/2)} \tag{16}$$

where the wave value $k = \omega \sqrt{[{}^{p}\rho v + {}^{m}\rho(1-v)]\left[\frac{v}{{}^{p}s_{11}} + \frac{1-v}{{}^{m}s_{11}}\right]^{-1}}$, $s_2 = v^{m} s_{11}^{B} + (1-v)^{p} s_{11}$, and effective permeability of piezomagnetic layer is given as:

$$\mu_{eff} = \frac{s_2 {}^{m}s_{11}^{B} {}^{m}\mu_{11} kL}{({}^{m}s_{11}^{B} + {}^{m}g_{11}^{2} {}^{m}\mu_{11})kLs_2 + 2{}^{m}g_{11}^{2} {}^{m}\mu_{11} {}^{p}s_{11}(1-v)\tan(kL/2)}$$

b. *ME Effect at Bending Mode of EMR*

A key drawback for ME effect at longitudinal mode is that the frequencies are high of the order of hundreds of kHz for nominal sample dimensions. The eddy current losses for the magnetostrictive phase can be quite high at such frequencies, in particular for transition metals and alloys and earth rare alloys such as Terfenol-D, resulting in an inefficient ME energy conversion. In order to reduce the operating frequency, one must therefore increase the laminate size that is inconvenient for any applications. An alternative for getting a strong ME coupling is the resonance enhancement at bending modes of the composite (Xing et al. 2006; Zhai et al. 2008). Under the open circuit condition, the ME voltage coefficient at bending modes of EMR can be given as (Petrov et al. 2009):

$$\alpha_{E,31} = \frac{E_3}{H_1} = -\frac{\int_{z_0-p_t}^{z_0} {}^{p}E_3 dz}{tH_1 {}^{P}\varepsilon_{33}} \tag{17}$$

where E_3 and H_1 are the average electric field induced across the sample and applied magnetic field.

Nanostructures in the shape of wires, pillars, and films are important for increased functionality in miniature devices. The models must take into account the clamping effect of substrate, flexural deformations, and the contribution of lattice mismatch between composite phases and substrate to ME coupling. Theoretical modeling of the size- and shape-dependent characterization of the ME effect in composites, particularly down to nanoscale dimensions, has been reviewed by Bichurin et al. (2010b, c) and is be discussed here.

4.8 Application of ME Composite

Promising applications for ME composite include magnetic field sensors, transducer, spintronics, and voltage tunable RF/microwave device. ME devices can be categorized into several groups based upon the mechanisms listed in Table 4.8. The development of ME devices and applications have been summarized and reviewed in several review articles (Nan et al. 2008; Ma et al. 2011;Scott 2012; Srinivasan and Sun 2012). Here, we discuss two important ME devices: (1) ME magnetic field sensor and (2) ME energy harvester.

Table 4.8 Classification of different multiferroic devices (Srinivasan and Sun 2012)

ME coupling	Physical mechanism	ME devices
Direct ME coupling	*H* control of electric polarization	Magnetic sensors
		Current sensors
		Energy harvesters
		Transformers
Converse ME coupling	*E* control of magnetization switching	Spintronics, including ME random access memory
	E control of permeability μ	Voltage tunable inductors, tunable band-pass filters, phase shifters
	E control of spin wave	Voltage tunable filters, tunable resonators, phase shifters
No ME coupling needed	High μ and high ε	Compact antennas, and other RF/ microwave devices

ME magnetoelectric

4.8.1 ME Magnetic Field Sensors

ME composites consisting of piezoelectric and magnetostrictive components have a considerable potential as highly sensitive magnetic field sensors to replace superconducting quantum interference device (SQUID) for weak field detection due to the low cost and no requirement of cryogenic condition. The practical usefulness of a magnetic sensor is determined not only by the output signal of the sensor in response to an incident magnetic field, but also by the equivalent magnetic noise generated in the absence of an incident field. The challenge of fabricating an ME composite with a high α_E and a low equivalent magnetic noise has restricted the realization of ME magnetic sensors.

The highest ME coefficient α_E has been reported by Wang et al. in a multi-push–pull mode Metglas/PMN–PT sensor (Fig. 4.32). Under optimal DC magnetic bias, α_E reaches a maximum value of 52 V cm^{-1} Oe^{-1} (Fig. 4.33). An extremely low equivalent magnetic noise of 5.1 pT Hz$^{-1/2}$ was found at 1 Hz for these Metglas/PMN–PT sensors (Fig. 4.45), which is very close to the predicted value of 4.2 pT Hz$^{-1/2}$. Thin film ME composites are of special interest due to their potential for miniaturization that is essential for sensor arrays or for spatially resolved measurements. They are also of interest due to the possibility of avoiding magnetic bias fields by using exchanged biased magnetostrictive multilayers, and due to their perfect coupling between the two phases avoiding any gluing processes. Piorra et al. reported a magnetic field sensor based on the ME thin film composites with interdigitated electrodes (IDE) that had a limit of detection of 2.6 pT/Hz$^{1/2}$ at the mechanical resonance frequency (f_r=974.4 Hz). These ME thin film composites are fabricated on silicon cantilevers using PZT films with interdigitated transducer electrodes on the top side and FeCoSiB amorphous magnetostrictive thin films on the backsides. Figure 4.46 shows the sketch of cantilever with interdigitated electrodes and plate capacitor-type electrodes.

Fig. 4.45 Measured and estimated equivalent magnetic noise of the multi-push–pull mode Metglas/PMN–PT sensor over the frequency range of 0.125 Hz<f<100 Hz. (Reprinted with permission from Wang et al. (2011). Copyright, 2011, Wiley)

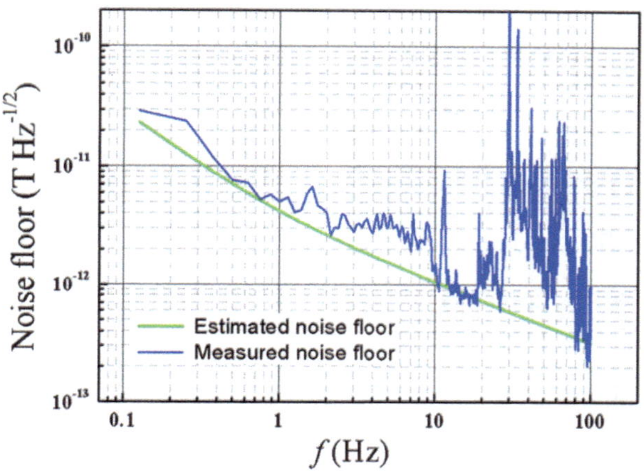

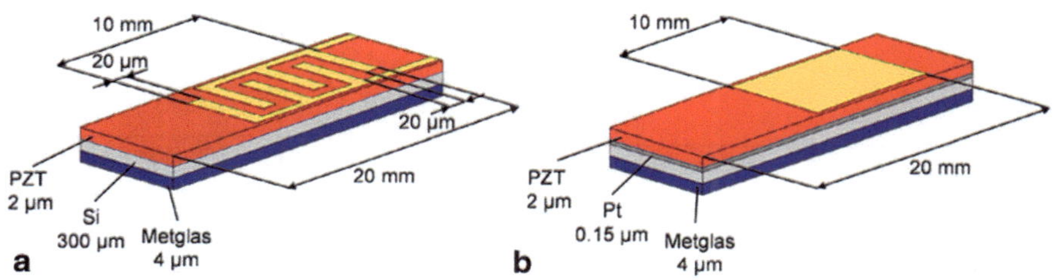

Fig. 4.46 Illustration of cantilever designs. **a** Cantilever with interdigitated electrodes, 4 μm Metglas on backside. **b** Cantilever with plate capacitor-type electrodes, 4 μm Metglas on backside. (Reprinted with permission from Piorra et al. (2013). Copyright, 2011, AIP)

The induced ME voltage is proportional to the electrode spacing. The interdigitated electrode layout can overcome this limitation because the spacing of the IDE can be freely defined by the lithography process and is typically ten times the film thickness. Further, as mentioned earlier, the effective piezoelectric coefficient will be the d_{33} coefficient which is approximately twice that of the d_{31} of the plate capacitor geometry. There are two things that need to be mentioned here: (1) Since the IDE electrode design requires a poling process resulting in an antiparallel poling direction between the neighboring electrodes, ferroelectric materials instead of aluminum nitride (AlN) are used as the piezoelectric thin films. (2) For the IDE layout, the ferroelectric layer should be deposited on a nonconducting material to avoid a short circuit of the electric field. Therefore, magnetostrictive and piezoelectric layers need to be deposited on the opposite sides of a nonconducting cantilever as shown in Fig. 4.46.

Figure 4.47 shows the ME response of the PZT ME composite with an IDE. An ME coefficient of approximately $\alpha_{ME}=200$ V/cm Oe and an ME voltage coefficient of approximately $\alpha=0.4$ V/Oe was measured at the mechanical resonance frequency ($f_r=974.4$ Hz). Compared to the plate capacitor design, which shows an

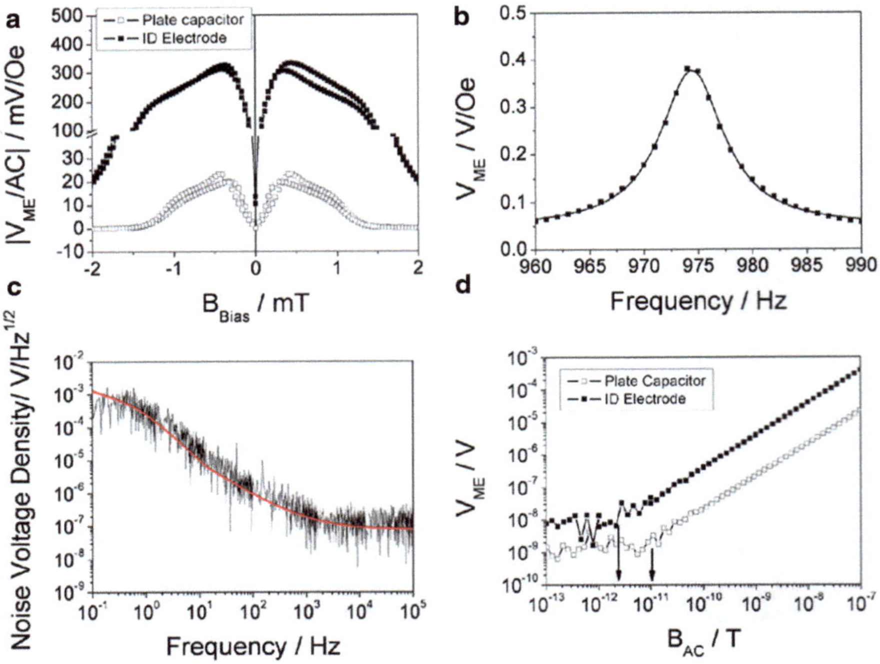

Fig. 4.47 **a** ME voltage V_{ME} versus bias field B_{bias}. **b** ME voltage V_{ME} versus frequency for cantilever with interdigitated electrodes. **c** Measured (*black curve*) and calculated (*red curve*) noise voltage density versus frequency for a cantilever with an IDE at the output of the amplifier. **d** ME voltage V_{ME} versus magnetic AC field B_{ac}. The *arrow* indicates the calculated limit of detection. (Reprinted with permission from (Piorra et al. 2013). Copyright, 2011, AIP)

ME voltage coefficient of approximately of $\alpha = 0.02$ V/Oe and an ME coefficient of approximately $\alpha_{ME} = 100$ V/cm Oe, the ME voltage coefficient is enhanced by a factor of approximately 20 (Fig. 4.47a). Figure 4.47c shows a comparison of the measured and calculated noise density spectrum for the IDE sensor. The equivalent magnetic noise density can be calculated to be $B_{noise} = 2.6$ pT/√Hz. This value is in good agreement with the sensitivity measurements which show a limit of detection of 2.3 pT/√Hz (Fig. 4.47d). The sensitivity measurement of the PZT plate capacitor sensor shows limit of detection of approximately 10 pT/√Hz.

As mentioned earlier, the practical usefulness of a magnetic sensor is determined not only by the output signal of the sensor in response to an incident magnetic field but also by the equivalent magnetic noise generated in the absence of an incident field. The reduction of equivalent magnetic noise is very important to achieve high signal noise ratio (SNR) in order to improve the sensitivity of ME magnetic sensor. Wang et al. have reviewed the equivalent magnetic noise of ME laminate sensors, analyzed the noise contributions, and summarized noise reduction strategies (Wang et al. 2014). There are two types of noise: intrinsic and extrinsic. Extrinsic noise sources exist in the environment (examples include vibration and thermal noise), which can often be reduced or eliminated by rejection techniques. However, it is not possible to reject the intrinsic noise sources from ME materials (dielectric loss noise, N_{loss}, and leakage resistance noise, N_R) and circuits (thermal noise from feedback resistor, N_{Rf}, current noise from amplifier, N_i, and the transfer function noise, Nv).

The noise charge density (C $Hz^{-1/2}$) from ME materials can be given as:

$$N_{loss} = \sqrt{\frac{4kTC \tan\delta}{2\pi f}} \quad (18)$$

$$N_R = \frac{1}{2\pi f}\sqrt{\frac{4kT}{R}} \quad (19)$$

where C is capacitance, R is resistance of ME composite, k is Boltzmann's constant, T is the temperature in kelvin, $\tan\delta$ is the dielectric loss, and f is the frequency in hertz.

The noise charge density (C $Hz^{-1/2}$) from circuits can be given as:

$$N_{R_f} = \frac{1}{2\pi f}\sqrt{\frac{4kT}{R_f}} \quad (20)$$

$$N_i = \frac{1}{2\pi f^{i_n}} \quad (21)$$

$$N_v = \frac{e_n\left|1 + Z / Z_f\right|}{\left|H(s)\right|} \quad (22)$$

where i_n is the current noise density and e_n is the voltage noise density of the low-noise amplifier, Z and Z_f are the impedance of ME composite and the feedback circuits.

Thus, the total noise charge density ($C\,Hz^{-1/2}$) can be expressed as:

$$\text{noise charge density } (\text{CHz}^{-1/2}) = \sqrt{N_R^2 + N_{loss}^2 + N_{R_f}^2 + N_i^2 + N_v^2} \quad (23)$$

and the equivalent magnetic noise ($T\,Hz^{-1/2}$) can then be written as:

$$\text{equivalent magnetic noise } (\text{THz}^{-1/2}) = \frac{1}{\alpha_Q}\sqrt{N_R^2 + N_{loss}^2 + N_{R_f}^2 + N_i^2 + N_v^2} \quad (24)$$

From above equation, the strategies to reduce the equivalent magnetic noise of an ME sensor include: (i) reducing the charge amplifier noise (via low-noise detection methods), (ii) reducing the ME laminate noise (via reducing dielectric loss

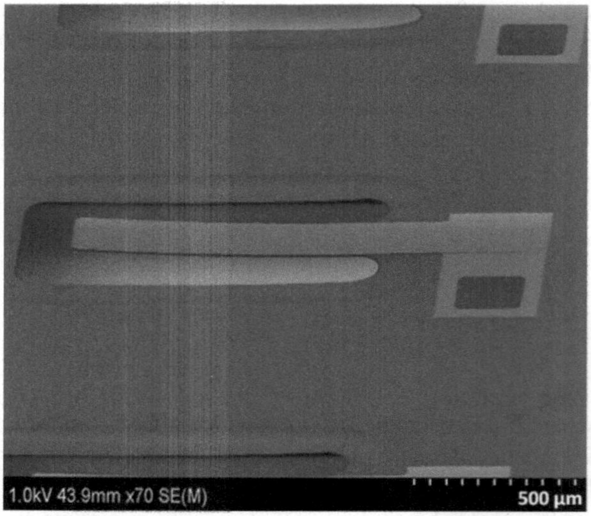

Fig. 4.48 SEM image of an ME energy harvester. (Reprinted with permission from Onuta et al. (2011). Copyright, 2011, AIP)

and increasing the DC resistance of piezoelectric phase), (iii) increasing the charge coefficient without introducing additional noise (via increasing the piezomagnetic coefficient of magnetostrictive materials), and (iv) increasing the charge coefficient (via sensor arrays; Wang et al. 2014).

The contribution of each noise source was significantly dependent on the frequency. At a frequency of $f = 1$ Hz, the total noise charge density was dominated by the ME laminate noise, while the noise contributions from the circuit was much smaller (Wang et al. 2014). To reduce the ME laminate noise, piezoelectric materials with high piezoelectric response and low dielectric loss (such as single crystal) are preferred. Further, as discussed earlier, optimization of interfacial bonding condition (epoxy bonding, co-firing) and effective poling (IDE) during fabrication process are critical factors to suppress the dielectric loss, thereby achieving low noise.

4.8.2 ME Energy Harvesters

Onuta et al. (2011) reported a highly efficient miniaturized energy harvesters using all-thin-film ME structures on Si-micromachined cantilevers. The ME layers consist of a magnetostrictive $Fe_{0.7}Ga_{0.3}$ thin film and a PZT piezoelectric thin film. The thin film heterostructure was fabricated on a Si substrate with a plasma-enhanced chemical vapor deposited (PECVD) silicon oxide/nitride/oxide (ONO) stack. A 20-nm/100-nm Ti/Pt layer was sputtered at 430 °C to form the bottom electrode of the piezoelectric layer. A ~500-nm PZT layer was spun by a standard solgel process. A 35-nm Pt buffer layer was sputtered at 305 °C on the PZT layer and then a 500-nm $Fe_{0.7}Ga_{0.3}$ layer was sputtered at room temperature. To fabricate the cantilever beam structure, a four-mask photolithographic process was used to make the free-standing cantilevers having 950 μm length and 200 μm width as shown in Fig. 4.48.

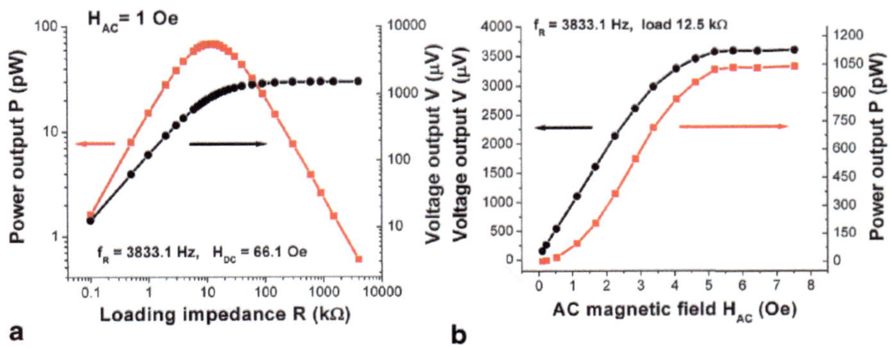

Fig. 4.49 **a** Power output and RMS voltage as a function of loading impedance at resonance frequency; **b** dependence of the output voltage and power on the AC magnetic field at resonance. RMS root mean square. (Reprinted with permission from Onuta et al. (2011). Copyright, 2011, AIP)

Figure 4.49a shows the results of energy harvesting measurements from a single all-thin-film multiferroic device. The power output $(P = V_{\mathrm{RMS}}^2 / R_{\mathrm{load}})$ has a peak that occurs for a load impedance of 12.5 kΩ. To determine the power density of the multiferroic harvester, an effective volume (950 μm × 200 μm × 0.5 μm) determined by taking into account the freestanding length of the cantilever and the thickness of the poled (PZT) piezoelectric film was used. The measured peak power density was 0.7 mW/cm^3 (root mean square, RMS). This value is similar to the other reported harvested power densities of bulk laminate devices at 1 Oe (RMS). Figure 4.49b shows the AC magnetic field dependence of the harvested voltage and power. The harvested voltage and power output increases with the increase in AC magnetic field before arriving the saturation plateau. The saturation plateau in both power and voltage with respect to increasing AC magnetic field was due to the saturation of the internal stress of magnetic origin that involves the rotation of magnetization vector (Onuta et al. 2011).

Yuan et al. proposed a dual-phase (vibration and magnetic) energy harvesting system utilizing a self-biased ME response (Zhou et al. 2013). High ME voltage output was found to be generated from this structure under the applied AC magnetic field (H_{ac}) in the absence of DC bias (H_{dc}). The voltage output was further enhanced with the co-existence of external vibration. Figure 4.50 shows the performance of the dual-phase energy harvester working under both mechanical and magnetic condition at various resonance frequencies. It can be clearly noticed that the output voltage under the dual-phase mode was improved over a wide range of frequency, which implies that both power density and efficiency were enhanced through additive effect. The increase in voltage output at the first three resonance frequencies under the dual-phase mode was found to be 2.4, 35.5, and 360.7 %.

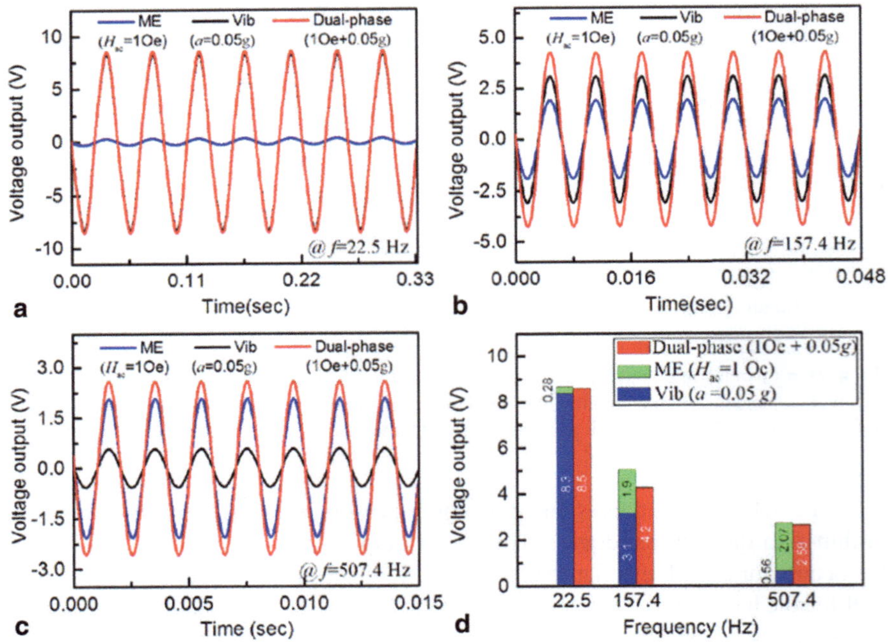

Fig. 4.50 Performance of the dual-phase energy harvester working under both mechanical and magnetic condition at various resonance frequencies: **a** 22.5 Hz, **b** 157.4 Hz, and **c** 507.4 Hz. **d** Comparison of the EH voltage output at single mode and dual-phase mode under different driving frequency. *EH* energy harvesting (Reprinted with permission from Yuan et al. (2013). Copyright, 2013, AIP)

4.9 Summary and Future Perspective

Multiferroic composites have recently stimulated an increasing number of research activities owing to the significant technological promise in the design of novel multifunctional devices. In this chapter, we reviewed the materials issues for developing high-performance ME composites/hybrids. The chapter covers the development of each material's constituent (piezoelectrics and magnetostriction), the design of the composite structure, interface strain coupling between piezoelectric and magnetostrictive phases, and the fabrication process of several important hybrid materials systems. The scaling effect of the ME composite was discussed because multiferroic ME nanostructures have become an important topic for microelectronic devices.

Although much progress has been made in the past decade, there are still several important issues that need to be resolved. These include:

1. Materials Synthesis
   The bulk ME composite exhibits a large ME effect above room temperature, however, from the application viewpoint, ME composite films are more desirable because of easy on-chip integration. Currently, most ME composite films

reported in the literature have quite low magnitude of ME coefficients compared to their bulk counterparts. New material design and fabrication processes are required to grow desired ME nanocomposites with controlled composition, atomic arrangement, and interface.

2. Coupling mechanism in ME nanocomposite/hybrid
   In nanostructured films, many open questions remain regarding the ME coupling mechanism. Is there a critical dimension below which the ME coupling will disappear or change its nature? What is the role of interfaces? How do ferroelectric and ferromagnetic domains interact with each other in the presence and absence of a field? These questions and other related issues need to be addressed for further advancing the progress in nanocomposites.
3. Applications
   Although some potential applications and prototypes devices have been demonstrated, challenges still remain in terms of comparative advantage, reliability, and cost-effectiveness. For example, most of the conventional ME composites need a DC bias to obtain the maximum ME response, which in turn results in problems of device size and electromagnetic interference. Thus, a novel design is desired to achieve a large α_{ME} in the absence of a DC magnetic field (called self-biased composites), which will pave the way for on-chip devices utilization of ME phenomenon.

Acknowledgment Y.Y. acknowledges the financial support from Office of Basic Energy Science, Department of Energy. S.P. would like to acknowledge the support from Office of Naval Research through Center for Energy Harvesting Materials and Systems (CEHMS).

References

Bichurin MI, Petrov VM, Srinivasan G (2003) Theory of low-frequency magnetoelectric coupling in magnetostrictive-piezoelectric bilayers. Phys Rev B 68: 054402

Bichurin MI, Petrov VM, Averkin SV, Filippov AV (2010a) Electromechanical resonance in magnetoelectric layered structures. Phys Solid State 52(10):2116–2122

Bichurin MI, Petrov VM, Averkin SV, Liverts E (2010b) Present status of theoretical modeling the magnetoelectric effect in magnetostrictive-piezoelectric nanostructures. Part I: low frequency and electromechanical resonance ranges. J Appl Phys 107: 053904

Bichurin MI, Petrov VM, Averkin SV, Liverts E (2010c) Present status of theoretical modeling the magnetoelectric effect in magnetostrictive-piezoelectric nanostructures. Part II: magnetic and magnetoacoustic resonance ranges. J Appl Phys 107: 053905

Brown WF, Hornreich RM, Shtrikman S (1968) Upper bound on the magnetoelectric susceptibility. Phys Rev 168(2):574–577

Cho KH, Park CS, Priya S (2010) Effect of intensive and extensive loss factors on the dynamic response of magnetoelectric laminates. Appl Phys Lett 97: 182902

Clarke R, Whatmore WR (1976) The growth and characterization of Pzt. J Cryst Growth 33:29–38

Damjanovic D (1998) Ferroelectric, dielectric and piezoelectric properties of ferroelectric thin films and ceramics. Rep Prog Phys 61(9):1267–1324

Deng C, Zhang Y, Ma J, Lin Y, Nan C-W (2008) Magnetoelectric effect in multiferroic heteroepitaxial BaTiO3-NiFe2O4 composite thin films. Acta Mater 56(3):405–412

Dong SX, Li JF, Viehland D (2003) Longitudinal and transverse magnetoelectric voltage coefficients of magnetostrictive/piezoelectric laminate composite: theory. IEEE Trans Ultrason Ferroelectr Freq Control 50(10):1253–1261
Dong SX, Li JF, Viehland D (2004) A longitudinal-longitudinal mode TERFENOL-D/Pb(Mg1/3Nb2/3)O-3-PbTiO3 laminate composite. Appl Phys Lett 85(22):5305–5306
Dong SX, Zhai JY, Li JF, Viehland D (2006) Near-ideal magnetoelectricity in high-permeability magnetostrictive/piezofiber laminates with a (2–1) connectivity. Appl Phys Lett 89: 252904
Engdahl G (2000) Handbook of giant magnetostrictive materials. Academic Press, San Diego
Guo R, Cross LE, Park SE, Noheda B, Cox DE, Shirane G (2000) Origin of the high piezoelectric response in PbZr1-xTixO3. Phys Rev Lett 84(23):5423–5426
Harshe GR (1991) Magnetoelectric effect in piezoelectric-magnetostrictive composites, vol 9204210. The Pennsylvania State University, Ann Arbor, pp 299–99
Hill NA (2000) Why are there so few magnetic ferroelectrics? J Phys Chem B 104(29):6694–6709
Huan Y, Wang XH, Fang J, Li LT (2014) Grain size effect on piezoelectric and ferroelectric properties of BaTiO3 ceramics. J Eur Ceram Soc 34(5):1445–1448
Islam RA (2008) Composition-microstructure-property relationships in dual phase bulk magnetoelectric composite, vol 3310386. The University of Texas at Arlington, Ann Arbor, pp 244
Islam RA, Bedekar V, Poudyal N, Liu JP, Priya S (2008) Magnetoelectric properties of core-shell particulate nanocomposites. J Appl Phys 104: 104111
Israel C, Mathur ND, Scott JF (2008) A one-cent room-temperature magnetoelectric sensor. Nat Mater 7(2):93–94
Jaffe B (1971) Piezoelectric ceramics. Academic Press, London, p ix, 317
Laletin VM, Paddubnaya N, Srinivasan G, De Vreugd CP, Bichurin MI, Petrov VM, Filippov DA (2005) Frequency and field dependence of magnetoelectric interactions in layered ferromagnetic transition metal-piezoelectric lead zirconate titanate. Appl Phys Lett 87: 222507
Leontsev SO, Eitel RE (2010) Progress in engineering high strain lead-free piezoelectric ceramics. Sci Technol Adv Mat 11: 044302
Liu WF, Ren XB (2009) Large piezoelectric effect in Pb-Free ceramics. Phys Rev Lett 103: 257602
Liu D, Yan YK, Zhou HP (2007) Synthesis of micron-scale platelet BaTiO3. J Am Ceram Soc 90(4):1323–1326
Liu R, Zhao Y, Huang R, Zhao Y, Zhou H (2010) Multiferroic ferrite/perovskite oxide core/shell nanostructures. J Mater Chem 20(47):10665–10670
Lotgering FK (1959) Topotactical reactions with ferrimagnetic oxides having hexagonal crystal structures—I. J Inorg Nucl Chem 9(2):113–123
Ma J, Shi Z, Nan C-W (2007) Magnetoelectric properties of composites of single Pb(Zr, Ti)O-3 rods and Terfenol-D/epoxy with a single-period of 1–3-type structure. Adv Mater 19(18):2571–2573
Ma J, Hu JM, Li Z, Nan CW (2011) Recent progress in multiferroic magnetoelectric composites: from bulk to thin films. Adv Mater 23(9):1062–10687
Mandal SK, Sreenivasulu G, Petrov VM, Srinivasan G (2010) Flexural deformation in a compositionally stepped ferrite and magnetoelectric effects in a composite with piezoelectrics. Appl Phys Lett 96: 192502
Martin LW (2010) Engineering functionality in the multiferroic BiFeO3—controlling chemistry to enable advanced applications. Dalton Trans 39(45):10813–10826
Messing GL, Trolier-McKinstry S, Sabolsky EM, Duran C, Kwon S, Brahmaroutu B, Park P, Yilmaz H, Rehrig PW, Eitel KB, Suvaci E, Seabaugh M, Oh KS (2004) Templated grain growth of textured piezoelectric ceramics. Crit Rev Solid State 29(2):45–96
Molegraaf HJA, Hoffman J, Vaz CAF, Gariglio S, van der Marel D, Ahn CH, Triscone J-M (2009) Magnetoelectric effects in complex oxides with competing ground states. Adv Mater 21(34):3470
Nan CW (1993) Physics of inhomogeneous inorganic materials. Prog Mater Sci 37(1):1–116
Nan CW, Li M, Huang JH (2001) Calculations of giant magnetoelectric effects in ferroic composites of rare-earth-iron alloys and ferroelectric polymers. Phys Rev B 63: 144415
Nan CW, Bichurin MI, Dong SX, Viehland D, Srinivasan G (2008) Multiferroic magnetoelectric composites: historical perspective, status, and future directions. J Appl Phys 103: 031101

Newnham RE (2005) Properties of materials: anisotropy, symmetry, structure. Oxford University Press, Oxford
Newnham RE, Skinner DP, Cross LE (1978) Connectivity and Piezoelectric-Pyroelectric composites. Mater Res Bull 13(5):525–536
Nye JF (1985) Physical properties of crystals: their representation by tensors and matrices. Clarendon Press; Oxford University Press, Oxford
O'Dell TH (1963) The field invariants in a magnetoelectric medium. Phila Mag 8:411–418
Onuta T-D, Wang Y, Long CJ, Takeuchi I (2011) Energy harvesting properties of all-thin-film multiferroic cantilevers. Appl Phys Lett 99: 203506
Park CS, Priya S (2011) Cofired magnetoelectric laminate composites. J Am Ceram Soc 94(4):1087–1095
Park SE, Shrout TR (1997) Ultrahigh strain and piezoelectric behavior in relaxor based ferroelectric single crystals. J Appl Phys 82(4):1804–1811
Park C-S, Ryu J, Choi J-J, Park D-S, Ahn C-W, Priya S (2009) Giant Magnetoelectric coefficient in 3–2 nanocomposite thick films. Japanese J Appl Phys 48: 080204
Petrov VM, Srinivasan G, Bichurin MI, Galkina TA (2009) Theory of magnetoelectric effect for bending modes in magnetostrictive-piezoelectric bilayers. J Appl Phys 105: 063911
Piorra A, Jahns R, Teliban I, Gugat JL, Gerken M, Knoechel R, Quandt E (2013) Magnetoelectric thin film composites with interdigital electrodes. Appl Phys Lett 103: 032902
Priya S, Islam R, Dong S, Viehland D (2007) Recent advancements in magnetoelectric particulate and laminate composites. J Electroceram 19(1):149–166
Priya S, Ryu J, Park CS, Oliver J, Choi JJ, Park DS (2009) Piezoelectric and magnetoelectric thick films for fabricating power sources in wireless sensor nodes. Sensors 9(8):6362–6384
Randall CA, Kim N, Kucera JP, Cao WW, Shrout TR (1998) Intrinsic and extrinsic size effects in fine-grained morphotropic-phase-boundary lead zirconate titanate ceramics. J Am Ceram Soc 81(3):677–688
Ryu J, Carazo AV, Uchino K, Kim HE (2001) Piezoelectric and magnetoelectric properties of lead zirconate titanate/Ni-Ferrite particulate composites. J Electroceram 7(1):17–24
Ryu H, Murugavel P, Lee JH, Chae SC, Noh TW, Oh YS, Kim HJ, Kim KH, Jang JH, Kim M, Bae C, Park JG (2006) Magnetoelectric effects of nanoparticulate Pb(Zr0.52Ti0.48)O-3-NiFe2O4 composite films. Appl Phys Lett 89: 102907
Sandlund L, Fahlander M, Cedell T, Clark AE, Restorff JB, Wun-Fogle M (1994) Magnetostriction, elastic moduli, and coupling factors of composite Terfenol-D. J Appl Phys 75(10):5656–5658
Schileo G (2013) Recent developments in ceramic multiferroic composites based on core/shell and other heterostructures obtained by sol-gel routes. Prog Solid State Chem 41(4):87–98
Scott JF (2012) Applications of magnetoelectrics. J Mater Chem 22(11):4567–4574
Shi Z, Nan CW, Zhang J, Cai N, Li JF (2005) Magnetoelectric effect of Pb(Zr, Ti)O-3 rod arrays in a (Tb, Dy)Fe-2/epoxy medium. Appl Phys Lett 87: 012503
Spaldin NA, Fiebig M (2005) The renaissance of magnetoelectric multiferroics. Science 309(5733):391–392
Srinivasan G (2010) Magnetoelectric Composites, pp 153–78. in Annual Review of Materials Research, Vol 40, Annual Review of Materials Research. Annual Reviews, Palo Alto, 2010
Srinivasan G, Sun NX (2012) Voltage control of magnetism in multiferroic heterostructures and devices. SPIN 02(03):1240004
Srinivasan G, Rasmussen ET, Gallegos J, Srinivasan R, Bokhan YI, Laletin VM (2001) Magnetoelectric bilayer and multilayer structures of magnetostrictive and piezoelectric oxides. Phys Rev B 64: 214408
Srinivasan G, Rasmussen ET, Levin BJ, Hayes R (2002) Magnetoelectric effects in bilayers and multilayers of magnetostrictive and piezoelectric perovskite oxides. Phys Rev B 65: 134402
Uchino K (2010) Ferroelectric devices, 2nd ed. CRC, Boca Raton, p xix, 347
Uchino K, Sadanaga E, Hirose T (1989) Dependence of the crystal-structure on particle-size in Barium-Titanate. J Am Ceram Soc 72(8):1555–1558
Vaz CAF, Hoffman J, Anh CH, Ramesh R (2010) Magnetoelectric coupling effects in multiferroic complex oxide composite structures. Adv Mater 22(26–27):2900–2918

Wang KF, Liu JM, Ren ZF (2009) Multiferroicity: the coupling between magnetic and polarization orders. Adv Phys 58(4):321–448

Wang YJ, Gray D, Berry D, Gao JQ, Li MH, Li JF, Viehland D (2011) An extremely low equivalent magnetic noise magnetoelectric sensor. Adv Mater 23(35):4111

Wang YJ, Gray D, Berry D, Li MH, Gao JQ, Li JF, Viehland D (2012) Influence of interfacial bonding condition on magnetoelectric properties in piezofiber/Metglas heterostructures. J Alloys Compd 513:242–244

Wang YJ, Gao JQ, Li MH, Shen Y, Hasanyan D, Li JF, Viehland D (2014) A review on equivalent magnetic noise of magnetoelectric laminate sensors. Philos Trans R Soc a-Math Phys Eng Sci 372: 20120455

Wilkie WK, Bryant RG, High JW, Fox RL, Hellbaum RF, Jalink A, Little BD, Mirick PH (2000) Low-cost piezocomposite actuator for structural control applications. pp. 323–334. in smart structures and material 2000: industrial and commercial applications of smart structures technologies, Vol. 3991. Proceedings of the Society of Photo-Optical Instrumentation Engineers (Spie). Edited by J. H. Jacobs, 2000.

Xing Z, Dong S, Zhai J, Yan L, Li J, Viehland D (2006) Resonant bending mode of Terfenol-D/steel/Pb(Zr, Ti)O3 magnetoelectric laminate composites. Appl Phys Lett 89: 112911

Yan Y, Cho K-H, Priya S (2012a) Piezoelectric properties and temperature stability of Mn-doped Pb(Mg1/3Nb2/3)-PbZrO3-PbTiO3 textured ceramics. Appl Phys Lett 100: 132908

Yan YK, Wang YU, Priya S (2012b) Electromechanical behavior of 001-textured Pb(Mg1/3Nb2/3)O-3-PbTiO3 ceramics. Appl Phys Lett 100: 192905

Yan Y, Zhou Y, Gupta S, Priya S (2013a) Fatigue mechanism of textured Pb(Mg1/3Nb2/3)O-3-PbTiO3 ceramics. Appl Phys Lett 103: 082906

Yan YK, Cho KH, Maurya D, Kumar A, Kalinin S, Khachaturyan A, Priya S (2013b) Giant energy density in 001 -textured Pb(Mg1/3Nb2/3)O-3-PbZrO3-PbTiO3 piezoelectric ceramics. Appl Phys Lett 102: 042903

Yan YK, Zhou Y, Priya S (2013c) Giant self-biased magnetoelectric coupling in co-fired textured layered composites. Appl Phys Lett 102: 052907

Yan Y, Zhou Y, Priya S (2014) Enhanced electromechanical coupling in Pb(Mg1/3Nb2/3)O-3-PbTiO3 < 001 > (C) radially textured cylinders. Appl Phys Lett 104: 012910

Zhai J, Xing Z, Dong S, Li J, Viehland D (2008) Thermal noise cancellation in symmetric magnetoelectric bimorph laminates. Appl Phys Lett 93: 072906

Zhang SJ, Li F (2012) High performance ferroelectric relaxor-PbTiO3 single crystals: status and perspective. J Appl Phys 111: 031301

Zhang SJ, Xia R, Shrout TR (2007) Lead-free piezoelectric ceramics vs. PZT? J Electroceram 19(4):251–257

Zheng H, Wang J, Mohaddes-Ardabili L, Wuttig M, Salamanca-Riba L, Schlom DG, Ramesh R (2004) Three-dimensional heteroepitaxy in self-assembled BaTiO3-CoFe2O4 nanostructures. Appl Phys Lett 85(11):2035–2037

Zhou Y, Apo DJ, Priya S (2013) Dual-phase self-biased magnetoelectric energy harvester. Appl Phys Lett 103: 192909

Chapter 5
Clay/Polymer Nanocomposites: Processing, Properties, and Applications

Vasanth Chakravarthy Shunmugasamy, Chongchen Xiang and Nikhil Gupta

Abstract Clay/polymer nanocomposites have been extensively studied in recent years. The present state of the art for these materials is summarized in this chapter. The development of fabrication methods for these composites is very challenging because the platelets of nanoclay exist in the form of clusters, which need to be dispersed in the matrix resin in order to obtain any benefit from the high surface area of nanoclay. Incorporation of only a small weight fraction (1–5 %) of nanoclay in polymers provides significant benefits in the properties of composites. Several tensile, flexural, and thermal properties are found to increase by 30–40 % due to the presence of nanoclay in the composite compared to the properties of the neat resin. Entrapment of air porosity at higher nanoclay content can lead to reversal of the trends and can actually reduce the mechanical properties of nanocomposites. Theoretical models have been developed to estimate the properties of nanocomposites by accounting for the microstructure that may include clustered, intercalated, or exfoliated nanoclay. The benefit in mechanical properties obtained from incorporating nanoclay is much greater than that can be achieved with microscale reinforcement at the same loading levels. The current applications of clay/epoxy nanocomposites are in the area of automotive moldings and fire retardant coatings.

Keywords nanoclay · nanocomposite · mechanical properties · nanocomposite processing · nanocomposite applications · molecular simulation · glass transition temperature · exfoliation

5.1 Introduction

Polymer matrix nanocomposites have been extensively studied in recent years with an aim to achieve properties that are beyond the realm of microcomposites. Nanocomposites have shown remarkable improvement in mechanical, electrical,

N. Gupta (✉) · V. C. Shunmugasamy · C. Xiang
Composite Materials and Mechanics Laboratory,
Mechanical and Aerospace Engineering Department,
New York University Polytechnic School of Engineering, Brooklyn, NY 11201, USA
e-mail: ngupta@nyu.edu

C.-S. Kim et al. (eds.), *Hybrid and Hierarchical Composite Materials*,
DOI 10.1007/978-3-319-12868-9_5

and thermal properties compared to the bulk materials and composites reinforced with microsized particles and fibers. Although it is difficult to assign a strict definition to what is a nanoscale reinforcement, typically materials that have at least one dimension smaller than 100 nm are classified into this category. Some of the dimensions of nanoscale reinforcements may actually be very large. For example, carbon nanotubes and nanofibers have been fabricated in several meters length. Nanoclay and graphene have a platelet structure where only the platelet thickness is in a few nanometers but the other two dimensions are considerably larger. Similarities in the structure of graphene and clay in terms of platelet structures can help in understanding the structural effects of such reinforcement. One-dimensional reinforcements such as nanotubes and nanofibers are also extensively studied for reinforcing capabilities. One of the main areas of interest in carbon-based nanomaterial is that their incorporation can make thermally and electrically conductive composites using nonconductive matrix resins such as epoxy or polypropylene resins.

Clay, a natural mineral, has been used in fabricating nanocomposites. Clay is abundantly available in the nature at low cost, which makes it attractive in making high performance composite materials. This chapter is focused on nanoclay-reinforced polymer matrix composites, with an aim of understanding the structure of nanoclay and understanding the effect of nanoclay reinforcement on mechanical properties of clay/epoxy nanocomposites. One of the considerations in fabrication of clay-reinforced nanocomposites is that they can also be reinforced with microfibers or particles to create hybrid composite materials that have two different reinforcements at two different length scales. Such hybrid composites can provide high mechanical properties in addition to functional properties such as thermal or electrical conductivity. Understanding the properties of nanocomposites and developing structure-property correlations are key factors in development of multiscale-reinforced hybrid composites.

Some of the polymers commonly used as matrix for clay/polymer nanocomposites include epoxy, polystyrene, polypropylene, polyurethane, nylon, and polyethylene (Pavlidou and Papaspyrides 2008; Tjong 2006; Alexandre and Dubois 2000). Although various polymers have been used as the matrix to prepare polymer matrix nanocomposites, the discussion in this chapter emphasizes on epoxy matrix nanocomposites, because of the large body of literature that is available on clay/epoxy nanocomposites. The mechanisms of reinforcement attributed to nanoclay under different loading conditions are discussed.

5.2 Applications of Nanoclay Composites

Research and development of nanoclay-reinforced composites in industry and research labs/academia have resulted in several composites that have already found industrial applications. Some common applications of nanoclay-reinforced polymer matrix composites are summarized in Table 5.1. A comprehensive listing of early applications of clay/epoxy nanocomposites is presented in a review article (Gao

Table 5.1 Examples of applications of nanoclay-reinforced polymer matrix composites

Product	Usage
Mitsubishi gas chemical—nylon nanocomposite (Europe) (Kang 2010)	Alcoholic beverage containers
Honeywell—nanoclay composites (Korea) (Kang 2010)	Alcoholic beverage containers
LG Chem. Ltd.—high-barrier monolayer blow-molded container of HDPE with 3–5 wt% nanoclay (Korea) (Kang 2010)	Chemical (toluene) storage
InMat Inc.—elastomeric nanocomposite barrier coating (Kang 2010)	Wilson Double Core tennis ball
Nanoclay–nylon coatings (Chinnamuthu and Murugesaboopathi 2009)	Beverage glass bottles
Nanocor polymer composites (Lan)	Military food packaging
Nanoclay–polymer coatings (Fischer 2003)	Temperature and ultraviolet ray protection coating
Starch–clay nanocomposite (Fischer 2003)	Hydrophobic starch (biocompatible and biodegradable) membrane

HDPE high-density polyethylene

2004). The earliest known application (year 1991) is attributed to Toyota's use of clay/nylon-6 nanocomposite in drive belt covers (Edser 2002). Similar applications were then developed by other automakers. Nylon matrix nanoclay composites are used in food packaging and the automotive industry. The diffusion rate of gases such as carbon dioxide and oxygen in nanocomposites can be considerably lower than the matrix resin because the nanoclay platelets act as barriers (Kang 2010). Applications such as beverage and liquid containers that require low diffusion rates of gases to maintain the quality of the contents for longer time benefit from such properties. Enhanced environmental performance of nanoclay-reinforced composites makes them suitable for the packaging industry (Lan et al. 2001).

Nanoclay-reinforced polymer matrix composites are also used in the automotive sector as parts for cars and trucks. Nanoclay-reinforced polypropylene is used to make car seat backs on the Acura TL (2004). Nanoclay-reinforced thermoplastic olefin composites are used to make components such as step-assists for the Chevrolet Astro van (2002), body side molding for the Chevrolet Impala (2004), and the cargo bed on the General Motors Hummer (2005) (Kang 2010). Superior wear and impact resistance of clay reinforced polymers compared to the neat resin is useful in such applications.

5.2.1 Flame Retardant Materials

Flame retardancy is another application where nanoclay-reinforced polymer matrix composites are finding utilization (Kiliaris and Papaspyrides 2010). Studies have characterized the flame retardancy of nanoclay epoxy composites with added flame retarding agents such as organophosphorous and aluminum trihydrate (Kiliaris and Papaspyrides 2010). The addition of the flame retardants to the polymer matrix usually accompanies a decrease in the mechanical property of the composite (Markarian 2005). A portion of the flame retardant can be replaced with nanoclay, which will also act as a flame retardant while retaining the mechanical property of the composite. The total heat release of 122 MJ/m^2 for 10 wt% of aluminum trihydrate/epoxy composite decreased to 115 MJ/m^2 with the addition of 5 wt% of nanoclay and 5 wt% of aluminum trihydrate to the epoxy matrix (Schartel et al. 2006).

5.2.2 Drug Delivery System

Recently, nanoclay composites have attracted attention for their possible uses in drug delivery systems. Nanoclay composites have been studied for their ability to control drug delivery in order to improve effectiveness and reduce side effects. Montmorillonite (MMT) clay has large specific surface area, exhibits good adsorption characteristics and possess good cation exchange capacity (CEC), leading to good drug-carrying capability. Several recent articles are available on the utilization of nanoclay as an efficient drug delivery tool (Rodrigues et al. 2013; Suresh et al. 2010; Patel et al. 2006; Ha and Xanthos 2011). Nanoclay has been used to supply propranolol hydrochloride (PPN), an antihypertension drug (Seema and Datta 2013). MMT dispersed in a biodegradable polymer matrix was used as sustained release carrier for PPN, which allowed for more effective release of the drug into the intestines.

Polyurethane-based MMT clay-reinforced nanocomposites have been used for the controlled release of dexamethasone acetate (ACT), which is an anti-inflammatory medication to treat or prevent allergic reactions (Da Silva et al. 2009). The chemical structure of the drug remained unchanged and the ACT did not promote any modification in the morphology of the polymeric matrix. The drug was shown to be completely dissolved within the nanocomposites. Based on in vivo study, the nanocomposites exhibited a high degree of biocompatibility because they induced mild inflammatory response in the tissue adjacent to the materials, which was completely resolved in a short period of time (14 days).

5.3 Components

5.3.1 Clay

Clay comprises layered silicates stacked one on top of the other held together by van der Waals forces (Nguyen and Baird 2006). Clay can be classified into four different groups based on the arrangement of the layered structure.

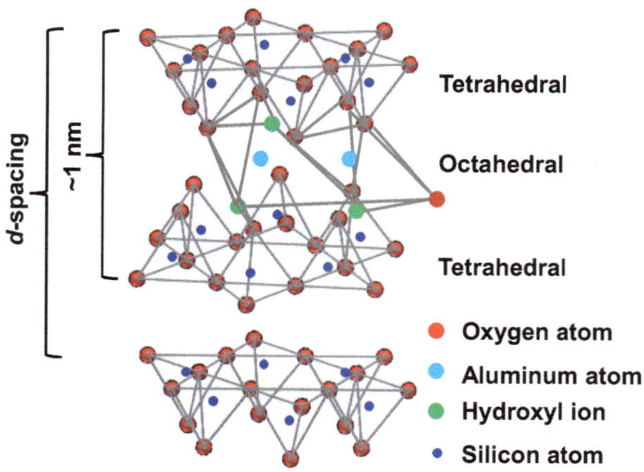

Fig. 5.1 Structure of nanoclay (montmorillonite)

5.3.1.1 Group 1:1—Kaolin Group

This group is distinguished by the arrangement of one octahedral alumina layer ($Al_2(OH)_4$) bonded to one tetrahedral silica layer (Si_2O_5). This group is dimorphic (comprising two sheet minerals) and is represented by $Al_2Si_2O_5(OH)_4$ (Uddin 2008). Some examples of this group are kaolinite, nacrite, and halloysite. These clays have the same chemical representation while having different structure and are referred as polymorphs.

5.3.1.2 Group 2:1—Smectite Group

This group is characterized by the arrangement of two tetrahedral silica layers sandwiching an octahedral alumina layer (Uddin 2008). Examples of this group include MMT, mentioned in Sect. 5.2, talc, laponite, and saponite. This group of clay can be expanded (Farris 2014), which makes it suitable as fillers for composites.

The MMT is the most commonly used clay mineral that is used as reinforcement in polymer matrix composites (Azeez et al. 2013; Sinha Ray and Okamoto 2003; LeBaron et al. 1999). MMT clay is chemically represented as $M_x(Al_{4-x}Mg_x)Si_8O_{20}(OH)_4$, where M represents a monovalent cation and x the degree of isomorphous substitution which can range between 0.5 and 1.3 (Sinha Ray and Okamoto 2003). MMT clay belongs to the smectite group of clays and has a trimorphic structure comprising an edge-shared octahedral alumina layer enclosed between two tetrahedral layers of silica (Azeez et al. 2013), as shown in Fig. 5.1. This arrangement of the MMT clay platelet structure is about 1 nm thick. The structure of the clay extends two dimensionally in a plane and is stacked one on top of the other, along the third axis. The *d*-spacing of the nanoclay is defined as the sum of platelet thicknesses along with the distance between the adjacent layers of the platelets. The distance between the MMT nanoclay platelets can be extended considerably and this can be utilized to impart a reinforcing effect, when used in composites.

5.3.1.3 Group 2:1—Illite Group

This group is also characterized by the arrangement of two tetrahedral silica layers sandwiching an octahedral alumina layer (Uddin 2008). An example of this group is mineral illite. This clay group contains potassium, calcium, or magnesium interlayer cations, which prevent the ingress of water molecules, thereby preventing the clays from expanding (Farris 2014).

5.3.1.4 Group 2:2—Chlorite Group

This group comprises an alternating arrangement of tetrahedral silica and octahedral alumina layers (Azeez et al. 2013). Examples of this group are chlorite, amesite, and chamosite. The general composition of this group can be expressed as $(Mg, Fe, Al)_6[(SiAl)_4O_{10}](OH)_8$. The 2:2 notation signifies that two layers of Si-O are followed by two layers of Al-O or (Mg, Fe)-O.

5.3.1.5 Surface Modification

In general, the clays in their natural form are hydrophilic in nature and are incompatible with most polymers. Surface modification and functionalization of clay helps in improving the compatibility with polymers and helps in dispersing the clay platelets in polymer matrix composites (Tcherbi-Narteh et al. 2013). The clays are treated with ammonium or phosphonium ions which replace the inorganic cations present between two adjacent platelets (Alexandre and Dubois 2000; Azeez et al. 2013). The common commercially available organically modified nanoclays are Nanomer I.28E and Cloisite 30B, which are modified by octadecyl trimethyl ammonium and methyl, tallow, bis-2-hydroxyethyl, and quaternary ammonium, respectively (Yasmin et al. 2006). The clay surface becomes hydrophobic after subjecting them to these organic modifications, which makes them compatible with epoxy resins. A review article is available for more details on organic modification of nanoclay (de Paiva et al. 2008).

5.3.2 Epoxy

Epoxy resin is a class of thermosetting polymer that is characterized by the presence of two or more oxirane rings (epoxide group) in its molecular structure (Ratna 2009). Thermosetting resins require curing of the polymer. The curing of the polymer is catalyzed using a hardener (amines, thiols, and alcohols) (Ratna 2009), which results in the formation of cross-linking secondary bonds. The type of hardener used depends on the curing temperature, required physical and chemical properties, and processing methodology utilized. Upon curing, the epoxy resins do not produce volatile products, despite the presence of a volatile solvent (Bhatnagar and Salomone 1996).

One of the most commonly used thermosetting resins is the diglycidyl ether of bisphenol A (DGEBA). DGEBA is commonly synthesized using bisphenol A and epichlorohydrin. The cross-linked nature of the resin helps in attaining high strength, good high-temperature properties, high corrosion tolerance, and good insulation properties (Ratna 2009). Epoxies are commonly used as the matrix for structural composites and as bonding agents and adhesives, due to their high strength. Resistance to environmental degradation helps epoxies to be used in marine environment as paint binder for naval structures. The good electrical insulation enables their use in the electronics industry for insulation in motors, generators, and transformers.

5.3.3 Clay/Epoxy Morphology

The micron-sized aggregates in the clay can have high aspect ratio of 10:1. The aggregates contain a stacked sequence of nanosized platelets (each ~ 1 nm) and are held together by weak van der Waals forces. The platelets have high stiffness that is reported to be about 170 GPa (Luo and Daniel 2003). The dispersion of the platelets from the aggregated micron-sized particles of the nanoclay helps in imparting high stiffness to the polymer matrix composites. The morphology of epoxy matrix nanoclay composites can be classified into three types:

1. Aggregated microcomposite—the nanoclay aggregates exist as such in the matrix with no penetration by the polymer, as illustrated in Fig. 5.2a
2. Intercalated nanocomposite—the polymer penetrates into the clay structure and increases the interplatelet spacing (Giannelis 1996), as illustrated in Fig. 5.2b and
3. Exfoliated nanocomposite—the polymer penetrates into the clay structure and the individual nanoclay platelets are separated and distributed in the matrix resin, as shown in Fig. 5.2c.

The *d*-spacing is defined as the sum of the platelet thickness and the interlayer distance between the platelets. With the intercalation and exfoliation of the clay, the *d*-spacing increases progressively. Analyzing Tables 5.2, 5.3, and 5.4 for the *d*-spacing values reported for intercalated composites, it is observed that the *d*-spacing values are 1–4 nm for intercalation of the nanoclay, and in the case of exfoliation, the *d*-spacing values are >7 nm. Wide-angle X-ray diffraction (WAXD) is commonly used to observe the intercalation and exfoliation of the clay in the clay/polymer nanocomposites. Transmission electron microscope (TEM) images of nanoclay-reinforced polypropylene are shown in Fig. 5.3 (Nguyen and Baird 2007). Supercritical carbon dioxide was used as processing aid in fabricating this sample.

Complete exfoliation and homogenous dispersion of the nanoclay is hard to achieve, which is a major limitation in synthesizing clay-reinforced nanocomposites. The homogenous distribution of the nanoclay in epoxy has been analyzed by observing the microstructure of the composites using scanning electron microscope (SEM) and TEM micrograph images. The interaction between the nanoclay and

Table 5.2 Studies on nanoclay/epoxy composites utilizing mechanical mixing (MM) method to prepare the composite

Reference	Composite fabrication technique	Morphology
Chan et al. (2011b)	Vacuum + spinning (MM)	Nanoclay clusters (<200 nm)
Ferreira et al. (2013)	MM	Intercalated + exfoliated
Guevara- Morales and Taylor (2014)	MM	Cloisite: intercalated (d=3.25 nm) Nanomer: intercalated (d=3.45 nm)
(Ho et al. 2006)	MM	Nanoclay acted as nanodispersion in the composite (no peak shift observed in XRD spectra)
Huskić et al. (2013)	MM	(d=1.5–3.5 nm)
Jumahat et al. (2012)	MM	Intercalated (d=2.4–8 nm)
Kusmono and Mohd Ishak (2013)	MM	Clay: (d=2.49 nm) 3 wt% clay: exfoliated 6 wt% clay: intercalated
Lu et al. (2004)	MM, HEHM, HEHM + ball milling	HEHM + ball milling: Exfoliated (d=>8.8 nm)
Park and Jana (2003)	Magnetic mixing (MM)	Exfoliation observed for the following combinations: (curing agent + clay type + curing temp. (°C) + curing time (h)) DDS+1 and 3+200+2 D230+1 and 3+100+3 D230+1 and 3+125+3 PACM+1 and 3+100+3 PACM+1 and 3+125+3 All the combinations for 5 wt% of clay in Epon 828 matrix
Qi et al. (2006)	MM	*MMT-Na+* (d=1.2 nm) Epoxy/(2–10 wt.%) MMT-Na^+ - Tactoids *MMT-30b* (d=1.8 nm) Epoxy/(2 wt.%) MMT-30B: (d=3.3 nm) Epoxy/(5 wt.%) MMT-30B: (d=3.2 nm): intercalated Epoxy/(10 wt.%) MMT-30B: (d=3.4 nm) *MMT-I30E* (d=2.2 nm) Epoxy/(2&10wt.%) MMT-I30E: (no peak) Epoxy/(5 wt.%) MMT-I30E: (no peak)—exfoliated *MMT-CPC* (d=2 nm) Epoxy/(2wt.%) MMT-CPC: (d=5.6 nm) Epoxy/(5 wt.%) MMT-CPC: (d=2.9 nm): intercalated Epoxy/(10 wt.%) MMT-CPC: (d=2.9 nm)
Saber-Samandari et al. (2007)	Centrifugal mixing (MM)	Hand mixing: agglomeration Centrifugal mixing: intercalated + exfoliated
Shi et al. (1996)		3A-mont-1x: intercalated 3A-mont-3x: exfoliated

Table 5.2 (continued)

Reference	Composite fabrication technique	Morphology
Tcherbi-Narteh et al. (2013)	Magnetic stirring technique (MM)	*Nanomer* I.28E (*d*=0.79 nm) Epoxy/nanomer I.28E (*d*=1.32 nm): intercalated *Cloisite 10A* (*d*=1.19 nm) Epoxy/cloisite 10A (*d*=2.72 nm): intercalated *Cloisite 30B* (*d*=1.14 nm) Epoxy/Cloisite 30B (*d*=2.44 nm): intercalated
Wang et al. (2006)	MM	Exfoliated
Wang and Pinnavaia (1994)	Magnetically stirred (MM)	Exfoliated
Zaman et al. (2011)	MM	*Sodium clay* (*d*=1.2 nm) Epoxy/sodium clay: intercalated *eth-clay* (*d*=1.24 nm) Epoxy/eth-clay: intercalated (*d*=1.42 nm) *m27-clay* (*d*=1.65 nm) Epoxy/m27-clay: exfoliated (*d*=1.77 nm) *xtj-clay* (*d*=1.65 nm) Epoxy/xtj-clay: exfoliated (*d*=1.78 nm) *Sequence for exfoliation*: eth-clay m27-clay xtj-clay
Ku et al. (2013)	MM	Not reported
Ku and Trada (2013)	MM	Nanoclay cluster
Guevara-Morales and Taylor (2014)	MM	*C30B* (*d*=1.77 nm) 5 wt% C30B/epoxy: intercalated (*d*=3.25 nm) *I.28E* (*d*=2.26 nm) 5 wt% I.28E/epoxy: intercalated (*d*=3.45 nm)
Dai et al. (2005)	MM	Intercalated (*d*=3.7 nm)
Huskić et al. (2013)	MM	*sMMT 0.3*: (*d*=1.49 nm) 5 wt% sMMT 0.3/Epoxy: (*d*=1.36 nm) *sMMT 0.5*: (*d*=1.63 nm) 5 wt% sMMT 0.5/epoxy: (*d*=1.50 nm) *sMMT 1.0*: (*d*=2.00 nm) 5 wt% sMMT 1.0/epoxy: (*d*=1.92 nm) *sMMT 1.5*: (*d*=2.10 nm) 5 wt% sMMT 1.5/epoxy: (*d*=2.00 nm) *Nanofil 2*: (*d*=1.96 nm) 5 wt% Nanofil 2/epoxy: (*d*=2.94 nm) *Nanofil 8*: (*d*=3.52 nm) 5 wt% Nanofil 8/epoxy: (*d*=3.57 nm)
Lu and Nutt (2003)	MM	3 wt% Cloisite 30B/epoxy: intercalated (*d*=5.43 nm) 5 wt% Cloisite 30B/epoxy: intercalated (*d*=5.60 nm) 7 wt% Cloisite 30B/epoxy: intercalated (*d*=4.85 nm) 10 wt% Cloisite 30B/epoxy: intercalated (*d*=4.57 nm)

XRD X-ray diffraction, *MM* mechanical mixing, *MMT* montmorillonite, *temp.* temperature, *HEHM* high-speed emulsifying and homogenizing mixer *sMMT 0.3* silylated MMT "0.3" represents amount of aminosilane (g)

Table 5.3 Studies on nanoclay/epoxy composites utilizing mechanical mixing (MM) and ultrasonication (US) methods to prepare the composite

Reference	Composite fabrication technique	Morphology
Lam and Lau (2006)	MM + US	Not reported
Chau (2012)	MM + US	Not reported
Kaya et al. (2008)	MM + US	*MMT*: (d=1.43 nm) *OMMT*: (d=1.81 nm) Clay/epoxy nanocomposite: intercalated
Liu et al. (2004)	MM + US	*Cloisite 93A*: (d=2.58 nm) 1 wt% Cloisite 93A/epoxy: exfoliated (absence of basal reflection corresponding to (001) plane) 2 and 4 wt% Cloisite 93A/epoxy: intercalated (d=3.78 nm)
Sodeifian et al. (2012)	MM + US	*Cloisite 30B*: 1.85 nm Cloisite 30B/epoxy: premixed: 2.66 nm Cloisite 30B/epoxy: premixed + sonication: 3.95 nm
Basara et al. (2005)	MM + US	*CloisiteNa*$^+$ (d=1.17 nm) 3 wt% CloisiteNa$^+$/epoxy: intercalated (d=1.54 nm) *Cloisite30B* (d=1.84 nm) 3 wt% Cloisite30B/epoxy: intercalated + exfoliated (d=3.82 nm)
Silan et al. (2012)	MM + US (slurry compounding process)	Exfoliated
Wang et al. (2005b)	MM + US (slurry compounding process)	Clay—(d=1.21 nm) Clay/epoxy: Exfoliated (d=1.22 nm)
Wang and Qin (2007)	MM + US	Exfoliated (absence of basal reflection corresponding to (001) plane)
Wang et al. (2005a)	MM + US	Epoxy/93A2.5: intercalated Epoxy/S-clays: exfoliated

Table 5.4 Studies on nanoclay/epoxy composites utilizing shear mixing (SM) and ultrasonication (US) methods to prepare the composite

Reference	Composite fabrication technique	Morphology
Yasmin et al. (2006)	Three-roll milling (SM)	Nanomer/epoxy: intercalated (d=3.6 nm) Cloisite/epoxy: exfoliated (d=~8 nm)
Yasmin et al. (2003)	Three-roll milling (SM)	Nanomer/epoxy: intercalated (d=3.5 nm) Cloisite/epoxy: exfoliated+intercalated (d=~7 nm)
Chen et al. (2008b)	High and low SM + US	Low SM: intercalated (larger aggregates) (d=15 nm) High SM: intercalated (smaller aggregates) (d=15 nm) High SM + US: exfoliated (d=32 nm)
Chen and Tolle (2004)	High SM + US	Exfoliated (d=~32 nm)
Di Gianni et al. (2008)	US	Intercalated + exfoliated (d=~2.95 nm)

Table 5.4 (continued)

Reference	Composite fabrication technique	Morphology
Lam et al. (2005)	US	No change
Sarathi et al. (2007)	SM	*Organophillic* MMT clay: (*d*=1.7 nm) Epoxy + (1,3 & 5 wt.% clay): exfoliated Epoxy + (10 wt.% clay): intercalated
Al-Qadhi et al. (2013)	SM + MM	*Nanomer I.30E*—(*d*=2.2 nm) Nanomer I.30E/epoxy—intercalated + exfoliated (*d* 7 nm)
Miyagawa et al. (2004b)	US	Cloisite 30B/uncured epoxy—intercalated (*d*=3.76 nm) 2.5 and 5 wt.% Cloisite 30B/epoxy—intercalated + exfoliated (absence of basal reflection orresponding to (001) plane) 7.5 wt.% Cloisite 30B/epoxy—intercalated (*d*=4.91 nm) 10 wt.% Cloisite 30B/epoxy—intercalated (*d*=4.53 nm)
Miyagawa et al. (2004a)	US	Same as above
Lingaiah et al. (2008)	SM	Exfoliated
Zunjarrao et al. (2006)	US and SM	Shear mixing: exfoliated (absence of basal reflection corresponding to (001) plane) Ultrasonication: intercalated
Koerner et al. (2006)	Process 1: SM Process 2: SM + US Process 3: SM + US + compounding	*I30.E*—(*d*=3.75 nm) 3 wt% I30.E/Epoxy: Process 1: poorly dispersed (*d*=12.5 nm) Process 2: poorly dispersed (*d*=12.5 nm) Process 3: poorly dispersed (*d*=22 nm) *Cloisite 30A*—(*d*=3.5 nm) 3 wt% Cloisite 30A/Epoxy: Process 1: intercalated, poorly dispersed (*d*=3.39 nm) Process 2: intercalated, poorly dispersed (*d*=3.39 nm) Process 3: exfoliated, well dispersed (*d*=3.3 nm)

SM shear mixing, *US* Ultrasoniaction

nanocomposite are shown in Fig. 5.4. The results from the FTIR spectrum can be analyzed as follows:

1. The range of 1100–1000 cm^{-1} can be assigned to Si–O–R based on the asymmetric Si–O–C stretching which shows the Si in nanoclay reacting with epoxy.
2. In the range of 1250–1200 cm^{-1}, the deformation of symmetric CH_3 occurs and denotes the presence of Si–CH_2–R.

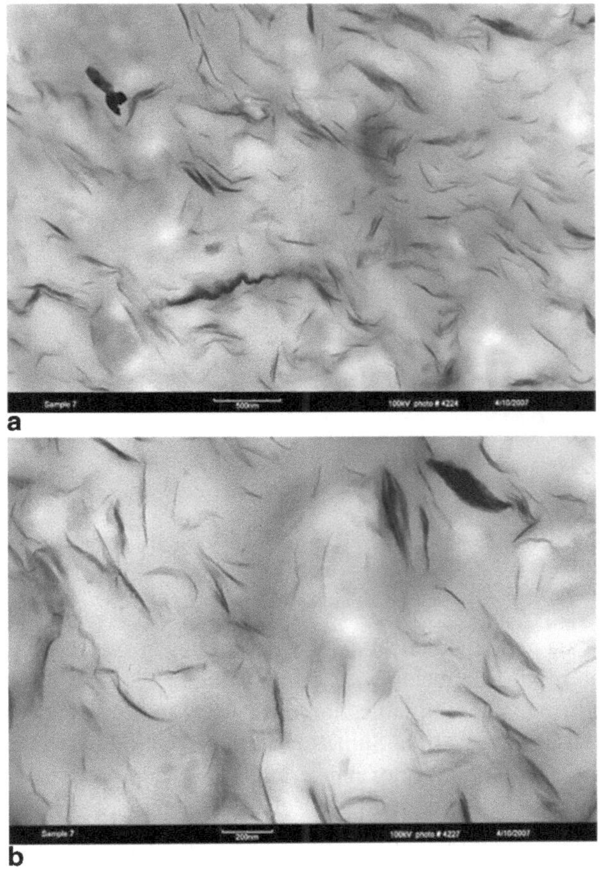

Fig. 5.3 TEM image of 6.6 wt% nanoclay-reinforced polypropylene matrix composite at two different magnifications (Nguyen and Baird 2007)

The observation from the XPS spectra on clay/epoxy nanocomposites, shown in Fig. 5.5, is explained in Table 5.5. It is observed that silicon present in the nanoclay reacts with carbon and oxygen in the epoxy and aluminum present in the nanoclay reacts with oxygen in epoxy polymer.

5.4 Fabrication of Clay/Epoxy Nanocomposites

Exfoliation presents enormous surface area on the clay platelets for the polymer molecules to attach. This effect leads to a rapid increase in the viscosity of the resin as the exfoliated nanoclay content increases and bonds with the resin. However, various studies have shown different degrees of intercalation and exfoliation. Several processing techniques are available for synthesizing clay/polymer nanocomposites such as melt intercalation (Azeez et al. 2013, Barick and Tripathy 2011), in situ polymerization (Sinha Ray and Okamoto 2003), and solution casting

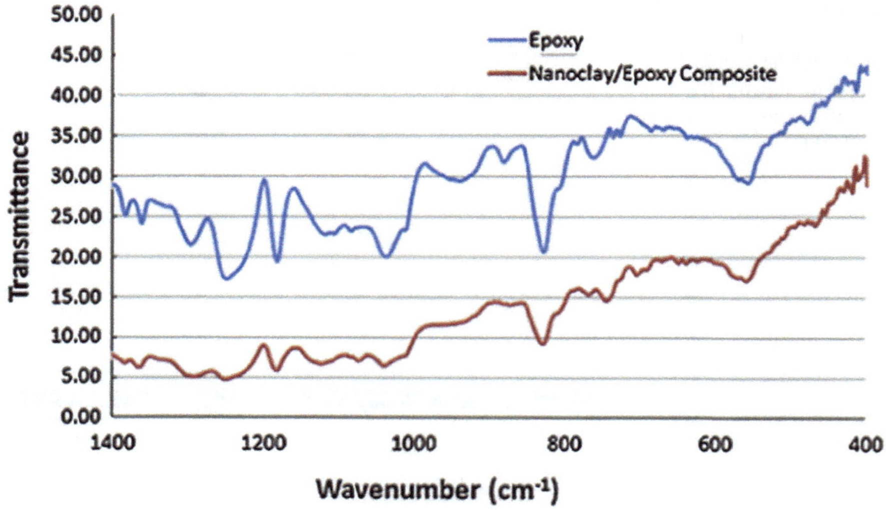

Fig. 5.4 FTIR spectroscopy for neat epoxy and nanoclay (5 wt%) epoxy composite. (Chan et al. 2011a)

(Vaia and Giannelis 1997). Most studies in the existing literature are concentrated on mixing nanoclay with epoxy using mechanical mixing (hand mixing, magnetic stirring, spinning table, and centrifugal mixing), shear mixing (three-roll milling), and ultrasonication. Sometimes a combination of two techniques is used to achieve the desired quality of nanocomposites. Table 5.2, 5.3, and 5.4 provide a comprehensive listing of studies on clay/epoxy nanocomposites with the method used for synthesis and the effect on the nanoclay dispersion. Most of the available methods for clay/polymer nanocomposites can be divided into two categories as discussed below.

5.4.1 Mechanical Processing Methods

Mechanical mixing methods have been extensively used for mixing nanoclay in polymers. Most of these methods rely on creating high shear forces in the resin to slide the nanoclay platelets with respect to each other in order to achieve exfoliation. Mechanical mixing using simple drill press setups using high shear impellers has been used for this purpose. The main advantages of this method are low cost of the setup and the possibility of processing large batches of nanocomposites. However, the mechanics of exfoliation is complex and complete exfoliation is often not achieved. This method is also simple to modify and heaters can be added to conduct the processing at higher temperatures because addition of nanoclay increases the resin viscosity and makes mixing more difficult as the nanoclay is continuously added. Mixing impeller design, mixing velocity, mixture viscosity, and mixing time are the important parameters in these methods.

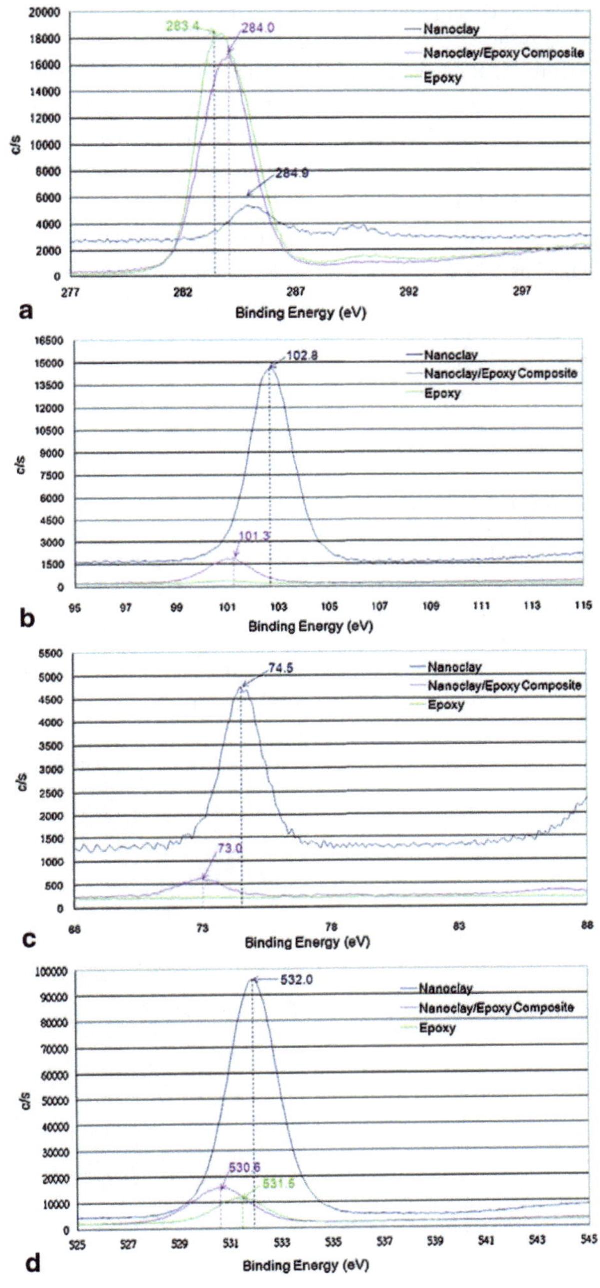

Fig. 5.5 X-ray photoelectron spectroscopy for **a** carbon 1s, **b** silicon 2p core, **c** Aluminum 2p core, and **d** oxygen 1s core of nanoclay, epoxy–nanoclay composite, and epoxy resin. The XPS is obtained from Chan et al. (2011a)

Three-roll milling is another widely used method to prepare intercalated and exfoliated nanoclay/polymer composites by taking advantage of the shear forces produced between the rolls. The three-roll mills consist of a feed roll, a center roll, and an apron roll. A schematic of the three-roll mill is shown in Fig. 5.6a and an ac-

Table 5.5 Description of the XPS spectra (Fig. 5.5) of nanoclay, nanoclay/epoxy composites, and epoxy resin

Figure	Binding energy (eV)	Remark (Chan et al. 2011a)
Fig. 5.5a	283.4	Represents the energy corresponding to Carbon 1s core level of the aromatic ring
	284.9	Represents carbon atoms (C–C, C=C and C–H)
Fig. 5.5b	101.3	Represents elemental SiC for epoxy: nanoclay composite
	102.8	Represents elemental Si–O for nanoclay
Fig. 5.5c	73.0	Represents peak centers of aluminum 2p state for epoxy-nanoclay composite—Al from nanoclay reacting with epoxy
	74.5	Represents peak centers of aluminum 2p state for nanoclay
Fig. 5.5d	531.5	Represents –COO–C_6H_4–
	530.6	Represents Si–O
	532.0	Represents C–O

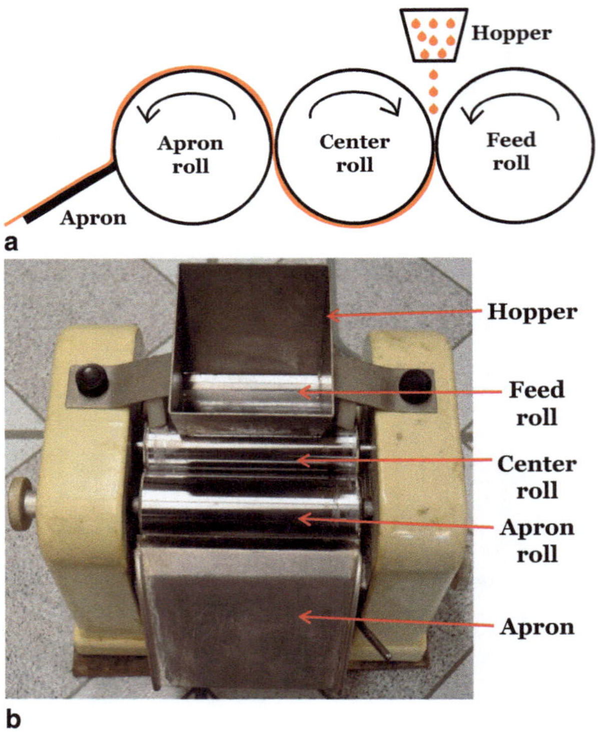

Fig. 5.6 **a** Schematic of three-roll mill (shear mixing) method for preparing nanoclay epoxy composite (Yasmin et al. 2006) and **b** the actual setup

tual setup is shown in Fig. 5.6b. The gap between the rolls is a few microns in order to generate high shear forces in the flowing nanoclay–resin mixture. The mixture is first prepared using hand mixing or mechanical mixing to wet the nanoclay clusters with the resin and then this mixture is fed through a hopper between the feed and center roll to create a thin layer around the rolls. The mixing happens due to the

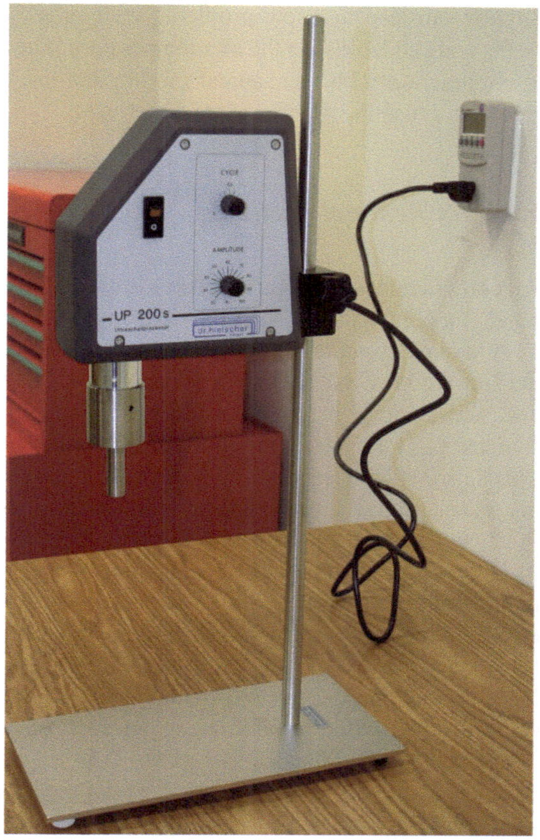

Fig. 5.7 Ultrasonic mixing equipment

shear forces generated by the differences in the speed between the rollers. The mixture is removed using the apron from the apron roll and is fed back to the hopper. This action is performed as a closed loop over a period of time to obtain intercalation or exfoliation in the nanoclay (Yasmin et al. 2006). The mechanism of shear mixing has been studied in detail in the published literature. Thermoplastic matrix composites have been widely processed using single- and twin-screw extruders and brabenders, which rely on shear forces to disperse particles.

Ultrasonic vibrations are widely used for dispersing nanoclay in polymers. An example of a commercially available ultrasonicator is shown in Fig. 5.7. The equipment uses an ultrasonic probe that is submerged in a mechanically mixed nanoclay-polymer mixture. The ultrasonic vibrations generated through the probe create intense localized mixing action leading to nanoclay exfoliation. Intense ultrasonic vibration leads to cavitation in the liquids, which can disperse a very high amount of energy locally when the bubbles collapse under extreme pressure. Cavitation can lead to very high localized velocities, pressures, and temperatures in the resin, which are responsible for dispersing nanoclay platelets. While cavitation is beneficial and effective for nanoclay dispersion, the localized nature of effects in this process can lead to a few problems such as self-polymerization of thermosetting resins and sudden rise in the mix temperature. The ultrasonication may be employed in conjunction with mechanical mixing and use of cold baths around the mixing chamber to use the process more effectively.

5.4.2 Chemical Methods

Several chemical methods are also available for dispersing nanoclay in resins and obtaining composites with exfoliated nanoclay.

A method commonly utilized to prepare exfoliated nanoclay epoxy composite is the slurry compounding process (Wang et al. 2005b; Silan et al. 2012). In this process, the nanoclay is first dispersed in deionized water and subjected to stirring and sonication. The suspension is then added to acetone to prepare a clay-acetone slurry. This slurry is blended with epoxy and the acetone is removed by evaporating it using a vacuum oven. The epoxy nanoclay mixture is cured to prepare the composite. A schematic representation of the slurry compounding process is shown in Fig. 5.8.

Other chemical methods work on modifying the surface of nanoclay with species that are chemically compatible with the resin. The bonding between the outer layers with the resin combined with rigorous mixing leads to separation of the surface layer from the rest of the cluster and the process is repeated until complete exfoliation is obtained. Chemical methods may work in conjunction with mechanical or ultrasonic mixing to facilitate penetration of the solvent or the reactive species inside the nanoclay cluster.

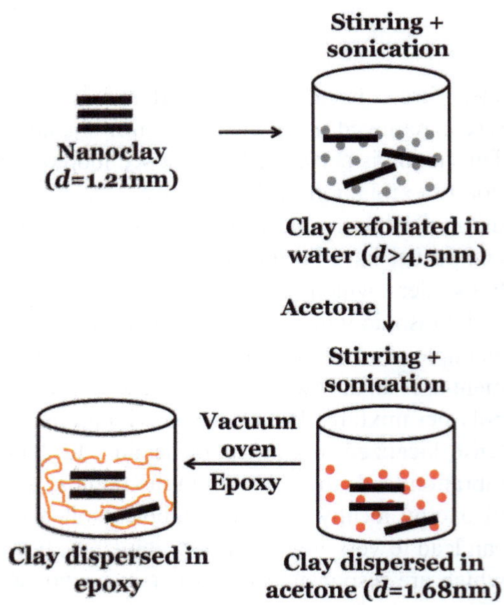

Fig. 5.8 Schematic representation of slurry compounding process to prepare nanoclay-epoxy composite. (Wang et al. 2005b)

5.5 Properties of Clay/Epoxy Nanocomposites

A vast body of literature is available on the properties of clay/epoxy nanocomposites. Representative data sets are compared here to understand the possible improvement in the properties of nanocomposites compared to the neat resin due to the presence of nanoclay.

5.5.1 Tensile Properties

Mechanical properties such as modulus and strength of nanoclay-reinforced epoxy matrix composites tested under tensile loading are shown in Fig. 5.9a and b, respectively. The tensile modulus and strength values are normalized with the matrix epoxy resin values and are presented in Fig. 5.10a and b, respectively. The addition of nanoclay content (wt%) in the epoxy matrix has resulted in improvement in the tensile modulus. Most of the studies have obtained up to 1.7 times increase in the tensile modulus in comparison to the neat epoxy, with up to 10 wt% addition of nanoclay. This can be attributed to the addition of high stiffness (170 GPa (Luo and Daniel 2003)) nanoclay platelets in the epoxy matrix and the reduced mobility of the polymer due to extensive bonding with the nanoclay platelets.

In comparison to the change in modulus, the tensile strength is not significantly affected by the addition of nanoclay. At high content (>5 wt%) of nanoclay, studies (Ho et al. 2006, Yasmin et al. 2003) have observed entrapment of air voids in the nanocomposite during the fabrication process, which leads to early failure, thereby causing lower strength. Intercalation and exfoliation of the nanoclay in the composite provide different stiffening mechanisms. A study using Nanomer I.28E and CloisiteB in epoxy nanocomposites observed 1.2 and 1.4 times improvement in the tensile modulus, in comparison to the neat epoxy resin (Yasmin et al. 2006). The Nanomer I.28E-epoxy composite showed an intercalated microstructure and Cloisite 30B-epoxy showed an exfoliated microstructure. The variation of tensile modulus based on the organic modification of the clay can be observed in a study by Basara et al. (2005). The study was performed with natural Na-MMT (Cloisite Na^{+}) and organically modified MMT (Cloisite 30B) using epoxy matrix. The study observed 10 and 17% improvement in the tensile modulus for Cloisite Na^{+} and Cloisite 30B composite, respectively, in comparison to the epoxy resin.

The trends seen in normalized strength and modulus are also very illustrative because they demonstrate the potential of increase in the properties of composite compared to the neat matrix. Figure 5.10a shows that the modulus of nearly all composites is higher than the matrix material. However, the strength of composites seems to peak at 3–5 wt% and then decline at further nanoclay addition. The density of nanoclay is higher than the polymer matrix; therefore, higher nanoclay contents lead to heavier composites with lower tensile strength. Entrapped voids, which may be stabilized by the nanoparticles, are considered one of the factors responsible for such trend. Difficulty in complete exfoliation of nanoclay is also a factor that leads to such trend.

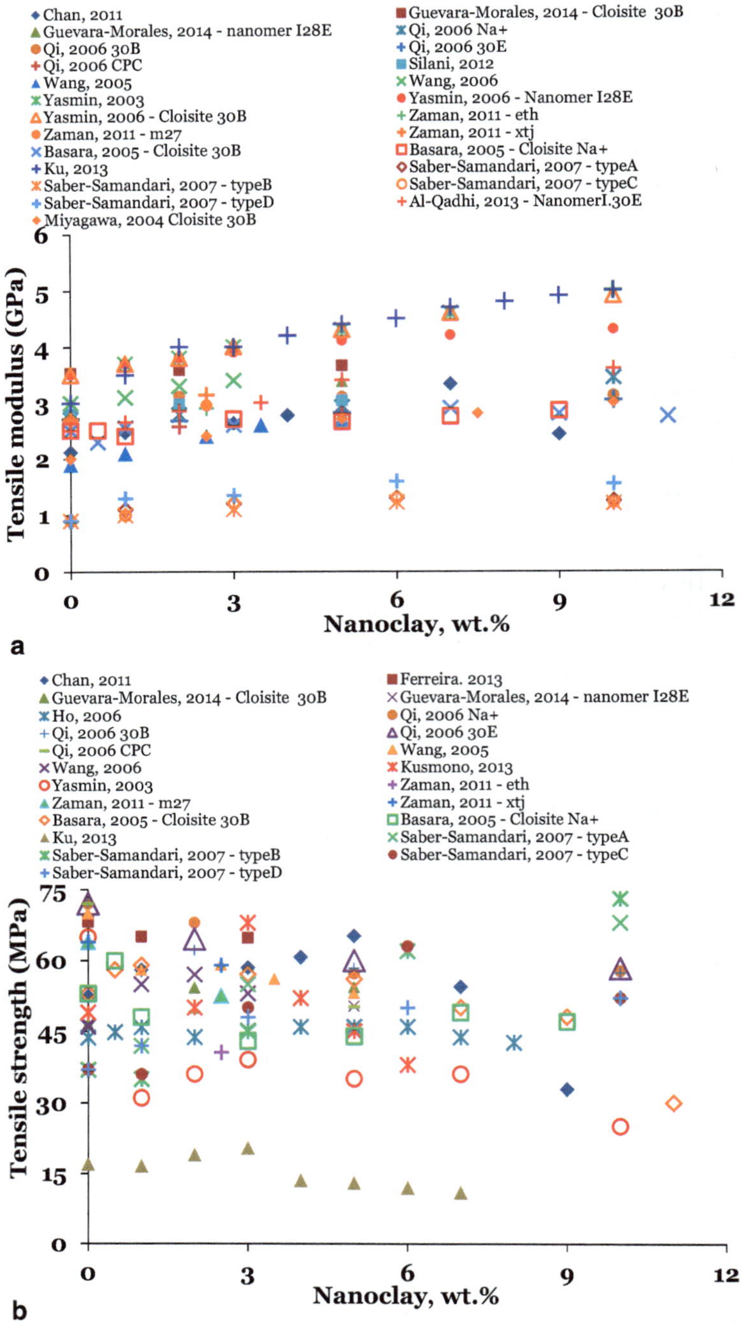

Fig. 5.9 Variation of tensile **a** modulus and **b** strength with respect to the nanoclay content in epoxy matrix composites. (Yasmin et al. 2003, 2006; Silan et al. 2012; Ho et al. 2006; Basara et al. 2005; Zaman et al. 2011; Miyagawa et al. 2004a; Chan et al. 2011b; Ferreira et al. 2013; Guevara-Morales and Taylor 2014; Kusmono and Mohd Ishak 2013; Qi et al. 2006; Saber-Samandari et al. 2007; Wang et al. 2006; Ku and Trada 2013; Wang et al. 2005a; Al-Qadhi et al. 2013)

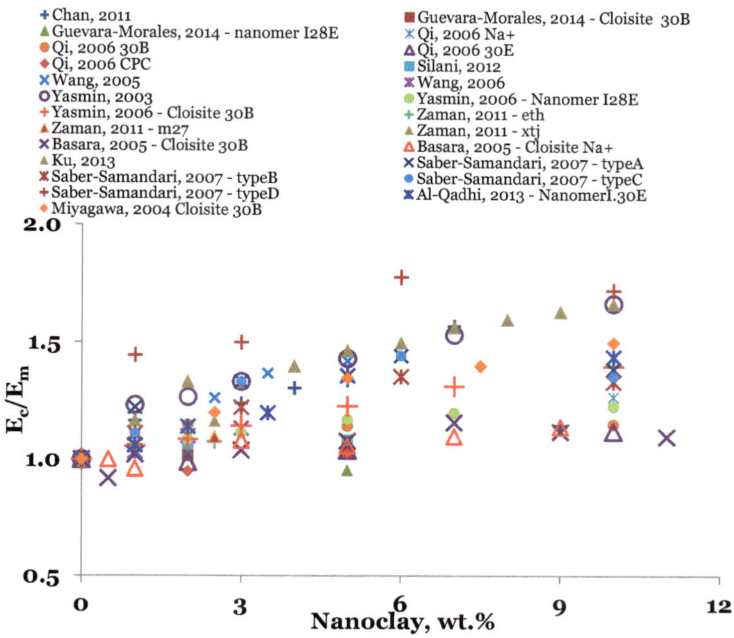

a

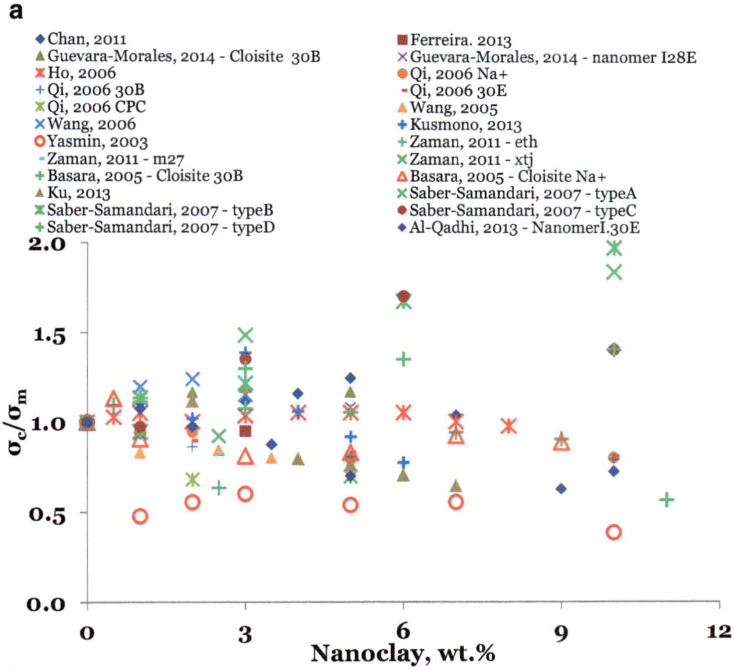

b

Fig. 5.10 Variation of normalized tensile **a** modulus and **b** strength (normalized with respect to the neat epoxy), with respect to nanoclay content in epoxy matrix composites. (Yasmin et al. 2003; 2006, Silan et al. 2012; Ho et al. 2006; Basara et al. 2005; Zaman et al. 2011; Miyagawa et al. 2004a; Chan et al. 2011b; Ferreira et al. 2013; Guevara-Morales and Taylor 2014; Kusmono and Mohd Ishak 2013; Qi et al. 2006; Saber-Samandari et al. 2007; Wang et al. 2006; Ku and Trada 2013; Wang et al. 2005a; Al-Qadhi et al. 2013)

5.5.2 Flexural Properties

The variation of the flexural modulus and strength with respect to the nanoclay weight fraction is presented in Fig. 5.11 for data extracted from various studies. The same data are replotted in Fig. 5.12 in the form of values normalized with those of the matrix resin. The flexural modulus of the nanocomposite has been observed to improve by up to 1.4 times in comparison to the neat resin. Similar to the tensile modulus, the flexural modulus is also dependent on the composite morphology. Zunjarrao et al. (2006) studied the effect of processing methodologies on the flexural modulus. They utilized shear mixing and sonication techniques to prepare the composites. The shear mixing technique resulted in exfoliation of the nanoclay. thereby stiffening the composite (Fig. 5.12). The exfoliated morphology is observed to improve the flexural strength of the nanoclay epoxy composite while intercalated or agglomerated morphology showed no improvement. A study by Lu et al. (2004) prepared composites using mechanical mixing and shear mixing (ball milling). The ball-milling preparation method yielded an exfoliated composite morphology which resulted in 8% improvement in the flexural strength.

The trends observed in flexural properties are qualitatively similar to those observed for tensile properties. Such similarity is observed because most of the data are available for composites containing brittle matrix materials such as epoxy resins. When tested under flexural conditions, such specimens tend to fail from the tensile side. Therefore, their tensile and flexural properties show similar trends. The flexural strength is also found to peak around 3 wt% nanoclay and goes down below the strength of neat matrix above 6 wt% nanoclay in the composite. These comparison graphs are useful in identifying the compositions that can provide the maximum benefit in mechanical properties.

5.5.3 Glass Transition Temperature

The glass transition temperature (T_g) for a thermosetting resin is defined as the temperature at which the polymer goes to the flow region and the modulus drops to a low value. Identifying the glass transition temperature is crucial as it governs the useful temperature limit of the polymer and its composites for practical applications. Exfoliation of nanoclay in the polymer matrix can provide stiffness to the matrix and restricts the motion of the polymer chains, which helps in improving T_g. The variation of T_g with respect to the nanoclay weight fraction is presented in Fig. 5.13a, and the T_g of nanocomposites normalized with the neat epoxy T_g is presented in Fig. 5.13b. A clear trend in the variation of T_g with respect to the clay content is not observed in the studies have characterized T_g. However, studies have shown that the exfoliation of the clay in epoxy matrix results in improvement in the T_g. Zaman et al. analyzed the surface-modified clay epoxy composite (Zaman et al. 2011). The clay modified using Jeffamine® XTJ502 in the epoxy matrix composite resulted in an exfoliated morphology which showed an increase of 6% in the T_g, in comparison to the neat resin.

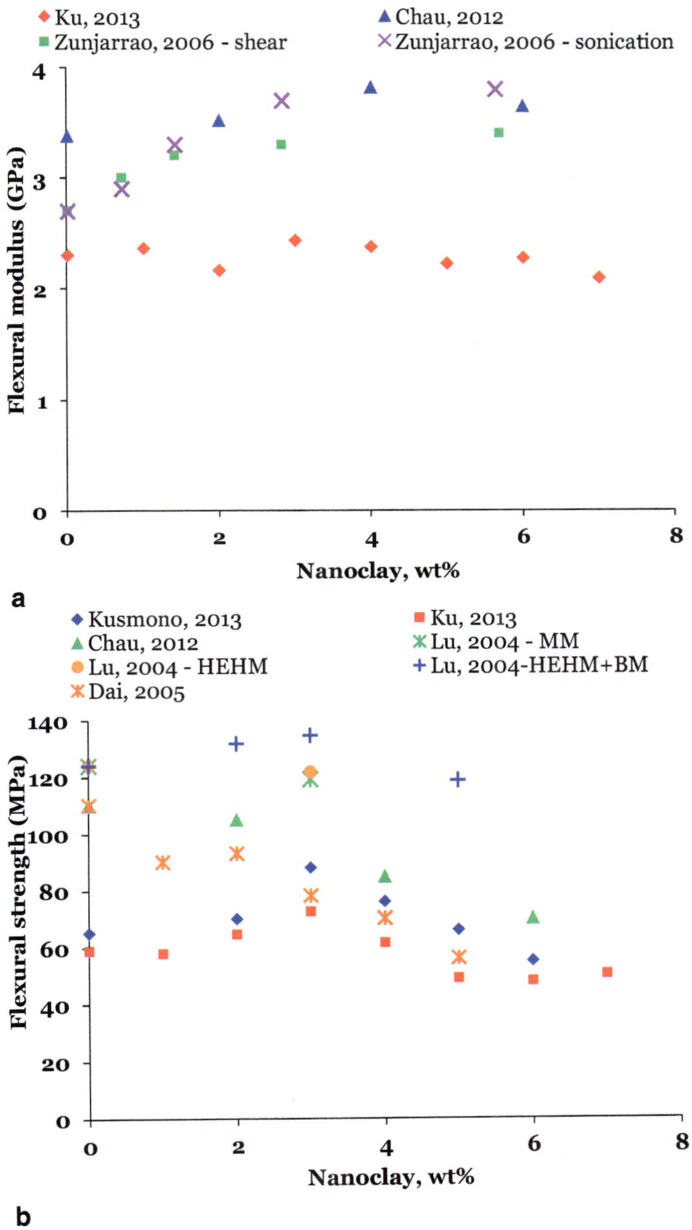

Fig. 5.11 Variation of flexural **a** modulus and **b** strength with respect to the nanoclay content in epoxy matrix composites (Zunjarrao et al. 2006; Lu et al. 2004; Kusmono and Mohd Ishak 2013; Ku and Trada 2013; Dai et al. 2005; Chau 2012)

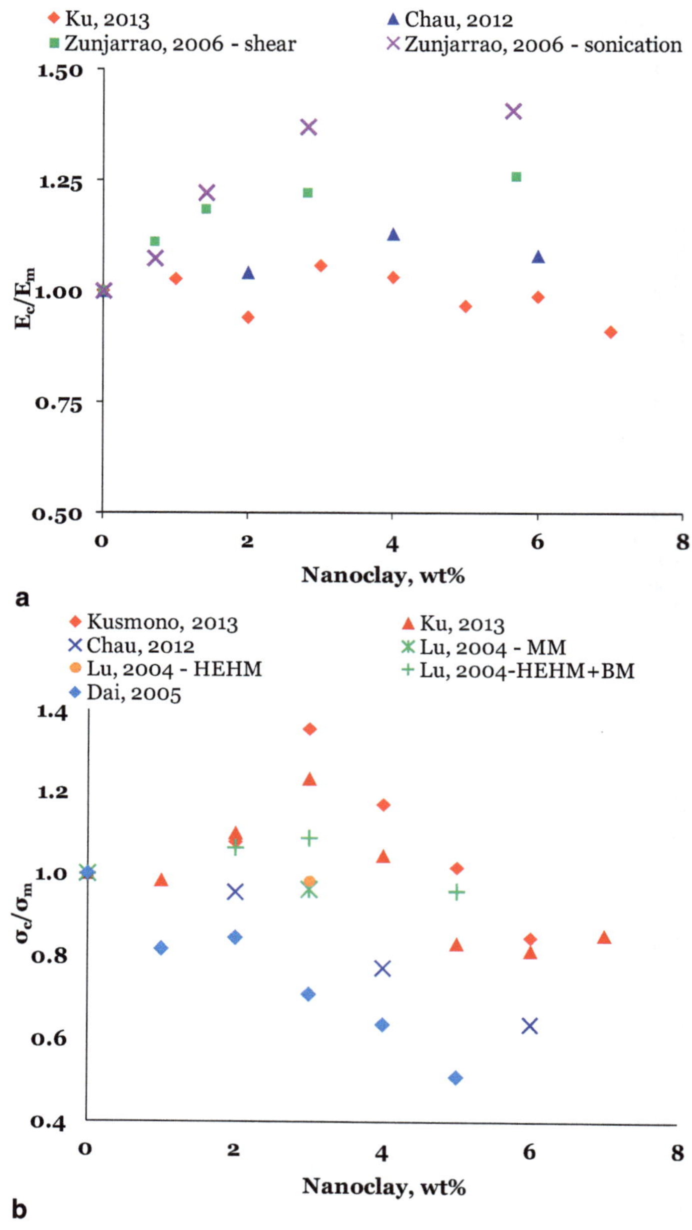

Fig. 5.12 Variation of normalized flexural **a** modulus and **b** strength (normalized with respect to the neat epoxy), with respect to nanoclay content in epoxy matrix composites (Zunjarrao et al. 2006; Lu et al. 2004; Kusmono and Mohd Ishak 2013; Ku et al. 2013; Dai et al. 2005; Chau 2012)

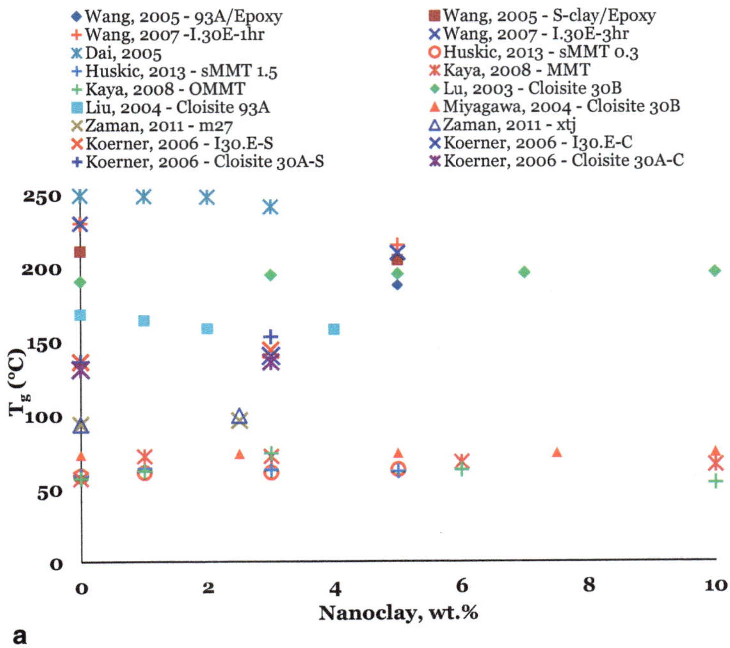

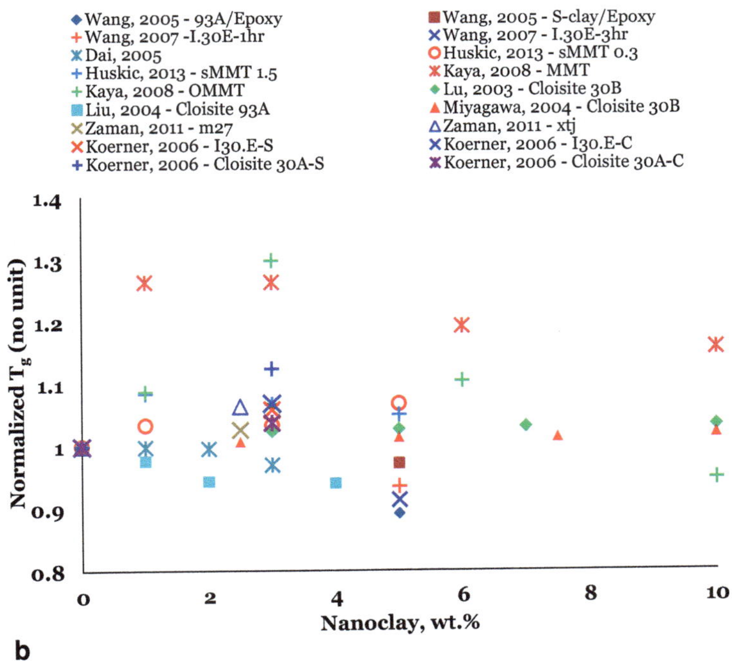

Fig. 5.13 Variation in nanoclay–epoxy composites' **a** glass transition temperature and **b** normalized T_g (normalized with the neat epoxy resin T_g) with respect to nanoclay weight fraction. S and C in the study by Koerner (2006) represents sonicated and compounded (Zaman et al. 2011; Miyagawa et al. 2004a; Huskić et al. 2013; Dai et al. 2005; Lu and Nutt 2003; Kaya et al. 2008; Liu et al. 2004; Wang and Qin 2007; Wang et al. 2005a; Koerner et al. 2006)

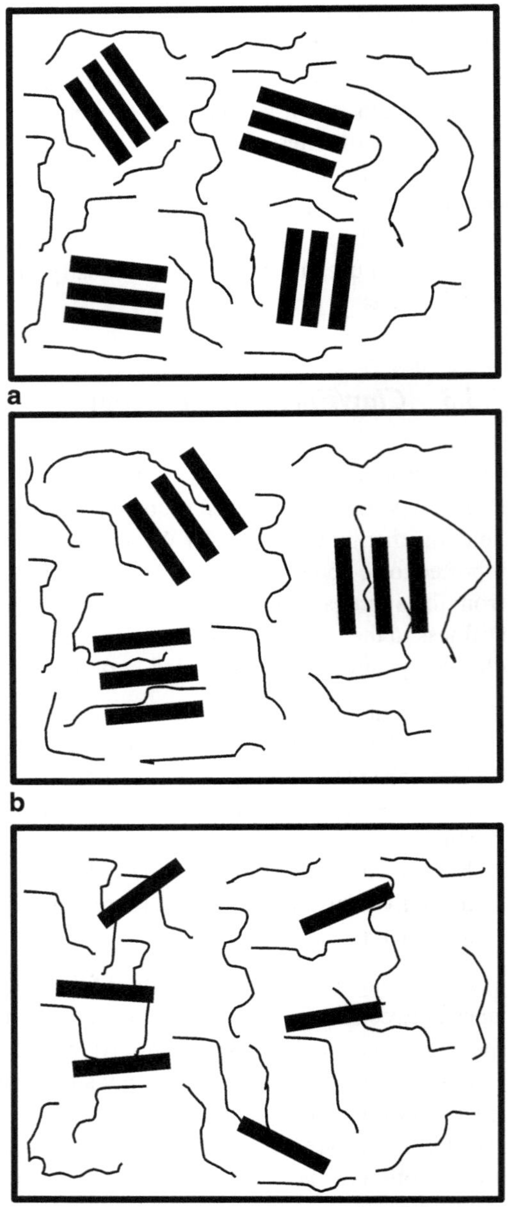

Fig. 5.2 Dispersion of clay platelets in polymer matrix composites showing **a** aggregated microcomposite, and **b** intercalated, and **c** exfoliated nanoscale composites, respectively. (Alexandre and Dubois 2000)

epoxy matrix has been studied using techniques such as X-ray photoelectron spectroscopy (XPS) and Fourier transform infrared spectroscopy (FTIR) (Chan et al. 2011a). The XPS and FTIR techniques are used to analyze the chemical surface composition and the conversion of functional groups for the epoxy matrix nanocomposites (Chan et al. 2011a). The FTIR spectroscopy results for a clay/epoxy

5.5.4 *Transport Properties*

Nanoclay-reinforced polymer composites are used to prepare gloves, protective devices, and are also used in food packaging industry because they inhibit the transport of gas molecules.

A study on nanoclay-reinforced carboxylated nitrile butadiene rubber observed that the nanoclay–polymer affinity hindered the permeability of the solvents. The permeability coefficient of nanoclay/rubber composites containing 3 phr (parts per hundred of rubber) of nanoclay was found to be 5.44×10^{-7} g/cm-s for permeation of ethyl acetate (Aliabadi et al. 2014). This permeability coefficient is 40 % lower in comparison to the neat rubber material. This shows that the addition of nanoclay helps in lowering the permeability coefficient, thereby enabling the material to be used to produce safety devices such as gloves.

5.6 Modeling and Simulations

Several modeling and simulation approaches can be found published for nanoclay/epoxy composites. The capabilities of predicting the properties of composites allow determining the compositions that can be the most suitable for a given application. Some of the modeling and simulation studies are summarized below.

5.6.1 *Flexural Modulus*

Nanoclay epoxy composites were characterized for flexural properties using a four point bend method (Zunjarrao et al. 2006). The results were compared with estimates obtained using the Hashin–Shtrikman (Hashin and Shtrikman 1963) and Norris (1990) models. The Young's modulus for a two-phase material is obtained using (Zunjarrao et al. 2006)

$$E_x^* = \frac{9K_x^* G_x^*}{3K_x^* + G_x^*} \tag{5.1}$$

where K_x^* and G_x^* are the estimated bulk and the estimated shear moduli of the composite, x can be either U or L representing upper and lower bounds, respectively. The bulk (K) and shear (G) moduli for the upper and lower bounds are given by (Hashin and Shtrikman 1963)

$$K_U^* = K_2 + V_1 \left(\frac{1}{K_1 - K_2} + \frac{3V_2}{3K_2 + 4G_2} \right)^{-1} \tag{5.2}$$

$$K_L^* = K_1 + V_2\left(\frac{1}{K_2 - K_1} + \frac{3V_1}{3K_1 + 4G_1}\right)^{-1} \tag{5.3}$$

$$G_U^* = G_2 + V_1\left(\frac{1}{G_1 - G_2} + \frac{6(K_2 + 2G_2)V_2}{5G_2(3K_2 + 4G_2)}\right)^{-1} \tag{5.4}$$

$$G_L^* = G_1 + V_2\left(\frac{1}{G_2 - G_1} + \frac{6(K_1 + 2G_1)V_1}{5G_1(3K_1 + 4G_1)}\right)^{-1} \tag{5.5}$$

where V is the volume fraction and the subscript 1 and 2 represent the matrix and the inclusion, respectively. The bulk (K) and the shear (G) moduli for isotropic oblate spheroid inclusions in a matrix given by the Norris model are represented as (Norris 1990)

$$K = K_1 + \frac{4}{9}V_2\left(\chi\frac{\pi}{8}\frac{3-4\upsilon_1}{G_1(1-\upsilon_1)} + \frac{1}{G_1}\frac{1-\upsilon_2}{1+\upsilon_2}\right)^{-1} \tag{5.6}$$

$$G = G_1 + \frac{1}{15}V_2\left(\chi\frac{\pi}{8}\frac{3-4\upsilon_1}{G_1(1-\upsilon_1)} + \frac{1}{G_2}\frac{1-\upsilon_2}{1+\upsilon_2}\right)^{-1} + \frac{2}{5}V_2\left(\chi\frac{\pi}{16}\frac{7-8\upsilon_1}{G_1(1-\upsilon_1)} + \frac{1}{G_2}\right)^{-1} \tag{5.7}$$

where χ is the aspect ratio and is given by $\chi = c/a$ and υ is the Poisson's ratio. The parameters c and a are the polar and the equatorial radii of the inclusion, respectively. The bulk and the shear moduli are given in terms of the Young's modulus (E) and the Poisson's ratio (υ) as

$$K = \frac{E}{3(1-2\upsilon)} \tag{5.8}$$

$$G = \frac{E}{2(1+\upsilon)} \tag{5.9}$$

The Young's modulus of the epoxy and clay are taken as 2.7 and 167 GPa, respectively (Zunjarrao et al. 2006). The Poisson's ratio of the epoxy and clay are taken as 0.3 and 0.23, respectively (Zunjarrao et al. 2006). The predicted values from the Hashin–Shtrikman and Norris models along with the experimental results are presented in Fig. 5.14. The experimental values are observed to be between the Hashin–Shtrikman upper and lower bounds. The modulus values increase up to 2%, and are close to the values obtained with $\chi = 1/200$, indicating an exfoliated morphology. At nanoclay content >2%, the variation of the modulus is estimated with $\chi = 1/7$ and $\chi = 1/4.5$, indicating that the composites behave as a two-phase material.

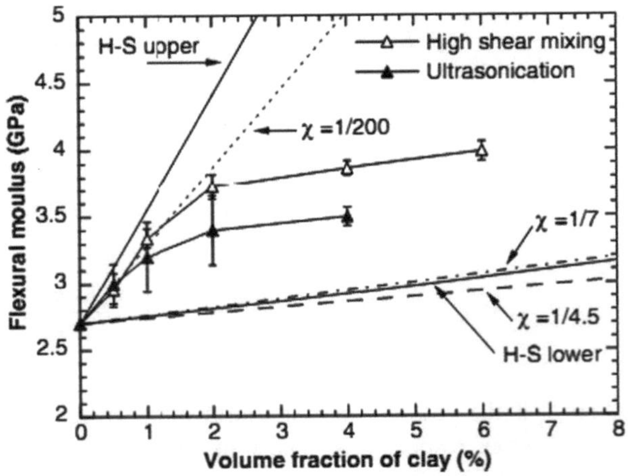

Fig. 5.14 Variation of nanoclay–epoxy composite flexural modulus with varying content of nanoclay (Zunjarrao et al. 2006). The results are compared with estimations obtained from Hashin–Shtrikman (1963 and Norris (1990)

5.6.2 Tensile Modulus

The variation of the tensile modulus with respect to the varying clay content is modeled using the Halpin–Tsai model and the Tandon–Weng model (Tandon and Weng 1986) in a study on clay/epoxy nanocomposites (Miyagawa et al. 2004a). The Halpin-Tsai model for predicting the modulus (E) is given by (Miyagawa et al. 2004a)

$$E_L = E_m \frac{1+\eta_L \xi V_c}{1-\eta_L V_c} \tag{5.10}$$

$$E_T = E_m \frac{1+2\eta_T V_c}{1-\eta_T V_c} \tag{5.11}$$

where

$$\eta_L = \frac{(E_c/E_m)-1}{(E_c/E_m)+\zeta} \tag{5.12}$$

$$\eta_T = \frac{(E_c/E_m)-1}{(E_c/E_m)+2} \tag{5.13}$$

where V is the volume fraction and the subscript L, T, m, and c refer to the longitudinal, transverse, matrix, and clay, respectively. The parameter ζ is given by (Miyagawa et al. 2004a)

$$\zeta = \frac{2\alpha_c}{3} = \frac{2l_c}{3t_c} \tag{5.14}$$

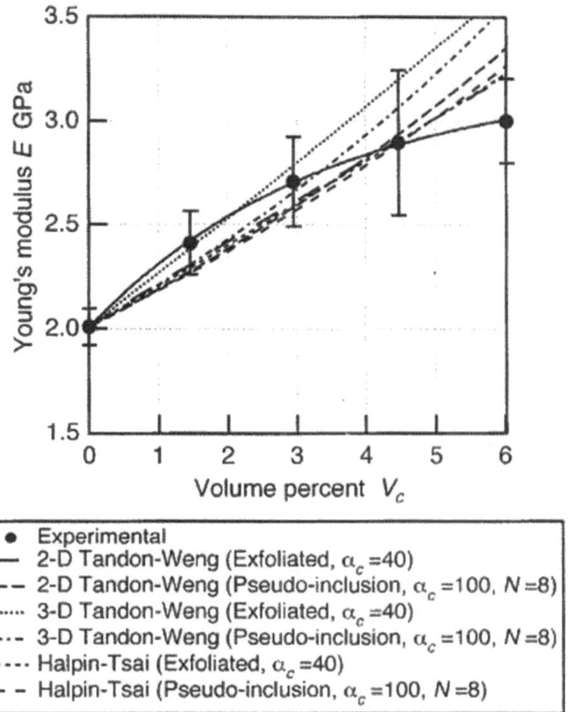

Fig. 5.15 Variation of nanoclay–epoxy composite tensile modulus with respect to the nanoclay content (Miyagawa et al. 2004a). The results are compared with prediction made using Tandom–Weng and Halpin-Tsai

where α_c , l_c, and t_c represent the aspect ratio, length, and thickness of the exfoliated clay platelets. The modulus of nanocomposites containing randomly oriented clay platelets is given by

$$E_n = \frac{3}{8}E_L + \frac{5}{8}E_T \tag{5.15}$$

The detailed expression for the Tandon–Weng model is given in (Tandon and Weng 1986). The predictions of the models are compared with the experimental data in Fig. 5.15. The modulus of epoxy and nanoclay are taken as 2.50 and 170 GPa, respectively (Miyagawa et al. 2004a). The Poisson's ratios of epoxy and nanoclay are taken as 0.416 and 0.3, respectively (Miyagawa et al. 2004a). The models show predictions close to the experimental tensile modulus values obtained for clay/epoxy nanocomposites. It should be noted in the figure that the standard deviations in experimental data are very large so that a wide range of predictions can be fit on this data set.

Luo and Daniel modeled the tensile modulus of clay/epoxy nanocomposites using a three phase model accounting for the partial exfoliation and intercalation of the nanoclay (Luo and Daniel 2003). The modulus of epoxy and nanoclay are taken as 3.1 and 176 GPa, respectively. The Poisson's ratios of epoxy and nanoclay are taken as 0.35 and 0.25, respectively. The epoxy matrix and the clay embedded in the matrix are assumed to be isotropic and the properties of intercalated clay structure are evaluated by assuming them as parallel nanolayers. The volume fraction of the nanoclay (V_c) in epoxy, assuming no intercalation or exfoliation, is given by

$$V_c = \frac{w/\rho_c}{w/\rho_c + (1-w)/\rho_m} \tag{5.16}$$

where ρ_c and ρ_m are the clay and matrix density and w is the weight fraction of clay. The exfoliated (V_e) and intercalated (V_i) clay volumes are represented as

$$V_e = V_c r_e \frac{t}{d_0} \tag{5.17}$$

$$V_i = V_c(1-r_e)\frac{d}{d_0} \tag{5.18}$$

where r_e is the exfoliated clay volume fraction, t the layer thickness (1 nm), d the layer spacing, and d_0 is the initial layer spacing (1.85 nm). The effective composite stiffness (C^*) properties were obtained using Mori–Tanaka method

$$C^* = C_1 + V_2\left((C_2 - C_1)A\right) \tag{5.19}$$

where C_1 and C_2 are the matrix phase and the inclusion phase stiffness tensors. The concentration tensor A is given by

$$A = A^{(dil)}\left[V_1 I + V_2\left\{A^{(dil)}\right\}\right]^{-1} \tag{5.20}$$

where $A^{(dil)}$ is given for a dilute solution using Eshelby's solution

$$A^{(dil)} = \left[I + SC_1^{-1}(C_2 - C_1)\right]^{-1} \tag{5.21}$$

where I is the fourth-order unit tensor and S is the fourth-order Eshelby's tensor. The results obtained experimentally are compared with the model and are also compared with the Voigt and Reuss models in Fig. 5.16. The exfoliation ratio and the d-spacing of the nanoclay are obtained using TEM micrographs and are taken as 10% and 4.8 nm (Luo and Daniel 2003). From Fig. 5.16, it could be observed that the predicted values are in close agreement with the experimental values.

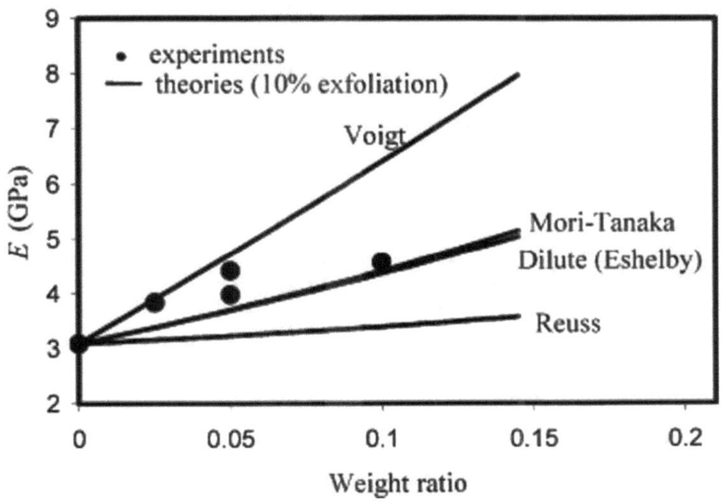

Fig. 5.16 Variation of nanoclay-epoxy composite tensile modulus with respect to the nanoclay content (Luo and Daniel 2003). The results are compared with estimates made using Mori–Tanaka and dilute theory

5.6.3 Molecular Simulation Studies

To determine the property of the composites, the properties of the nanoclay needs to be understood. However, the clay platelets do not possess long-range order and are not perfect crystals. The direct measurement of the properties of clay platelets is very challenging. Because of the size scale of the clay platelets and the presence of complex clay-polymer interactions, molecular simulation methods can be very useful for gaining insight into the properties of nanoclay-reinforced composites (Chen et al. 2008a). Studies on atomistic simulation on organoclay-polymer matrix composites are concentrated in two main groups: (a) studies concentrating on understanding the clay structure, and (b) studies concentrating on the interaction between the clay and the polymers. A brief overview on molecular dynamic simulation studies on the computation of elastic modulus and the bending of the nanoclay layers is presented in this work. For further knowledge and understanding, the reader is encouraged to refer to the existing literature (Suter et al. 2009, 2007; Mazo et al. 2009; Scocchi et al. 2007; Katti et al. 2012) and the references cited therein.

The elastic modulus of the nanoclay platelet was evaluated using large-scale molecular dynamic simulations on MMT clay (Suter et al. 2007). The number of atoms used in the study range between 6752 and 9,495,000. An example of the nanoclay sheet model is shown in Fig. 5.17. CLAYFF force field was utilized in the simulation and the metal–oxygen interactions were evaluated using the Lennard–Jones (6–12) potential. The simulation was run for at least 1 ns at 300 K at a pressure of 1 atm, using an isobaric–isothermal ensemble. The elastic properties were evaluated by applying uniaxial tension and compression to the cell, by changing the length of

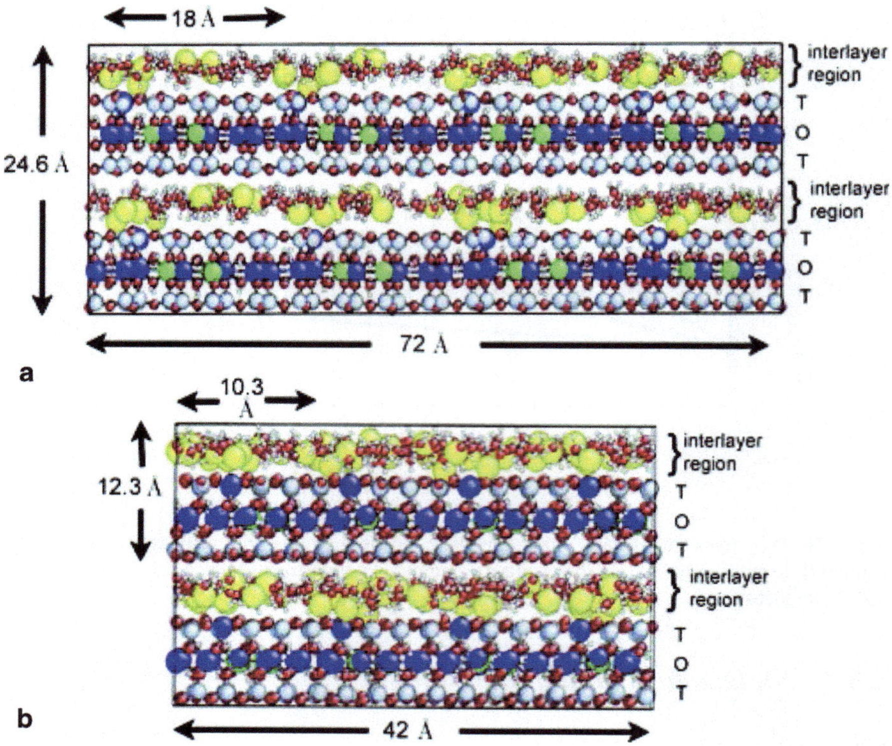

Fig. 5.17 Image of molecular dynamics simulation of montmorillonite clay containing 6752 atoms, **a** in *yz* plane and **b** *xz* plane. Each clay sheet comprises two tetrahedral (T) and one octahedral sheet (O). (Reprinted with permission from Suter et al. 2007. Copyright (2007) American Chemical Society)

the cell in the direction of the deformation. The stress was evaluated based on the applied deformation using

$$\sigma_{kl} = \frac{1}{V}\sum_{i=1}^{N}\left(\frac{m_i}{2}v_i v_i + \frac{1}{2}\sum_{j=1}^{N}F_i r_{ij}\right)_{kl} \tag{5.22}$$

where V is the volume, v_i is the velocity, m_i is the mass, F_i is the force on the *i*th particle, and $r_{ij}=(r_i\text{-}r_j)$ with r_i is the position. The stress–strain curves for the loading applied in the *x* and *y*-direction is shown in Fig. 5.18. The average Young's modulus in the *x*- and *y*-direction is calculated to be 172 and 182 GPa. The Young's modulus evaluated from the molecular simulation studies is in good agreement with the values commonly used in the literature.

The effect of temperature on the stiffness coefficients of nanoclay have been studied using molecular dynamic simulation (Mazo et al. 2008). A parallelepiped containing 45 unit cells (1800 atoms) was constructed and periodic boundary condi-

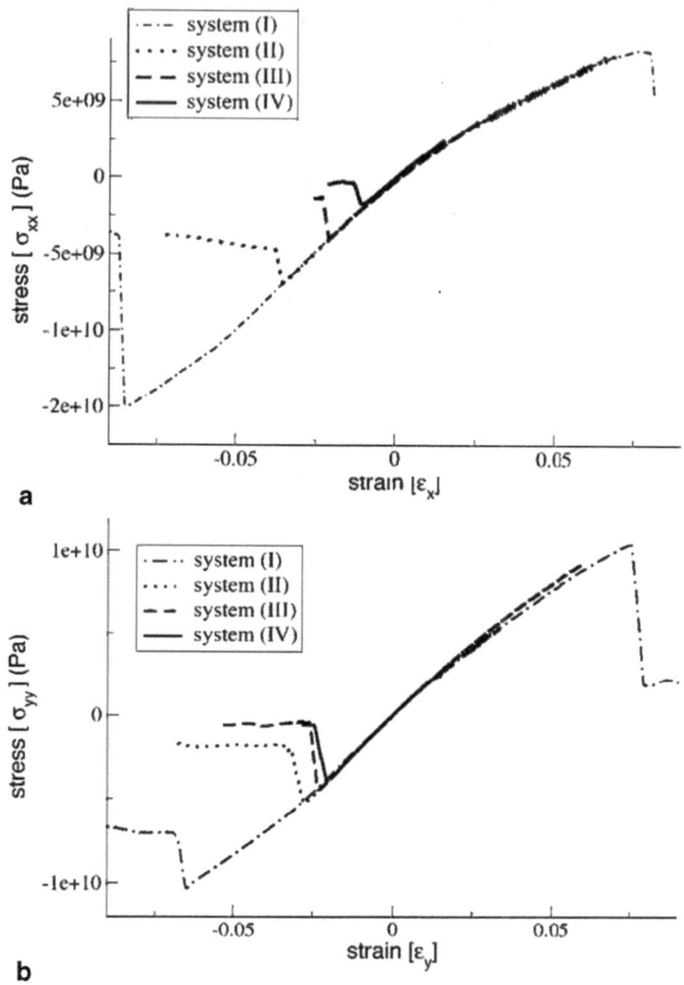

Fig. 5.18 Stress–strain plot from the molecular simulation study on nanoclay with the strain applied in **a** *x*- and **b** *y*-direction. (Reprinted with permission from Suter et al. (2007). Copyright (2007) American Chemical Society)

tions were applied on all three directions. The energy of the system was computed using CLAYFF force field and the long-range Coulomb interactions were computed using smooth particle Ewald method. The deformation of the cell was executed by changing the size of the cell in the desired direction. The 36 compliance constants (S_{ij}) were calculated based on the strain applied as

$$\varepsilon_i = S_{i1}\sigma_1 + S_{i6}\sigma_6 (i = 1, 2, ..., 6) \quad (5.23)$$

$$\varepsilon_i = S_{i2}\sigma_2 + S_{i6}\sigma_6 (i = 1, 2, ..., 6) \quad (5.24)$$

Table 5.6 Elastic coefficients evaluated at three different temperatures. (Adapted with permission from (Mazo et al. 2008). Copyright (2008) American Chemical Society)

T (K)	C_{11}	C_{22}	C_{33}	C_{44}	C_{55}	C_{66}
100	470	437	268	56	73	127
300	442	399	253	54	70	120
700	413	381	229	50	65	113

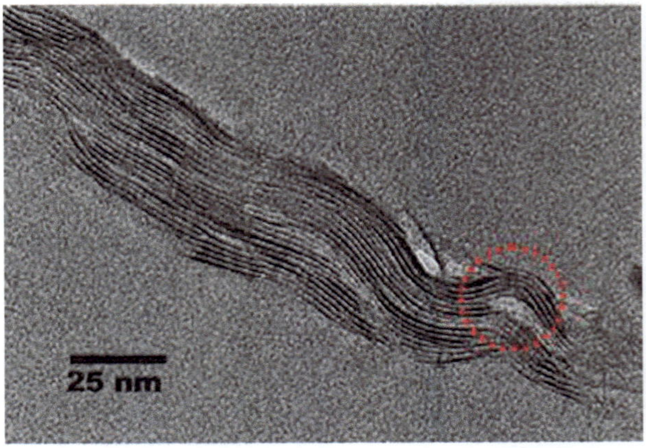

Fig. 5.19 Transmission electron micrograph of nanoclay-reinforced epoxy matrix composites showing bent nanoclay. (Reprinted with permission from Fu et al. (2011). Copyright (2011) American Chemical Society)

$$\varepsilon_i = S_{i6}\sigma_6 (i = 1, 2, ..., 6) \quad (5.25)$$

$$\varepsilon_i = S_{i3}P_3 + S_{i6}\sigma_6 (i = 1, 2, ..., 6) \quad (5.26)$$

$$\varepsilon_i = S_{i4}P_4 + S_{i6}\sigma_6 (i = 1, 2, ..., 6) \quad (5.27)$$

$$\varepsilon_i = S_{i5}P_5 + S_{i6}\sigma_6 (i = 1, 2, ..., 6) \quad (5.28)$$

where ε_i and, σ_1 and σ_6 are measured during the simulations and P_3, P_4, and P_5 are external applied stresses. The elastic stiffness coefficients are calculated as $C_{ij} = S_{ij}^{-1}$and is shown in Table 5.6. The modulus is shown to decrease with increase in temperature. The nanoclay stiffness in the *x*- and *y*-directions (in the plane of the platelet) are higher than the stiffness in the *z*-direction (normal to the platelet).

Nanoclay embedded in polymer matrix materials such as epoxy can bend during the manufacturing process. The bending can occur due to the out-of-plane or vertical forces acting on the platelets. Another possible cause for the platelets to bend is, when the in-plane compressive forces exceed a certain threshold (Cygan et al. 2009). A transmission electron micrograph of nanoclay-reinforced epoxy composite showing bent nanoclay platelets is shown in Fig. 5.19. The linear properties of the nanoclay composites have been studied in detail, however the properties of the bent nanoclay system are difficult to study by experimental methods. The problem has therefore only been studied by molecular simulations (Fu et al. 2011). The bending of the nanoclay layers results in changes in the bond angles and lengths. Also, the alkali ions shift when the nanoclay layers bend resulting in changes in the bending energy. To analyze the bending energy associated with the bending of the nanoclay layers, molecular dynamic simulations was conducted on MMT clay. A $12.9 \times 2.7 \times 1$ nm cell was created to model large and bendable layers. The periodic boundary condition was employed only in the *y*-direction, to create isolated clay layers. Alkali cations were equally distributed on either side on the clay layers. Phyllosilicate force field embedded in a polymer-consistent force was employed in the simulation. The straight nanoclay layers were inserted between carbon wraps possessing a radius of curvature. The bend energy (E_b) of a single clay layer was calculated as

$$E_b = E_t - E_1 - E_c \tag{5.29}$$

where E_t is the total energy of the bent layer and the two wraps, E_1 is the energy of the corresponding linear clay layer without wraps, and E_c is the energy due to the van der Waals interaction between the bent aluminosilicate clay layer and the carbon wraps. The bending energy evaluated for a single layer of nanoclay increases from 0 to 10 mJ/m^2 for the bending radius of 20 nm (Fu et al. 2011). For smaller bending radius on the order of 10 nm, the bending energy of a single layer was calculated to be 50–100 mJ/m^2. The specific stored energy in the clay aluminosilicate layers, per unit mass and unit volume, possessing a radius of curvature of 6–8 nm are 40 kJ/kg and 100 kJ/dm^3. These energy densities values are in the magnitudes of one fourth of the energy density of lead acid batteries and higher than the energy density of supercapacitors (Cygan et al. 2009). The bending flexibility of the nanoclay can be exploited to be used as energy storage devices, in addition to being used as structural reinforcements.

Summary

Low cost and abundant availability of clay have been viewed as significant advantages in favor of developing high performance clay/polymer nanocomposites. Both thermoplastic and thermosetting resin matrix nanocomposites have been studied in the existing literature. The layered structure of clay can be exfoliated to create enormous surface area where matrix resins can be bonded to provide nanocomposites with increased mechanical properties. However, the dispersion of nanoclay is difficult and expensive. Several mechanical and chemical processes have been developed to effectively exfoliate nanoclay and obtain high quality composites. However, a review of existing data reveals that the potential of clay/polymer nanocomposites has not been fully realized and most properties are improved by 30–40 % compared to those of neat resin. Such improvement is obtained at low nanoclay content in the composites (1–5 wt%). At higher nanoclay loading levels, the exfoliation becomes progressively more difficult and entrapment of air in the composite during the mixing process increases and the advantages in the mechanical properties diminish. Tensile, flexural, and thermal properties of nanocomposites have been extensively studied in the available literature. Several innovative modeling approaches are now available in the literature. One of the main challenges for these schemes is to account for the presence of clustered, intercalated, and exfoliated nanoclay contents. The estimated properties are found to be within Hashin–Shtrikman bounds for the nanocomposites.

Industrial applications exist for clay/epoxy nanocomposites, while some new applications are also being exploited. Clay/epoxy nanocomposites provide improved corrosion protection, so that it might find applications in modern aircraft anticorrosive coatings (Tomić et al. 2014). Clay/epoxy nanocomposites have been widely used for structural adhesive applications, because of its potential to improve adhesive performance, practicality, and lower cost (Sancaktar and Kuznicki 2011). Most of the current applications are in automobiles, which now extensively use polymers inside the vehicle. Dashboards, parts of seat structures, and fixtures are examples of current applications. New applications are continuously being developed for nanoclay composites.

Acknowledgments The authors would like to thank the US Army Research Laboratory Cooperative Agreement W911NF-11-2-0096 and the Office of Naval Research grant N00014-10-1-0988 for supporting the work. The authors thank Steven E. Zeltmann for his help with the manuscript preparation.

References

Alexandre M, Dubois P (2000) Polymer-layered silicate nanocomposites: preparation, properties and uses of a new class of materials. Mater Sci Eng R Rep 28(1–2):1–63

Aliabadi MM, Naderi G, Shahtaheri SJ, Forushani AR, Mohammadfam I, Jahangiri M (2014) Transport properties of carboxylated nitrile butadiene rubber (XNBR)-nanoclay composites; a promising material for protective gloves in occupational exposures. J Environ Health Sci Eng 12(1):51–58

Al-Qadhi M, Merah N, Gasem ZM (2013) Mechanical properties and water uptake of epoxy–clay nanocomposites containing different clay loadings. J Mater Sci 48(10):3798–3804

Azeez AA, Rhee KY, Park SJ, Hui D (2013) Epoxy clay nanocomposites—processing, properties and applications: a review. Compos Part B: Eng 45(1):308–320

Barick AK, Tripathy DK (2011) Effect of organically modified layered silicate nanoclay on the dynamic viscoelastic properties of thermoplastic polyurethane nanocomposites. Appl Clay Sci 52(3):312–321

Basara C, Yilmazer U, Bayram G (2005) Synthesis and characterization of epoxy based nanocomposites. J Appl Polym Sci 98(3):1081–1086

Bhatnagar MS (1996) Epoxy resins (overview). In: Salomone JC (ed) Polymeric materials encyclopedia. CRC Press, Boca Raton, pp 1–11

Chan M-L, Lau K-T, Wong TT, Cardona F (2011a) Interfacial bonding characteristic of nanoclay/polymer composites. Appl Surf Sci 258(2):860–864

Chan M-L, Lau K-T, Wong T-T, Ho M-P, Hui D (2011b) Mechanism of reinforcement in a nanoclay/polymer composite. Compos Part B: Eng 42(6):1708–1712

Chau DV (2012) A study on water absorption and its effect on strength of nano organoclay-epoxy composites. J Appl Sci 12(18):1939–1945
Chen C, Tolle TB (2004) Fully exfoliated layered silicate epoxy nanocomposites. J Polym Sci Part B: Polym Phys 42(21):3981–3986
Chen B, Evans JRG, Greenwell HC, Boulet P, Coveney PV, Bowden AA, Whiting A (2008a) A critical appraisal of polymer—clay nanocomposites. Chem Soc Rev 37:568–594
Chen C, Benson-Tolle T, Baur JW, Putthanarat S (2008b) Processing—morphology regulation of epoxy/layered-silicate nanocomposites. J Appl Polym Sci 108(5):3324–3333
Chinnamuthu CR, Murugesaboopathi P (2009) Nanotechnology and agroecosystem. Madras Agric J 96(1–6):17–31
Cygan RT, Greathouse JA, Heinz H, Kalinichev AG (2009) Molecular models and simulations of layered materials. J Mater Chem 19(17):2470–2481
Dai F, Xu Y-H, Zheng Y-P, Yi X-S (2005) Study on morphology and mechanical properties of high-functional epoxy based clay nanocomposites. Chin J Aeronaut 18(3):279–282
Da Silva GR, Ayres E, Orefice RL, Moura SAL, Cara DC, Cunha ADS (2009) Controlled release of dexamethasone acetate from biodegradable and biocompatible polyurethane and polyurethane nanocomposite. J Drug Target 17(5):374–383
Di Gianni A, Amerio E, Monticelli O, Bongiovanni R (2008) Preparation of polymer/clay mineral nanocomposites via dispersion of silylated montmorillonite in a UV curable epoxy matrix. Appl Clay Sci 42(1–2):116–124
Edser C (2000) Auto applications drive commercialization of nanocomposites (2002). Plast Addit Compd 4(1):30–33
Farris DW Clay minerals. http://earth.usc.edu/~dfarris/Mineralogy/17_ClayMinerals.pdf. Accessed 29 Jan 2014
Ferreira JAM, Borrego LP, Costa JDM, Capela C (2013) Fatigue behaviour of nanoclay reinforced epoxy resin composites. Compos Part B: Eng 52:286–291
Fischer H (2003) Polymer nanocomposites: from fundamental research to specific applications. Mater Sci Eng: C 23(6–8):763–772
Fu Y-T, Zartman GD, Yoonessi M, Drummy LF, Heinz H (2011) Bending of layered silicates on the nanometer scale: mechanism, stored energy, and curvature limits. J Phys Chem C 115(45):22292–22300
Gao F (2004) Clay/polymer composites: the story. Mater Today 7(11):50–55
Giannelis EP (1996) Polymer layered silicate nanocomposites. Adv Mater 8(1):29–35
Guevara-Morales A, Taylor AC (2014) Mechanical and dielectric properties of epoxy—clay nanocomposites. J Mater Sci 49(4):1574–1584
Ha JU, Xanthos M (2011) Drug release characteristics from nanoclay hybrids and their dispersions in organic polymers. Int J Pharm 414(1–2):321–331
Hashin Z, Shtrikman S (1963) A variational approach to the theory of the elastic behaviour of multiphase materials. J Mech Phys Solids 11(2):127–140
Ho M-W, Lam C-K, Lau K-T, Ng DHL, Hui D (2006) Mechanical properties of epoxy-based composites using nanoclays. Compos Struct 75(1–4):415–421
Huskić M, Žigon M, Ivanković M (2013) Comparison of the properties of clay polymer nanocomposites prepared by montmorillonite modified by silane and by quaternary ammonium salts. Appl Clay Sci 85:109–115
Jumahat A, Soutis C, Mahmud J, Ahmad N (2012) Compressive properties of nanoclay/epoxy nanocomposites. Procedia Eng 41:1607–1613
Kang H-Y (2010) A review of the emerging nanotechnology industry: materials, fabrications, and applications. September 2010: Sacramento pp 8–15
Katti DR, Katti KS, Raviprasad M, Gu C (2012) Role of polymer interactions with clays and modifiers on nanomechanical properties and crystallinity in polymer clay nanocomposites. J Nanomater 2012:1–15
Kaya E, Tanoğlu M, Okur S (2008) Layered clay/epoxy nanocomposites: thermomechanical, flame retardancy, and optical properties. J Appl Polym Sci 109(2):834–840

Kiliaris P, Papaspyrides CD (2010) Polymer/layered silicate (clay) nanocomposites: an overview of flame retardancy. Prog Polym Sci 35(7):902–958

Koerner H, Misra D, Tan A, Drummy L, Mirau P, Vaia R (2006) Montmorillonite-thermoset nanocomposites via cryo-compounding. Polymer 47(10):3426–3435

Ku H, Trada M (2013) Tensile properties of nanoclay reinforced epoxy composites. In: Epaarachchi JA, Lau AK-T, Leng J (Eds) Fourth International Conference on Smart Materials and Nanotechnology in Engineering, Proceedings of SPIE, Gold Coast, 10–12 July 2013, pp 1–5

Ku H, Fung S, Islam M, Li X, Chong Y (2013) A pilot study on flexural properties of epoxy nanoclay reinforced composites. J Reinf Plast Compos 32(2):96–104

Kusmono, Wildan MW, Mohd Ishak ZA (2013) Preparation and properties of clay-reinforced epoxy nanocomposites. Int J Polym Sci 2013:1–7

Lam CK, Lau KT (2006) Localized elastic modulus distribution of nanoclay/epoxy composites by using nanoindentation. Compos Struct 75(1–4):553–558

Lam C-K, Lau K-T, Cheung H-Y, Ling H-Y (2005) Effect of ultrasound sonication in nanoclay clusters of nanoclay/epoxy composites. Mater Lett 59(11):1369–1372

Lan T. Nanocomposite materials for packaging applications. http://www.nanocor.com/tech_papers/Antec-Nanocor-Tie%20Lan-5–07.pdf. Accessed 27 Jan 2014

Lan T, Cho J, Liang Y, Qian J, Peter M (2001) Application of nanomer in nanocomposites: from concept to reality. In: Proceedings of Nanocomposites 2001. 25–27 June 2001. Chicago, pp 1–10

LeBaron PC, Wang Z, Pinnavaia TJ (1999) Polymer-layered silicate nanocomposites: an overview. Appl Clay Sci 15(1–2):11–29

Lingaiah S, Sadler R, Ibeh C, Shivakumar K (2008) A method of visualization of inorganic nanoparticles dispersion in nanocomposites. Compos Part B: Eng 39(1):196–201

Liu T, Tjiu WC, Tong Y, He C, Goh SS, Chung T-S (2004) Morphology and fracture behavior of intercalated epoxy/clay nanocomposites. J Appl Polym Sci 94:1236–1244

Lu H, Nutt S (2003) Restricted relaxation in polymer nanocomposites near the glass transition. Macromolecules 36(11):4010–4016

Lu H-J, Liang G-Z, Ma X-Y, Zhang B-Y, Chen X-B (2004) Epoxy/clay nanocomposites: further exfoliation of newly modified clay induced by shearing force of ball milling. Polym Int 53(10):1545–1553

Luo J-J, Daniel IM (2003) Characterization and modeling of mechanical behavior of polymer/clay nanocomposites. Compos Sci Technol 63(11):1607–1616

Markarian J (2005) Automotive and packaging offer growth opportunities for nanocomposites. Plast Addit Compd 7(6):18–21

Mazo MA, Manevitch LI, Gusarova EB, Shamaev MY, Berlin AA, Balabaev NK, Rutledge GC (2008) Molecular dynamics simulation of thermo-mechanical properties of montmorillonite crystal. 1. Isolated clay nanoplate. J Phys Chem B 112(10):2964–2969

Mazo MA, Manevich LI, Balabaev NK (2009) Molecular dynamics simulation of thermo-mechanical properties of montmorillonite crystal. Nanotechnol Russia 4(9–10):676–699

Miyagawa H, Rich MJ, Drzal LT (2004a) Amine-cured epoxy/clay nanocomposites. II. The effect of the nanoclay aspect ratio. J Polym Sci Part B: Polym Phys 42(23):4391–4400

Miyagawa H, Rich MJ, Drzal LT (2004b) Amine-cured epoxy/clay nanocomposites. I. Processing and chemical characterization. J Polym Sci Part B: Polym Phys 42(23):4384–4390

Nguyen QT, Baird DG (2006) Preparation of polymer–clay nanocomposites and their properties. Adv Polym Technol 25(4):270–285

Nguyen QT, Baird DG (2007) An improved technique for exfoliating and dispersing nanoclay particles into polymer matrices using supercritical carbon dioxide. Polymer 48(23):6923–6933

Norris AN (1990) The mechanical properties of platelet reinforced composites. Int J Solids Struct 26(5–6):663–674

de Paiva LB, Morales AR, Valenzuela Díaz FR (2008) Organoclays: properties, preparation and applications. Appl Clay Sci 42(1–2):8–24

Park JH, Jana SC (2003) Mechanism of exfoliation of nanoclay particles in epoxy—clay nanocomposites. Macromolecules 36(8):2758–2768

Patel HA, Somani RS, Bajaj HC, Jasra RV (2006) Nanoclays for polymer nanocomposites, paints, inks, greases and cosmetics formulations, drug delivery vehicle and waste water treatment. Bull Mater Sci 29(2):133–145
Pavlidou S, Papaspyrides CD (2008) A review on polymer—layered silicate nanocomposites. Prog Polym Sci 33(12):1119–1198
Qi B, Zhang QX, Bannister M, Mai YW (2006) Investigation of the mechanical properties of DGEBA-based epoxy resin with nanoclay additives. Compos Struct 75(1–4):514–519
Ratna D (2009) Handbook of thermosetting resins. iSmithers Rapra, Shropshire, pp. 1–385
Rodrigues LAdS, Figueiras A, Veiga F, de Freitas RM, Nunes LCC, da Silva Filho EC, da Silva Leite CM (2013) The systems containing clays and clay minerals from modified drug release: a review. Coll Surf B: Biointerfaces 103:642–651
Saber-Samandari S, Khatibi AA, Basic D (2007) An experimental study on clay/epoxy nanocomposites produced in a centrifuge. Compos Part B: Eng 38(1):102–107
Sancaktar E, Kuznicki J (2011) Nanocomposite adhesives: mechanical behavior with nanoclay. Int J Adhes Adhes 31(5):286–300
Sarathi R, Sahu RK, Rajeshkumar P (2007) Understanding the thermal, mechanical and electrical properties of epoxy nanocomposites. Mater Sci Eng: A 445–446:567–578
Schartel B, Knoll U, Hartwig A, Pütz D (2006) Phosphonium-modified layered silicate epoxy resins nanocomposites and their combinations with ATH and organo-phosphorus fire retardants. Polym Adv Technol 17(4):281–293
Scocchi G, Posocco P, Fermeglia M, Pricl S (2007) Polymer–clay nanocomposites: a multiscale molecular modeling approach. J Phys Chem B 111(9):2143–2151
Seema, Datta M (2013) Clay—polymer nanocomposites as a novel drug carrier: synthesis, characterization and controlled release study of propranolol hydrochloride. Appl Clay Sci 80–81:85–92
Shi H, Lan T, Pinnavaia TJ (1996) Interfacial effects on the reinforcement properties of polymer—organoclay nanocomposites. Chem Mater 8(8):1584–1587
Silan, M, Ziaei-Rad S, Esfahanian M, Tan VBC (2012) On the experimental and numerical investigation of clay/epoxy nanocomposites. Compos Struct 94(11):3142–3148
Sinha Ray S, Okamoto M (2003) Polymer/layered silicate nanocomposites: a review from preparation to processing. Prog Polym Sci 28(11):1539–1641
Sodeifian G, Nikooamal H, Yousefi A (2012) Molecular dynamics study of epoxy/clay nanocomposites: rheology and molecular confinement. J Polym Res 19(6):1–12, 9897
Suresh R, Borkar S, Sawant V, Shende V, Dimble S (2010) Nanoclay drug delivery system. Int J Pharm Sci Nanotechnol 3:901–905
Suter JL, Coveney PV, Greenwell HC, Thyveetil M-A (2007) Large-scale molecular dynamics study of montmorillonite clay: emergence of undulatory fluctuations and determination of materials properties. J Phys Chem C 111(23):12
Suter JL, Anderson RL, Greenwell HC, Coveney PV (2009) Recent advances in large-scale atomistic and coarse-grained molecular dynamics simulation of clay minerals. J Mater Chem 19:2482–2493
Tandon GP, Weng GJ (1986) Average stress in the matrix and effective moduli of randomly oriented composites. Compos Sci Technol 27(2):111–132
Tcherbi-Narteh A, Hosur MV, Triggs E, Jeelani S (2013) Effects of surface treatments of montmorillonite nanoclay on cure behavior of diglycidyl ether of bisphenol A epoxy resin. J Nanosci 2013:1–12
Tjong SC (2006) Structural and mechanical properties of polymer nanocomposites. Mater Sci Eng: R: Rep 53(3–4):73–197
Tomić MD, Dunjić B, Likić V, Bajat J, Rogan J, Djonlagić J (2014) The use of nanoclay in preparation of epoxy anticorrosive coatings. Prog Organ Coat 77(2):518–527
Uddin F (2008) Clays, nanoclays, and montmorillonite minerals. Metall Mater Trans A 39(12):2804–2814
Vaia RA, Giannelis EP (1997) Lattice model of polymer melt intercalation in organically-modified layered silicates. Macromolecules 30(25):7990–7999

Wang MS, Pinnavaia TJ (1994) Clay-polymer nanocomposites formed from acidic derivatives of montmorillonite and an epoxy resin. Chem Mater. 6(4):468–474

Wang J, Qin S (2007) Study on the thermal and mechanical properties of epoxy—nanoclay composites: the effect of ultrasonic stirring time. Mater Lett 61(19–20):4222–4224

Wang K, Chen L, Wu J, Toh ML, He C, Yee AF (2005a) Epoxy nanocomposites with highly exfoliated clay: mechanical properties and fracture mechanisms. Macromolecules 38(3):788–800

Wang K, Wang L, Wu J, Chen L, He C (2005b) Preparation of highly exfoliated epoxy/clay nanocomposites by "slurry compounding": process and mechanisms. Langmuir 21(8):3613–3618

Wang L, Wang K, Chen L, Zhang Y, He C (2006) Preparation, morphology and thermal/mechanical properties of epoxy/nanoclay composite. Compos Part A: Appl Sci Manuf 37(11):1890–1896

Yasmin A, Abot JL, Daniel IM (2003) Processing of clay/epoxy nanocomposites by shear mixing. Scr Mater 49(1):81–86

Yasmin A, Luo JJ, Abot JL, Daniel IM (2006) Mechanical and thermal behavior of clay/epoxy nanocomposites. Compos Sci Technol 66(14):2415–2422

Zaman I, Le Q-H, Kuan H-C, Kawashima N, Luong L, Gerson A, Ma J (2011) Interface-tuned epoxy/clay nanocomposites. Polymer 52(2):497–504

Zunjarrao SC, Sriraman R, Singh RP (2006) Effect of processing parameters and clay volume fraction on the mechanical properties of epoxy-clay nanocomposites. J Mater Sci 41(8):2219–2228

Part II
Hierarchical Composites

Chapter 6
Medical Applications of Hierarchical Composites

Manoj Kumar R, Khelendra Agrawal and Debrupa Lahiri

6.1 Introduction

Nature always fascinates scientists and engineers with a variety of complex hierarchical morphologies, resulting in their excellent properties. It is tempting to mimic them by novel processing routes and synthetic material systems (Dong et al. 2007). The idea of a hierarchical composite structure was conceived from the numerous examples available in nature. The best and extensively discussed illustration of this category is the structure of bone. An ideal combination of properties is available in bone for its intended performance, i.e., the right proportion of stiffness, strength, fracture toughness, and low density. This is possible only through its complex hierarchical composite structure composed of mineral crystals and protein fibers spanning from the nanoscale to macroscale range (Landis 1995). Development of such synthetic well-engineered hierarchical structures for sophisticated medical applications has inspired a community of great researchers toward some unique findings. Composites have proven to be a great material system to achieve highly complex hierarchical structures with properties to meet the medical demand. Combining two or more different natural or synthetic biomaterials with novel fabrication processes often helps realize such a material system. With the advancement in technology and current focus on nanotechnology, it has become possible to engineer the composite with complex hierarchical structure containing different scales of morphology to resemble natural living hard tissues.

Inspired by the great variety of interesting properties, material researchers have explored the novel functionalities of hierarchical composite structures. In this chapter, attempts were made to describe composite materials and fabrication of three-dimensional (3D) hierarchical structures. The composites were of different material variety, including various ceramics, collagen, and inorganic/organic materials sys-

D. Lahiri (✉) · M. Kumar R · K. Agrawal
Biomaterials and Multiscale Mechanics Lab, Department of Metallurgical and Materials Engineering, Indian Institute of Technology Roorkee, Roorkee 247667, India
e-mail: dlahifmt@iitr.ac.in

C.-S. Kim et al. (eds.), *Hybrid and Hierarchical Composite Materials*,
DOI 10.1007/978-3-319-12868-9_6

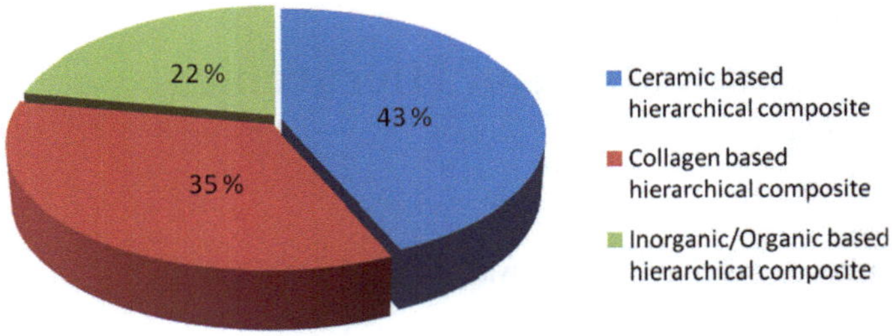

Fig. 6.1 Classification of hierarchical composite with respect to number of publications on each category as listed in *scopus.com* till May 2014

tems. Diverse synthesis techniques were used, including biomineralization (Zhang et al. 2003; Bradt et al. 1999), replica molding (Kim et al. 2012, 2013), rapid prototyping (Huang 2013), solvent casting (Bae et al. 2014), and others. The major challenge in these research endeavors still remains to some extent, toward mimicking the natural hard tissue properties and structure. Hence, it is necessary to review and highlight different strategies for developing these hierarchical composites, with their advantages and limitations.

This chapter provides an overview of the different hierarchical composites and their synthesis route to tailor the structure and properties required for medical and medicinal applications (Feng et al. 2002; Xiu et al. 2007; Liao 2004; Murugan and Ramakrishna 2005). How different combinations of materials in hierarchical composite structures can be applied to medical applications will be covered. Most of the studies on such structures are related to developing the material system for scaffold and tissue engineering. The composites are divided into the following three categories: ceramic-based, collagen-based, and inorganic/organic-based material systems, followed by the discussion on their processing techniques. As shown in Fig. 6.1, more studies have been carried out for ceramic-based structures, followed by those of collagen based. This observation clearly indicates the majority of interest being oriented towards hard tissue engineering. In addition, unique biological and mechanical properties of composite systems resulting from their structure and its suitability in targeted applications are discussed. Special emphasis is given on medical applications of such structures in tissue engineering and drug delivery. A brief discussion on modeling of mechanical properties for hard tissue replacement is also addressed. Lastly, the current challenges and the future directions of research on hierarchical composites for medical applications are described.

6.2 Types of Hierarchically Structured Composites in Medical Applications

Hierarchical multiscale structures are abundant in nature. Some naturally occurring hierarchical composites are: lotus leaves, spider silks, kidney's glomerules, bones, and many others, which posses a variety of fascinating properties due to their micro- and nanoscale material structures. It has been a challenge for materials scientists to design artificial structures with similar functionalities (Bae et al. 2014). This challenge has been facilitated by remarkable advancements in nanotechnology and diverse fabricating methodologies. Therefore, it is suitable to summarize the different advanced composite material systems used to tailor the hierarchical structures for medical applications. At each length scale of these composite structures, the components are combined together with specific interactions and are designed to serve highly specific functions, while the total architecture helps to shape the composite into a unique structural system allowing for hierarchical performance (Baer and Hiltner 1992). Researchers have tried to create synthetic composites to mimic the natural biological structure with the help of basic concepts from materials chemistry and physics along with some input from engineering design. Different combinations of composite materials have been tested for this purpose, which can be broadly classified as ceramic-based, polymer-based, and inorganic/organic polymer-based composites. These efforts mostly involve mimicking nature, which necessitates gathering thorough knowledge of the mineralized and living tissues found in nature (Oliveira et al. 2003).

6.2.1 *Ceramics-Based Hierarchically Structured Composite*

Hard tissues of vertebrates, like bone and teeth, typically have a 3D complex hierarchical structure with an intimate composite of the inorganic mineral phases (hydroxyapatite, (HA)) and organic matrices like collagen and nanocollagen protein fibrils as depicted in Fig. 6.2 (Imai and Tatara 2003; Meyers et al. 2008). Synthetic ceramics, e.g., tricalcium phosphate, HA, and other suitable bioactive (ability to form bone apatite like material on their surfaces) calcium phosphate derivatives can

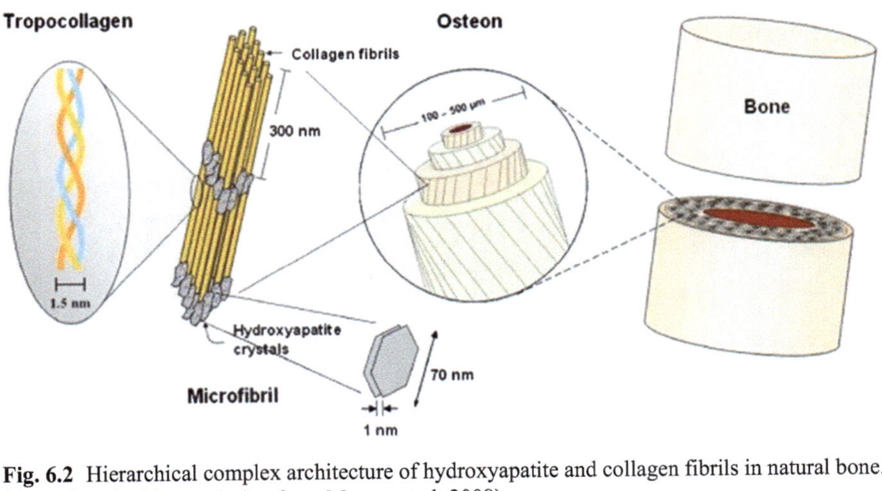

Fig. 6.2 Hierarchical complex architecture of hydroxyapatite and collagen fibrils in natural bone. (Reproduced with permission from Meyers et al. 2008)

be a potential bioceramic material system for the replacement of the mineral part of a synthetic hard tissue structure, since they can easily replicate the bone structure, maintaining the biocompatible behaviors with most of the related cell types. Moreover, the osteoconductive (ability of the material to provide a template for new bone formation) properties and chemistry of such a synthetic structure allows the formation of new bone on its surface (Qiu 1999). Hence, bioceramic-based composites with appropriate 3D geometry are the effective orthopedic biomaterials for the regeneration of hard tissues. In this regard, researchers have tried bioceramic-based composite systems with different ceramic materials like calcium phosphate derivatives, bioglass, and mesoporous bioactive glasses (MBG) with different organic polymers like collagen, cellulose, and other synthetic polymer with a variety of novel fabrication processes such as periodic precipitation, biomimetic approach, 3D printing and plotting as well as mineralization and sol–gel route to develop the 3D bone-like hierarchical structure (Bradt et al. 1999; Chang et al. 2001; Yokogawa et al. 1998; Murphy 2002; Girija et al 2004). However, tailoring ceramic-based polymer composites with hierarchical structures similar to natural bone (mineralization of extracellular matrix) are still challenging. In this section, attempt has been made to explore the different bioceramic material system used to fabricate the hierarchical composite structure for medical applications.

The preparation of various hierarchical structures of calcium phosphate and organic polymer nanocomposites is currently being carried out by many researchers (Imai and Tatara 2003; Murphy 2002; Girija et al 2004; Du et al. 1999, 2000; Yuan 2001; Murphy 2000). By adding organic polymer, e.g., gelatin, with calcium phosphate in different solution preparation conditions, such as concentrations of precursors and gel density, the architecture of the composite can be easily controlled according to the application. This unique composite hierarchy construction was fabricated through the calcification of a poly(acrylic acid) (PAA) hydrogel (Furuichi et al. 2006). Macroscopically, the calcification was achieved by the diffusion of calcium ions in an organic hydrogel-containing phosphate solution. Micrometer-scale layered structures consisting of HA nanoscale crystals in the opaque products are found (Fig. 6.3). The porous HA structure acts as an adsorbent as well as provides

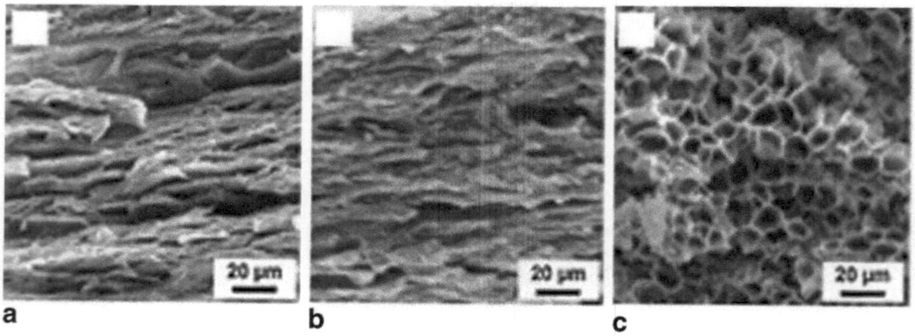

Fig. 6.3 Scanning electron microscope (*SEM*) images of calcined porous structures fabricated at standard conditions by various gel densities with **a** 5 g, **b** 7 g, and **c** 10 g poly(acrylic acid) (*PAA*) in 45 cm^3 water. (Reproduced with permission from Furuichi et al. 2006)

strength to the scaffold. The mechanical properties and structure of this laminated HA–polymer composite are similar to natural bone and thus can serve as a potential replacement (Furuichi et al. 2006; Wu and Luo 2011; Yun et al. 2011).

Homogeneous nanocomposites of HA and organic collagen hydrogel were developed using the in situ precipitation technique through dual template-driven reaction and freeze-drying method (Shen et al. 2011). After the multilevel freeze-drying process, the products exhibited hierarchical architecture of foams with primary macroporosity and sub microporosity depicted in Fig. 6.4. The ultrafine nanoscale HA particles were homogeneously distributed in the organic collagen. The foam scaffold, with interconnected porous hierarchical structure, provided superior mechanical properties and enhanced bioactivity. The collagen hydrogel acts as an organic template, provides an aqueous 3D compartment for the precipitation reaction of calcium phosphate ions, and also gives a template-driving force under the regulation of pH. Controlling the pH of the template helps in controlling the nucleation and size of the HA phase, leading to the structural change of the composite.

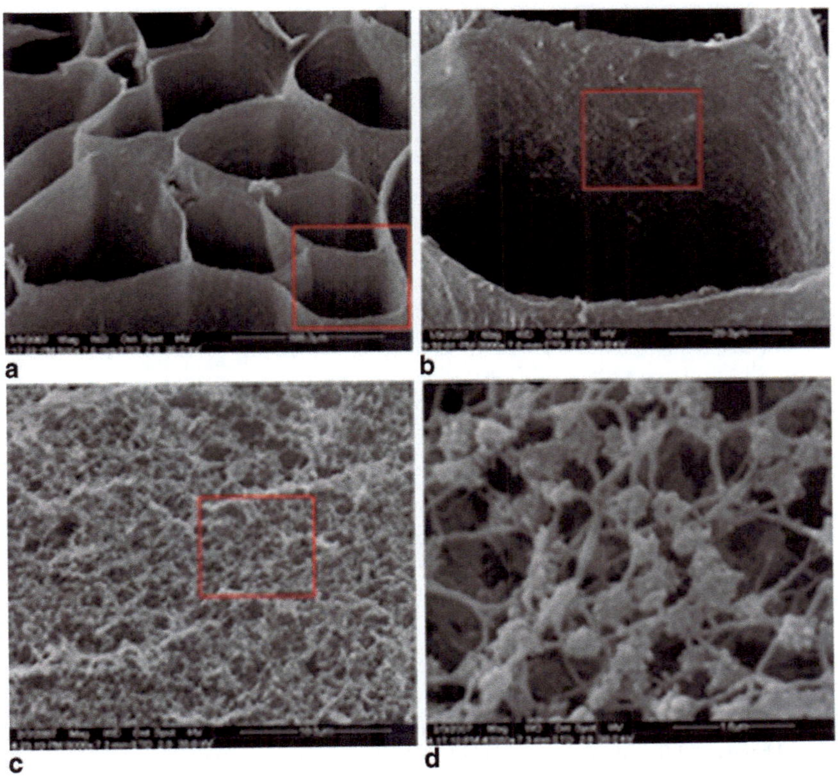

Fig. 6.4 SEM images of HA/collagen nanocomposite scaffold, *insets* are enlarged in the following figures. **a** 500X, **b** 2000X, **c** 5000X, **d** 40,000X. *SEM* scanning electron microscope, *HA* hydroxyapatite. (Reproduced with permission from Shen et al. 2011)

Success of this composite system inspired the fabrication of diverse homogeneous inorganic/organic nanocomposites (Chang et al. 2001; Yuan 2001; Shen et al. 2011).

Among other approaches, the laminated calcium phosphate PAA composite with a hierarchically layered structure was successfully prepared using a periodic precipitation technique. This was achieved through a nonequilibrium reaction system with the diffusion of reactants in an organic polymer matrix. Porous HA was bent microscopically by removing the organic components by sintering the composites at 700°C in air. The structural design of the products was altered from a layered structure into a cellular one with equiaxed porosities by controlling the concentration of the reactants depicted in Fig. 6.5 (Imai and Tatara 2003). In another effort, bone-like nano-HA/collagen (nHAC) biocomposite was developed by mineralizing the type I collagen sheet. Basic in vitro (*test conducted in an artificial environment*) and in vivo (*test conducted in a living organism*) studies revealed the effective bioactive and biodegradable property of this composite, but it failed to give

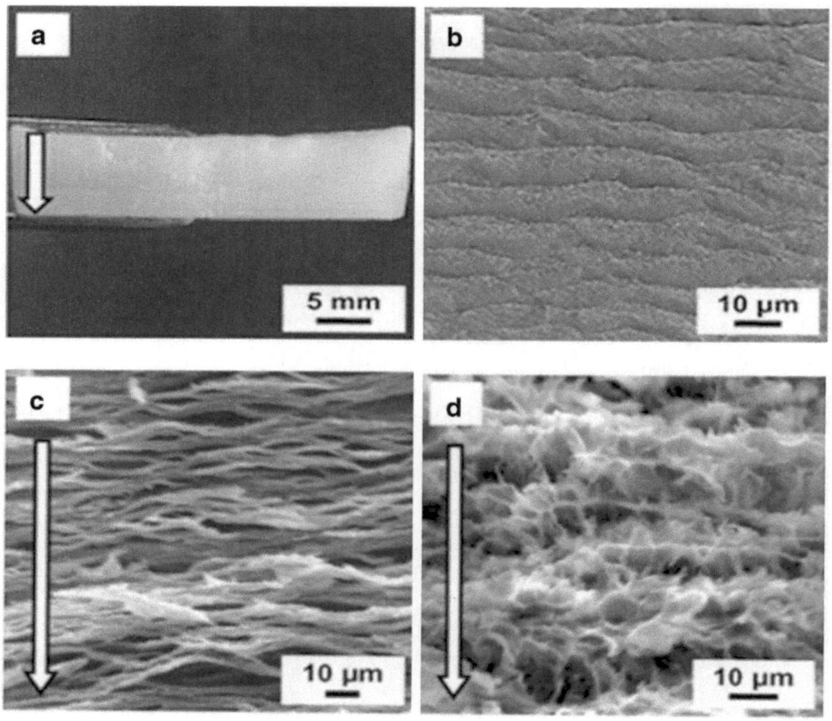

Fig. 6.5 **a** Cross-sectional image of solidified gel **b** represents the FESEM image of hierarchically laminated bands of diffused calcium ions $[Ca^{2+}]=4.0$ M (molar concentration) and phosphate ions $[PO_4^{2-}]=1.2$ M) in the gel matrix. **c** and **d** Porous HAp consisting of grooved thin films with cellular structure obtained in the lower concentration of precursor. *FESEM* field emission scanning electron microscope. (Reproduced with permission from Imai and Tatara 2003)

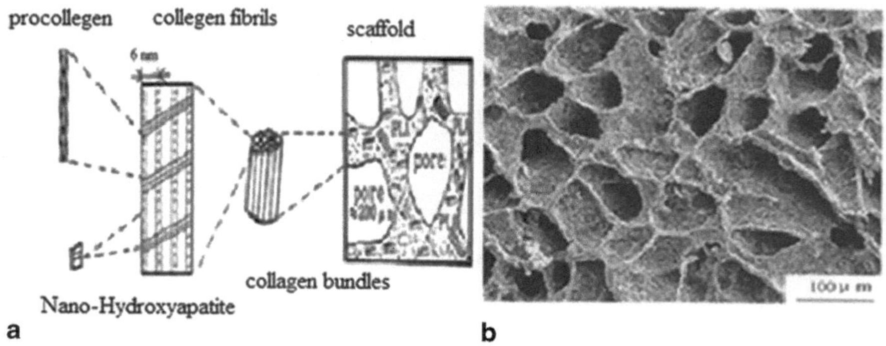

Fig. 6.6 **a** Schematic of nHAC/PLA nanocomposite with hierarchical structure **b** SEM micrograph of actual hierarchical scaffold. *nHAC* nano-HA/collagen, *PLA* poly(lactic acid), *SEM* scanning electron microscope. (Reproduced with permission from Liao 2004)

reasonable mechanical strength (Du et al. 1999, 2000). In order to get better mechanical properties, as well as improving the forming ability of the material, a novel composite of nHAC/poly(lactic acid) (nHAC/PLA) was explored. This composite structure possesses similar composition and hierarchical microstructure to natural bone and can be easily fabricated using a biomimetic strategy (Fig. 6.6). nHA and collagen assembled into mineralized fibril and 3D porous scaffold materials imitate the microstructure of cancellous bone. The cell culture and in vivo experiments established the potential bioactivity and its potential use for bone tissue engineering (Liao 2004).

New generation of biomaterials, such as MBG, have been extensively used in the bone restoration due to their excellent bioactivity, degradation, and drug delivery ability (Wu and Luo 2011), though the intrinsic brittleness and low mechanical strength of MBG restrict its use in 3D scaffolds construction. To overcome these limitations, a modified 3D printing technique with polyvinyl alcohol (PVA) as a binder is employed. Hierarchical pore architectures of 3D printed multifunctional MBG scaffolds showed high compressive strength and reduced brittleness due to the addition of PVA as a binder and also good mineralization ability (calcium phosphate crystal formation from bone-forming cells) for bone regeneration (Fig. 6.7). In another effort, MBG and concentrated alginate ceramic-based nanocomposite scaffold were fabricated via 3D plotting. The scaffold structure, well organized with nanochannels and controlled macro–micro pores, was found to be very effective for tissue engineering and drug delivery applications (Luo et al. 2013). Systematic study of MBG/alginate composites explored the potential mechanical strength, apatite mineralization ability, and cytocompatibility. MBG/alginate paste provides good processing ability in the 3D plotting process and the incorporation of MBG particles in the composite significantly enhances the mechanical and alkaline phosphate activity of the scaffolds (Luo et al. 2013). Finally, MBG ceramic-based composite studies paved the way for solving issues that normally exist with ceramic-based biodegradable composite scaffolds, e.g., uncontrollable pore architecture, poor strength, brittleness, and the constraint of second sintering (Wu and Luo 2011).

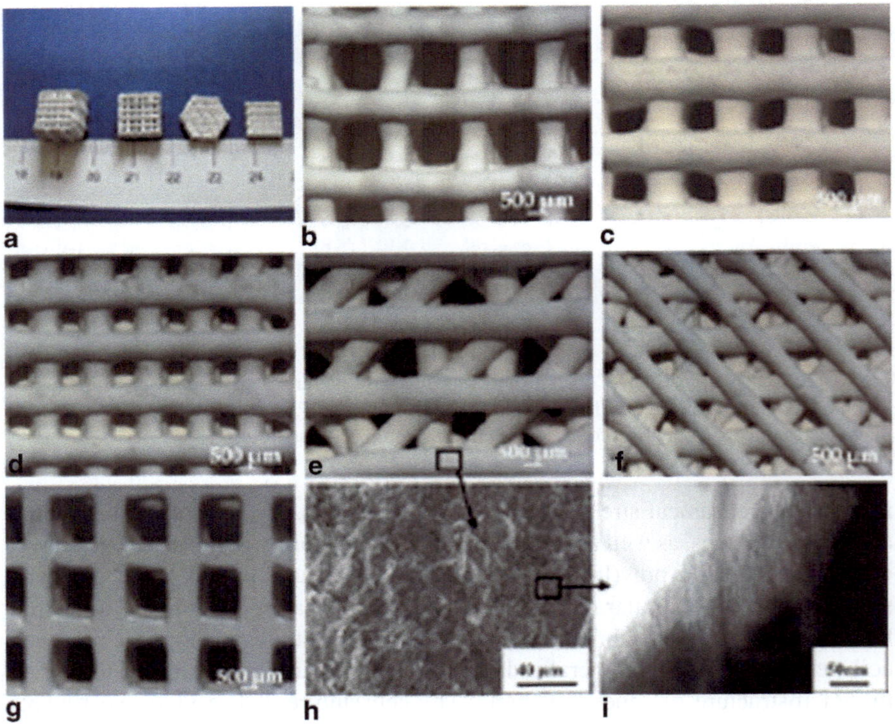

Fig. 6.7 **a** Mesoporous bioactive glasses (*MBG*) scaffolds with different pore morphology and microstructure showing varying pore sizes of **b** 1307±40 μm, **c** 1001±48 μm and **d** 624±40 μm; and **d–f** diverse morphologies. **g** Pore morphology of the bottom side of the MBG scaffolds, which remain open. **h** SEM image of the microstructure of pore walls. **i** TEM micrographs of pore walls with well-ordered mesopore channel structure. *MBG* mesoporous bioactive glasses, *SEM* scanning electron microscope, *TEM* transmission electron microscope. (Reproduced with permission from Wu and Luo 2011)

Hierarchical 3D scaffolds with well-arranged and interconnected pores at multiple length scales were fabricated with bioactive glass (BG)/poly ε-caprolactone (PCL) using three different combinations of synthetic methods (Yun et al. 2011). These methods include the following: (1) evaporation-induced self-assembly process in the presence of nonionic triblock copolymer, $EO_{100}PO_{65}EO_{100}$ (F127) as a template, (2) salt leaching (NaCl templates), and (3) rapid prototyping. The triblock copolymer templating technique introduced nanopores, the salt-leaching technique produced micropores with sizes varying from ten to hundreds of micrometers, and the rapid prototyping technique introduced giant pores (more than 100 μm) in the 3D scaffolds, as observed in Fig. 6.8. This composite material showed excellent bioactivity, biocompatibility, and better mechanical properties aiding in bone regeneration (Yun et al. 2011).

All the above discussion reveals the considerable research efforts from the past two decades to mimic living hard tissue through ceramic-based hierarchical com-

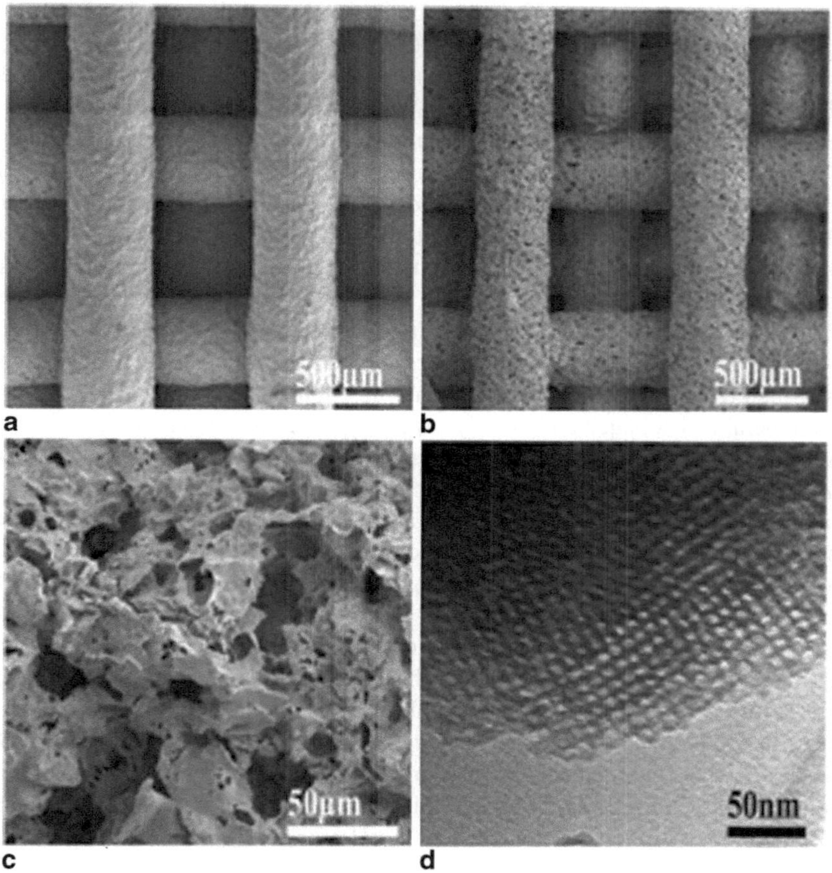

Fig. 6.8 SEM images of **a** MBG 0.6, **b** MBG 0.6Na1, **c** enlarged image of (**b**), **d** TEM image of MBG. *MBG* mesoporous bioactive glasses, *SEM* scanning electron microscope, *TEM* transmission electron microscope. (Reproduced with permission from Yun et al. 2011)

posite structures. These efforts are mainly driven to achieve high mechanical properties along with other functionalities, e.g., bone mineralization, drug delivery, etc. But a significant challenge still exists towards achievement of complex architectures with desired mechanical properties. Further innovations in material systems and processing techniques can meet the medical demand for hard tissue replacement in the future.

6.2.2 Collagen-Based Hierarchically Structured Composite

Collagens are natural polymers that forms collagen fibers and are abundantly distributed within the living tissues including bone, ligaments, tendons, skin, and blood vessel walls. The fibrillar protein collagen type I is one of the common structural elements found in a variety of skeletal bone tissues and the protein for controlling the cell functions. Recently, it has been treated as the key material for tissue engineering and regenerative medicine due to its exclusive biocompatibility, osteoinductivity, and good attachment to cells, and its ductile properties, which help to increase the fracture toughness of the composite as well as its relative abundance in living tissue (Stegemann 2013). To mimic the hierarchical morphology of the natural hard tissues, collagen-based composites are the appropriate choice for material researchers. This subsection will feature collagen-based hierarchical composites, mainly collagen and calcium phosphate derivatives with different fabrication routes, and their effects on the structure are addressed (Stegemann 2013; Zhang et al. 2004; Weadock et al. 1983; Roche and Ronzière 2001).

There are significant interests in replacing hard tissues like bone, with collagen-based hierarchical structures. The intricate hierarchical structure of natural bone with mineralized collagen and orderly deposited HA minerals within it, leads these tissues to take a part in special roles in supporting different functions (Weiner and Wagner 1998). Controlling the nucleation and growth of calcium phosphate (CaP) minerals through the collagen matrix in an aqueous media, and developing hierarchical structure of nanofibrils that exactly resembles the bone tissue is a challenging step toward the synthesis of collagen-based composites (Zhang et al. 2003; Clarke 1993). In this structure, organic and inorganic nanophases have specific relationship with respect to each other to balance the perfect proportion of biological and mechanical properties. In such structures, organic phases like collagen, increase the cell attachment, biocompatibility, and fracture toughness. Inorganic phases like CaP, increase the osteoconductivity and strength. Zhang et al. (2003) have designed the hierarchical structure to resemble the extracellular matrix of bone through self-assembly of nanofibrils of mineralized collagen. Collagen and calcium phosphate can easily self-assemble when the pH of the collagen-containing calcium and phosphate solution changes. HA crystals grow on collagen fibril surfaces with their *c*-axis oriented along the longitudinal axis of the fibrils as shown in Fig. 6.9. In a separate effort, Bradt et al. (1999) have revealed that the addition of small polyaspartate in the reaction mixture can significantly improve the attachment between the collagen fibrils and the calcium phosphate crystals. The authors noted that this was possible due to the polyaspartate getting adsorbed on to the calcium phosphate surface, inhibiting clustering of the calcium phosphate. Less clustering increases the surface energy that helps in bonding the CaP to the collagen fibrils better, through electrostatic interactions (Fujisawa et al. 1996). Kinetics of the calcium phosphate and fibril formation have a positive effect with polyaspartate concentration in the solution (Bradt et al. 1999).

Recently, researchers have demonstrated naturally calcified biominerals can be remineralized, replacing the Ca compound crystals with amorphous silica. This process produces a composite with interpenetrating collagen and silica phases possessing a unique hierarchical structure (Niu et al. 2013). The authors have also mentioned the possibility of fabricating collagen–apatite–silica multiphase composites.

Hierarchical tertiary structure and biocompatibility of collagen type I make it favorable as templates for many organic–inorganic composite materials. Natural collagen and mineral biocomposites gives better strength and toughness unlike a reconstituted collagen, due to the incidence of different scale of hierarchy in the

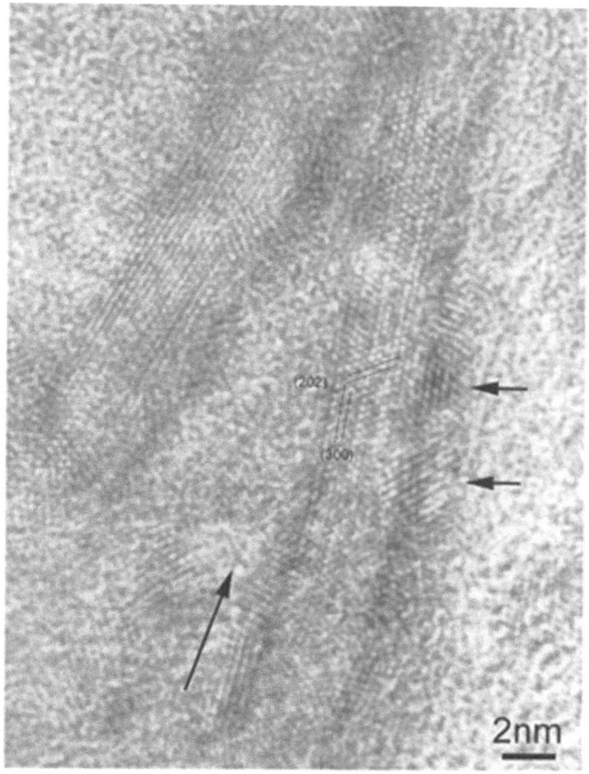

Fig. 6.9 HRTEM image of mineralized collagen fibrils. Collagen fibrils are in longitudinal direction as indicated by the *long arrow*. Two *short arrows* show HA nanocrystals. *HA* hydroxyapatite. (Reproduced with permission from Zhang et al. 2003)

organic components (Ehrlich 2010; Weiner et al. 2006). Natural biominerals, containing collagen-based composites, demonstrated different range of nano- to macroscales of sophisticated complex hierarchical architectures. These complex hierarchical architectures are responsible for the biominerals' potential mechanical property and other metabolic functions.

The complex spatial organization found in natural tissues with combinational organic and inorganic phases are far superior than what materials researchers are able to currently achieve by the contemporary engineering techniques (Niu et al. 2013). Still, collagen-based biocomposites are one of the high-performance engineering materials to mimic natural tissue. However, with novel processing routes being explored, researchers are hopeful to match the complex organization of natural bone structure.

6.2.3 Inorganic/Organic Polymer-Based Hierarchical Structured Composites

A large number of inorganic/organic polymer-based composites have been synthesized with various compositions and processing techniques to achieve the 3D hierarchical morphology for medical applications (Chen et al. 2011; Sanchez et al. 2011; Shi et al. 2013). Proper organization of one- and two-dimensional building blocks of organic/

inorganic components leads to the formation of complex hierarchically structured nanocomposites that can take full advantage of each kind of material (Zhao et al. 2012). Asymmetric hollow microspheres with polystyrene-ethylacrylate (PSt-EA) semi-spherical cores and Ni-silicate hierarchical porous shells, aimed for drug and biomolecule delivery, were fabricated successfully by the combination of three processes, (a) emulsifier-free polymerization, (b) modified Stober method (to synthesize the core/shell PSt-EA@SiO_2 microspheres), and (c) in situ hydrothermal conversion reaction (Fig. 6.10). Hierarchical Ni-silicate and hollow spherical PSt-EA composites

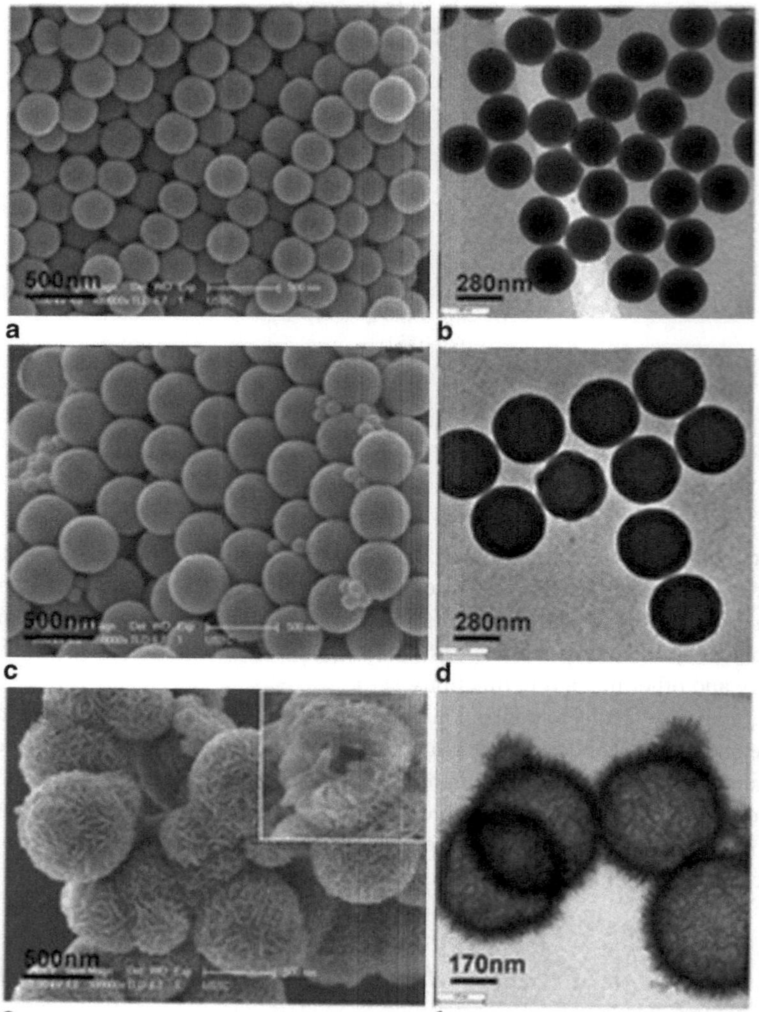

Fig. 6.10 SEM and TEM images of **a–b** PSt-EA microspheres; **c–d** PSt-EA-SiO_2 microspheres **e** hydrothermal microspheres and **f** hollow Ni-silicate microspheres. *SEM* scanning electron microscope, *TEM* transmission electron microscope. (Reproduced with permission from Zhou et al. 2013)

were fabricated through calcinations of asymmetric PSt-EA/Ni-silicate at 500°C for 5 h (Zhou et al. 2013). The large porous and nanostructured morphology of inorganic/organic materials have attracted attention as a carrier vehicle for many biological molecules and antibiotics due to their unique properties, e.g., high pore volume, high surface area, and well-defined architecture (Dergunov 2011).

Many studies have proven that controlled architecture of organic and inorganic hybrid composite materials has high potential for medical applications because of their ideal biocompatibility, biodegradability, and superior mechanical properties (Kakizawa et al. 2004; Epple et al. 2010; Zhang and Kataoka 2009). Among these, a hybrid structure of calcium carbonate ($CaCO_3$) and a rational choice of biocompatible polymer (usually selected based on their mechanical, thermal behavior, and chemical stability) have attracted attention for drug carrier vehicle applications (Zhang and Kataoka 2009; Sokolova and Epple 2008). Shi et al. (2013) have successfully developed hierarchical poly (N-isopropylacrylamide) and calcium carbonate micro/monohybrid composites through in situ biomineralization reaction for smart drug delivery applications. Here, the nucleation and growth of inorganic crystals occurs at the interface of the organic substrate in biological conditions.

The recent contender in this category of organic–inorganic composites are multifunctional foams with hierarchical porosity, made up of polymer and inorganic composites, which have shown promising results toward the applications in biomedicine and tissue engineering (Dong et al. 2007; Su et al. 2012; Singh et al. 2010). Hierarchically organized foams were successfully developed by synthesizing bioactive PVA sepiolite composites via freeze-drying process (Wicklein et al. 2013).

Apart from their well-organized multiscale structure, organic–inorganic composites are instrumental in their unique functionalities to solve many leading medical challenges, like controlled drug delivery, new bone formation, and properly imitating the skeletal bone structure (Sanchez and Julian 2005).

6.3 Processing of Hierarchical Composites for Medical Applications

In the processing of hierarchical composites, the architecture and synthesis of functional materials have attracted attention from researchers across the globe. Their main focus lies in organizing the nanostructures of functional materials over multiple length scales (Cölfen and Mann 2003). Several processing methods or routes are used in the development of hierarchical composite structures. Some of the important routes include sol–gel processing (Kreetachat et al. 2013), rapid prototyping, layer-by-layer growth (Jan and Kotov 2007), solvent-evaporation induced self-assembly (EISA) technique (Zhang et al. 2003), coprecipitation (Gee and Hong 2003), alkaline treatment of collagen matrix (Rocha et al. 2002), calcification of HA-containing hydrogel (Furuichi et al. 2006), as well as thermal decomposition of organometallic compounds. The most conventional composite fabrication methods employed are mechanical mixing and coprecipitation (Zhang et al. 2003; Yin et al.

2000; Wei and Ma 2004a; Taguchi et al. 1999 Doi et al. 1996). However, these techniques suffered from two common limitations. First, inorganic particles cannot be distributed well in the organic matrices at the nanolevel using these techniques. Moreover, the scaffold synthesized from these methods cannot meet various required functions because of their single-sized porous structure. These disadvantages limit their applications in hard tissue repair and replacement.

Natural bone consists of a complex nanofibril system with an intricate hierarchical structure of mineralized collagen. To exploit this property of bone, some research groups tried and succeeded in depositing HA in a collagen matrix through a self-assembly route (Kischi and Itoh 2001). Also, few researchers prepared HA collagen nanocomposite through crystallization of HA on collagen fibers from the acid–base reaction using $CaHPO_4$ as a HA precursor (Miyamoto et al. 1998). However, it has been observed that these HA collagen composites lack the compositional and structural similarities to natural bone.

Among the methods mentioned above, the sol–gel route and the self-assembly technique have attracted much attention recently because of their unique processing route and ease of fabrication. The interfacial sol–gel approach is one of these which was proven to be the most efficient in developing hierarchical structures (Antonietti 2001). This technique not only mimics and replicates natural biomaterials, but is also helpful in the generation of an intricate structural profile (Mann 2001). In this route, the initial precursors (sol ingredients) undergoe surface gelation and then later react with biological elements like proteins, polysaccharides, and functional bio-residues, to develop advanced materials with predetermined chemical and physical properties like molecular weight, surface energy, hydrophilicity, shape, and structure (Dong et al. 2007). Some researchers also suggested the application of biomimetic approach in designing the hierarchical composite structures (Liao 2004). Basically, the approach was more useful in the preparation of calcium phosphate organic polymer nanocomposites with adequate control over the morphology of calcium phosphate crystals. Research studies reported that the HA-organic polymer nanocomposites hold a structural similarity to the natural HA-collagen composite present in real bone (Furuichi et al. 2006). Also, bone-like HA coatings over the various organic substrates have been achieved by mimicking the biological process in a simulated body fluid (SBF) environment (Tanahashi 1994; Michalske et al. 1990; Rhee et al. 2000). Reports show that the apatite development was restricted only up to the surface of the polymer fiber, when exposed to the precursor solution in the process of fabricating apatite–polymer fiber composite by soaking in SBF (Furuichi et al. 2006). An alternative soaking approach carried out at an ambient temperature using calcium and phosphorous solutions, resulted in a 3D structured HA formation in the matrix of organic hydrogel (Taguchi et al. 1998). Some researchers have applied a PAA matrix containing phosphate ions for the diffusion of calcium ions, to achieve hierarchical organized structures of HA (Furuichi et al. 2006).

Hierarchical 3D porous architectures, produced by rapid prototyping (RP) techniques, are found to posses controlled and regular pores sizes, shapes, and interconnectivity (Jan and Kotov 2007; Landers et al. 2002; Hutmacher et al. 2004). Most of the RP processes, including 3D plotting processes are conducted under mild environment conditions, so that biological components, such as proteins and even living cells can be printed simultaneously by using an appropriate material (Fedorovich et al. 2008;

Maher et al. 2009). In this process, printing is carried out with the help of a robotic arm. The number and shape of pores along with their spatial distribution in the scaffold are controlled by computer-aided design (CAD)/computer-aided manufacturing (CAM) modeling of the movements of robots. The wide range of materials already used for 3D printing includes BG and PCL (Sobral et al. 2011; Shor et al. 2007).

Bioactive glass ceramics were developed for fabricating scaffolds, which were easily chemically bonded with living hard tissue by forming a HA layer (Hench 1971; 2006; Vallet-Regí et al. 2003; Habraken et al. 2007). A recently developed MBG show good biocompatibility and promotes new bone formation when loaded with osteogenic agents. This was due to its unique pore architecture, which was comprised of large specific surface area, controllable pore size, as well as pore structure (Yan et al. 2004; Shi et al. 2006; Yun et al. 2007; Izquierdo-Barba et al. 2008a; 2008b).

Yun et al. (2011) have developed 3D porous scaffolds based on MBG using triblock copolymer $EO_{20}PO_{70}EO_{20}$ (P123) or $EO_{100}PO_{65}$- EO_{100} (F127), and polyurethane (PU) as co-templates for achieving nanoporosity less than 10 nm in diameter and giant porosity with diameter greater than 100 μm simultaneously. But all these processing methods lack adequate control over interconnecting pores and structures, which is crucial for bone tissue and drug delivery applications.

Another effort in synthesizing MBG/PCL composite scaffolds by the solvent-casting–particulate-leaching method, also results in poor interconnectivity of the pores in the structure (Li et al. 2008). Yun et al. (2011) reported an improved approach in which 3D nano–macro–giant-porous MBG/PCL composite scaffolds were synthesized using a combination of the sol–gel, polymer templating, salt leaching, and RP techniques with a gantry robotic deposition apparatus. The schematic representation of experimental procedure is been shown in Fig. 6.11. In this fabrication technique, NaCl was mixed with the gel paste before the 3D printing process. The purpose of adding salt was to induce large-sized pores in the scaffold. This gel paste was extruded over the surface of a cold polymer matrix substrate to form the scaffold, followed by the removal of NaCl using deionized water (Yun et al. 2011).

Some researchers used a biopolymer deposition mechanism in the past to fabricate alginate scaffolds from low concentration alginate sols (0.1–3 % w/v) (Khalil

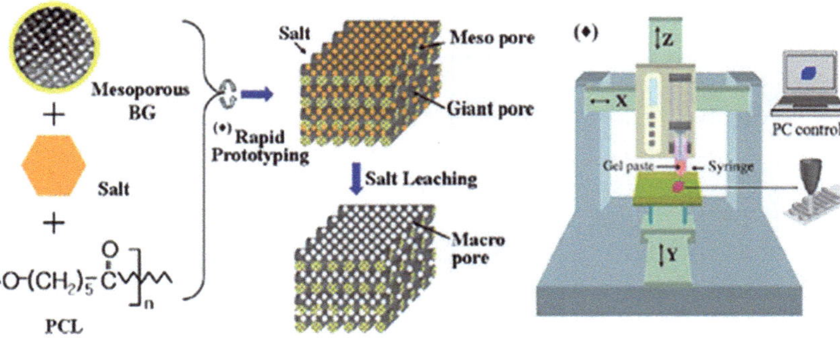

Fig. 6.11 A schematic of processing routes for synthesizing inorganic–organic composite scaffolds with meso–macro–giant pores in a hierarchical 3D structure. (Reproduced with permission from Yun et al. 2011)

et al. 2005; Khalil and Sun 2009). They attempted to fabricate two to three layers of alginate scaffolds, but even with 30 layers, this scaffold lacked the stiffness to hold its 3D structure and thereby exhibited poor mechanical strength. Also, the internal pores were closed in these scaffolds which adversely affected the drug delivery (Khalil et al. 2005). Despite this, it was reported that alginates have some positive characteristics like good biocompatibility and controlled degradation as well as the ability to cross link with divalent cations (Augst et al. 2006). To overcome the limitations related to porosity and strength, an improved processing route was suggested by Luo et al. (2013). They developed a composite material by combining alginate with MBG. For fabrication, the alginate powder was mixed with MBG powder in a fixed mass ratio and stirred in a PVA and water solution to form a paste, which was used for 3D printing at room temperature. The scaffolds were designed in two kinds of patterns XX (X direction in the first layer and Y direction in the second layer) and XXYY (first two layers in the X direction and next two layers in the Y direction). Figure 6.12 shows the variation of the pore structure. The final scaffolds were then

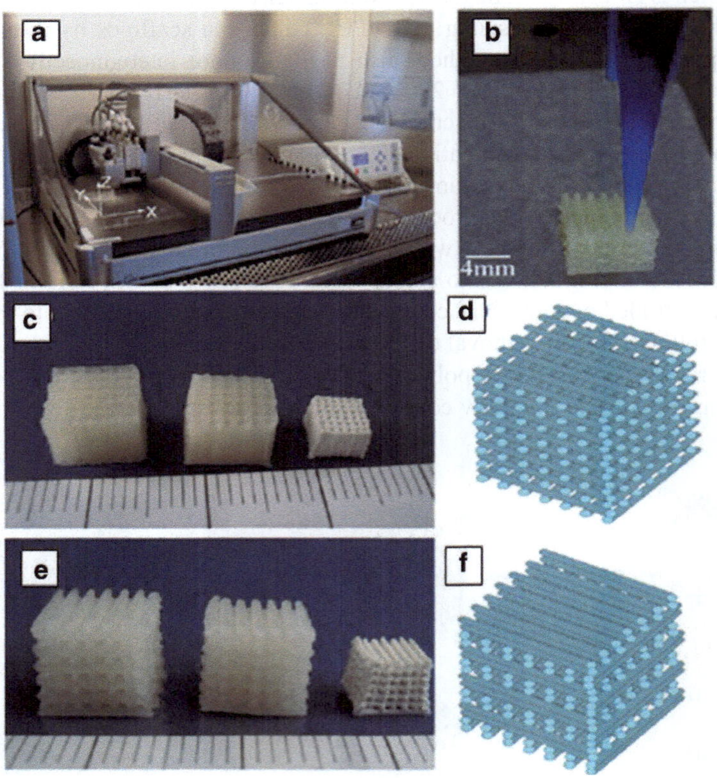

Fig. 6.12 **a** 3D plotting machine; **b** scaffold being plotted. Scaffolds (30% MBG/alginate) and CAD models with **c, d** XY pattern **e, f** and XXYY pattern. Scaffolds are shown in the wet state after cross-linking in 500 mM $CaCl_2$ solution and in the dry state. *MBG* mesoporous bioactive glasses, *CAD* computer-aided design. (Reproduced with permission from Luo et al. 2013)

dipped into an aqueous solution of $CaCl_2$ for 10 h, washed in deionized water, and dried at room temperature.

The researchers in this field have tried different processing routes with different combination of composite materials to develop the ideal material that exactly mimics the living hard tissue with their unusual shape and complexity. But designing a hierarchically organized structure with controllable length scales and pore sizes with adequate mechanical strength is a major challenge. With advancements in the field of materials science and chemistry there is a definite possibility to derive newer processing routes with the thorough understanding of capable materials, which resembles living hard tissues.

6.4 Properties of Hierarchical Composites

Mechanical and biological properties are the most critical part of designing hierarchical composites for medical applications. With advanced processing techniques, it is fairly possible to construct hierarchical structures with different length scales. In general, the macrostructure impart mechanical strength, while the nanocounterpart helps in biological activity. In this subsection, effort has been made to explain the mechanical and biological properties of various hierarchical composites and evaluate their potential use in medical applications.

6.4.1 *Mechanical Properties*

Over the past decades, significant research effort has been made in achieving an ideal combination of properties like bone, and researchers have engineered different processing routes to develop the hierarchical composite for this purpose. The nHAC/PLA-based composite structure, fabricated by the biomimetic route has shown improvement in the mechanical strength and the formability (Liao 2004). The compressive strength, evaluated for a cylindrical sample (9 mm diameter and 15 mm length based on ASTM D5024-95a) was found to increase with PLA concentration, even when the porosity content was maintained similar for all compositions. However, the elastic modulus value showed a different trend and was not in accordance with compressive strength for the respective compositions. The researchers envisioned the clinical application of nHAC/PLA-based composites by tailoring the properties through PLA concentrations. A composite of HA and PAA has demonstrated a Young's modulus and tensile strength of 3.7 and 73.1 MPa, respectively, when evaluated by the three-point bending test (Furuichi et al. 2006). These properties are close to those of natural bone (Young's modulus: 0.3–24 GPa; tensile strength: 100–220 MPa), indicating the potential use of such structures in orthopedics.

In regards to MBG/PCL composites scaffolds having hierarchical structure, the compressive modulus of this composite having MBG to PCL weight ratio (0.6:1) is

4.3±0.7 MPa (Khalil et al. 2005). But when the concentration of NaCl granules increased in the MBG/PCL composites to improve hardness as well as to enhance molding capability, the compressive modulus decreased. MBG0.6Na0.5 and MBG0.6Na1 scaffolds have shown compressive modulus to be 3.3±0.5 and 2.2±0.3 MPa, respectively. However, the compressive modulus of MBG0.6Na1 was found to be more than the value for the MBG/PCL composite scaffolds (0.5±0.1 MPa) prepared by the salt leaching method. Researchers also mentioned that increasing the amount of NaCl not only eased the formation of macropores in the scaffolds but also increased the porosity which is responsible for the decrease in the mechanical strength. The dynamic mechanical properties, viz., damping factor and storage modulus (dynamic stiffness) for the scaffold, were also evaluated. Macro porous composite scaffolds, with ~84 % porosity, offer a significant increase in the value of the damping factor and a clear decrease in the value of the storage modulus. MBG0.6Na1 with superior viscoelastic behavior is found to be suitable for replacing cartilages.

As mentioned earlier, 3D printing had been a very attractive fabrication technique for synthesizing scaffolds with hierarchical structures. A newer approach was proposed by Luo et al. (2013) for synthesizing a composite scaffold combining MBG and alginate powder in different proportions with different 3D printing orientations, namely XY and XXYY. The compressive strength of the composite was tested before and after immersion in SBF for a different time period. It was estimated from the test results that the compressive strength and compressive modulus of the plotted alginate scaffolds increased with increasing MBG concentrations (Fig. 6.13). The composite having 50 % concentration of MBG had higher compressive strength and modulus compared to those of the pure alginate scaffold for both kinds of plotted patterns XY and XXYY, respectively for a particular time period. The presence of open pores in the Z direction of XXYY scaffolds showed an adverse effect on its strength when compared to XY scaffolds. Longer incubation periods in SBF also reduced the mechanical strength of the scaffold (Luo et al. 2013). In another effort, MBG scaffolds, prepared through 3D printing and incorporating polyvinyl alcohol as a binder, have shown excellent mechanical properties (Wu and Luo 2011). For a MBG scaffold with ~ 1000 μm pore sizes, made by 3D printing, compressive strength and modulus were 16.10±1.53 and 155.13±14.89 MPa, respectively, and as the percentage of deformation increased the compressive strength also increased (Fig. 6.14). In other words, the compressive strength of 3D printed MBG scaffolds was 200 times greater than that of the polyurethane template type. The authors reported that it was easy to handle and utilized the 3D printed scaffolds on the basis of its mechanical properties, which was also comparable to the average compressive strength of human trabecular bone (2–12 MPa). On the other hand, researchers have developed inorganic materials like metal, glass, ceramic, and organic components (polymers) composites that can be architectured into hierarchical structures, and their structure property correlation can be easily tailored according to the desired applications. In such composites, the organic part contributes in shaping and better processing as well as providing the templating effect, and the inorganic components facilitate the mechanical and thermal stability (Sanchez et al. 2005). Hence, a majority of hierarchical composites are being developed for replacing bone tissue with the mechanical properties of such structures being the most critical aspect.

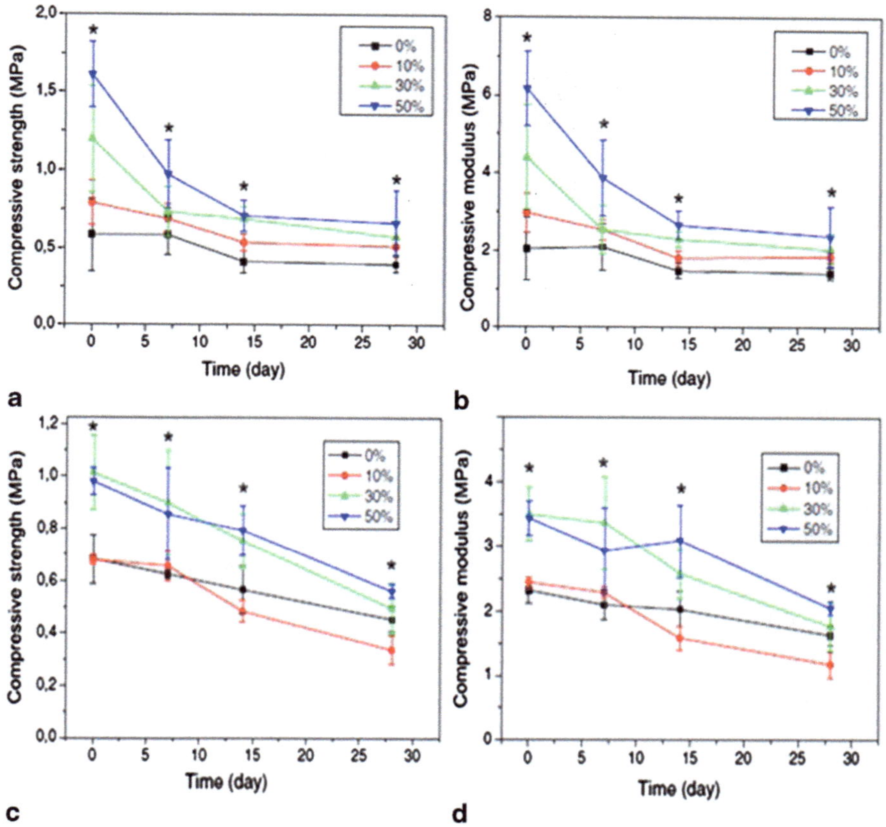

Fig. 6.13 The compressive strength and modulus of the (**a, b**) XY and XXYY (**c, d**) scaffolds after soaking in SBF for 2 h, 7, 14 and 28 days. The percentages in the legends denote % MBG in alginate scaffold. *MBG* mesoporous bioactive glasses, *SBF* simulated body fluid. (Reproduced with permission from Luo et al. 2013)

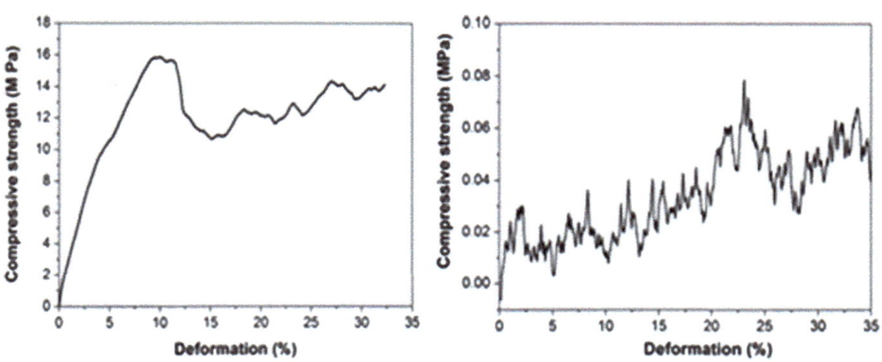

Fig. 6.14 Compressive strength of **a** 3D printed MBG scaffolds and **b** MBG scaffolds made by polyurethane templating. *MBG* mesoporous bioactive glasses. (Reproduced with permission from Wu and Luo 2011)

6.4.2 Biological Properties

Being a material for medical applications, the biological evaluation of hierarchical composite structures is a must. Hence, most of the available literatures include some discussion on biological properties of such composites and the role of its structural hierarchy. The evaluation would depend on the nature of targeted application. The majority of hierarchical composites are used in hard tissue engineering. This requires checking the compatibility of the composite structure with bone forming cells and ability of integration with the neobone formed. The composites developed for drug delivery should be checked for kinetics of drug delivery in vitro followed by in vivo, along with their compatibility in the human body. A summary of such evaluations with their outcomes in deciding the efficiency of the hierarchical composite structure in their targeted application is given below.

Researchers have found nHACollagen/PLA composite (Liao 2004) fulfilling most of the requirements as a suitable underlying layer for bone tissue engineering. This composite has the advantage of high biocompatibility as well as the nHAC promoting osteogenesis with controlled cell function. The morphological changes (mainly spindle shape to polygonal) of osteoblasts were observed in the nHAC/PLA composite, which was responsible for adhesion, proliferation, and maturation. The in vitro cell culture studies revealed spreading of cells in the interconnected pores of the composite which results in the smooth surface of the composite due to biodegradation. The biodegradation of scaffold did not change the pH value of the media and hence did not affect the growth of osteoblasts.

PLA is reported to restrict the acid leakages from the composite during degradation (Liao 2004). The possible reason is the presence of nanomineralized collagen in the nHAC/PLA composite. The in vivo studies, using rabbits as animal subjects, quick healing of large segmental defects was observed after 12 weeks using a scaffold. Thus, nHAC/PLA composites were found to be an appropriate bone substitute material, that eased new bone formation on its porous surface by facilitating osteoblasts attachment and providing a suitable environment for collagen and bone mineral deposition.

The biological behavior of MBG/PCL (Yun et al. 2011) composite scaffold was studied by Yun et al. both in vitro and in vivo. In earlier literature, it was reported that the MBG scaffold structure facilitated easy bone formation due to the diffusion of ions (Yun et al. 2007;Izquierdo-Barba et al. 2008a). The in vitro result in a SBF environment on MBG0.6Na1 scaffold showed the formation of HA nanoparticles over its surface, which covered the surface at an exposure of 24 h. The Ca^{2+} level and pH value for the PCL scaffolds did not change during immersion periods. But for the MBG0.6Na1 scaffold, both the Ca^{2+} concentration and pH increased initially while soaking was done for 8 h and then repeatedly decreased and increased with further soaking. The rapid release of Ca^{2+} and changing pH values in SBF indicate rapid HA deposition and thereby indicating superior bone formation ability compared to PCL scaffolds. The scaffold with macropores created by salt leaching also enhanced the bioactive properties in MBG0.6Na1 scaffolds. The in vitro test results were evaluated by studying human bone marrow stromal cell response for MBG0.6 and MBG0.6Na.1 scaffolds for the period of 15 days. The cell test results showed favorable cell behavior including better proliferation, osteogenesis, and similar cell numbers/counts for all the tested scaffolds with no significant differences in cell attachment efficiencies.

Researchers have emphasized the fact that bio scaffolds used for bone replacements offer improved osteoblasts growth in the SBF environment due to apatite formation which also further influences the in vivo response (Wu et al. 2007; Xu et al. 2008; Kokubo and Takadama 2006; Chou et al. 2005). MBG scaffolds, fabricated through 3D printing possess excellent apatite mineralization ability in SBF (Wu and Luo 2011). When soaked for 1 or 3 days, platelet-like crystals were found over the scaffold. The authors also commented that the use of a small amount of PVA solution, which acts as a binder during printing of 3D scaffolds, does not show any adverse effect on the bioactivity. MBG scaffolds display rapid release of ion in a biological environment when soaked for 7 days, which has been considered an important factor responsible for rapid degradation of the scaffold.

The ion-release behavior of MBG/alginate composite scaffolds and pure alginate scaffolds soaked in SBF for a period of 42 days, showed an increase in concentration of Ca^{2+} and silicate ions in the medium (Luo et al. 2013). But the concentrations of phosphate ions decreased with exposure time for all types of scaffolds (pure alginate and the MBG/alginate composite). The possible reason is precipitate formation on the products when combined with the released Ca^{2+} ions from calcium. The osteogenic differentiation was also evaluated over a period of 1 week by seeding human bone marrow-derived mesenchymal stem cells (hBMSC) onto a MBG/alginate scaffold after incubation. The results showed the incorporation of the MBG significantly improved the apatite-forming ability of alginate scaffolds in SBF and cell culture medium. This was further responsible for higher ALP activity and cell response over MBG/alginate composite surface compared to pure alginate scaffolds (Luo et al. 2013).

Efforts have also been made in the preparation of hollow mesoporous silica spheres (HMSs) with controllable pore sizes (Chen et al. 2010b). Inorganic nanocrystal core has been developed through selective etching process by Chen et al. (2011). In such mesoporous structures, the interior part is etched away, while keeping the outer shell intact from the unique solid silica core (Chen et al. 2010a). The hierarchical pore size distribution can be varied by parameters like reaction time, temperature, precursor concentrations, etc. It was reported that if the pores of varying sizes can coexist and maintain their identity at different parts of the structure, then molecules of different sizes could be diffused and used for size-dependent drug release (Lopez-Noriega et al. 2009).

However, as most of the hierarchical composites in medical application are being developed with the aim of application in orthopedics, the mechanical and biological properties must include the suitability in the targeted application.

6.5 Medical Applications of Hierarchical Composite

Virtually all living matters in the universe could be seen as a complex hierarchical structure, since this spatial scale architecture in the living organism helps efficiently perform all biological, chemical, and mechanical functions. Synthesized or artificial biomaterials with similar living structures are suitable to replace the natural functions efficiently in different in vivo conditions. Hence, the hierarchical composite are being explored for various medical applications. Major applications of such manmade hierarchical composites are found in tissue engineering, mainly to reinstate the damaged hard tissues, e.g., bone and teeth. Another opportunity is established as drug carrier vehicle, since it permits the extended release of a drug in a controlled manner. Further, use of such structures in cancer therapy is also discussed in this section by stressing the unique capability of hierarchical composite structures (Bae et al. 2014; n.d., http://en.wikipedia.org/wiki/Biomaterial).

6.5.1 Tissue Engineering

Application of hierarchical composite in tissue engineering mostly includes developing material system to replace the hard tissues, like bone. The scaffold or implant should have comparable properties to the hard tissue in their targeted application in order to work efficiently. Some of the mechanical properties of the bone are human cortical bone–elastic modulus 5–10 GPa, compressive strength 130–210 MPa, tensile strength 90–190 MPa, human cancellous bone—compressive strength 2–38 MPa.

Living hard tissues are a perfect organization of diverse microscale cells and nanoscale extracellular matrix (ECM) in a multiscale hierarchical structure. Native ECM has quite a complex structure, which gives the structural support to cells as well as controls the cell attachment. Presently, replacement and regeneration of hard tissues are performed through different therapies like autograft, allograft, and other artificial implants or scaffold made of metals, bioceramics, and polymers (Goldberg 1992;Costantino 1994). However, each of these therapies has their own limitations. Autografts (grafting tissues from the same patient) have the drawback of insufficient supply and donor morbidity as well as increase in the surgery time. Allografts (grafting tissue from a donor) and xenografts (interspecies grafting) are associated with the lack of immune response and the risk of disease transmission. Further, synthetic materials do not possess hierarchical structure in their original form, and as a result often fails to support the full functionality of the replaced tissue. Hence, engineering of biomimetic artificial 3D tissue scaffolds, incorporating multiple length scale structures for an effective cell attachment and growth, as well as formation in perfect shape and maintaining functionalities is of great importance in tissue engineering (Lutolf 2005; Bettinger 2009; Bonfiglid 1972).

Polymer-based synthetic ECM nanostructure was successfully developed via replica molding, nanoimprinting, and nanopattering methods. In this approach, nanoscale molds of biodegradable polymers, such as ultraviolet-curable polyurethane acrylate and polyethylene glycol were used to produce the fine ECM-like structure, which serves as a new platform to study the cell interaction (Kim et al. 2012, 2013;Bae et al. 2014). Researchers have cultured fibroblast (Kim 2012), muscle cells (Zhao et al. 2009), and stem cells (Yim 2010) on artificial nanoscale ECM and shown promising results on cellular response in the form of morphological change, adhesion, and proliferation of the nanoscale patterns as compared to the microscale patterns. The observations reveal that the development of artificial hierarchical composite scaffolds with optimized nano- and microscale structural geometries are necessary for tissue engineering (Teixeira 2003, 2006;Rebollar et al. 2008). Multiscaled porous architecture within the 3D scaffold helps mass transfer and cell migration through micropores. In addition, it gives the mechanical stability for structural support. The nanofeatures define rigidity and nanotopography of ECM, which allows for the cell attachment. Functional nanomaterials composite of HA and collagen were deposited onto preformed microscale patterns to engineer the combined micro/nanofeatures within the scaffolds (Bae et al. 2014;Tan and Saltzman 2004). Nanohydroxyapatite (nHAP)/polymer 3D scaffolds composite with highly controlled pore architecture was found to greatly improve the mechanical property and protein adsorption capability, due to higher surface energy and area of nanoparticles. By varying the thermally induced phase separation parameters, e.g., concentration of the polymer solution, nearly 90% of the interconnected porosity and controlled nano- to micropore sizes were achieved. Nanosized HA addition decreases the porosity slightly but greatly affects the protein adsorption, cell attachment and differentiation to form new tissue. HA, distributed within the pore walls and forming a loose fibrous structure, enhances the surface to volume ratio of the scaffold than in the solid pore walls, which greatly affects the protein adsorption as depicted in Fig. 6.15 (Wei and Ma 2004a). Recently, hierarchical MBG and alginate composite scaffolds with well-ordered nanochannels and interconnected micropores were produced by the rapid prototyping technique (Luo et al. 2013). This technique offers a great solution to the fabrication of 3D scaffolds similar to the native tissue with mechanical strength, apatite mineralization ability, and alkaline phosphate activity comparable to the human bone (Luo et al. 2013). Some other studies (Singh et al. 2010;Zorlutuna et al. 2012;Ngiam et al. 2009;Deng et al. 2012) have also strongly supported the strength of hierarchical polymer and ceramic-based composite for tissue engineering applications. All these findings reveal that hierarchical structures can effectively provide the biomimetic microenvironment, improving cell attachment, interaction, and growth of tissues inside the scaffold, which is the first step toward functional tissue engineering.

6.5.2 *Drug Delivery Agent*

Over the past few decades, extensive research efforts have been directed toward developing better drug carrier vehicles (Liu and Webster 2010;Pham et al. 2002). The intention was not only to fight critical and incurable diseases through targeted delivery of drugs but also to fight the bacterial infection around the implantation site for

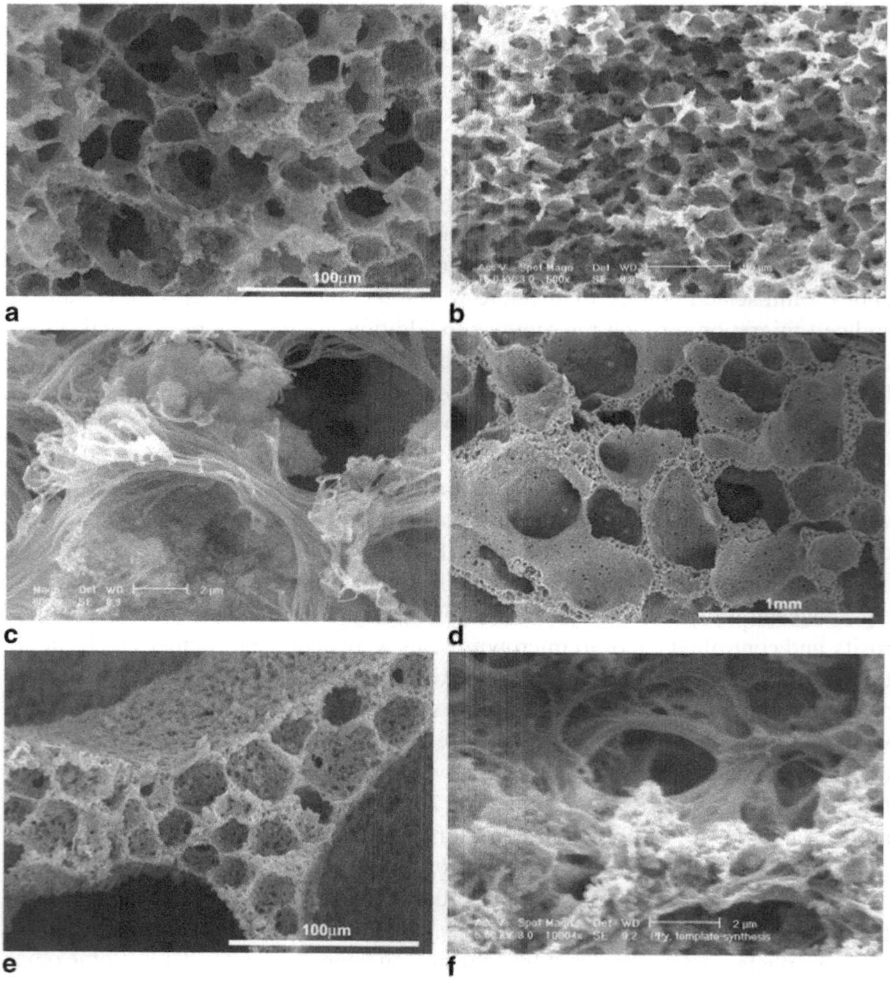

Fig. 6.15 SEM micrographs of nHAP/PLLA (30:70) scaffold showing the effect of different composition of solvents. **a** Dioxane:water=95:5, ×500. **b, c** Dioxane:water=90:10, ×500, ×8000. **d, e, f** Dioxane:water=87:13, ×45, ×500, ×10,000. *SEM* scanning electron microscope, *nHAP* nanohydroxyapatite, *PLLA* poly-L-lactide. (Reproduced with permission from Wei and Ma 2004a)

implants and scaffolds that aid in wound healing and accelerating the total curing process (Aviv 2007). Drug carrier vehicles should have the potential ability to incorporate the drug physically or chemically and to carry the drug until it reaches the target site. The perfect vehicle should also possess the ability to release the drug in controlled and sustained manner throughout the desired period (Bose et al. 2011). Hierarchical structured composites are an attractive vehicle for effective drug loading and are an ideal carrier system due to their multiscale porous architecture. The interconnected micropores have a high surface energy, hence drug encapsulation is easier due to improved attachment. Through degradation of the composite structure, sustained drug release can be achieved in a controlled manner. Hierarchical micro–nanostructured composites with organic and inorganic phases, such as $CaCO_3$ and polymer hybrids are explored as smart drug carriers due to their ideal mechanical strength, controlled release behavior, and biodegradability (Sokolova and Epple 2008).

Research has been carried out to develop the novel self-assembled drug carrier system by mineralizing $CaCO_3$ on polymer templates which has shown sustained release of doxorubicin in a controlled manner for bone cancer treatment (Lee 2010;Wakayama et al. 2006). Recently, Shi et al. (2013) have prepared the hierarchical poly (N-isopropylacrylamide; PNIPAA) and $CaCO_3$ composite via self-assembly biomineralization reaction. Sodium poly(styrene sulfonate) (PSS) and PAA were used to control the nucleation and morphology of polymorph of $CaCO_3$ microcrystal by interaction between PSS and Ca^{2+}, which helps in the formation of hierarchical micro/nanohybrid composites (Jin et al. 2009). This composite was synthesized by adding $(NH_4)_2CO_3$ to the initial solution of PSS and $CaCl_2$. The inorganic salts react to form $CaCO_3$, which incorporates PNIPAA nanoparticles, resulting in formation of the hybrid composite micro particles (Jin et al. 2009). Vitamin B_2 (VB_2) release behavior was examined in different pH states and thermal conditions for this composite (Fig. 6.16). The drug release rate at 37 °C was 82 % and decreased to 67 % at 25 °C. At higher temperatures, shrinkage of PNIPAAm nanogel compacts the structure of the composite resulting in an accelerated drug release. In an acidic environment, increase in the dissolution of $CaCO_3$ microparticles resulted in a relatively high drug release. Introduction of $CaCO_3$ microspheres helps in controlling the permeation of encapsulated VB_2 and effectively managing the sustained release profiles for tissue application.

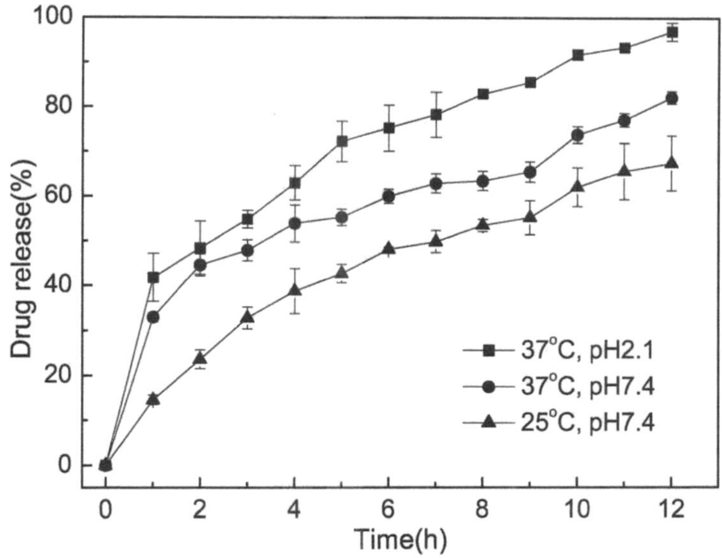

Fig. 6.16 Drug release profiles of the hybrid composites produced with 2.37 mM of Ca^{2+} using PSS as crystal growth additive as a function of pH and temperature. *PSS* poly(styrene sulfonate). (Reproduced with permission from Jin et al. 2009)

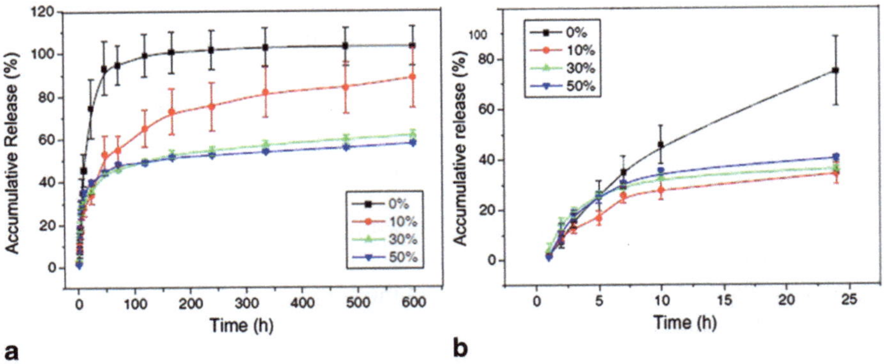

Fig. 6.17 Dexamethasone drug release profile of pure MBG/algnate composite scaffolds in SBF **a** more than 25 days **b** at the early stage (24 h). Percentages in the legend denote MBG content. *SEM* scanning electron microscope, *SBF* simulated body fluid. (Reproduced with permission from Luo et al. 2013)

MBG and concentrated alginate composite scaffold with well-ordered porous architecture, fabricated via 3D printing, have shown sustained drug delivery to reduce the chance of post-surgery infection. In pure alginate scaffolds, the dexamethasone drug delivery kinetics was very rapid with a high initial burst, but MBG/alginate composite scaffolds showed a slower and sustained release profile (Fig. 6.17). The addition of MBG increases the osteoinductivity and bioactivity of the scaffold, hence fast apatite-film formation takes place on the surface, and controls the drug delivery. By varying the MBG content in the composite, drug release kinetics as well as the mechanical property of the scaffolds can be tailored according to the tissue engineering application (Luo et al. 2013;Wu et al. 2010).

In other efforts, researchers have tried a hierarchical porous nanoparticles-based drug delivery system for cancer therapy and tumor treatment. Inorganic nanoparticles with porous hierarchical shells have gained much more attention due to their surface chemistry, biocompatibility, lack of toxicity, and high transfection efficiency (process of deliberately introducing nucleic acids into cells) (Bose and Tarafder 2012). Wei et al. (2008) have developed hierarchical hollow spherical $CaCO_3$ particles and used these particles effectively as a pH sensitive anticancer drug carrier. Electrostatic interaction between negatively charged inner mesoporous silica shell/sphere and positively charged cationic surfactant hexadecyl trimethyl ammonium bromide (C16TAB) creates the second shell. These two adjacent shells create the hierarchical pore structure. Combining the second shells with initial mesoporous shells increases pore volumes, as well as the surface energy of the particle, which helps in holding the drug firmly. The double-shelled mesoporous silica spheres can be used as drug carrier vehicles for cancer therapy due to its high drug loading capacity and highly biocompatible nature. Further, deposition of a second mesoporous shell can control the sustained release of drugs (Chen et al. 2011). Currently, the lack of studies and the complex fabrication techniques limits extensive clinical usage of these hierarchical structured nanoparticles for drug delivery applications.

6.6 Composite Modeling and Simulation

As the structure becomes complicated, modeling becomes a very useful tool for predicting the mechanical behavior of any heterogeneous structure. However, with increasing amount of complexity in the structure, the available computational capabilities are often constrained with the inherent limitation. This aspect seems to explain the scarcity of available literature on modeling and simulation studies of hierarchical composites for medical applications. Few literatures, available on related topics, mostly deal with the structure-based property and failure aspects of bone, which is one of the most efficient hierarchical composite structures by nature. This section summarizes few such studies that show the potential of modeling and simulation for predicting the performance of hierarchical composite structures.

The organic and inorganic composition of bone tissue and its hierarchical structure is more intricate than most of the artificial engineering composites. Hence, basic understanding may be achieved through employing the modeling and simulation of collagenous matrix and mineral crystals (Fratzl et al. 2004). Mechanical property and failure behavior of bone are completely dependent on the property of its constituent phases as well as their structural relation at various length scales of hierarchy (Landis 1995). Efforts have been made to explain the elastic property of whole bone using composite rules of mixtures of each element (Lees and Davidson 1977). Dong (2009) have proposed two-dimensional plane strain probabilistic finite elemental model to examine the mechanism of microdamage progression from an initial defect in lamellar structure of mineral–collagen composite in tension. The probabilistic failure model and simulation demonstrated significant effect of micro damage progression from an interfacial debonding around an initial defect (Fig. 6.18). Debonding of mineral and collagen interfaces with an applied load helps in scattering of stress and the reduction in the crack progression rate

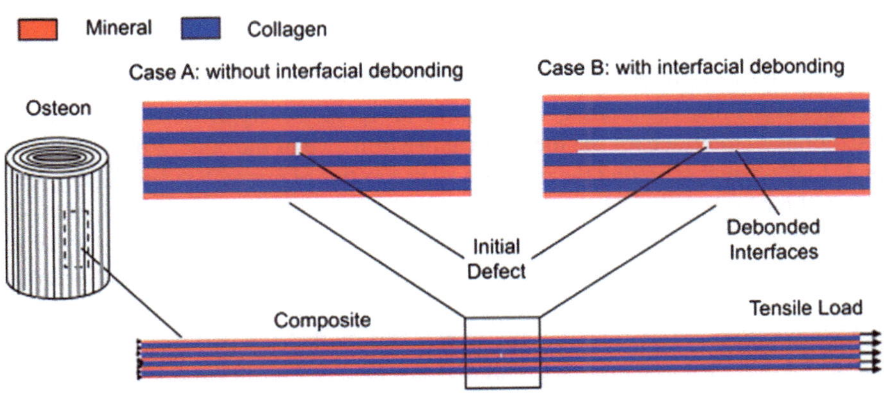

Fig. 6.18 Graphical illustration of finite element method for a mineral–collagen composite, without and with interfacial debonding between the mineral and collagen layers around the initial defect. (Reproduced with permission from Dong 2009)

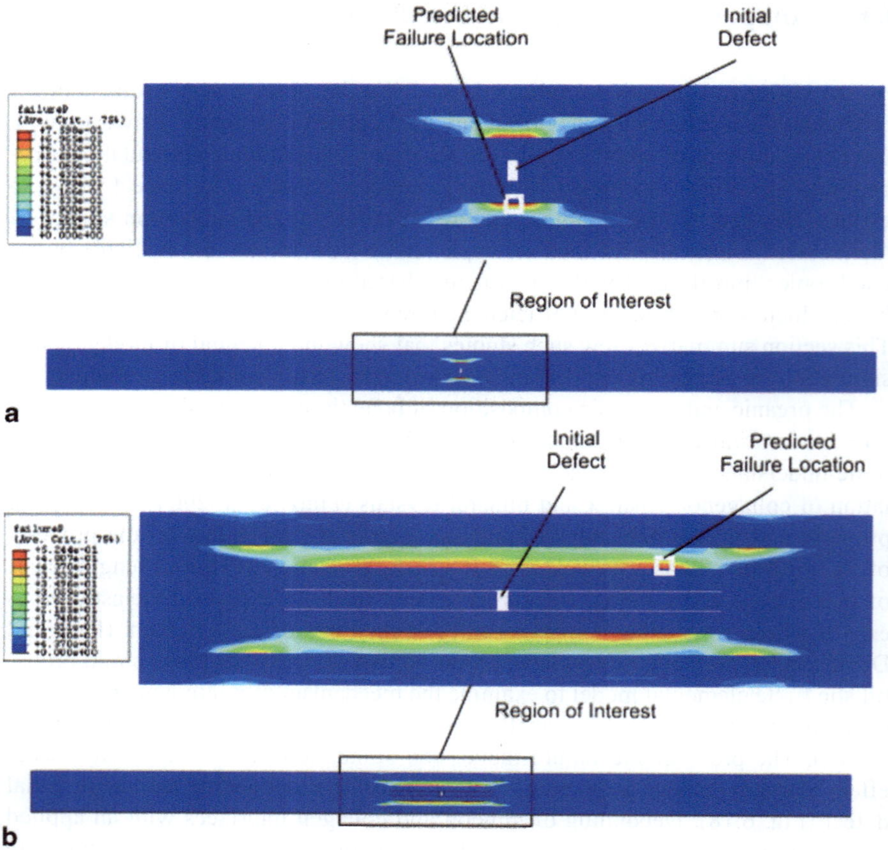

Fig. 6.19 Failure possibility distribution in the mineral–collagen composite **a** without interfacial debonding and **b** with initial defect. (Reproduced with permission from Dong 2009)

from the initial damage site. This deviates the probability of failure from the plane of initial defect through the mechanism of micro damage progression as shown in Fig. 6.19. Finally, research results have revealed the significance of the mineral–collagen interface for the mechanical performance of the bone. Efforts are also made to analyze and predict the mechanical properties of mineralized collagen fibril and the bone at the nanostructure level through analytical and computational models (Hamed 2010;Ghanbari and Naghdabadi 2009;Hambli and Barkaoui 2012;Yuan et al. 2011;Barkaoui and Hambli 2011). However, the immensely heterogeneous structure of bone with different phases and structures across different length scales make it complicated to construct the model and predict the performance (Hamed and Jasiuk 2012). Further in-depth studies are required to thoroughly arrange and feed the properties and geometrical arrangements of mineralized crystal and collagen to develop the ideal model of bone from the nano- through the microscale level.

6.7 Challenges and Future Direction

Looking to the near future, hierarchical composites will be one of the material systems that provide the most suitable combination of mechanical and biological properties to support the present and future medical challenges. Research in composites with advanced materials, and understanding the nanoscience and technology to design innovative hierarchical structures (from nano- to millimeter length scale) through advanced synthesis routes would help in realizing this dream.

New generation of bioceramic-based composites with a properly designed 3D hierarchical structure have been identified for bone regeneration and tissue engineering applications due to their excellent osteoconductivity and bioactivity. But, inherent brittleness and low mechanical strength of such structures limit their clinical application. In addition, the essential high temperature sintering processes lead to uncontrolled interconnected porosity, resulting in poor drug delivery and cellular response of the scaffolds. Notable research has been performed on collagen-based hierarchical composites to exactly mimic the complicated natural extracellular matrix-like structure. However, unsatisfactory results in mimicking intricate hierarchical structures with mineralized collagen and orderly deposited HA minerals, leaves room for further research.

Inorganic and organic material-based composites have been introduced as creative alternatives to the market to take the full advantage of the distinctly different properties of each kind of material system. However, the major constraint in this category has been the lack of a suitable fabrication technique to synthesize the perfect hierarchical architecture. Recent advances in technology have made available, methods like solvent casting, particle leaching, freeze drying, rapid prototyping approaches. These methods have enabled the processing of hybrids with artificial structures with fascinating multifunctionality. Researchers are actively engaged in efforts to overcome the present limitations and to take the full advantage of these techniques.

A majority of the related literature is on bone tissue engineering. Hierarchical composites with 3D porous scaffolds are important for scaffold-based bone tissue engineering applications. They provide the mechanical support and bioactivity to repair bone defects and to regenerate bone functions. While nanostructure offers great interactions with living cells and improves biological activities, microstructure plays a role in strengthening the mechanical functionalities. Therefore, constructing the potential scaffolds with enhanced bioactivity and mechanical property is a challenging task, and depends on the proper selection of biomaterials and fabrication techniques to control the interconnected multiscale pore architecture.

Recently, hierarchical composites have been used in drug delivery applications because of their ability to incorporate the drug in the structure physically or chemically, and to carry forward until it reaches the target site. Significant challenges still remain in the development of an ideal carrier of drugs over an extended period of time.

Several examples of different materials hierarchical composite with their challenges and future direction have been discussed. But these possibilities only represent a small fraction of the tip of the iceberg, which is yet to be explored.

Conclusions

Nature has many creatures with highly organized hierarchical architecture ranging from the nano- to macro-length scale. An in-depth understanding of the key role of natural hierarchy is essential for developing artificial replacements. During the past few decades, a considerable effort has been made to construct the ideal hierarchical structure through different composite approaches to meet the advanced medical demand, which has achieved partial success. Material scientists are still searching for the most suitable composite materials system to develop artificial biomimetic structures.

Directing the nature-inspired technologies toward the development of hierarchical composites with permutation and combination of different material systems through innovative processing routes, and transitioning the achievements from the laboratory level into the real world by counterbalancing the limitations is an attractive and challenging job. This overview was aimed at identifying the direction and scope of the development of artificial hierarchical composite structures with the best possible functionalities for biomedical applications.

References

Antonietti RA (2001) Sol–gel nanocoating: an approach to the preparation of structured materials. Chem Mater 13:3272–3282

Augst AD, Kong HJ, Mooney DJ (2006) Alginate hydrogels as biomaterials. Macromol Biosci 6:623–633

Aviv MI (2007) Gentamicin-loaded bioresorbable films for prevention of bacterial infections associated with orthopedic implants. J Biomed Mater Res A 83:10–19

Bae WG, Kim HN, Park S-H, Jeong HE Suh K-Y (2014) 25th anniversary article: scalable multiscale patterned structures inspired by nature: the role of hierarchy. Adv Mater 26(5):675–700

Baer E, Hiltner A Morgan RJ (1992) Biological and synthetic hierarchical composites. Phys Today 45(10):60

Barkaoui A, Hambli R (2011) Finite element 3D modeling of mechanical behavior of mineralized collagen microfibrils. J Appl Biomater Biomech 9(3):199–205

Bettinger CJ, Langer R, Borenstein JT (2009) Engineering substrate topography at the micro-and nanoscale to control cell function. Angew Chem Int Ed Engl 48:5406–5415

Bonfiglid M, Jeter WS (1972) Immunological responses to bone. Clinical Orthop Relat Res 87:19–27

Bose S, Tarafder S (2012) Calcium phosphate ceramic systems in growth factor and drug delivery for bone tissue engineering: a review. Acta Biomater 8:1401–1421

Bose S, Tarafder S, Edgington J, Bandyopadhyay A (2011) Calcium phosphate ceramics in drug delivery. JOM 63(4):93–98

Bradt J-H et al (1999) Biomimetic mineralization of collagen by combined fibril assembly and calcium phosphate formation. Chem Mater 11:2694–2701
Chang MC, Ikoma T, Kikuchi M, Tanaka J (2001) Preparation of a porous hydroxyapatite/collagen nanocomposite using glutaraldehyde as a crosslinkage agent. J Mater Sci Lett 20:1199–1201
Chen Y, Chen H, Guo L, He Q, Chen F, Zhou J, Feng J, Shi J (2010a) Hollow/rattle-type mesoporous nanostructures by a structural difference-based selective etching strategy. ACS Nano 4:529–539
Chen Y, Chen H, Zhang S, Chen F, Zhang L, Zhang J, Zhu M, Wu H, Guo L, Feng J, Shi J (2010b) Multifunctional mesoporous nanoellipsoids for biological bimodal imaging and magnetically targeted delivery of anticancer drugs. Adv Funct Mater 21(2):270–278.
Chen Y, Chen H, Ma M, Chen F, Guo L, Zhang L, Shi J (2011) Double mesoporous silica shelled spherical/ellipsoidal nanostructures: synthesis and hydrophilic/hydrophobic anticancer drug delivery. J Mater Chem 21:5290–5298
Chou YF, Huang W, Dunn JC, Miller TA, Wu BM (2005) The effect of biomimetic apatite structure on osteoblasts viability, proliferation, and gene expression. Biomaterials 26:285–295
Clarke KI, Graves SE, Wong ATC, Triffitt JT, Francis MJO, Czernuszka JT (1993) Investigation into the formation and mechanical properties of a bioactive material based on collagen and calcium phosphate. J Mater Sci: Mater Med 4(2):107–110
Cölfen H, Mann S (2003) Higher-order organization by mesoscale self-assembly and transformation of hybrid nanostructures. Angew Chem Int Ed 42(21):2350–2365
Costantino PD, Friedman CD (1994) Synthetic bone graft substitutes. Otolaryngol Clin North Am 27(5):1037–1074
Deng M, James R, Laurencin CT, Kumbar SG (2012) Nanostructured polymeric scaffolds for orthopaedic regenerative engineering. NanoBioscience I(11):3–14
Dergunov SA, Pinkhassik E.(2011) Synergistic co-entrapment and triggered release in hollow nanocapsules with uniform nanopores. J Am Chem Soc 133(49):19656–19659
Doi Y, Horiguchi T, Moriwaki Y, Kitago H, Kajimoto T, Iwayama Y (1996) Formation of apatite–collagen complexes. J Biomed Mater Res 31:43–49
Dong Q, Su H, Cao W, Zhang D, Guo Q, Lai Y (2007) Synthesis and characterizations of hierarchical biomorphic titania oxide by a bio-inspired bottom-up assembly solution technique. J Solid State Chem 180:949–955
Dong XN, Guda T, Millwater HR, Wang X.(2009) Probabilistic failure analysis of bone using a finite element model of mineral–collagen composites. J Biomech 42(3):202–209
Du C, Cui F, Zhu XD, de Groot K.(1999) Three-dimensional nano-HAp/collagen matrix loading with osteogenic cells in organ culture. J Biomed Mater Res 44:407–415
Du C, Cui FZ, Zhang W, Feng QL, Zhu XD, de Groot K (2000) Formation of calcium phosphate/collagen composites through mineralization of collagen matrix. J Biomed Mater Res 50:518–527
Ehrlich H (2010) Biological materials of marine origin: invertebrates. Biologically-inspired systems. Springer, Dordrecht, pp 25–122
Epple M, Ganesan K, Heumann R, Klesing J, Kovtun A, Neumann S, Sokolova V (2010) Application of calcium phosphate nanoparticles in biomedicine. J Mater Chem 20(1):18–23
Fedorovich NE, De Wijn JR, Verbout AJ (2008) Three-dimensional fiber deposition of cell-laden, viable, patterned constructs for bone tissue printing. Tissue Eng Part A 14:127–133
Feng L et al (2002) Super-hydrophobic surfaces: from natural to artificial. Adv Mater 14:1857–1860
Fratzl P et al (2004) Structure and mechanical quality of the collagen–mineral nano-composite in bone. J Mater Chem 14(14):2115–2123
Fujisawa R et al (1996) Acidic amino acid-rich sequences as binding sites of osteonectin to hydroxyapatite crystals. Biochim Biophys Acta 1292(1):53–60.
Furuichi K, Oaki Y, Ichimiya H et al (2006) Preparation of hierarchically organized calcium phosphate–organic polymer composites by calcification of hydrogel. Sci Technol Adv Mater 7(2):219–225

Gee SH, Hong YK (2003) Synthesis and aging effect of spherical magnetite (Fe_3O_4) nanoparticles for biosensor applications. J App Phys 93:7560
Ghanbari J, Naghdabadi R (2009) Nonlinear hierarchical multiscale modeling of cortical bone considering its nanoscale microstructure. J Biomech 42(10):1560–1565
Girija EK, Yokogawa Y, Nagata F (2004) Apatite formation on collagen fibrils in the presence of polyacrylic acid. J Mater Sci Mater Med 15(5):593–599
Goldberg VM (1992) Natural history of autografts and allografts. In: Older MWJ (ed) Bone implant grafting. Springer, London, pp 9–12
Habraken WJ, Wolke JG, Jansen JA (2007) Ceramic composites as matrices and scaffolds for drug delivery in tissue engineering. Adv Drug Deliv Rev 59:234–248
Hambli R, Barkaoui A (2012) Physically based 3D finite element model of a single mineralized collagen microfibril. J Theor Biol 30(1):28–41
Hamed EY (2010) Multiscale modeling of elastic properties of trabecular bone. 6th World Congress of Biomechanics (WCB 2010)
Hamed E, Jasiuk I (2012) Elastic modeling of bone at nanostructural level. Mater Sci Eng R 73(3):27–49
Hench LS (1971) Bonding mechanism at the interface of ceramics prosthetic materials. J Biomed Mater Res Symp 2 5:117–141
Hench LL (2006) The story of bioglass. J Mater Sci Mater Med.17:967–978
Huang YK (2013) Rapid prototyping of a hybrid hierarchical polyurethane-cell/hydrogel construct for regenerative medicine. Mater Sci Eng: C 33(6):3220–3229
Hutmacher DW, Sittinger M, Risbud MV (2004) Scaffold-based tissue engineering: rationale for computer-aided design and solid free-form fabrication systems. Trends Biotechnol 22:354–362
Imai H, Tatara S (2003) Formation of calcium phosphate having a hierarchically laminated architecture through periodic precipitation in organic gel. Chem Commun (Camb) 15:1952–1953
Izquierdo-Barba I, Arcos D, Sakamoto Y et al (2008a) High-performance mesoporous bioceramics mimicking bone mineralization. Chem Mater 20(9):3191–3198
Izquierdo-Barba I, Colilla M, Vallet-Regí M (2008b) Nanostructured mesoporous silicas for bone tissue regeneration. J Nanomater 2008 106970.
Jan E, Kotov N (2007) Successful differentiation of mouse neural stem cells on layer by layer assembled single walled carbon nanotube composite. Nano Lett 7:1123–1128
Jin Y et al (2009) (Protamine/dextran sulfate)$_6$ microcapules templated on biocompatible calcium carbonate microspheres. Colloids Surfaces A: Physicochem Eng Asp 342(1):40–45
Kakizawa Y et al (2004) Size-controlled formation of a calcium phosphate-based organic–inorganic hybrid vector for gene delivery using poly (ethylene glycol)-block-poly (aspartic acid). Adv Mater 16:699–702
Khalil S, Sun W (2009) Bioprinting endothelial cells with alginate for 3D tissue constructs. J Biomech Eng 131:111002
Khalil S, Nam J, Sun W (2005) Multi-nozzle deposition forconstruction of 3D biopolymer tissue scaffolds. Rapid Prototyp J 11:9–17
Kim HN (2012) Effect of orientation and density of nanotopography in dermal wound healing. Biomaterials 33(34):8782–8792
Kim HN et al (2012) Patterning methods for polymers in cell and tissue engineering. Ann Biomed Eng 40:1339–1355
Kim HN et al (2013) Nanotopography-guided tissue engineering and regenerative medicine. Adv Drug Deliv Rev 65:536–558
Kischi M, Itoh S (2001) Self-organization mechanism in a bone-like hydroxyapatite/collagen nanocomposite synthesized in vitro and its biological reaction in vivo. Biomaterials 22:1705–1711
Kokubo T, Takadama H (2006) How useful is SBF in predicting in vivo bone bioactivity? Biomaterials 27:2907–2915
Kreetachat T, Kruenate J, Suwannahong K (2013) Preparation of Tio2/Bio-composite film by sol-gel method in VOCs photocatalytic degradation process. Appl Mech Mater 390:552–556

Landers R, Hübner U, Schmelzeisen R, Mülhaupt R (2002) Rapid prototyping of scaffolds derived from thermoreversible hydrogels and tailored for applications in tissue engineering. Biomaterials 23:4437–4447

Landis WJ (1995) The strength of a calcified tissue depends in part on the molecular structure and organization of its constituent mineral crystals in their organic matrix. Bone 16:533–544

Lee HJ (2010) Spatially mineralized self-assembled polymeric nanocarriers with enhanced robustness and controlled drug-releasing property. Chem Commun 46(3):377–379

Lees S, Davidson CL (1977) The role of collagen in the elastic proper-ties of calcified tissues. J Biomech 10:473–486

Li X, Shi J, Dong X, Zhang L, Zeng H (2008) A mesoporous bioactive glass/polycaprolactone composite scaffold and its bioactivity behavior. J Biomed Mater Res A 84:84–91

Liao SS et al (2004) Hierarchically biomimetic bone scaffold materials: nano-HA/collagen/PLA composite. J Biomed Mater Res Part B: Appl Biomater 69(2):158–165

Liu H, Webster TJ (2010) Ceramic/polymer nanocomposites with tunable drug delivery capability at specific disease sites. J Biomed Mater Res A 93(3):1180–1192

Lopez-Noriega A, Ruiz-Hernandez E, Stevens SM et al (2009) Mesoporous microspheres with doubly ordered core-shell structure. Chem Mater 21:218–20

Luo Y, Wu C, Lode A, Gelinsky M (2013) Hierarchical mesoporous bioactive glass/alginate composite scaffolds fabricated by three-dimensional plotting for bone tissue engineering. Biofabrication 5(1):015005

Lutolf MP, Hubbell JA (2005) Synthetic biomaterials as instructive extracellular microenvironments for morphogenesis in tissue engineering. Nat Biotechnol 23(1):47–55

Maher PS, Keatch R, Donnelly K (2009) Construction of 3D biological matrices using rapid prototyping technology. Rapid Prototype J 15:204–210

Mann S (2001) Biomineralization. Oxford University Press, New York

Meyers MA, Chen P-Y, Lin AYM, Seki Y (2008) Biological materials: structure and mechanical properties. Prog Mater Sci 53(1):1–206

Michalske TA, Bunker BC, Keefer KD (1990) Mechanical properties and adhesion of hydrated glass surface layers. J Non-Cryst Solids 120(1–3):138–151

Miyamoto Y, Ishikawa K, Takechi M et al (1998) Basic properties of calcium phosphate cement containing atelocollagen in its liquid or powder phases. Biomaterials 19:707–715

Murphy WL, Mooney DJ (2002) Bioinspired growth of crystalline carbonate apatite on biodegradable polymer substrata. J Am Chem Soc 124(9):1910–1917

Murphy WL, Kohn DH, Mooney DJ.(2000) Growth of continuous bonelike mineral within porous poly (lactide-co-glycolide) scaffolds in vitro. J Biomed Mater Res 50(1):50–58

Murugan R, Ramakrishna S (2005) Development of nanocomposites for bone grafting. Compos Sci Technol 65:2385–2406

(n.d.) http://en.wikipedia.org/wiki/Biomaterial

Ngiam M et al (2009) The fabrication of nano-hydroxyapatite on PLGA and PLGA/collagen nanofibrous composite scaffolds and their effects in osteoblastic behavior for bone tissue engineering. Bone 45(1):4–16

Niu LN et al (2013) Biomimetic silicification of demineralized hierarchical collagenous tissues. Biomacromolecules 14(5):1661–1668

Oliveira AL, Mano JF, Reis RL (2003) Nature-inspired calcium phosphate coatings: present status and novel advances in the science of mimicry. Curr Opin Solid State Mater Sci 7(4–5):309–318

Pham HH, Luo P, Génin F, Dash AK (2002) Synthesis and characterization of hydroxyapatite-ciprofloxacin delivery systems by precipitation and spray drying technique. AAPS PharmSciTech 3(1):1–9

Qiu PD (1999) Bioactive ceramics: the effect of surface reactivity on bone formation and bone cell function. Biomaterials 20:2287–2303

Rebollar E et al (2008) Proliferation of aligned mammalian cells on laser-nanostructured polystyrene. Biomaterials 29:1796–1806

Rhee S-H, Lee JD, Tanaka J (2000) Nucleation of hydroxyapatite crystal through chemical interaction with collagen. J Am Ceram Soc 83:2890–2892

Rocha LB, Goissis G, Rossi MA (2002) Biocompatibility of anionic collagen matrix as scaffold for bone healing. Biomaterials 23:449–456
Roche S, Ronzière M et al (2001) Native and DPPA cross-linked collagen sponges seeded with fetal bovine epiphyseal chondrocytes used for cartilage tissue engineering. Biomaterials 22:9–18
Sanchez C, Julián C, Belleville P et al (2005) Applications of hybrid organic–inorganic nanocomposites. J Mater Chem 15:3559–3592
Sanchez C et al (2011) Applications of advanced hybrid organic–inorganic nanomaterials: from laboratory to market. Chem Soc Rev 40(2):696–753
Shen X, Chen L et al (2011) A novel method for the fabrication of homogeneous hydroxyapatite/collagen nanocomposite and nanocomposite scaffold with hierarchical porosity. J Mater Sci Mater Med 22:299–305
Shi QH, Wang JF, Zhang JP et al (2006). Rapid-setting, mesoporous, bioactive glass cements that induce accelerated in vitro apatite formation. Adv Mater 18:1038
Shi J et al (2013) Micro/nanohybrid hierarchical poly (N-isopropylacrylamide)/calcium carbonate composites for smart drug delivery. J Appl Polym Sci 129(2):577–584
Shor L, Güçeri S, Wen X et al (2007) Fabrication of three-dimensional polycaprolactone/hydroxyapatite tissue scaffolds and osteoblast-scaffold interactions in vitro. Biomaterials 28:5291–5297
Singh R et al (2010) Hierarchically structured titanium foams for tissue scaffold applications. Acta Biomater 6(12):4596–4604
Sobral JM, Caridade SG, Sousa RA et al (2011) Three-dimensional plotted scaffolds with controlled pore size gradients: effect of scaffold geometry on mechanical performance and cell seeding efficiency. Acta Biomater 7:1009–1018
Sokolova V, Epple M (2008) Inorganic nanoparticles as carriers of nucleic acids into cells. Angew Chem Int Ed Engl 47:1382–1395
Stegemann BD (2013) Strategies for directing the structure and function of three-dimensional collagen biomaterials across length scales. Acta Biomater
Su B-L, Sanchez C, Yang X-Y (2012) Hierarchically structured porous materials: from nanoscience to catalysis, separation, optics, energy, and life science. Wiley, Weinheim
Taguchi T, Aiko K, Akashi M (1998) Hydroxyapatite formation on/in poly(vinyl alcohol) hydrogel matrices using a novel alternate soaking process. Chem Lett 27(8):711–712
Taguchi T, Kishida A, Akashi M (1999) Apatite formation on/in hydrogel matrices using an alternate soaking process. II. Effect of swelling ratios of poly(vinyl alcohol) hydrogel matrices on apatite formation. J Biomater Sci Polym Ed 1D:331–339
Tan J, Saltzman WM (2004) Biomaterials with hierarchically defined micro-and nanoscale structure. Biomaterials 25(17):3593–3601
Tanahashi M et al (1994) Apatite coating on organic polymers by a biomimetic process. J Am Ceram Soc 77(11):2805–2808
Teixeira AI et al (2003) Epithelial contact guidance on well-defined micro-and nanostructured substrates. J Cell Sci 116(10):1881–1892
Teixeira AI et al (2006) The effect of environmental factors on the response of human corneal epithelial cells to nanoscale substrate topography. Biomaterials 27(21):3945–3954
Vallet-Regí M, Ragel CV, Salinas AJ (2003) Glasses with medical applications. Eur J Inorg Chem 2003(6):1029–1042
Wakayama H et al (2006) CaCO3/biopolymer composite films prepared using supercritical CO2. Ind Eng Chem Res 45:3332–3334
Weadock K, Olson R, Silver FH (1983) Evaluation of collagen crosslinking techniques. Biomater Med Devices Artif Organs 11:293–318
Wei GB, Ma PX (2004a) Structure and properties of nano-hydroxyapatite/polymer composite scaffolds for bone tissue engineering. Biomaterials 25:4749–4757
Wei W et al (2008) Preparation of hierarchical hollow CaCO3 particles and the application as anticancer drug carrier. J Am Chem Soc 130(47):15808–15810
Weiner S, Traub W (1986) Organization of hydroxyapatite crystals within collagen fibrils. FEBS Lett 206:262–266

Weiner S, Wagner HD (1998) The material bone: structure-mechanical function relations. Annu Rev Mater Sci 28:271–298
Weiner S et al (2006) Mineralized biological materials: a perspective on interfaces and interphases designed over millions of years. Biointerphases 1(2)12–14
Wicklein B et al (2013) Hierarchically structured bioactive foams based on polyvinyl alcohol–sepiolite nanocomposites. J Mater Chem B 1(23):2911–2920
Wu C, Luo Y (2011) Three-dimensional printing of hierarchical and tough mesoporous bioactive glass scaffolds with a controllable pore architecture, excellent mechanical strength and mineralization ability. Acta Biomater 7:2644–2650
Wu C, Chang J, Zhai W et al (2007) A novel bioactive porous bredigite (Ca(7)MgSi(4)O(16)) scaffold with biomimetic apatite layer for bone tissue engineering. J Mater Sci Mater Med 18:857–864
Wu C et al (2010) Bioactive mesopore-glass microspheres with controllable protein-delivery properties by biomimetic surface modification. J Biomed Mater Res A 95(2):476–485
Xiu Y et al (2007) Hierarchical silicon etched structures for controlled hydrophobicity/superhydrophobicity. Nano Lett 7:3388–3393
Xu S, Lin K, Wang Z et al (2008) Reconstruction of calvarial defect of rabbits using porous calcium silicate bioactive ceramics. Biomaterials 29:2588–2596
Yan X, Yu C, Zhou X et al (2004) Highly ordered mesoporous bioactive glasses with superior in vitro bone-forming bioactivities. Angew Chem Int Ed Engl 43:5980–5984
Yim EK et al (2010) Nanotopography-induced changes in focal adhesions, cytoskeletal organization, and mechanical properties of human mesenchymal stem cells. Biomaterials 31(6):1299–1306
Yin YJ, Zhao F, Yao KD et al (2000) Preparation and characterization of hydroxyapatite/chitosan–gelatin network composite. J Appl Polym Sci 77:2929–2938
Yokogawa Y et al (1998) Calcium phosphate compound-cellulose fiber composite material prepared in soaking medium at 36.5–60 °C. J Mater Res 13(4):922–925
Yuan XA (2001) Formation of bone-like apatite on poly (L-lactic acid) fibers by a biomimetic process. J Biomed Mater Res 57(1):140–150
Yuan F et al (2011) A new model to simulate the elastic properties of mineralized collagen fibril. Biomech Model Mechanobiol 10(2):147–160
Yun HS, Kim S, Hyeon YT (2007) Highly ordered mesoporous bioactive glasses with Im3m symmetry. Mater Lett 61:4569–4572
Yun H, Kim S, Hyeon YT (2011) Bioactive glass–poly (ε-caprolactone) composite scaffolds with 3 dimensionally hierarchical pore networks. Mater Sci Eng C 31:198–205
Zhang M, Kataoka K (2009) Nano-structured composites based on calcium phosphate for cellular delivery of therapeutic and diagnostic agents. Nano Today 4(6):508–517
Zhang W, Liao SS, Cui FZ (2003) Hierarchical self-assembly of nano-fibrils in mineralized collagen. Chem Mater 15:3221–3226
Zhang L-J, Feng X-S, Liu H-G, Qian D-J, Zhang L et al (2004) Hydroxyapatite/collagen composite materials formation in simulated body fluid environment. Mater Lett 58:719–722
Zhao Y et al (2009) Fabrication of skeletal muscle constructs by topographic activation of cell alignment. Biotechnol Bioeng 102(2):624–631
Zhao M-Q et al (2012) Hierarchical nanocomposites derived from nanocarbons and layered double hydroxides-properties, synthesis, and applications. Adv Funct Mater 22 (4):675–694
Zhou Y et al (2013) Asymmetric PSt-EA/Ni-silicate hollow microsphere with a hierarchical porous shell. J Mater Chem B 1(10):1414–1420
Zorlutuna P et al (2012) Microfabricated biomaterials for engineering 3D tissues. Adv Mater 24(12):1782–1804

Chapter 7
Electrochemical Hierarchical Composites

Yu Zhao, Lele Peng and Guihua Yu

Abstract Composites with hierarchical structures are considered to be one of the most prevailing materials for various electrochemical applications as they constitute a complex architecture that provides a large internal surface area and enables synergistic effects for electrochemical reactions at the interface. Considering novel nanoscale properties with large-scale processability and affordable cost, hierarchically structured composites could potentially provide the new and transformative approaches to meet the challenges for the modern society: enabling powerful electrochemical devices for renewable energy conversion and storage and environmental monitoring. This chapter reviews the applications of hierarchically structured composites for energy storage, energy conversion, and environmental monitoring. In the section of energy storage, supercapacitors and batteries are intensively reviewed with emphasis on the design, synthesis, and performance evaluation of the hierarchically structured composites; in the section of energy conversion, photoelectrochemical cells and fuel cells are introduced; and in the section of environmental monitoring, we introduce some sensing devices based on the hierarchically structured composites. The fundamental understanding of the structure–property relationship between these hierarchically structured materials and their performances in electrochemical devices will further promote the design of new electrochemical materials with unprecedented properties.

Keywords Electrochemical devices · Hierarchical composite · Energy storage · Energy conversion · Sensing

7.1 Introduction

Driven by novel properties that nanostructures possess and emergent needs for developing low-cost and highly efficient devices for electrochemical (EC) device applications, hierarchically nanostructured materials that are of geometric complex-

---

G. Yu (✉) · Y. Zhao · L. Peng
Materials Science and Engineering Program and Department of Mechanical Engineering,
The University of Texas at Austin, Austin, TX 78712, USA
e-mail: ghyu@austin.utexas.edu

C.-S. Kim et al. (eds.), *Hybrid and Hierarchical Composite Materials*,
DOI 10.1007/978-3-319-12868-9_7

ity or composed of multiple constituents and related EC devices have become a promising class of candidates for the next generation, environmentally sustainable technologies. Therefore, research efforts focused on these materials with the goal of improving scientific understanding at unprecedented length scales are of considerable interest (Ding et al. 2006; Gao et al. 2012; Gao and Wang 2002; Lao et al. 2002; Whang et al. 2003). The hierarchical structures can be categorized into either (i) structures with nanoscale building blocks extended into more than one dimension or (ii) structures with multiple components. In the first category, successful material manipulation results in the production of a useful and complex hierarchy of nanoscale building blocks. In particular, the complexity in extended dimensions brings about advantages such as high surface area; this is important for electronic device manufacturing. However, materials hierarchy may refer to more than just geometric arrangement. In the second category, different combinations of dissimilar units may lead to multiple functionalities exhibiting other complex properties in the resulting hierarchical materials.

Materials with hierarchical structures have been extensively investigated in EC devices such as energy storage, energy conversion, and sensors. Hierarchical structures can provide large surface areas for reaction, interfacial transport, or dispersion of active sites at different length scales of pores and shorten diffusion paths or reduce diffusion effects (Li et al. 2012). By rational design of nanostructured architectures and tuning of chemical compositions, EC properties such as reactivity, EC properties, and light harvesting capabilities can be greatly enhanced (Coleman et al. 2006; Giannelis 1996; Yang et al. 2009; Chen et al. 2009, 2011; Zeng et al. 2005; Veedu et al. 2006). For instance, the combined clay/polymer, carbon nanotube (CNT)/polymer, or graphene/polymer nanocomposites show significantly improved mechanical, electrical, and energy-absorbing properties (Giannelis 1996; Zeng et al. 2005; Veedu et al. 2006; Bai et al. 2011; Liu et al. 2011b). For instance, the electrical conductivity of hierarchical three dimensional (3D) composite constructed from aligned nanotubes grown on carbon fiber cloth is six orders of magnitude higher than that of its 2D counterpart (Veedu et al. 2006). The much-improved through-thickness electrical conductivity observed in the 3D composites would potentially impart to these structures an electrical sensing capability for structural health monitoring during crack propagation or delamination. Tailored assembly of 1D CNTs and 2D graphene into 3D architectures can further improve the performance of graphitic carbon materials in nanoelectronics, sensors, and energy storage/conversion devices (Zhang et al. 2010; Li et al. 2011; Lee et al. 2011; Fan et al. 2010). Hierarchical structures can also act as host materials to stabilize or to incorporate other active components. Novel hierarchical carbonaceous materials acting as electrode scaffold or catalyst supporter for fuel cell applications show good electrical conductivity, excellent chemical, mechanical, and thermal stabilities. An example can be found in the 3D sandwiched CNT/graphene electrode matrix (Fan et al. 2010). The significant improvement of the EC performance of such electrodes is mainly attributed to the unique sandwich structure of graphene and CNTs. The CNTs act as a structural buffer for the large volume expansion of loaded pseudocapacitive materials during the reduction–oxidation (redox) reaction, and have good electrical contact with the materials upon cycling. The introduction of CNTs can

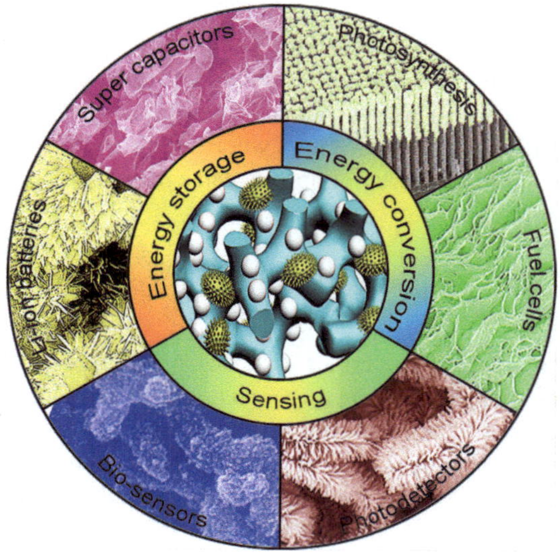

Fig. 7.1 Illustration of the potential applications of hierarchically structured materials/composites on electrochemical devices

also provide diffusion path on the surface of graphene, facilitating the diffusion and migration of electrolyte inside the electrode. The applications of hierarchically structured materials in photoelectrochemical cells present significant advantages to increase the efficiency/cost ratio by enhancing the effective optical path and significantly decreasing the probability of charge recombination. For instance, 1D hierarchical nanostructures have been demonstrated to be efficient in photoelectrochemical cell applications because they can decouple the directions of light absorption and charge carrier collection. When the lifetime of the minority carrier is short, the minority carrier undergoes recombination in bulk before reaching the semiconductor/electrolyte junction. However, due to the small radius of nanowires, the minority carrier can diffuse to the surface before recombination, which potentially increases the charge separation efficiency, particularly, when the minority carrier diffusion length is comparable to the radius of the nanowire.

In this chapter, electrochemical applications of hierarchically structured materials such as energy storage, energy conversion, and sensing (Fig. 7.1) are discussed. Their applications in energy storage, such as supercapacitors and secondary batteries, is reviewed first. Their applications in energy conversion technologies such as electrode scaffold/catalyst for photoelectrochemical cells and fuel cells are then commented on.

7.2 Applications in Energy Storage

Energy storage is accomplished by devices or physical media that store some form of energy to perform some useful operation at a later time. Energy storage methods can be for example: (1) chemical H_2 or hydrocarbon storage, (2) biological storage such as glycogen or starch, (3) electrochemical storage such as batteries or supercapacitors, (4) mechanical storage such as compressed air energy storage, and (5) thermal storage such as ice storage and stem accumulators. This section focuses primarily on electrochemical energy storage (EES) devices, such as Li-ion batteries (LIB) and supercapacitors where hierarchically structured porous materials contribute to the important improvements in energy storage performance and efficiency.

7.2.1 Supercapacitors

EES systems are critical to a variety of applications ranging from portable electronics, to hybrid electric vehicles, and to stationary power and energy management. Owing to the capability to deliver high power output and extremely long cycle life, electrochemical capacitors (ECs), also known as supercapacitors, have attracted considerable attention recently since high-power performance becomes one of the most critical bottlenecks in EES systems. This section describes the most recent progress in the development of electrode materials for EC technology, with a particular focus on hierarchical composites and nanostructured materials that combine carbon-based materials with pseudocapacitive metal oxides or conducting polymers for achieving high-performance ECs. This section starts with an overview of current EC technologies, followed by a brief description of energy storage mechanisms for different types of EC materials, and focus on the development of hierarchical composites for effective electrochemical utilization. Future research directions and challenges for next-generation ECs are discussed.

Owing to the discovery of new electrode materials, and the design of new hierarchical composites that combine faradaic and capacitive electrodes, EC technology in the past decade has experienced an impressive development by means of improving the power performance and cyclability. Hierarchical composites have become even more important for EES, given their unusual electrical, mechanical, and surface properties due to the multifunctional capability and morphology. The promising advantages and associated disadvantages of the electrode materials have been discussed and reviewed previously (Arico et al. 2005). The key advantages for electrode materials with hierarchical structure arise from short charge/mass transport distance, large surface area exposed to electrolyte, and new reactions not possible in their bulk phase. Main disadvantages include increased undesirable reactions at the electrode/electrolyte interface due to large surface area and potentially more complex synthesis and higher manufacturing cost (Yu et al. 2013).

7.2.1.1 Basic Working Mechanisms of ECs

Depending on the energy storage mechanisms, ECs can be classified into two general categories: electrochemical double-layer capacitors (EDLCs) and pseudocapacitors. In a conventional EDLC, energy is stored by moving charge carriers, typically electrons, from one electrode to another. The charge separation generates a potential between the two electrodes, which can be harnessed in an external circuit. The total energy stored in this fashion increases with either the amount of charge stored or the potential between the electrodes. The amount of charge stored per unit voltage is essentially a function of the size, the distance, and the material properties of the electrodes and the electrolytes. Currently, the most common and commercially available ECs are EDLCs where electrical energy is stored by electrostatic accumulation of charges in the electric double-layer at all electrode/electrolyte interfaces is schematically shown in Fig. 7.2a. Electrochemical double-layer capacitors (EDLCs) can provide extremely high power density and excellent life span due to their fast and near-surface EC process. In EDLCs, carbon-based active electrode materials with high surface area are commonly used (Conway 1999). The specific capacitance of EDLCs using carbon-based materials is typically 100–250 F g^{-1} because the charges stored in EDLCs are often limited by electrical charge separation at the interface of the electrode materials and the electrolyte, and by the operating voltages that are dominated by the applicable potential window of the electrolyte. In contrast, pseudocapacitors use fast and reversible redox reactions at the surface of electroactive materials for charge storage as schematically shown in Fig. 7.2b. Pseudocapacitors store electrical energy faradaically by electron charge transfer between electrode and electrolyte. This is accomplished through electrosorption, redox reactions, and intercalation processes, termed pseudocapacitance. A pseudocapacitor has a chemical reaction at the electrode, unlike EDLCs where the electrical charge storage is stored electrostatically with no interaction between the electrode and the ions. Compared with EDLCs, though at some cost of power density and

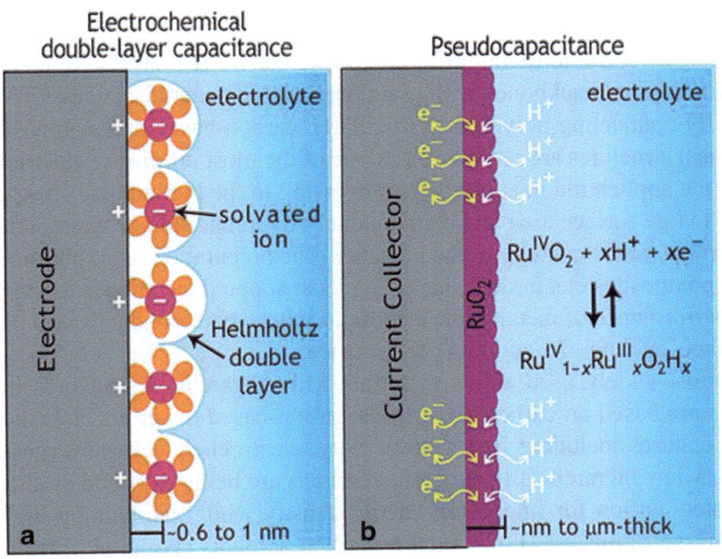

Fig. 7.2 Schematic of two different charge-storage mechanisms via **a** electrochemical double-layer capacitance (*EDLC*) or **b** redox reactions-based pseudocapacitance. (Reproduced with permission Long et al. (2011) Copyright 2011, Materials Research Society)

cycle life, pseudocapacitors can increase specific capacitance and energy density because of the nature of the faradaic process (redox reactions) involved in energy storage. Typical active pseudocapacitive materials include transition metal oxides (TMOs) such as RuO_2, Fe_3O_4, NiO, and MnO_2 (Cottineau et al. 2006; Bélanger et al. 2008), and electronically conducting redox polymers such as polyanilines (PAnis), polypyrroles (PPys), and polythiophenes (Naoi and Morita 2008). The typical specific pseudocapacitance of faradaic electrodes reaches 300–1000 F g^{-1}, exceeding that of EDLCs using double-layer charge storage. In addition to EDLCs and pseudocapacitors, "hybrid capacitor" is a special type in capacitor system. The hybrid capacitors usually combine one battery-type faradaic electrode (as energy source) with the other capacitive electrode (as power source) in the same EC cell (Simon and Gogotsi 2008), in which the battery-like electrode provides high energy density and the EDLC electrode enables high power capability. This type of ECs generally shows higher capacitance as well as the energy density when compared with EDLCs (Chen et al. 2010a; Wu et al. 2010b; Sivakkumar et al. 2007), but the limited cyclability of the faradaic electrodes is a significant drawback of these hybrid capacitor devices. Overall, the key to achieving high power and energy-density hybrid ECs with long cycle life is to explore novel electrode materials with rational design of material combination and structure/morphology/size tuning, and to explore proper electrolytes that can operate at high voltages and provide sufficient ionic conductivity and electrochemical stability.

7.2.1.2 Hierarchical Carbon-Based Materials for ELDCs

The electrochemical performance of ECs can be achieved by introducing pseudocapacitance effects into hierarchical composite electrode systems combining hierarchical structured backbones with pseudocapacitive materials, such as TMOs and electrically conducting polymers. Particularly, carbon-based nanocomposites with hierarchical structures are recognized as one of the most promising electrode materials for the application in supercapacitors owing to the hierarchical structures that provide a large surface area and high conductivity in addition to synergistic effects for electrochemical reactions at the interface. The preparation of carbon-based hierarchical composites electrode is one of the most adopted techniques to improve the device performance of metal oxide electrodes whose poor conductivities limit their capacitance, cycling life, and rate performance.

The most widely used active materials with hierarchical structures in ELDC electrode are based on carbon since these carbon-based materials offer the advantageous features including high conductivity, electrochemical stability, and open porosity. Many hierarchically structured carbons are being explored as alternatives to activated carbon for improving energy density while maintaining high power density in the ultimate EDLCs, such as carbon aerogels (Wu et al. 2013b; Biener et al. 2011; Li et al. 2010), carbide-derived porous carbons (Chmiola et al. 2010; Heon et al. 2011), onion-like carbons (Pech et al. 2010; Portet et al. 2007), and graphene (Liu et al. 2010a; Zhu et al. 2011; Stoller et al. 2008). Xu and coworkers developed a facile, green, and template-free route toward sponge-like carbonaceous hydrogels and aerogels by using crude biomass, watermelon, as the carbon source (Fig. 7.3a; Wu et al. 2013b). The sponge-like carbonaceous hydrogels consisted of both a global catenulate carbon network and a carbon nanofiber cross-linked carbon nanosphere network that together constitute the 3D porous structure. The pores created by the interconnected networks varied from several nanometers to micrometers, and the average pore diameter was measured to be 45.8 nm. The presence of a large number of micropores allowed the efficient diffusion/mass transfer of liquid/gas phase of other components, leading to potential applications such as adsorbents as well as substrates. The obtained 3D flexible carbonaceous gels incorporated with Fe_3O_4 nanoparticles after calcination showed excellent EC performance with capacitance up to 330 F g^{-1} and stable cycling stability. Gogotsi et al. and Simon et al.

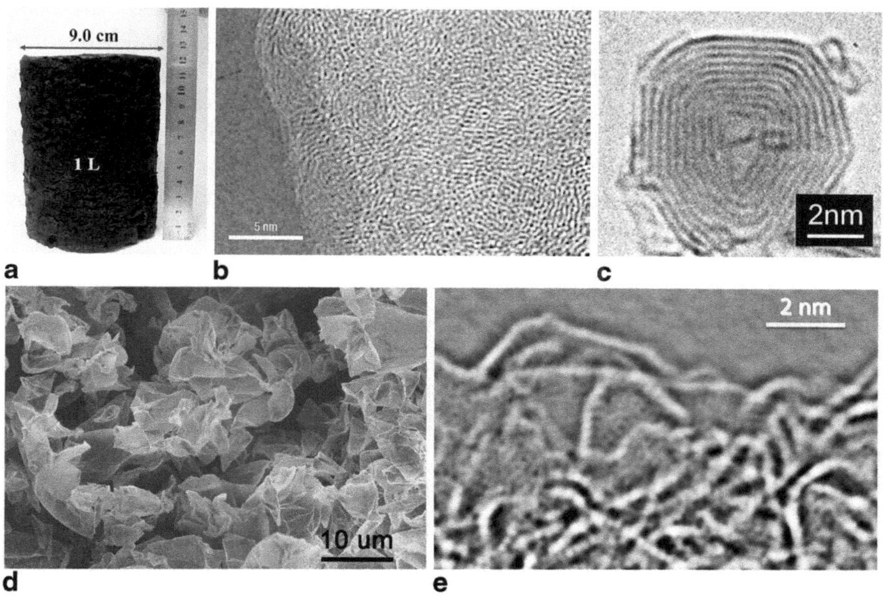

Fig. 7.3 Hierarchically structured carbon materials as active materials for electrochemical double-layer capacitance (EDLCs): **a** Digital image of carbonaceous hydrogel monoliths. (Reproduced with permission from Wu et al. (2013b) Copyright 2013, American Chemical Society). **b** Typical transmission electron microscopy (*TEM*) image of a disordered microporous carbon (SiC-derived carbon; Reproduced with permission from Simon and Gogotsi (2008) Copyright 2008, Macmillan Publishers Limited). **c** TEM image of onion-like carbon. (Reproduced with permission from Pech et al. (2010) Copyright 2010, Macmillan Publishers Limited). **d** Scanning electron microscopy (*SEM*) image of curved graphene sheets. (Reproduced with permission from Liu et al. (2010a) Copyright 2010, American Chemical Society). **e** Reconstructed high-resolution transmission electron microscopy (*HRTEM*) image from the edge of KOH-activated microwave-exfoliated graphene oxide (*GO*) sample. The in-plane carbon atoms are clearly resolved, and a variety of *n*-membered carbon rings can be seen. Substantial curvature of the single-carbon sheets is visible, with the in-plane crystallinity being preserved. (Reproduced with permission Zhu et al. (2011) Copyright 2011, American Association for the Advancement of Science)

recently developed several unique nanostructured carbons, such as carbide-derived carbons (Fig. 7.3b; Simon and Gogotsi 2008) and onion-like carbons (Fig. 7.3c; Pech et al. 2010) for microscale supercapacitors which exhibited good volumetric capacity and excellent power density. The activated carbons are derived from carbon-rich organic precursors by carbonization (heat treatment) in inert atmosphere with subsequent selective oxidation in CO_2, water vapor, or KOH to increase the specific surface area and pore volume. A porous network in the bulk of the carbon particles is produced after activation; micropores (<2 nm in size), mesopores (2–50 nm), and macropores (>50 nm) can be created in carbon grains. Jang et al. reported the preparation of curved graphene sheets aiming at the full utilization of the highest intrinsic surface capacitance (Fig. 7.3d; Liu et al. 2010a). The obtained graphene-based electrodes showed a high specific energy density of 85.6 W h kg^{-1} and operating voltage >4 V even at room temperature with ionic liquids as electrolytes. Most recently, Ruoff and coworkers developed chemically activated highly porous graphite oxide with surface area up to 3100 m^2 g^{-1} (Fig. 7.3e; Zhu et al. 2011). The activation process etched the microwave-exfoliated GO and had generated a 3D distribution of what were referred to as mesopores and micropores in the porous materials literature. The activation with KOH yielded a continuous 3D network of pores of extremely small size, ranging from ~1 to ~10 nm, and a dense pore structure with a continuous 3D network of highly curved, predominantly atom-thick walls. Thus, it appeared that the chemical activation was not merely digesting the microwave-exfoliated GO but also dramatically restructuring it. The EDLC electrodes constructed with this carbon yielded high values of gravimetric capacitance and energy density both in organic and ionic liquid electrolytes. Despite these advancements made with the introducing of hierarchical structures, the specific energies of EDLCs are still limited by their reliance on double-layer capacitance as the primary energy storage mechanism. Furthermore, it is difficult to further promote the performance of ECs based on pure carbon. Alternatively, a drastic capacitance improvement can be realized by taking the advantage of pseudocapacitance effects through hybrid carbon with other pseudocapacitive materials.

7.2.1.3 Hierarchical Carbon–TMO Composites for Pseudocapacitors

For pseudocapacitors, there are two general types of pseudocapacitive materials: TMOs and electronically conducting polymers. In the former case, ruthenium oxide (Hu et al. 2006; Subramanian et al. 2004), manganese oxides (Bélanger et al. 2008; Yu et al. 2011a), iron oxides (Cottineau et al. 2006), and nickel oxide (Wang et al. 2008) are the most commonly known pseudocapacitive metal oxides; while in the latter case, PAnis, polythiophenes, (PPys), and other π-conjugated conductive polymers are typical electroactive polymers for redox pseudocapacitance (Naoi and Morita 2008). Owing to the high theoretical specific capacitance (1358 F g^{-1}) and good electronic conductivity (3×10^2 S cm^{-1}), RuO_2 is a widely studied pseudocapacitive material which exhibits distinct oxidation states accessible within a potential range of ~1.2 V in acidic aqueous electrolytes. The pseudocapacitive behavior of RuO_2 is generally ascribed to a series of fast and reversible electron transfer reactions coupled with adsorption of protons on the surface of RuO_2 electrodes, where Ru oxidation states can change from (II) up to (IV). Several factors such as the degree of particle crystallinity, particle size, and electrode architecture show effect on the pseudocapacitive performance of RuO_2, yielding specific pseudocapacitance of 200–1200 F g^{-1}. It is reported that a tubular arrayed porous structure of $RuO_2 \cdot xH_2O$ electrodes with uniform wall thickness of ca. 40 nm (Fig. 7.4a) yielded very high specific capacitance of ~ 1300 F g^{-1} (Hu et al. 2006), which was attributed to its tailored nanotubular array architecture for effective ion and electron transport besides its hydrous nature, enabling a high rate of proton exchange. Other RuO_2/carbon hierarchical structures have been investigated. For example, Sheu and co-workers demonstrated a novel type of RuO_2/MWCNTs hybrid electrode (Ye et al. 2005). TEM and SEM characterizations revealed that the RuO_2 film on the surface of the CNTs was composed of small crystalline grains hierarchically assembled into bundle-like microstructures. The capacitance of the electrodes was significantly increased from 0.35 to 16.94 mF cm^{-2}, indicating the enhanced capacitance by addi-

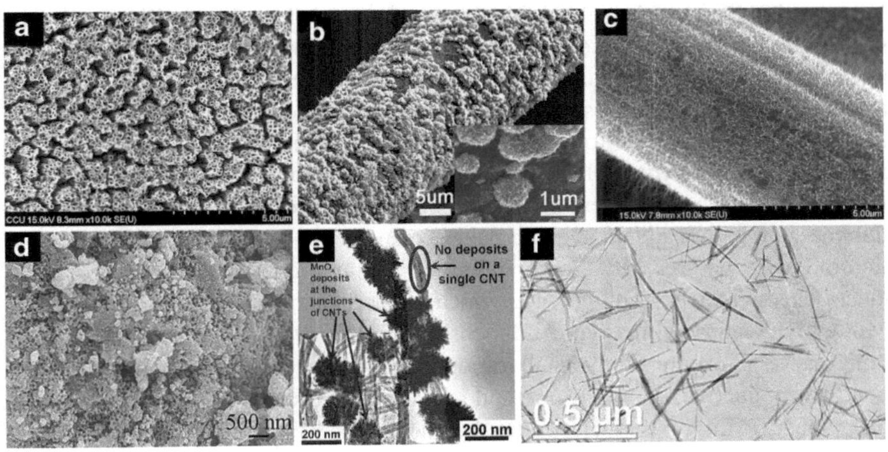

Fig. 7.4 High-performance hierarchically structured pseudocapacitive materials. **a** Scanning electron microscopy (*SEM*) image of $RuO_2 \cdot xH_2O$ nanotubular arrayed electrode. (Reproduced with permission from Hu et al. (2006) Copyright 2006, American Chemical Society). **b** SEM image of the energy textiles based on hybrid graphene/MnO_2 nanostructures as electrochemical capacitor (*EC*) electrodes. Inset shows the nanoflower structure of electrodeposited MnO_2 particles. (Reproduced with permission from Yu et al. (2011b) Copyright 2011, American Chemical Society). **c** SEM image of the needle-like α-MnO_2 nanofibers coated on carbon fiber fabric. (Reproduced with permission from Wu et al. (2010b) Copyright 2010, American Chemical Society). **d** SEM image of mesoporous carbon/MnO_2 hybrid structures with hierarchical pore structures. (Reproduced with permission from Peng et al. (2011) Copyright 2011, Springer-Verlag Berlin Heidelberg). **e** Transmission electron microscopy (TEM) images of the MnO_2/CNTs composite (Zhang et al. 2008) Copyright 2008, American Chemical Society. **f** TEM image of graphene oxide (GO)/MnO_2 hybrid composite. The single-layered GO sheet can be clearly distinguished from the background. MnO_2 nanoneedles are darker than others which seem to be enveloped by a thin film. (Reproduced with permission from Chen et al. (2010b) Copyright 2010, American Chemical Society)

tional pseudocapacitive RuO_2. Ramaprabhu et al. synthesized a series of hybrid materials of CNTs/RuO_2, CNTs/TiO_2, and CNTs/SnO_2 for pseudocapacitive electrodes via a chemical reduction method using functionalized CNTs and respective salts (Reddy and Ramaprabhu 2007). The specific capacitances of these electrodes are significantly increased compared to that of pristine multiwall nanotube (MWNT) electrodes.

Nevertheless, the high cost associated with the difficulties in large-scale production seriously limits the practical application of RuO_2. In comparison, MnO_2 is often considered as a practical candidate for pseudocapacitors due to its high theoretical specific capacitance (1370 F g^{-1}), rich abundance in nature, and environmental benignity. The charge storage mechanism for pseudocapacitive MnO_2 is based on surface adsorption of alkaline cations in the electrolyte as well as proton incorporation. Unfortunately, the theoretical capacitance of MnO_2 has seldom been achieved since its electronic conductivity is too low (~10^{-6} S cm^{-1}). Hierarchical composites, in which MnO_2 is combined with conductive matrix such as carbon and metal nanostructures, may provide a solution to improve the electrical conductivity and realize high specific capacitance of MnO_2-based electrodes. Yu et al. reported a solution-processed method to prepare graphene/MnO_2 nanostructured textiles (Fig. 7.4b; Yu et al. 2011b), which yielded a high specific capacitance up to 315 F g^{-1}. The solution-exfoliated graphene nanosheets could conformably be coated on 3D porous textiles support structures in large scale, and the controlled electrodeposition achieved high loading of pseudocapacitive MnO_2. To deposit graphene layers on textiles efficiently and uniformly, an ink solution with reasonable density and uniform size of exfoliated graphene is needed. Typical graphene ink solutions we used for textile coating yielded exfoliated graphene "nanosheets" with an average flake thickness of ~5 nm and an average flake dimension of ~0.6 μm, while the deposition of nanostructured MnO_2 on graphene-coated textiles was achieved through an electrochemical deposition process, which yielded nanoflower-shaped hierarchical MnO_2 architecture with a typical dimension of 300–800 nm and hundreds of random branches forming many small mesopores of size 5–30 nm. They further developed a "conductive wrapping" method to improve the supercapacitor performance of graphene/MnO_2-based electrodes by means of wrapping of graphene/MnO_2 nanostructures with CNTs or conducting polymer (Yu et al. 2011a). The specific capacitance of the electrodes has substantially increased by ~20 and ~45 %, with values as high as ~380 F g^{-1} achieved. Besides, these hierarchical composite electrodes have also exhibited excellent cycling performance with >95 % capacitance retention over 3000 cycles.

Other carbon-based hybrid materials with hierarchical structure have also been intensively studied. Carbon nanofibers have been used as the base-supporting materials to deposit MnO_2 forming the hierarchical structures of carbon/MnO_2. Li et al. demonstrated hybrid supercapacitor electrodes by coaxially coating MnO_2 thin films on a vertically aligned carbon nanofibers array (Liu et al. 2010b). This unique 3D brush-like hierarchical microstructure of carbon nanofibers/MnO_2 core–shell array shows a maximum specific capacitance of 365 F g^{-1}, a maximum specific energy density of ~32.5 W h kg^{-1}, and a maximum specific power density of ~6.2 kW kg^{-1}.

Wu et al. deposited needle-like α-MnO_2 nanofibers onto the conductive carbon fiber fabric for EC application (Fig. 7.4c) (Wu et al. 2010b). The average diameter and length of α-MnO_2 nanofiber coated on carbon fiber were approximately 20 nm and 1 μm, respectively. Some manganese oxide nanofibers were arranged roughly perpendicular to the surface of carbon fibers, whereas the manganese oxide nanofibers deposited on stainless steel tended to aggregate to form agglomerates. The α-MnO_2 nanofiber film with needle-like nanofibers is expected to have a positive effect on the capacitive behavior probably due to its high surface area and short ion diffusion path for EC reaction. The corresponding results revealed that protons intercalated into the interior region of the MnO_2 bulk in aqueous electrolytes, whereas the alkali metal cations only reached the near-surface region. The specific capacitance of MnO_2 nanofibers coated on carbon fabric could reach as high as 432 F g^{-1}. Besides, porous carbons such as carbon aerogels and mesoporous carbons have been used to produce carbon/MnO_2 hierarchical structures. Tong and coworkers prepared mesoporous MnO_2/carbon aerogel composites with mesoporous structure for use as EC electrodes, which achieved a high specific capacitance of 515.5 F g^{-1} (Pech et al. 2010). Lu et al. prepared carbon aerogel/Mn_3O_4 composite electrodes by a self-limiting anodic EC deposition of manganese oxide (Lin et al. 2011). Mn_3O_4 was grown in the form of thin nanofibers along the backbone of the carbon aerogel. The specific capacitance reached a maximum value of 503 F g^{-1} with only ~1 % capacitance decay for 6000 cycles. Lu et al. synthesized mesoporous carbon/MnO_2 hybrid structures with hierarchical pore structures (Fig. 7.4d) and controllable MnO_2 loading using a self-limiting growth method (Peng et al. 2011). The hierarchical carbon substrate showed the presence of macropores with diameter from 70 to 100 nm, and the mesoporous channels exhibited uniform diameters in the range of 6–8 nm. The hierarchical pore structure allows a tunable MnO_2 loading up to 83 wt%. The carbon/MnO_2 hybrid electrode with 70 wt% MnO_2 loading showed a maximum specific capacitance of 218 F g^{-1} and good capacitance retention. Moreover, CNTs/graphene and MnO_2 hybrid hierarchical structures were also frequently reported in recent years. Gu and coworkers prepared MnO_2 nanoflower/CNT array electrodes with hierarchical porosity (Fig. 7.4e; Zhang et al. 2008). The CNT array with thickness of 35 μm was formed by numerous densely packed and aligned CNTs with MnO_2 particles with 150 nm in diameter well dispersed within it. The specific surface area of the composite reached 234 m^2 g^{-1}, calculated from the specific surface area of MnO_2 particles (236 m^2 g^{-1}), CNTs array (201 m^2 g^{-1}), and the weight ratio of CNTs array in the composite (6.5 %). This binder-free electrode exhibited good rate capability with ~50 % capacity retention at 77 A g^{-1}, high capacitance of 199 F g^{-1} and 305 F cm^{-3}, and long cycle life with only 3 % capacity decay after 20,000 cycles. Zhu et al. synthesized a hybrid structure based on needle-like MnO_2 nanocrystals supported by GO (Fig. 7.4f; Chen et al. 2010b). A synergistic effect by the interaction between GO and MnO_2 was found to enhance the electrochemical performance of as-prepared hybrid electrodes. The GO/MnO_2 hybrid electrodes (~84 % capacitance retention from 197.2 to 165.9 F g^{-1}) shows improved electrochemical stability than that of MnO_2-only sample (69 % retention from 211.2 to 145.7 F g^{-1}) after 1000 cycles at current density of 200 mA g^{-1}. Similarly, Zhao et al. prepared hybrid materials of reduced GO (RGO) and MnO_2, which also exhibited enhanced capacitive performances with ~90% of original capacitance in the measured cycles than those of pure RGO and Na-typed birnessite (Na/MnO_2) sheets (Zhang et al. 2011).

7.2.1.4 Hierarchical Carbon-Conducting Polymer Composites for Pseudocapacitors

Conducting polymers such as PAni, PPy, and polythiophene have been also investigated as active electrode materials in energy storage systems (Pan et al. 2010). They are unique electroactive materials with large π-conjugation length and reversible redox and doping/dedoping reactions. However, conducting polymers such as pseudocapacitor electrode materials are limited by poor cyclability and fast capacitance decay at high current rates, presumably due to volume changes during the electrochemical operation and the accompanied decrease of their electrical conductivity. Moreover, the dopants may affect the chemical stability, hence the cycling stability of the conducting polymers. The composites of carbon and conducting polymers show synergistic effects that combine the advantages of both materials: conducting polymers provide superior pseudocapacitance, while hierarchical carbon scaffolds help conducting polymers to sustain from the strains upon cycling. Currently, one of the major research interests in this field is focused on tuning the hierarchical structures of the hybrid material. The studies on hybrid materials of conducting polymers and carbons with different chemical compositions, morphology, and phase structure are crucial for fundamental understanding of the interface between carbons and conducting polymers, which in turn provide the basis for future improvements of their device performance.

Hierarchical structures based on CNTs/porous carbon and conducting polymers were extensively studied with the expectation that the high conductivity and high specific surface area of CNTs can greatly enhance the ECs' performance of conducting polymers. Hammond and coworkers deposited hierarchical thin film electrodes of PAni nanofibers and functionalized MWCNTs by layer-by-layer (LbL) assembly for ECs (Fig. 7.5a; Hyder et al. 2011). Highly stable cationic PAni nanofibers, synthesized from the rapid aqueous phase polymerization of aniline, were assembled with carboxylic acid functionalized CNTs into LbL films. The LbL-PAni/CNTs film consisted of a nanoscale-interpenetrating network structure with well-developed nanopores. These LbL-PAni/CNTs films in lithium cell could store high volumetric capacitance of ~238 F cm^{-3} and high volumetric capacity of ~210 mA h cm^{-3}. In addition, rate-dependent galvanostatic tests showed that LbL-PAni/CNTs electrode could deliver both high power and high energy density of ~220 W h L^{-1} and ~100 kW L^{-1}, respectively. Wang et al. prepared the CMK-3/PAni hybrid nanomaterials, with ordered PAni nanofibers grown on the surface of ordered mesoporous carbons (Fig. 7.5b; Wang et al. 2006). The hybrid material exhibited a high specific capacitance of 900 F g^{-1} at current rate of 0.5 A g^{-1}. The capacitance retention of the hybrid was about 85% when the current rate increased to 5 A g^{-1}. The capacitance

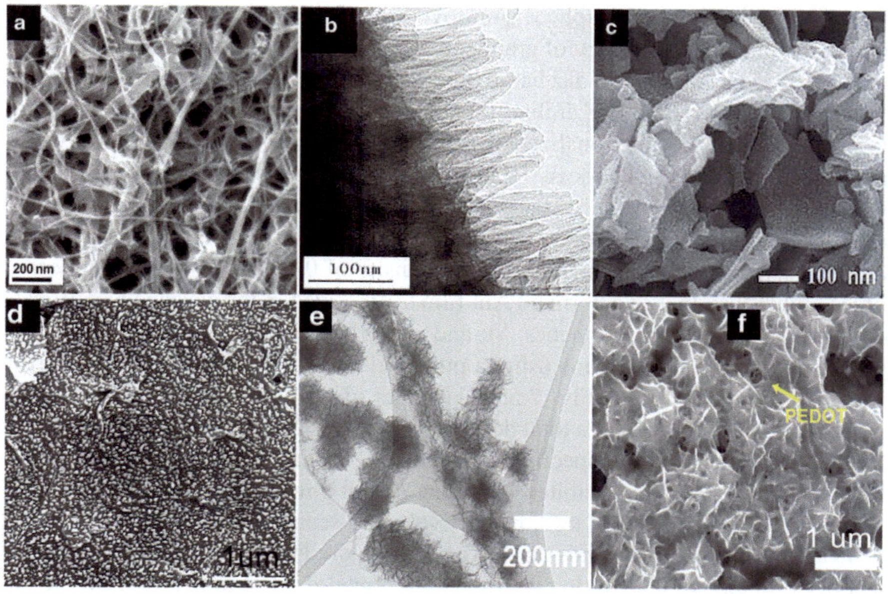

Fig. 7.5 Hierarchical carbon-conducting polymer composite pseudocapacitive materials. **a** Scanning electron microscopy (*SEM*) image of layer-by-layer polyaniline/carbon nanotubes (LbL-PAni/CNTs) thin film. (Reproduced with permission from Hyder et al. (2011) Copyright 2011, American Chemical Society). **b** Transmission electron microscopy (*TEM*) image of mesoporous carbon/PAni hybrid structure. Whisker-like structure is composed of a significant number of PAni thorns extending from the exterior of the mesoporous carbon into the interparticle open space, forming a loosely packed microstructure. (Reproduced with permission from Wang et al. (2006) Copyright 2006, WILEY-VCH Verlag GmbH & Co. KGaA, Weinheim). **c** SEM image of PAni/graphene hybrid structure from Zhao et al. (2011) Copyright 2011, the Royal Society of Chemistry). **d** SEM image graphene/PPy after 120 s electrodeposition. (Reproduced with permission from Davies et al. (2011) Copyright 2011, American Chemical Society). **e** TEM image of poly(3,4-ethylenedioxythiophene) polystyrene sulfonate (PEDOT-PSS)-dispersed MnO_2 nanospheres in situ grown on CNTs. (Reproduced with permission from Hou et al. (2010) Copyright 2010, American Chemical Society). **f** SEM image showing graphene/MnO_2/PEDOT:PSS nanostructures. (Reproduced with permission from Yu et al. (2011a) Copyright 2011, American Chemical Society)

remained ~95 % after 3000 cycles. It should be noted that the capacitance value is even higher than that of amorphous hydrated RuO_2 (840 F g^{-1}), indicating that this hybrid material could be a potential candidate for high-performance ECs to replace rare metal oxide RuO_2.

Graphene/conducting polymer hierarchically structured materials have also received considerable attention and represent one of the most promising directions for high-performance hybrid EC electrodes. Cheng et al. studied the supercapacitor performance of in situ polymerized PAni on carbon blacks, CNTs, and graphene nanosheets (Zhou et al. 2011). It was found that among all these formed structures, PAni/graphene hybrid showed the lowest internal resistance, the largest specific capacitance, as well as the highest cycling stability and rate capability, because of the high conductivity of graphene, improved interfacial contact of graphene/PAni due to the planar morphology of graphene, and therefore the stability to sustain the strain induced in the charge/discharge process. Shi et al. produced PAni nanosheets by electropolymerization of aniline monomer in the aqueous electrolyte containing sulfonated PAni-functionalized graphenes (Zhao et al. 2011). The as-prepared nanosheets have a thickness of ~10 nm. The resultant samples have a loosely stacked morphology on the surface of electrodes, giving a large effective surface area accessible to the electrolyte (Fig. 7.5c). The PAni nanosheets had a large specific capacitance of 372 F g^{-1} at a current density of 0.3 A g^{-1}, approximately five times higher than that of a normal PAni film synthesized in the absence of sulfonated PAni-functionalized graphenes. Yu and coworkers developed flexible, uniform graphene/PPy composite films using a pulsed electropolymerization technique for supercapacitor electrodes (Fig. 7.5d; Davies et al. 2011). A specific capacitance ~237 F g^{-1} was obtained for a total deposition time of 120 s, which is approximately four times higher than the specific capacitance of initial graphene film. This flexible supercapacitor film exhibited high energy and power densities with values of ~33 W h kg^{-1} and ~1184 W kg^{-1}, respectively, at a scan rate of 10 mV s^{-1}.

In addition to binary hybrids that have been explored to improve the electrochemical performance of conducting polymers and metal-oxides, ternary hybrid hierarchical structures have recently been explored as a new design that could integrate the advantages from all components: conducting carbon, pseudocapacitive metal oxides, and conducting polymers. Liu and coworkers designed a ternary hybrid material composed of MnO_2, CNTs, and PEDOT-PSS (Fig. 7.5e; Hou et al. 2010). Unique hierarchical MnO_2 architectures were successfully grown on a continuous functionalized CNTs network, and showed a tendency to strongly interact with the CNTs. No aggregations of the MnO_2 nanoparticles off CNTs scaffold were observed in the composite, indicating that the nucleation was predominantly on the exterior surfaces of CNTs, while PEDOT-PSS functioned as both the additional current collector and the binder material. In such a ternary composite, CNTs provided high surface for the deposition of hierarchical MnO_2 porous nanospheres, and improved the electrical conductivity and the mechanical stability of the composite; PEDOT-PSS functioned as an effective dispersant for MnO_2/CNTs structures and a binder material that improved the adhesion to the substrate and the connection among MnO_2/CNTs particles. The capacitance of the ternary composite electrode reached 427 F g^{-1}. Even at high mass loading and high concentration of MnO_2 of 60 %, the film still showed specific capacitance of ~200 F g^{-1}. The hierarchical electrode also exhibited impressive charge/discharge rate and good cycling stability, retaining over 99 % of its initial charge after 1000 cycles. Yu and coworkers developed a "3D conductive wrapping" method to improve the supercapacitor performance of graphene/MnO_2-based nanostructured electrodes (Fig. 7.5f; Yu et al. 2011a). They chose hybrid graphene/MnO_2 nanostructured textile electrodes in the studies as a platform to investigate the effect of conducting wrapping on improving the electrochemical performance of a graphene/MnO_2-based system. The graphene/MnO_2 textile electrodes were prepared by a two-step solution-based coating process, and the conductive wrapping with CNTs to form a graphene/MnO_2/CNT system was achieved by simply dipping graphene/MnO_2 CNT ink solution and subsequently drying. By 3D conductive wrapping of graphene/MnO_2 nanostructures with CNTs or conducting polymers, the specific capacitance of the electrodes has substantially increased by ~20 and ~45 %, respectively, with values as high as ~380 F g^{-1} achieved. Moreover, these ternary composite electrodes have also exhibited excellent cycling performance with >95 % capacitance retention over 3000 cycles. This 3D conductive wrapping approach presents a general direction for enhancing the device performance of metal oxide-based electrochemical supercapacitors and can be generalized for designing next-generation high-performance energy storage devices. These novel ternary hybrid systems take advantages of the synergistic effects from the hybrid structures, and make possible to effectively utilize the full potential of all the desired functions of each component. The new ternary hybrid system shows the advantages of high loading of active materials, excellent power and energy density, and long cycling life, and provides a new perspective for designing electrode architectures for next-generation high-performance ECs.

There have been tremendous experimental studies on synthesizing hierarchically structured carbons and carbon-based composites. Hierarchical porous carbons are of particular research interest. Kleitz and coworkers reported that increasing the hydrothermal temperature yielded larger pore sizes in mesoporous silica (KIT-6) and in its carbon replica (Kleitz et al. 2003). Vinu and Hartmann reported that mesoporous carbon materials with various textural properties such as specific surface area, specific pore volume, and pore diameters could be obtained at different temperatures by using SBA-15 as a template material (Vinu and Hartmann 2005; Hartmann and Vinu 2002). Lu et al. reported that the pore structures of the mesoporous carbon could be controlled by the amount of carbon precursor (Lu et al. 2003). Recently, Wu et al. synthesized mesoporous graphitic carbons by a typical chemical vapor deposition (CVD) process by using SBA-15 as the template and CH_4 as a precursor (Wu et al. 2012). It is even possible to obtain porous carbon materials with either a rod-type or a tube-type structure by using the same KIT-6 template material depending on the carbon source (Kleitz et al. 2003). In addition to experimental studies, nanostructured porous carbons have also been studied extensively using various computational techniques (Bandosz et al. 2003). By using the Car–Parrinello molecular dynamics (CPMD) technique, the local environment and the electronic structure of nanoporous carbon have been predicted (Galli et al. 1989, 1990; Marks et al. 1996a, b; Merchant et al. 2001). There are many other well-established empirical force fields to model carbon, such as Tersoff (1988), REBO (Brenner et al. 2002), ReaxFF (van Duin 2001), EDIP (Marks 2000), and LCBOP (Los and Fasolino 2003). Such models provide a better understanding of the molecular structure of porous carbon, which cannot be constructed solely based on experiments. These theoretical studies would potentially benefit the structural screening of proper carbon-related materials not only for supercapacitors but also for LIBs and fuel cells, catalyst supports, and materials for gas sensing and hydrogen storage.

7.2.2 Li-Ion Batteries

The fundamental difference between ECs and batteries lies in the charge storage mechanism. A typical battery stores charges in the bulk of the electrodes through faradaic reactions, whereas an EC stores charges near the surface (electrochemical double-layer capacitance or redox pseudocapacitance). Hence, batteries have higher specific energy (W h kg^{-1}), whereas ECs have higher specific power (W kg^{-1}). The former is useful when steady supply of energy is needed, while the latter finds applications where a burst of energy or high-frequency charge/discharge is needed, such as automotive regenerative braking and elevator operation. Rechargeable LIBs were first commercially introduced by the Sony Corporation through combining an $LiCoO_2$ cathode with a graphite anode in 1990. In the LIB configuration, lithium ions are incorporated/extracted between the cathode and anode during the discharging–charging processes (Fig. 7.6). The participants in the electrochemical reactions in an LIB are the negative and positive electrodes with the electrolyte providing a conductive medium for Li ions to move between the electrodes. Both electrodes allow Li ions to move in and out of their interiors. During insertion (or intercalation), Li ions move into the electrode. During the reverse process, extraction (or deintercalation), Li ions move back out. When an Li-ion-based cell is discharging, the positive Li ion moves from the negative electrode (usually graphite) and enters the positive electrode (lithium-containing compound). When the cell is charging, the reverse occurs. The exploration of higher specific-energy LIBs has recently attracted much research attention since they show potential applications for various high-end consumer electronic products (Wen and Li 2009). Some metals or metal alloys (e.g., Sn, Sb, Si, Bi, etc.) that react with lithium to form alloys are the most competitive candidates for new types of anode materials since they would provide a specific capacity much larger than that of conventional intercalation reactions (Arico et al. 2005; Kim et al. 2008; Chan et al. 2008b; Nam et al. 2006). Unfortunately, there are enormous volume changes during the alloying/dealloying processes in these materials, which would cause critical mechanical damage to the electrode, resulting in a marked loss of capacity during charge–discharge process (Besenhard et al. 2000).

It has been reported that the amount of reversibly incorporated lithium in the electrode materials and the voltage profile curves are affected by the structure, texture, and heteroatom content of the electrode materials (Flandrois and Simon 1999). To make those new electrode materials for practical use, challenges remain such as: low conductivity, large volume change of electrode material after lithium insertion, and the dissolution of active material during charge–discharge processes. All those challenges can be alleviated partially by constructing rationally designed hierarchical structures. Most of commercial LIBs use micrometer-sized particles as electrode materials. The low diffusion of lithium ions in solid bulk becomes the bottleneck in improving the battery's power density or rate performance. Designing hierarchical nanomaterials can alleviate this problem as the diffusion path of lithium ions will be reduced. Moreover, hierarchical nanostructures can accommodate volume change. The development of nanotechnology in the past decades has enabled excit-

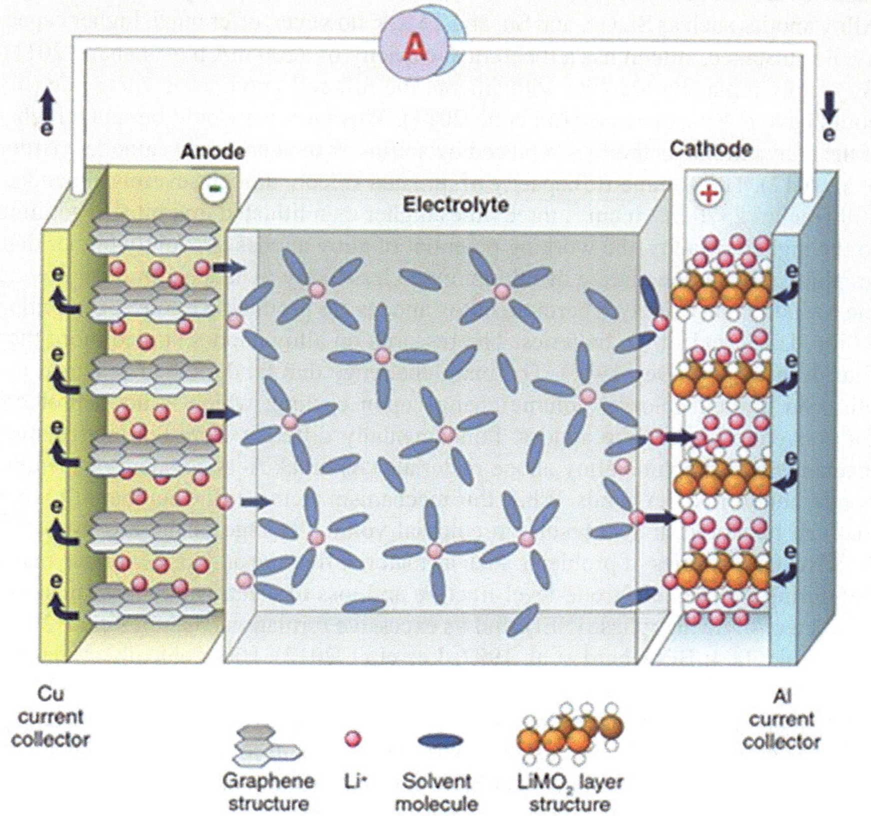

Fig. 7.6 Schematic of structure and working principle for Li-ion batteries (LIBs). The negative electrode is a graphitic carbon that holds Li in between the layers, whereas the positive electrode is an Li-intercalation compound. Both electrodes are able to reversibly insert and remove Li ions from their respective structures. The electrodes are separated by the electrolyte that transports Li ions between the electrodes. (Reproduced with permission (from Dunn et al. 2011) Copyright 2011, American Association for the Advancement of Science)

ing opportunities to design materials with desirable electronic, ionic, photonic, and mechanical properties. This development has also contributed to the advance in energy storage technologies. This section covers three aspects of recent research in hierarchically structured materials for LIBs: (i) hierarchical alloy anodes, (ii) hierarchical TMOs anodes, and (iii) hierarchical inorganic–organic composite anodes.

7.2.2.1 Hierarchical Alloy Anodes

State-of-the-art LIBs are composed of lithium metal oxide or phosphate cathodes and graphite anodes. To increase their energy density, electrode materials with higher capacity are needed. The graphite anode has a theoretical capacity of 372 mA h g^{-1}. Alloy anodes such as Si, Ge, and Sn, and TMOs, however, offer much higher capacity. For instance, silicon has a theoretical capacity of 4200 mA h g^{-1} (Zhang 2011). By simply replacing graphite with silicon, the full-cell gravimetric energy density could have a 40 % increase (Liu et al. 2014). This increase would be much higher if the conventional cathode is replaced by sulfur- or oxygen-based cathodes (Bruce et al. 2012). The volumetric capacity of lithiated silicon, though severely expanded, still reaches 2370 mA h cm^{-3}, three times higher than lithiated graphite. In addition to the higher capacity, the working potential of alloy anodes is slightly higher than graphite, making the plating of lithium metal less likely to happen, thus improving the overall battery safety. Therefore, alloy anodes are promising for next-generation Li-ion, Li–S, and Li–O_2 batteries. The research on alloy anodes started more than four decades ago (Dey 1971). The main challenge that hinders their practical application is their colossal volume change upon cycling, which is not a problem for conventional graphite anodes. Fundamentally different from the intercalation mechanism of graphite, alloy anode materials (A) break A–A covalent or metallic bonds, and form Li–A bonds. While this mechanism increases the number of Li ions that can be stored, it also results in colossal volume change of the host material, which causes a series of problems such as material fracture and loss of active material during cycling, electrode-level fracture and loss of electrical contact, unstable solid-electrolyte interphase (SEI), and its excessive formation (Beaulieu et al. 2001; Ryu et al. 2004; Besenhard et al. 1997; Lee et al. 2012). Hierarchically structured alloy anodes hosted on carbon matrix may be a viable solution to overcome the limitations of bulk particles. Hierarchal structure will provide the space needed for volume change and allow for fast transport of Li ions, and conducting carbon matrix will allow the improved solid/electrolyte interface formation, structural integrity, and high electrical conductivity. If properly engineered, the hierarchically structured alloy anodes can conserve size and shape on cycling, which is critical for industrial applications because commercial battery cells have very little volume available for anode expansion.

Yushin and coworkers reported a hierarchical bottom-up assembly method to prepare hierarchically structured Si–C nanocomposite (Magasinski et al. 2010). Si nanoparticles were grown on annealed carbon black, and further integrated the Si-coated carbon black particles into rigid spherical granules by CVD method (Fig. 7.7a). The size of the branches in the nanocarbon and the size of the deposited Si nanoparticles determine the pore size in the synthesized composite. The diameter of the composite particles is influenced by the carbon chemical vapor deposition (CVD) process parameters and by the size of the initial branched carbon particles. Therefore, the process allows control over the particle size, pore size, and composition of the composite. The obtained composite presented specific capacity based on Si of 3680 mA h g^{-1}. This high utilization of active Si is due to rapid Li ions into the particle bulk in this hierarchical structure. Besides, the pore available in this composite is large enough for the volume change of Si during Li insertion, and this can lead to the good cycling performance of this battery. Thus, the high capacity plus good stability should be contributed by special structure of this Si–C composite, and

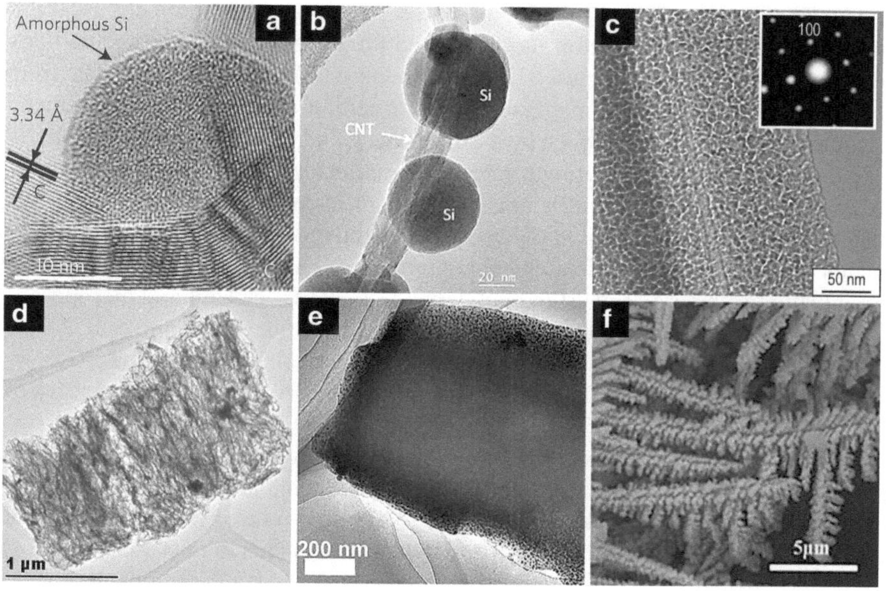

Fig. 7.7 Hierarchical alloy anodes for lithium ion batteries (LIB). Transmission electron microscopy (*TEM*) images of: **a** highly ordered graphitic structure of the carbon-black surface and the amorphous structure of Si. (Reproduced with permission from Magasinski et al. (2010) Copyright 2010, Macmillan Publishers Limited). **b** Single CNT covered with multiple silicon nanoclusters at defined spacing. (Reproduced with permission from Wang et al. (2011b) Copyright 2011, Elsevier B.V). **c** Nanoporous silicon powder. The pore size distribution is from several to tens of nanometers, and the remaining silicon scaffolding is as thin as several nanometers at points. The corresponding electron diffraction pattern indicates the single-crystalline nature of the porous particle. (Reproduced with permission from Zhao et al. 2012) Copyright 2012, the Royal Society of Chemistry). **d** Ge porous network architecture restructured from Ge nanowires after 100 charge/discharge cycles. (Reproduced with permission from Kennedy et al. (2014) Copyright 2014, American Chemical Society). **e** Sn nanoparticles (*NPs*) dispersed in N-doped porous carbon matrix. (Reproduced with permission from Zhu et al. 2014) Copyright 2014, American Chemical Society). **f** Dendritic Ni–Sn alloy. (Reproduced with permission from Zhuo et al. (2013) Copyright 2013, Elsevier B.V.)

this low-cost, safe and easy method is very promising for large-scale production of electrode materials. Kumta group also designed a Si/carbon hierarchical structure by a simple two-step CVD method (Fig. 7.7b; Wang et al. 2011b). This hierarchical structure is composed of vertically aligned Si/CNT arrays directly grown on metal current collector; therefore, no polymer binder is needed to make the electrode. The directly anchored Si/CNT on current collector can make a good electrical contact, which is beneficial for charge transfer. The CNTs function as a flexible mechanical support for strain release. The electrode demonstrates a reversible capacity of about 2000 mA h g^{-1} with capacity loss of 0.3 % per cycle. Zhao and coworkers used a wet-chemical etching method for the scale-up preparation of carbon-coated hierarchical microporous/nanoporous Si powders as an LIB anode (Fig. 7.7c; Zhao et al.

2012). The average size of the porous silicon powder is approximately 3–10 μm with wrinkled sidewalls uniformly distributed throughout the entire particle body to form a microporous structure. The gap among wrinkled sidewalls ranges from tens to several hundreds of nanometers. After carbon coating, the porous silicon particles are completely wrapped by a carbon layer, and the sidewall gap is partially filled. Compared with the growth rate of hierarchically structured Si/C composite, which is only several milligrams per hour or even lower based on the CVD method, this method can scale-up the yield of porous Si powder in gram scale. The etched Si powder shows a hierarchical microporous/nanoporous structure, alike the combination of microporous silicon obtained in HF/H_2O_2 etchant and mesoporous silicon prepared in $HF/AgNO_3$ etchant (Li and Bohna 2000; Hochbaum et al. 2009). After carbon intercalation inside the microholes/nanoholes of the porous Si powder, a stable capacity of 1500 mA h g^{-1} could be achieved with current density of 100 mA g^{-1}.

In addition to Si, hierarchically structured Ge, Sn, and Bi that react with lithium to form alloys are the other competitive candidates. Ryan and coworkers reported the formation of high-density Ge nanowire arrays grown directly from the current collector (Kennedy et al. 2014). An *ex situ* transmission electron microscopy (TEM) study has shown that the high performance can be attributed to the complete restructuring of the nanowires that occurs within the first 100 cycles to form a continuous porous network that is mechanically robust (Fig. 7.7d). Once formed, this restructured anode retains a remarkably stable capacity. The anodes retain capacities of ~900 mA h g^{-1} after 1100 cycles with excellent rate performance characteristics, even at very high discharge rates of 20–100 °C. Chen and coworkers reported the preparation of ultrasmall Sn nanoparticles embedded in nitrogen-doped porous carbon network (Fig. 7.7e; Zhu et al. 2014). Pyrolysis of Salen at 650 °C under Ar atmosphere was carried out to prepare N-doped porous Sn/C composite with the Brunauer, Emmett, and Teller (BET)-specific surface area that reached 280 m^2 g^{-1}. Sn nanoparticles with typical size of *ca.* 5 nm were homogenously embedded in the carbon framework. It should be noted that most of Sn nanoparticles were wrapped in the carbon network rather than exposed on the outer surface. As a result, the carbon matrix not only prevents the aggregation of tin particles but also avoids the exfoliation of Sn nanoparticles during lithiation/delithiation. The Sn/C composite showed an initial discharge capacity of 1014 mA h g^{-1} and a capacity retention of 722 mA h g^{-1} after 200 cycles at the current density of 0.2 A g^{-1}. Furthermore, a reversible capacity of ~480 mA h g^{-1} was obtained at higher current density of 5 A g^{-1}. The excellent electrochemical performance of Sn/C composite was attributed to the effective combination of ultrasmall Sn nanoparticles, uniform distribution, and porous carbon network structure, which simultaneously solved the major problems of pulverization, loss of electrical contact, and particle aggregation facing Sn anode. Chung and coworkers reported a highly porous 3D Ni–Sn alloy foam fabricated by electrodeposition accompanied by hydrogen evolution reaction (Fig. 7.7f; Zhuo et al. 2013). In their preparation, hydrogen bubbles function as a template in the electrodeposition of metals. 3D porous Ni–Sn alloy foams thus could be formed into self-supporting porous and/or dendritic structures by the dynamic hydrogen-template process. Close observation revealed that Ni was more likely to become aggregated particles and Sn to become needle-type dendrites. The size and numbers of the porous and/or dendritic structures were dependent on the concentrations of the metal ions and hydrogen ions and the applied over-potential. This alloy foam could evolve into porous and dendritic metal alloy structures. Dendritic $Ni_{50}Sn_{50}$ alloy foam exhibited high electrochemical capacity of 530 mA h g^{-1} at 1 C, and retained at 90.5 % of maximum capacity of discharge after 50 cycles. The highly porous structure of the Ni–Sn anode contributed to excellent cycle-life stability, providing facile accessibility to Li ions. These volumetric voids also act as mechanical bumpers preventing the structural disruption of Ni–Sn anodes.

Some hierarchical composites have shown promise in making a significant breakthrough toward commercialization. For example, a tin-based amorphous composite anode in Nexelion batteries made by Sony resulted from the deliberate combination of several material concepts such as active/inactive composite, intermetallic compound, and the use of carbonaceous material as a matrix phase. Despite the higher energy densities than carbonaceous anode materials, Li-alloying materials have generally been hindered by their poor cyclability due to huge volume changes during lithiation/delithiation reaction. Designing an anode material with hierarchical porosity addressing large volume changes and induced mechanical stresses therefore are suggested.

7.2.2.2 Hierarchical TMOs Anodes

TMOs have been studied as potential anode materials for LIBs owing to the stability and safety concerns, though their reversible capacities are relatively small compared to a graphite anode. The TMOs can be categorized into two groups based on the reaction mechanism: (i) insertion-type reaction of Li into the lattice of the electrode material without basic structural changes of the host material and (ii) conversion-type reaction with a complete decomposition of Li_2O and inactive transition metal (Park et al. 2010). In this section, TMO anode materials will be briefly discussed based on the type of the reaction mechanisms with Li.

Several TMOs including WO_2, MoO_2, and $Li_xFe_2O_3$ can store Li via insertion reaction but with low specific capacity and poor cyclability for Li insertion and extraction (Leroux et al. 1998; Dahn and McKinnon 1987; Sleigh and McKinnon 1991) Recently, TMOs as anode materials have gained a renewed interest due to the increased demand for new anode materials to fulfill requirements for large-scale batteries. In particular, spinel $Li_4Ti_5O_{12}$ has been widely studied as a promising anode to meet the requirements of LIBs for emerging applications such as electric vehicles and large-scale electrical energy storage systems (Ohzuku et al. 1995). Wang and coworkers prepared hierarchical $Li_4Ti_5O_{12}/TiO_2$ tubes composed of ultrathin nanoflakes via the calcination of the hydrothermal product of a porous amorphous TiO_2 precursor and lithium hydroxide monohydrate (Fig. 7.8a; Jiang et al. 2013). The hierarchical tubes exhibit a complex submicroscopic structure with regular structural imperfection, including a huge number of grain boundar-

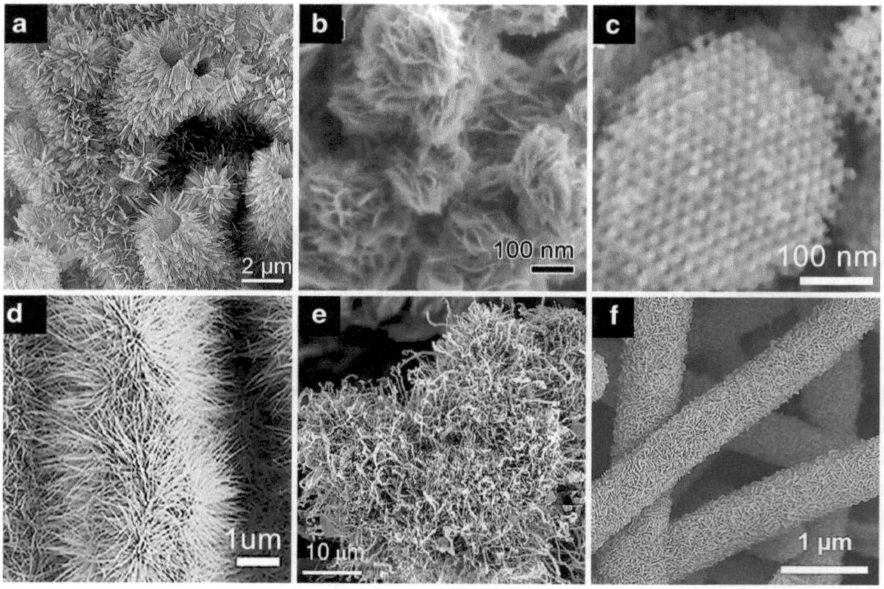

Fig. 7.8 Hierarchical transition metal oxide anodes for lithium ion batteries (LIBs). Scanning electron microscopy (*SEM*) images of: **a** $Li_4Ti_5O_{12}/TiO_2$ tube-like morphology assembled by interleaving ultrathin nanoflakes. (Reproduced with permission from Jiang et al. (2013) Copyright 2013, Macmillan Publishers Limited). **b** Self-assembled TiO_2–B nanosheets (Liu et al. 2012b) Copyright 2012, WILEY-VCH Verlag GmbH & Co. KGaA, Weinheim. **c** Mesoporous W-doped MoO_2 nanoparticles. (Reproduced with permission from Fang et al. (2012) Copyright 2012, the Royal Society of Chemistry). **d** $ZnCo_2O_4$ nanowire arrays growing on carbon cloth (Liu et al. 2012a) Copyright 2012, American Chemical Society. **e** Microscale hierarchical graphene–CNT–Fe_2O_3 3D nanostructure. (Reproduced with permission from Lee et al. (2013) Copyright 2013, American Chemical Society). **f** Tubular structures constructed by carbon-coated SnO_2 nanoplates. (Reproduced with permission from Zhang et al. (2013) Copyright 2013, WILEY-VCH Verlag GmbH & Co. KGaA, Weinheim)

ies and dislocations. The lithium ion storage property of these tubes showed initial discharge capacities of 420, 225, and 160 mA h g^{-1} at 0.01, 0.1, and 1.0 A g^{-1}, respectively. TiO_2 is also of great interest as an anode material due to its relatively larger theoretical capacity (335 mA h g^{-1} for $LiTiO_2$) as compared to $Li_4Ti_5O_{12}$ (175 mA h g^{-1}). Various TiO_2 polymorphs have been investigated and anatase TiO_2 and TiO_2–B have been reported to store and extract Li ions through the insertion reaction. Che and coworkers developed an approach for the self-assembly of titanium precursor chelated by $TiCl_4$ and ethylene glycol by solvothermal condensation in the presence of ammonia (Liu et al. 2012b). The obtained colloidal TiO_2–B nanosheets can be spontaneously assembled into piled architecture with abundant hierarchical pores (Fig. 7.8b). The petal-like layers with thickness of 5–10 nm were formed by stacking of several ultrathin nanosheets grown radically from core to outside with the inter layer spacing of ~1.2 nm. After calcination, the hierarchical structure with petal-like nanosheets morphology was still well preserved. Additionally, the calcined TiO_2–B shows the porous structure with a BET specific surface area of 151 $m^2\ g^{-1}$, pore volume of 0.278 $cm^3\ g^{-1}$, and pore size of ~3.6 nm. The reversible capacity reached 216 mA h g^{-1} and remained 200 mA h g^{-1} after 200 cycles with a capacity loss of 7.4% at a current rate of 10 C. The superior cycling performance testified the stability of hierarchical structure and the good accommodation to volume/strain changes during lithium insertion–extraction. MoO_2 with a distorted rutile structure can accommodate Li through the insertion reaction forming Li_xMoO_2 ($0<x<1$) with a theoretical capacity of about 209 mA h g^{-1} ($LiMoO_2$). Fang and coworkers synthesized ordered mesoporous W-doped MoO_2 by a nanocasting method (Fig. 7.8c; Fang et al. 2012). The formula for the resulting mesoporous W-doped MoO_2 can be expressed as $Mo_{0.82}W_{0.18}O_2$ based on the composition analysis of energy-dispersive X-ray spectroscopy. The mesoporous W-doped MoO_2 with pore wall thickness of *ca.* 8 nm showed a uniform mesoporosity with a specific surface area of 56 $m^2\ g^{-1}$ and a pore volume of 0.24 $cm^3\ g^{-1}$. The Li storage performance of mesoporous MoO_2 is significantly improved by tungsten doping, which exhibits a reversible capacity of 700 mA h g^{-1} at a current rate of 0.1 C with good cyclability. The W-doped MoO_2 material creates an optimized electrode configuration by combining the advantages of the high theoretical capacity of MoO_2 and the high electroactivity of WO_2, to give an overall performance that is better than that obtained by using either of these two components independently.

As mentioned above, reversible Li-ion storage could be achieved in TMOs (M=Fe, Co, Ni, Cu, etc.). These oxides were converted to a metallic state with the formation of Li_2O at the first lithiation, and reversibly returned to their initial state after the subsequent delithiation (Poizot et al. 2000). The number of electrons involved in the conversion reaction depends on the oxidation state of the transition metal, which is in contrast to the generally less than 1 for Li in the insertion reaction. This strategy to utilize all the possible oxidation states of the transition metal component enables attaining high Li storage capacities usually in the range of 400–1000 mA h g^{-1} with an average working potential of 1.8–2.0 V (vs. Li/Li^+). Shen and coworkers synthesized hierarchical $ZnCo_2O_4$ nanowire arrays on carbon cloth via hydrothermal route as binder-free anodes for LIBs (Fig. 7.8d; Liu et al. 2012a). $ZnCo_2O_4$ nanowire arrays/carbon composite fibers have uniform diameter of approximately 20 μm. Each composite fiber is composed of numerous highly ordered novel 3D $ZnCo_2O_4$ arrays with relatively high nanowire density grown on an individual carbon microfiber. The $ZnCo_2O_4$ nanowires were composed of small nanoparticles rather than conventional single-crystalline nanowire, and the diameter of $ZnCo_2O_4$ nanowire uniformly distributed in the range of 80−100 nm. The $ZnCo_2O_4$/carbon cloth electrodes showed a capacity about 1200 mA h g^{-1} after 160 cycles, and the capacity remained over 600 mA h g^{-1} at a high current rate of 5 °C. Oh and coworkers report a novel route via microwave irradiation to synthesize a bioinspired hierarchical graphene–nanotube–iron 3D nanostructure as an anode material in LIBs (Fig. 7.8e; Lee et al. 2013). The nanostructure comprises vertically aligned CNTs grown directly on graphene sheets along with shorter branches of CNTs stemming out from both the graphene sheets and the vertically aligned CNTs. In addition, functional iron-oxide nanoparticles decorated within the 3D hierarchical structure provides outstanding lithium storage characteristics, resulting in very high specific capacities. This composite material delivers a reversible capacity of ~1024 mA h g^{-1} even after prolonged cycling along with a Coulombic efficiency in excess of 99 %, which reflects the ability of the hierarchical network to prevent agglomeration of the iron-oxide nanoparticles. Lou and coworkers synthesized hierarchical tubular structures constructed by ultrathin carbon-coated SnO_2 nanoplates (Fig. 7.8f; Zhang et al. 2013), which simultaneously integrates the structural and compositional design rationales for high-energy anode materials based on the low-dimensional nanosized building blocks, hollow tubular structure, and carbon nanocoating. The hierarchically structured SnO_2@C manifests good cycling stability. The discharge capacity from the second cycle onwards shows relatively slow fading. After 50 cycles at a current density of 200 mA g^{-1}, a discharge capacity of 700 mA h g^{-1} could be retained, corresponding to about 80 % of the discharge capacity in the second cycle. However, despite their high specific capacities and good cyclability, these TMO materials are still not sufficient for commercial use due to their high working potentials as an anode, low Coulombic efficiency at the first cycle, and the large potential hysteresis between discharge and charge, which might cause a decrease of the power efficiency of the batteries.

7.2.2.3 Hierarchical Alloy/Conductive Polymer Composite Electrodes

Alloy-type and conversion-type anodes face significant challenges due to large volume expansion upon lithium insertion, which results in fracture and loss of electrical contact, and an unstable SEI growth on the surface of the electroactive materials. Recently, electrodes made from Si nanoparticle slurries have been investigated by several research groups as a potentially manufacturing-compatible route (Kim et al. 2008; Magasinski et al. 2010; Hu et al. 2011, 2012). Various polymer binders other than traditional polyvinylidene difluoride (PVDF), including polyacrylic acid (PAA), carboxy-methyl cellulose (CMC), poly(9,9-dioctylfluorene-co-Suorenone-comethylbenzoic acid) (PFFOMB), and alginate, have been found to enhance the cycle life (Liu et al. 2011a; Bridel et al. 2009; Verbrugge and Cheng 2008, 2009; Kovalenko et al. 2011). The cycle life improvement when using PAA, CMC, and alginate binders was attributed to the high polymer modulus values and the binding between the functional groups on the polymers with the surface oxide on the Si particles. In the case of the PFFOMB binder, an interesting concept was demonstrated that combined binding and conducting properties in the binder molecules (Beaulieu et al. 2001). Recently, Wu and Yu demonstrated that a unique feature of their approach for realizing excellent performance was the use of in situ polymerization, to form a bifunctional conformal coating that binds to the Si surface and also serves as a continuous 3D pathway for electronic conduction (Fig. 7.9a; Wu et al. 2013a). Conductive polymer hydrogels with hierarchical 3D porous microstructures and nanostructures were used as binders which could promote the transport of both electrons and ions owing to the availability of short diffusion paths and good 3D

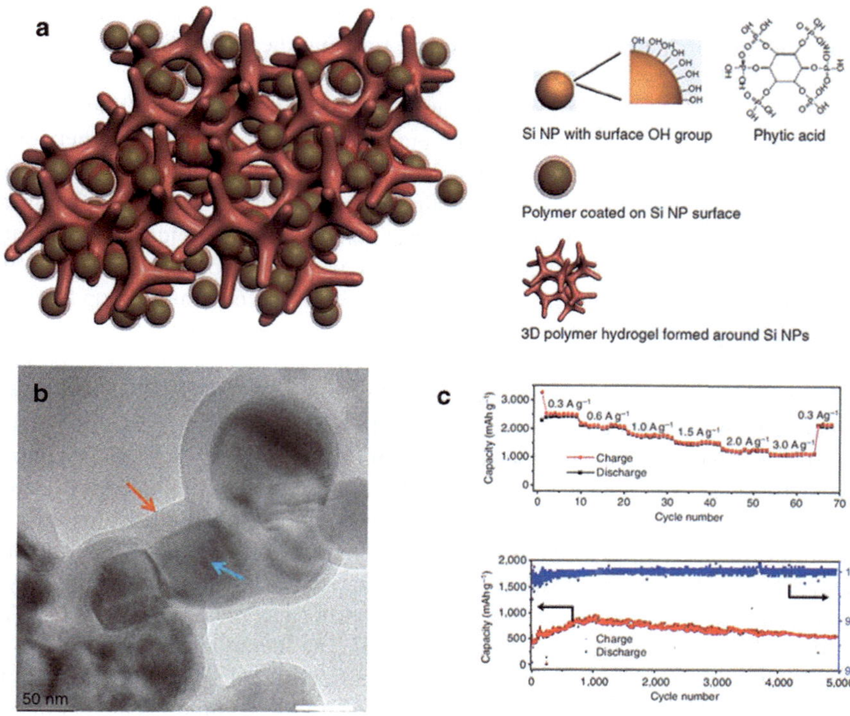

Fig. 7.9 **a** Schematic of 3D porous Si nanoparticle (SiNP)/conductive polymer hydrogel composite electrodes. Each SiNP is encapsulated within a thin polymeric coating layer and closely incorporated within the conductive PAni framework. The phytic acid functions as the crosslinker, which enhances the integration of SiNPs/conductive polymer framework and electric conductivity of the electrode. **b** Transmission electron microscopy (TEM) image shows Si nanoparticles (*blue arrow*) are coated with a uniform PAni polymer layer (*red arrow*). Scale bar, 50 nm. **c** Capacity of a Si nanoparticle–PAni electrode cycled at various current densities (*top*) and lithiation/delithiation capacity and CE of SiNP–PAni electrode cycled at current density of 6.0 A g^{-1} for 5000 cycles (*bottom*). (Reproduced with permission from Wu et al. (2013a) Copyright 2013, Macmillan Publishers Limited)

connectivity. Moreover, they could be synthesized by simply mixing two solutions, where one contained the initiator (oxidizer) and the other contained the crosslinker and the monomer. When Si nanoparticles were mixed into the solution during polymerization, the phosphoric acid groups in the phytic acid molecules could potentially bind with the SiO_2 on the Si particle surfaces via hydrogen bonding. This interaction could also result in the conformal coating of phytic acid molecules on the surface, which might further crosslink with aniline monomers during polymerization to generate a conformal conductive coating. Additionally, the negatively charged surface oxide might electrostatically interact with the positively charged

PAni doped by the phytic acid. All the above interactions result in the formation a 3D continuous-conducting hydrogel network, trapping the coated Si nanoparticles inside (Fig. 7.9b).

Typically, silicon nanoparticles have an average diameter of ca. 60 nm, while the dried polymer hydrogel consists of a hierarchical 3D porous foam-like network composed of dendritic nanofibers with typical diameters of 60–100 nm. As the PAni was formed in the presence of the Si nanoparticles, which are in intimate contact with the conductive polymer hydrogel matrix at both the microscopic and molecular level. Additionally, Si nanoparticles appear to be encapsulated by a conformal PAni coating due to the in situ polymerization (Fig. 7.9b). The continuous-conducting polymer hydrogel framework, which is directly connected to the current collector, provides channels for fast electron transport, thereby enabling outstanding rate capability. The capacity of a Si nanoparticle-PAni composite electrode varies from 2500 to 1100 mA h g^{-1} at charge/discharge rate from 0.3 to 3.0 A g^{-1}. At a high current density of 6.0 A g^{-1}, an electrode capacity of ca. 550 mA h g^{-1} was still retained after 5000 cycles, which corresponds to ca. 91 % capacity retention at current density of 6.0 A g^{-1} (Fig. 7.9c). The average CE of the composite electrode from the 2nd to 5000th cycle is 99.8 %, which is higher than that previously reported for silicon nanoparticles and nanowires (Yao et al. 2011; Chan et al. 2008a). In a similar way, Yu and coworkers reported a 3D ternary Si nanoparticles/conductive polymer/CNT hybrid anode material for LIBs (Liu et al. 2013a). The hierarchical conductive hydrogel framework with CNTs as the electronic fortifier offered a continuous electron transport network and high porosity to accommodate the volume expansion of Si particles. By 3D wrapping of silicon nanoparticles/single-wall CNTs with conducting polymer nanostructures, a greatly improved cycling performance is achieved with reversible discharge capacity over 1600 mA h g^{-1} and 86 % capacity retention over 1000 cycles at the current rate of 3.3 A g^{-1}. Moreover, above-described materials design for Si-based anodes may be extended to other alloy-type or conversion-type electrode material systems that experience large volume changes and unstable SEI formation upon cycling.

Hierarchical structures exhibit exceptional EC performance due to the greatly enhanced diffusion kinetics and structural stability for lithium storage. The integration of the hierarchical structure with electrochemically active materials arguably provides an effective way to generate high-performance electrode materials with great advantages over bulk and solid materials. It should be stressed that the performance evaluation of the electrodes should be carefully exploited by fully taking into account the important factors involved in materials and EC systems. A meaningful comparison can be only made based on comparable materials characteristics (e.g., the particle size, texture, and intrinsic EC properties) and similar conditions for the battery test such as the voltage window, current rate, cycle number, the kind and content of binder and electrolyte, and cell design.

7.3 Applications in Energy Conversion

Energy conversion concerns the transformation of energy from one form to another, for example, sunlight to chemical or electric energy, electricity to thermal and mechanical energy, and chemical to thermal energy and electricity. This section briefly introduces several conversion technologies used for converting sunlight to chemical and electric energy, and chemical energy to electricity. First, different conversion technologies using sunlight as energy sources such as photoelectrochemical cells for photochemical hydrogen production and photosynthesis are described, followed by hierarchically structured materials used for fuel cells.

7.3.1 Photoelectrochemical Cells

Solar energy, which is clean, renewable, and nearly inexhaustible, is our largest global renewable energy source, with the potential for providing orders-of-magnitude more energy than will be consumed annually (Hoffert et al. 1998). Solar conversion devices, however, are not yet ubiquitous because of the relatively large cost of energy produced this way when compared to traditional energy sources such as fossil fuels (Bierman and Jin 2009). Collecting and storing solar energy in chemical bonds, as nature accomplishes through photosynthesis, is a highly desirable approach to solving the energy challenge. Photolysis of water with semiconductor materials has been investigated as a clean and renewable energy conversion process by storing solar energy in chemical bonds such as hydrogen (Walter et al. 2010). Since Fujishima and Honda reported the capability of water splitting with TiO_2 (Fukishima and Honda 1972), metal oxide semiconductors have raised considerable research interest because of their stability under photoanodic conditions (Maeda and Domen 2011). A general approach for mimicking natural photosynthesis is to generate O_2 and H_2 separately via the "Z-scheme" (Wang et al. 2011a) using two inorganic semiconductors loaded with cocatalysts. When exposed to light illumination, the minority carriers in the two semiconductors are used to carry out the two individual reactions, while the majority carriers recombine at an ohmic contact between the materials (Bolton et al. 1985). Such an ohmic contact is an inorganic analogue of the electron transport chain in a chloroplast. Complex 2D and 3D nanostructures such as interlinked branches (Cho et al. 2011; Shi et al. 2011) and dendritic (Kay et al. 2006) nanostructures have shown enhanced efficiencies for solar water splitting from increased charge transport or charge separation. In order to bring the cost of solar energy in line with other sources, new science and new technology are needed. A variety of materials with hierarchical structures have been utilized to harvest solar energy that are cheaper than current technology while remaining efficient or could be even more efficient.

Yang and coworkers prepared a 3D hierarchical nanostructure of InGaN nanowires on Si wire arrays (Fig. 7.10a and b; Hwang et al. 2012). Single-phase InGaN nanowires grew vertically on the sidewalls of Si wires and acted as a high sur-

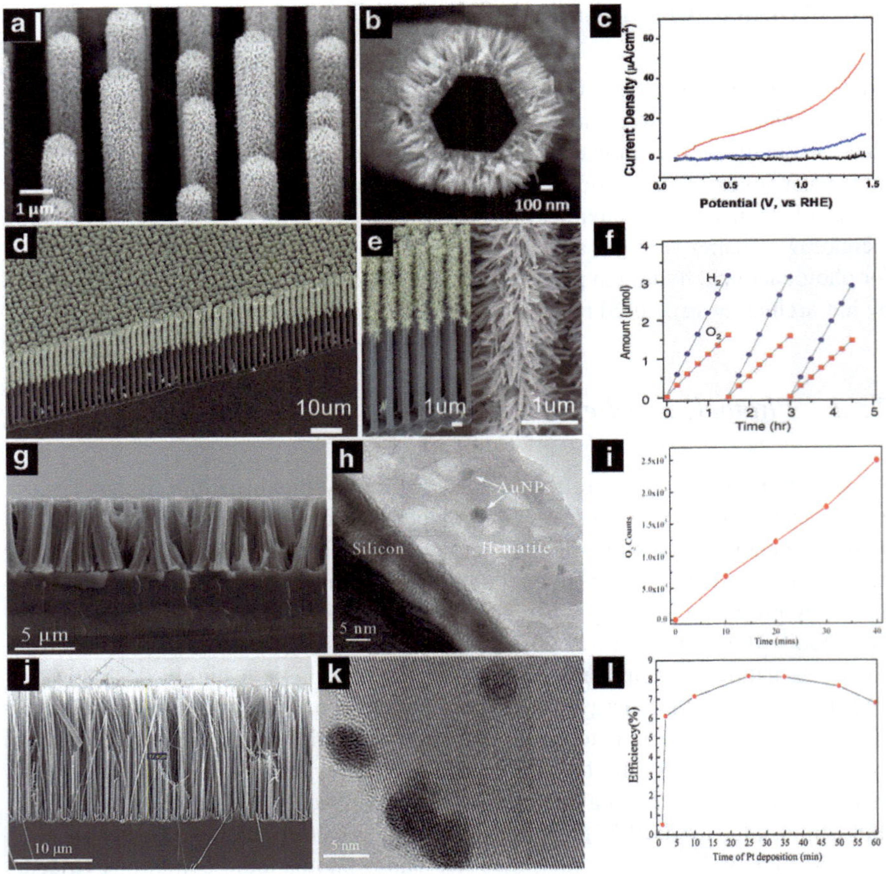

Fig. 7.10 Hierarchical nanostructures in photoelectrochemical cell application. **a, b** Scanning electron microscopy (*SEM*) images of hierarchical Si/$In_xGa_{1-x}N$ nanowire arrays on Si (111) substrate with $x=0.08-0.1$. A fractured wire reveals the cross section (**b**) showing that InGaN nanowires grow vertically from the six Si wire facets. **c** Photocurrent density plots of hierarchical Si/InxGa1–xN nanowire arrays (*red*) and planar Si/$In_xGa_{1-x}N$ (*blue*) under 100 W cm^{-2} without AM 1.5 filter show an increase in current for the hierarchical geometry. (Reproduced with permission from Hwang et al. (2012) Copyright 2012, American Chemical Society.) **d, e** SEM image of a Si/TiO_2 nanotree array at different magnifications. **f** The evolution of H_2 and O_2 gases measured by gas chromatography under simulated sunlight of 150 mW cm^{-2}. (Reproduced with permission from Liu et al. (2013b) Copyright 2013, American Chemical Society.) **g** SEM and **h** transmission electron microscopy (TEM) images of the Si nanowire (SiNW)/hematite–AuNPs photoanode. Si/hemitate core/shell nanowire array configuration integrates the advantages of enhanced light absorption, rapid carrier collection, and transport; additionally, it captures a wide solar spectrum since Si can absorb the long-wavelength part of the sunlight that is not absorbed by the hematite shell. Incorporation of plasmonic AuNPs enhances visible light absorption in the near-surface region of the hematite shell and shortens the diffusion distance of the photogenerated holes to the electrolyte. **i** Oxygen evolution as a function of time detected by gas chromatograph for a *p*-SiNW/hematite–AuNPs photoanode illuminated with AM 1.5 simulated sunlight. (Reproduced with permission from Wang et al. (2014) Copyright 2014, American Chemical Society). **j** Cross-sectional SEM image of SiNW array. **k** High-resolution TEM image of PtNP-decorated SiNW. **l** Effect of different amounts of PtNP loading via increasing deposition time on solar energy conversion efficiency. (Peng et al. 2009) Copyright 2009, American Chemical Society

face area photoanode for solar water splitting. EC measurements showed that the photocurrent density with hierarchical Si/InGaN nanowire arrays increased by five times compared to the photocurrent density with InGaN nanowire arrays grown on planar Si (Fig. 7.10c). Their results showed that Si/InGaN hierarchical nanostructures were a viable high surface area electrode geometry for solar water splitting. They further demonstrated a fully integrated system of nanoscale photoelectrodes assembled from hierarchical inorganic nanowire array for direct solar water splitting (Liu et al. 2013b). They synthesized the nanotree heterostructure that contains many nanotree heterostructures with Si nanowire (SiNW) trunks and TiO_2 nanowire branches (Fig. 7.10d). An individual nanotree showed a core–shell heterostructure with the photoanode of TiO_2 nanowires on the top and the photocathode of SiNW on the bottom (Fig. 7.10e). The high surface area of the TiO_2 nanowires provided abundant reactive sites for the sluggish O_2 evolution, while SiNW embedded underneath TiO_2 photoanode provided an ohmic contact for recombination of majority carriers and serves as a charge collector, which takes advantage of the high carrier mobility of single-crystalline SiNW. The 2:1 stoichiometry between H_2 and O_2 confirmed the water-splitting photoactivity, and the linear increase of gas concentrations revealed a stable catalytic performance (Fig. 7.10f).

Lee and coworkers reported a facile fabrication of 3D Si/hematite core/shell nanowire arrays decorated with gold nanoparticles and investigated their potential application for sunlight-driven solar water splitting (Wang et al. 2014). Wafer-scale aligned single-crystalline SiNWs arrays could be prepared on Si(100) wafers by metal-catalyzed electroless etching method in HF–$AgNO_3$ solution, which yielded SiNWs with diameters of 30–200 nm and lengths of a few micrometers (Fig. 7.10g) controlled via the etching time. Hematite shells incorporated with AuNPs were deposited on the surface of SiNWs by chemical wetting and annealing method using a mixed aqueous solution containing $Fe(NO_3)_3$ and $HAuCl_4$ precursors (Fig. 7.10h). When the photoanodes in the mixed configuration were immersed in aqueous solution of sodium phosphate in a quartz cell, copious amounts of gas bubbles dislodging from the electrode surface could be clearly observed under AM 1.5G illumination. Gas chromatography analysis confirmed that the collected gaseous product was oxygen and little hydrogen was produced due to the highly alkaline electrolyte solution. The amount of evolved oxygen versus time showed continuous oxygen evolution (Fig. 7.10i). They also used a similar wet-chemical etching method to produce high-density aligned *n*-type SiNW arrays (Fig. 7.10j) and further decorated the SiNWs with discrete 5–10 nm platinum nanoparticles (Fig. 7.10k) as photoanode of a photoelectrochemical cell (Peng et al. 2009). By controlling the coating time of PtNPs on SiNW sidewalls, a substantial enhancement in photoconversion efficiency and an apparent energy conversion efficiency of up to 8.14 % was obtained for photoelectrochemical cell in a liquid electrolyte containing Br^-/Br_2 redox couple (Fig. 7.10l). The improved performance demonstrated the superior ability of metal nanoparticles to modify the electronic properties of SiNWs, and the high potential of PtNPs-decorated SiNWs as a hybrid system for solar irradiation harvesting.

1D hierarchical nanostructures have been demonstrated to be efficient in photoelectrochemical cell applications because they can decouple the directions of light absorption and charge carrier collection. When the lifetime of the minority carrier is short, the minority carrier undergoes recombination in bulk before reaching the semiconductor/electrolyte junction. However, due to the small radius of nanowires, the minority carrier can diffuse to the surface before recombination, which potentially increases the charge separation efficiency, particularly, when the minority carrier diffusion length is comparable to the radius of the nanowire.

The search for earth-abundant materials with affordable cost and environmentally benign and photoelectrode stability remains challenging. The combinations of semiconductor materials, electrocatalysts, and cell configurations provide alternative ways for photoelectrolysis research. As solar fuels' research expands, standardizing both research methodologies and characterization techniques becomes paramount for accurate reporting, and ultimately helps to move the field forward into new areas of development and discovery. Recent efforts made on the development of efficient photoelectrolysis devices rely largely on controlling the size/shape and engineering the interface of microscale and nanoscale features of semiconductors and catalysts. The use of hierarchically structured photoelectrodes contributes to a more rational design of such devices because the control of the directionality of charge/hole movement and the control of the light absorption pathways can usually be achieved in wafer-scale. High aspect ratio photoelectrode surfaces resulting from such hierarchical structures also might allow for the use of less expensive catalysts with lowered catalytic activities. Due to an improvement in light absorption, charge separation, high surface area for electrochemical reactions, and the ability to fabricate hierarchical structured materials with high carrier mobility are attractive for photoelectrochemical energy conversion, of which the ultimate goal is to reduce the cost and improve the energy conversion efficiency, to enable large-scale adaptation of renewable energy. Fundamentally, this requires three processes to occur: absorption of light radiation to generate excited charge carriers, diffusion of charge carriers to an interface, and successful separation of these excited charge carriers. Proper materials and device architectures can facilitate these processes without the need for expensive manufacturing techniques. In addition, these systems must be scalable for large-scale deployment of renewable energy on a global scale.

7.3.2 Fuel Cell Electrode Scaffold/Catalyst

A fuel cell is an electrochemical cell that converts chemical energy from a fuel into electric energy in a constant temperature process. Electricity is generated from the reaction between a fuel supply and an oxidizing agent. Hydrogen is the most common fuel, but sometimes hydrocarbons are also used. Fuel cells require a constant source of fuel and oxygen/air to sustain the chemical reaction. The reactants flow into the cell, and the reaction products flow out of it, while the electrolyte remains within it. Fuel cells can operate continuously using wide range of fuels, including hydrogen, as long as the necessary reactant and oxidant flows are maintained. Fuel cell usually consists of an anode, a cathode, and an electrolyte that allows charges to move between the two sides of the fuel cell. At the anode, a catalyst oxidizes the fuel into positively charged ions and electrons. At the cathode, a catalyst turns the ions into the waste chemicals such as water or carbon dioxide. Fuel cells are classified by the type of electrolyte they use as the main difference among fuel cell types is the electrolyte. The reactions and processes that occur in the various fuel cell systems are summarized in Fig. 7.11 (Winter and Brodd 2004). Fuel cells are classified by the type of electrolyte they use, as the main difference among fuel cell types is the electrolyte. The lower temperature systems including alkaline fuel cells (Gülzow 1996), polymer electrolyte membrane fuel cells (Wang et al. 2011c), and phosphoric acid fuel cells (Bagotsky 2012) operate essentially on H_2 fuel, whereas the higher temperature systems, including molten carbonate fuel cells (Cherepy et al. 2005) and solid oxide fuel cells (Ormerod 2003), can also electrochemically oxidize carbon monoxide, which is advantageous when a hydrocarbon fuel is supplied to the fuel cell. For reasons of electrode activity, which translates into higher

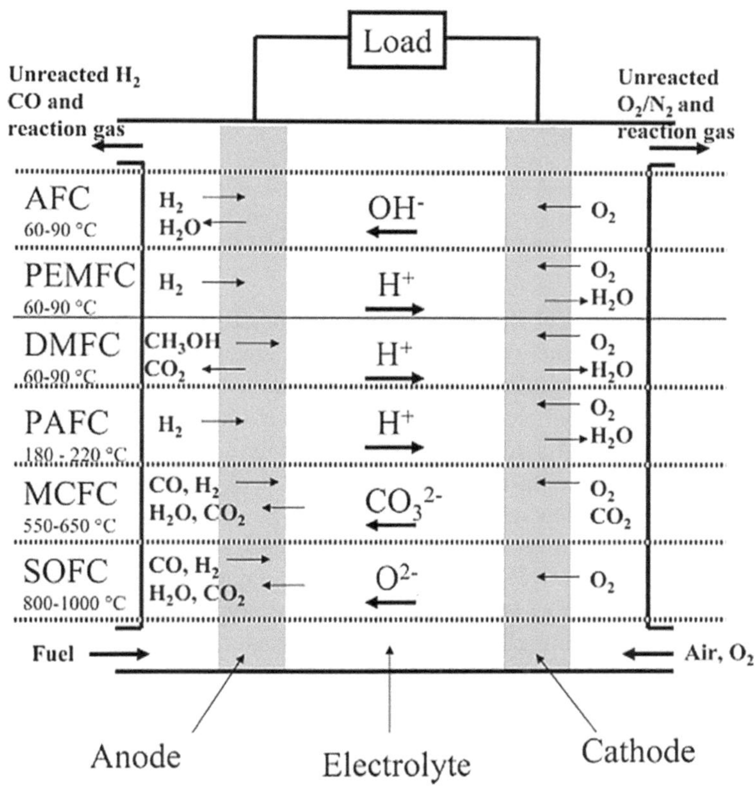

Fig. 7.11 Summary of the reactions and processes that occur in the various fuel cell systems: alkaline fuel cell (*AFC*), polymer electrolyte membrane fuel cells (*PEMFC*), direct methanol fuel cell (*DMFC*), phosphoric acid fuel cells (*PAFC*), molten-carbonate fuel cells (*MCFC*), and solid oxide fuel cell (*SOFC*). (Reproduced with permission from Winter and Brodd (2004) Copyright 2004, American Chemical Society)

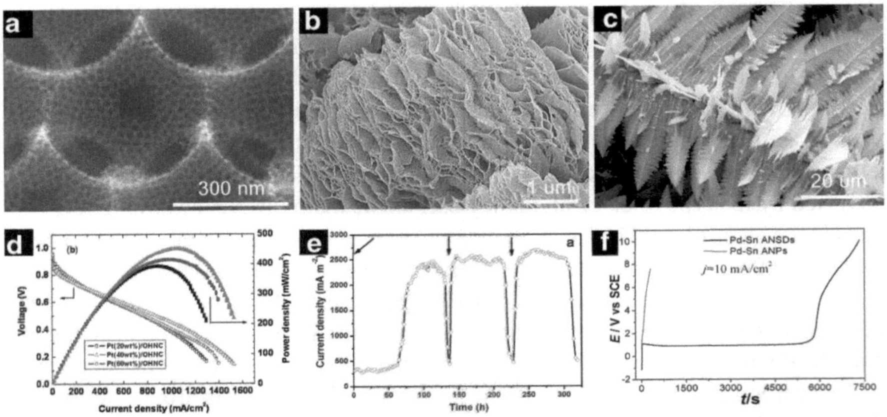

Fig. 7.12 **a** Scanning electron microscopy (SEM) image of ordered hierarchical nanostructured carbon (OHNC) fabricated through inverse replication of silica template as catalyst supporter for PEMFCs. **b** Polarization and power density plots at 60 °C for PEMFCs using OHNC-supported Pt catalyst cathodes with various Pt loadings. (Reproduced with permission from Fang et al. (2009) Copyright 2009, American Chemical Society). **c** SEM images of vacuum-stripped graphene scaffold as bioanode supporter for MFCs. **d** Constant load discharge curve of the MFC based on the biocompatible chitosan and vacuum-stripped graphene anode. (Reproduced with permission from He et al. (2012) Copyright 2012, American Chemical Society.) **e** SEM images of Pd–Sn dendrites prepared in 0.02 M $SnCl_2$ + 0.0025 M $PdCl_2$ + 0.01 M $C_6H_{11}O_7Na$ + 3.0 M HCl at 1.5 mA cm^{-2} for 60 min. **f** The chronopotentiometric curves of Pd–Sn ANSDs and Pd–Sn nanoparticles (*NPs*) in 1.0 M C_2H_5OH 1 1.0 M KOH at room temperature at 10 mA cm^{-2}. (Reproduced with permission from Ding et al. (2013) Copyright 2013, Macmillan Publishers Limited)

efficiency and greater fuel flexibility, higher temperature operation is preferred, but for portable (intermittent) power applications, lower temperature operation is typically favored, as it enables rapid start-up and minimizes stress due to thermal cycling. In addition, solid electrolyte systems can avoid the need to contain corrosive liquids; thus, solid oxide and polymer electrolyte fuel cells are preferred by many developers comparing to alkali, phosphoric acid, or molten carbonate fuel cells (Haile 2003; Brandon and Brett 2006).

Owing to the high power density, relatively quick start-up, rapid response to varying loading, and relatively low operating temperature, proton exchange membrane fuel cells (PEMFCs) are promising future energy sources for applications such as low/zero-emission electric vehicles, distributed home power generators, and power sources for small portable electronics (Springer et al. 1991; Toda et al. 1999). Hydrogen-fueled PEMFCs produce no pollutants, thus considered to be environmentally benign. The commercial viability of the PEMFCs technology is hindered by, for instance, poor kinetics of the oxygen reduction reaction and high cost of noble metal catalysts. Yu and coworkers reported an ordered hierarchical nanostructured carbon (Fig. 7.12a) fabricated through inverse replication of silica template to support high loading of Pt nanoparticles as cathode catalyst in polymer electrolyte membrane fuel cells (PEMFCs; Fang et al. 2009). The superb characteristics such as ordered hierarchical nanostructure, large surface area, and mesopore volume enable the hierarchical carbon cathode to support high loading (60 wt%) of PtNPs. The hierarchical carbon cathode exhibited enhanced catalytic activity toward oxygen reduction reaction and improved PEMFC performance. In addition, the hierarchical carbon cathode could operate well even at a low catalyst loading of 0.2 mg Pt/cm^2 (Fig. 7.12b), representing a significant breakthrough in fuel cell technology because of the expectable marked reduction in fabricating cost of the supported Pt catalyst and the corresponding system weight.

Microbial fuel cell (MFC) is capable of degrading organic waste and generating electricity simultaneously and hence has garnered strong interest as an alternative renewable power source (Lovley 2006). In MFC, the organic substrate is degraded by bacteria, and the electrons released during the metabolic process are transferred to the extracellular electrode, converting the chemical energy to electrical energy. However, the relatively low power density of MFC resulting from the sluggish electron transfer between bacteria cells and electrode limits its practical applications (Logan 2009). Graphene-based hierarchical materials prepared using the ice segregation induced self-assembly (ISISA) technique, with improved properties such as large specific surface area, good biocompatibility, and high electronic conductivity, are expected to realize a more efficient MFC anode (Vickery et al. 2009; Qian and Zhang 2011). He and coworkers reported the engineering of a novel hierarchically porous architecture as an efficient bioanode, consisting of biocompatible chitosan and vacuum-stripped graphene (CHI/VSG; He et al. 2012). With the hierarchical pores and unique VSG (Fig. 7.12c), the output current density of the microbial fuel cell (MFC) with chitosan and vacuum-stripped graphene (CHI/VSG) anode increases very slowly in the initial 70 h, a period of bacteria cells activation and colonization on the CHI/VSG anode, followed by a quick increase to a plateau of 2450 mA m^{-2} for 50 h, and then decreases sharply due to the depletion of glucose (Fig. 7.12d). With further glucose supplement, the current density quickly recovers to a value as high as 2550 mA m^{-2}, demonstrating the excellent stability of *Pseudomonas aeruginosa*-CHI/VSG anode for electricity harvesting in MFC. An optimized bioanode could deliver a remarkable maximum power density of 1530 mW m^{-2} in a mediator-less MFC, 78 times higher than a carbon-cloth anode.

Direct alcohol fuel cells (DAFCs) are generally accepted as promising future power sources in which the oxidation of methanol and ethanol is usually catalyzed by platinum (Chen et al. 2012; Zhou et al. 2010; Oezaslan et al. 2012; Wu et al. 2010a). However, its cost constitutes a major barrier to the practical application of those electrocatalysts (Gao et al. 2011a; Xia et al. 2012). Pt-free electrocatalysts might serve as alternative candidate toward the practical application of DAFCs (Lim et al. 2010). Recent studies have shown that Pd is highly electroactive for small organic molecule oxidation in basic media (Xu et al. 2007; Hong et al. 2011; Bianchini and Shen 2009). The addition of a second element, such as transition metal Sn, to Pd can obviously improve the overall electrocatalytic activities of Pd possibly due to the bimetallic promotional effect (Wei et al. 2008). Li and coworkers reported a facile and efficient electrodeposition approach for the controllable synthesis of Pd–Sn alloy nanosheet dendrites (ANSDs) with high surface area (Fig. 7.12e; Ding et al. 2013). These synthesized Pd–Sn ANSDs that exhibit high electrocatalytic activity and superior long-term cycle stability toward ethanol oxidation in alkaline media. The enhanced electrocataytic activity of Pd–Sn ANSDs may be attributed to Pd–Sn alloys, nanosheet dendrite-induced promotional effect, large number of active sites on dendrite surface, large surface area, and good electrical contact with the base electrode. Because of the simple implement and high flexibility, the proposed approach can be considered as a general and powerful strategy to synthesize the alloy electrocatalysts with high surface areas and open dendritic nanostructures. The chronoamperometric curves (Fig. 7.12f) of Pd–Sn ANSDs and Pd–Sn NPs in solution of 1.0 M C_2H_5OH + 1.0 M KOH at $^-$20 V for 800 s show that the Pd–Sn ANSDs have much higher catalytic activity than Pd–Sn ANPs. The polarization potential of Pd–Sn ANSDs is lower than that of Pd–Sn NPs at a high current density of 10 mA cm^{-2}, Pd–Sn NPs were poisoned immediately, but Pd–Sn ANSDs exhibit much longer sustaining stability time.

Fuel cell electrodes are more complex structures than battery electrodes. They serve three functions: (1) to ensure a stable interface between the reactant gas and the electrolyte, (2) to catalyze the electrode reactions, and (3) to conduct the electrons from or to the reaction sites. A significant problem is the control of the interface at the juncture of the reactant (gas) phase, the electrolyte medium, and the catalyzed conducting electrode, the so-called "three-phase boundary," where the electrolyte, electrode, and gas meet together. A stable three-phase boundary is critical to good performance and long operation. Therefore, the porosity and the wetting behavior with electrolyte and the electrode surface must be precisely adjusted. Meanwhile, it is critical to develop highly catalytic materials as well as optimize the electrode microstructure so that the performance in both the electrolytes and electrodes can be enhanced. Great progress has been made in developing preparation methods, electrode scaffolds/supporters, and down-selection approaches in recent years. Several simple and tunable methods have shown a superior ability to synthesize state-of-the-art catalysts. Some new hierarchical materials, such as nanostructured and mesostructured carbons, have been demonstrated to be feasible as catalyst supports, although their applications still face some challenges in terms of synthesis, metal loading, and electrode preparation.

7.4 Applications in Biosensing Devices

Biosensor is a kind of analytical device which is used for detection of an analyte, and it combines a biological component with a physicochemical detector. The most common example of biosensor is for the blood glucose detection, which has been commercially used. In this kind of biosensor, enzyme glucose oxidase is used to break down blood glucose. Ever since the first-generation glucose enzyme electrode was developed by Clark and Lyons in 1962, more and more attentions have been drawn into this field due to the promising applications in medical diagnosis, bioprocess monitoring, diabetes management, environmental monitoring, and so on

(Heller and Feldman 2008, 2010; Wang 2008). The most popular biosensors investigated are based on redox enzymes coupled with amperometric detection. Some kinds of conducting polymers, such as PPy, PAni, and polyphenol have been adopted for constructing enzyme sensors. The three roles of conducting polymers within enzyme-based sensors are: (1) to physically trap the enzymes onto a substrate, (2) to function as electron mediators, shuttling electrons between the electrode surface and the enzyme, and (3) to transduce a binding event. Recently, some researchers found that hierarchical structures of polymers are even more suitable for constructing such kinds of biosensors attributed to the advantages belonging to these special structures, such as high surface area, short charge diffusion length, and increased biocompatibility. What's more, the porous structure is beneficial for the infiltration of biochemical, thus leading to fast diffusion of these molecules. All these benefits make hierarchical structures more promising in the field of biosensing (Zhao et al. 2013).

Conducting-polymer-based sensors have been formulated in a variety of schemes. Conductometric sensors display changes in electrical conductivity in response to an analyte interaction. This is a particularly natural sensory scheme for electrically conducting polymers. These measurements are easily performed by connecting two adjacent electrodes (often microelectrodes) with a polymer film and then measuring the film's conductivity (or resistivity) as a function of analyte concentration. Conductometric studies can also be made while maintaining the system at a fixed electrochemical potential. Doping offers a powerful transduction mechanism since the polymer's conductivity often changes by multiple orders of magnitude with even a small amount of charge injection. The charged nature of the carriers also lends itself to interactions with the surrounding medium, thereby affecting the conductivity.

Among all kinds of biosensor, glucose sensors account for about 85 % of the total market (Wang 2008). Recently, a novel design based on 3D hierarchically porous PAni hydrogels was proposed as a kind of enzyme glucose sensor (Pan et al. 2012). The working principle of this sensor is that the enzymatic reaction of glucose oxidase and glucose can be monitored via electrochemical measurement of glucose oxidase/PAni hydrogel electrode (Fig. 7.13a). The formation of a mesh-like hydrogel network is due to the crosslinking effect that each phytic acid molecule can interact with more than one PAni chain. The gelation of the PAni hydrogel is of high efficiency, typically in minute scale (Fig. 7.13b). The surface morphology of the dehydrated PAni hydrogel is a foam-like nanostructure which is constructed with coral-like dendritic nanofibers with uniform diameters of 60–100 nm (Fig. 7.13c). The foam-like nanostructures are highly porous with hierarchical porosity on multiple length scales, from nanometer to micron-size pores (Fig. 7.13d and e). The nanoscale-interconnected conducting matrix and porosity offer greater effective surface areas than bulk materials and facilitate the transport of electrons and ions. The high surface area and interconnected conducting polymer hydrogels should be beneficial for high-performance chemical sensing. As a proof of concept, 3D hierarchically porous PAni hydrogels have been fabricated as enzyme glucose sensors. The working mechanism of the sensor is based on the redox properties of the GOx, where the enzymatic reaction of GOx and glucose is monitored via EC measure-

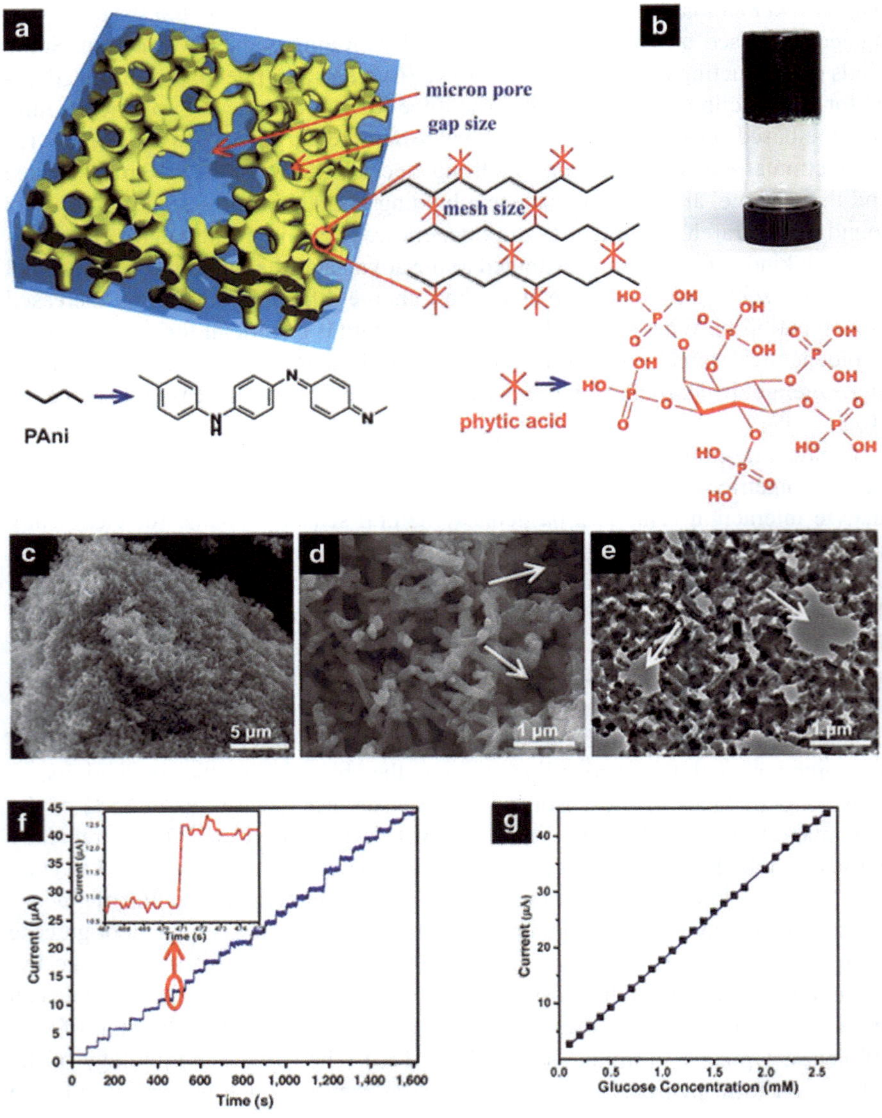

Fig. 7.13 Polyaniline (PAni) hydrogel as a 3D nanostructured electrode for glucose enzyme sensors. **a** Schematic of the 3D hierarchical microstructure of the gelated PAni hydrogel where phytic acid plays the role of dopant and crosslinker. **b** A photograph of the PAni hydrogel inside a glass vial. **c** Scanning electron microscopy (*SEM*) image of a piece of dehydrated hydrogel. **d** A higher magnification SEM image showing the interconnected network of dendritic PAni nanofibers. **e** A transmission electron microscopy (*TEM*) image showing the continuous nanostructured network of the dehydrated PAni hydrogel. **f** Amperometric response of the glucose oxidase (GOx)–PAni hydrogel-based enzymatic biosensor to successive addition of glucose in (phosphate buffer saline) PBS (0.1 M, pH 5.5). Inset shows a magnification of the seventh addition of glucose. **g** The corresponding calibration plot showing a linear amperometric response versus glucose concentration. (Reprinted with permission from Pan et al. (2012) Copyright 2012, National Academy of Sciences of the United States of America)

ment of the GOx–PAni hydrogel electrode (Dhand et al. 2011). It was observed that the current change of the PAni hydrogel was linearly correlated to the amount of glucose added into the buffer solution in the concentration range of 0.1–2.6 mM with a correlation coefficient as high as 0.9998 (Fig. 7.13f and g). Besides, the response time is only about 0.3 s and with an average sensing time of about 1.1 s when the concentration of glucose changes. This sensing time is comparable with the fastest response time of glucose sensors made with a single PAni nanojunction (~0.2 s) (Forzani et al. 2004), and is shorter than those based on PAni nanotube arrays (~3 s) (Wang et al. 2009). Moreover, the sensitivity of this PAni hierarchical hydrogel glucose sensor can reach 85.4 μA mM^{-1} cm^{-2}, and this result is even higher than those based on PPy, CNTs, or other PAni structures (Salimi et al. 2004; Uang and Chou 2003; Gao et al. 2011b). The fast response and high sensitivity are again attributed to the relatively short diffusion path that is due to the open channels of the hierarchical nanostructure and the continuously conductive framework of the PAni hydrogel.

For amperometric-type glucose sensors, GOx is the most commonly used enzyme for the specific recognition of glucose. GOx converts glucose into gluconolactone following reduction of the flavine adenine dinucleotide prosthetic group. The cofactor is then reoxidized during the second reaction, within which two electrons and two protons are transferred to molecular oxygen, yielding hydrogen peroxide (Li et al. 1998; Kim et al. 2011). As a result, the concentration of glucose can be monitored by detecting the formation of H_2O_2 (Fig. 7.14a). The developing of hierarchical conducting hydrogel matrix that combines GOx immobilization and the catalytic materials, such as PtNPs, for the electro-oxidation of hydrogen peroxide is necessary for constructing glucose sensors. The combination of the advantages of both hierarchical conductive polymer hydrogels and noble metal nanoparticles is expected to be an efficient route for increasing the sensitivity, selectivity, and electrode life of glucose sensors (Wu and Yin 2011; Xian et al. 2006; Zhou et al. 2005; Guo et al. 2009). Yu and coworkers reported an ultrahigh-sensitivity glucose sensor based on the PtNP/PAni hydrogel heterostructured electrodes, on which densely loaded monodispersed PtNPs were decorated onto the hydrogel matrix (Fig. 7.14b, c, d), acting as highly active catalysts for the electro-oxidation of hydrogen peroxide (Zhai et al. 2013). The hierarchically structured conducting polymer hydrogel played important roles in the dispersion of PtNPs and as the GOx immobilization matrix, and provided an electronically continuous 3D path for efficient charge collection. The current-time response of this heterostructure electrode showed rapid responses toward glucose concentration (Fig. 7.14e and f). The response time is only 3 s on average, and the steady-state current is also linear with the glucose concentration. The sensitivity can reach 96.1 μA mM^{-1} cm^{-2}, which is higher than those biosensors based on composites of Pt and conducting polymers. This high performance should be attributed to the advantages of both conducting polymer hydrogels and Pt nanoparticle catalyst. The porous structure of PAni hydrogel is beneficial to immobilize the enzyme and the penetration of soluble molecules, while the Pt nanoparticles are able to catalyze the decomposition of H_2O_2 generated during the enzymatic reaction. Moreover, this Pt nanoparticle–PAni hydrogel biosensor

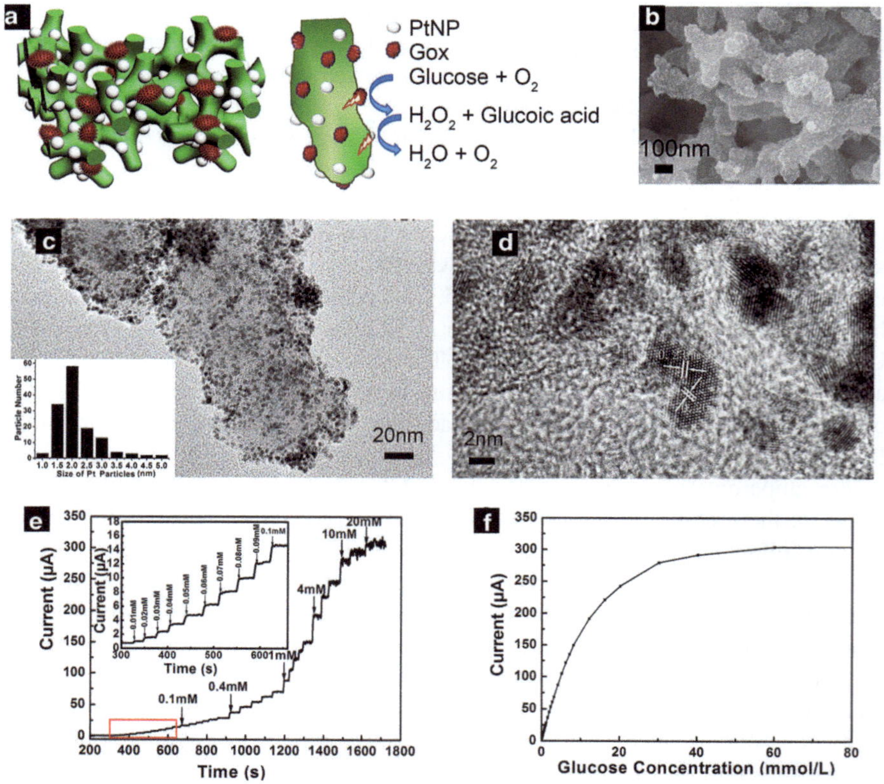

Fig. 7.14 **a** Schematic of the PtNP–PAni heterostructured hydrogel (*left*) and the reaction mechanism of the glucose sensor based on the PtNP–PAni hydrogel heterostructure (*right*). **b** Scanning electron microscopy (*SEM*) image of the PtNP-loaded PAni hydrogel. **c** Transmission electron microscopy (*TEM*) image indicating high-density PtNPs homogeneously loaded onto the PAni hydrogel matrix. Inset, statistical distribution of the PtNP sizes. **d** High-resolution transmission electron microscopy (*HRTEM*) image of the PtNPs on the PAni hydrogel. **e** Amperometric response of the PtNP–PAni hydrogel electrode after successive addition of glucose in 0.1 M PBS (pH ~5.6) at an applied potential of 0.56 V. **f** The calibration curve for glucose concentrations from 1 to 80 mM. (Reproduced with permission from Zhai et al. (2013) Copyright 2013 American Chemical Society)

showed good performance even when coupled with glassy carbon electrode without using expensive Pt electrode as substrate. It can be expected that the material cost will be further reduced by using this PtNP–PAni hydrogel and some other cheap substrates.

By efficiently reducing the diffusion length of the molecules and improving the charge transfer, the glucose sensors composed of the PtNP–PAni hydrogel heterostructure exhibited unprecedented performance with ultrahigh sensitivity, fast response, and a very low detection limit. The high-performance hydrogel-based heterostructure glucose sensors, combined with a low-cost, scalable microfabrication technique (such as ink-jet or screen printing), shows great potential for use in applications related to medical diagnosis, diabetes management, bioprocess monitoring, the beverage industry, and environmental monitoring.

7.5 Concluding Remarks

Hierarchical nanostructures have been extensively studied worldwide owing to their potential applications in many research and industrial areas. This chapter mainly summarizes the recent progress in hierarchically nanostructured materials developed for EC devices, particularly in energy storage, energy conversion, and EC biosensing. In each section, we elucidate the physics and chemistry behind the working mechanism and introduce the related hierarchical nanostructure design. The advantages from hierarchical assembly have been discussed in each specific application. It is worth pointing out that the nanostructure geometric arrangement and multifunctional coupling are two of the most important challenges from hierarchical structuring perspective, which if solved will enable considerable cost-effective, energy efficient, and multifunctional devices and solutions for practical applications in the relevant areas. Toward addressing these structuring and functioning challenges, it should be stressed that enhanced fundamental understanding and rationally deposited new principles or theories are in great need on the (dis-)similar nanostructure assembly, growth, and processing. On the other hand, the functional coupling and synergy between (dis-)similar nanostructures are worth tackling as the prerequisite understanding toward multifunctional nanostructure devices and systems suitable for various practical applications. Another concurrent need for hierarchical nanostructuring is the validation and improvement upon the green, scalable, and sustainable nanomaterials manufacturing, which is becoming a more and more important subject with further development of nanotechnology toward realistic technology platforms with novel functions and effective operations.

Currently, the primary challenges for EC technology are the limited energy density and the relatively high cost of ECs, hindering them being the energy storage technology of choice for many applications. Improving the energy density of ECs can be achieved by designing new structural electrode materials to combine EDL capacitance together with fast and highly reversible pseudocapacitance, by using superior electrolytes with broad EC voltage window, and by developing advanced hybrid EC systems to optimize the potential range for each electrode so as to widen the overall cell voltage. As for developing hybrid material system for EC electrodes, more research efforts should be placed on understanding two levels of interface, namely the interface between electrolyte and pseudocapacitor materials and the interface within the hybrid between carbon material and the pseudocapacitive material. The key challenge remains to enhancing the interactions between carbon and the pseudocapacitive material to improve the Faradic processes across the interface so as to achieve efficient pseudocapacitance in addition to the EDL capacitance, as well as to sustain the strain caused by phase/volume change for long cycle life.

The field of LIBs is of fundamental and technological importance to the large scientific and technology communities in materials science, electrochemistry, chemical engineering, computational chemistry, solid-state chemistry, physics, etc., because these battery systems are complicated and their development requires multidisciplinary skills and efforts. The lack of thermodynamic stability and the critical importance of passivation phenomena in the operation of these systems require rigorous studies of the correlation among composition, morphology structure, surface chemistry, intrinsic EC behavior, and thermal stability. Hierarchically structured materials have exhibited prospects of great potential for further applications. We can improve the EC performance of those materials by regulating the structure with more rational design and precise control, synthesizing more uniform products, as well as promoting its yield and realizing its uses in industrialization.

Research on developing hierarchically structured materials becomes more and more important owing to their practical potential in different areas from nanoscience to catalysis, separation, electronics, optics, optoelectronics, energy, environmental, and life science. Despite the great challenges in the fabrication of such materials, recent important discovery of the self-formation phenomenon of hierarchy (Su et al. 2011) provides the feasibility in their practical use and ease of synthesis of new advanced materials with tunable chemical compositions, controllable structures, and assembly behavior across broad length scales for desired multifunctionalities and high performance.

References

Arico AS, Bruce PG, Scrosati B, Tarascon J-M, Schalkwijk WV (2005) Nanostructured materials for advanced energy conversion and storage devices. Nat Mater 4:366–377

Bagotsky VS (2012) Fuel cells: problems and solutions, second edition, John Wiley & Sons, Inc., Hoboken, New Jersey.

Bai H, Li C, Shi GQ (2011) Functional composite materials based on chemically converted graphene. Adv Mater 23:1089–1115

Bandosz TJ, Biggs MJ, Gubbins KE, Hattori Y, Liyama T, Kaneko K, Pikunic J, Thomson KT (2003) Chemistry & physics of carbon: Volume 28, Marcel Dekker, New York.

Beaulieu LY, Eberman KW, Turner RL, Krause LJ, Dahn JR (2001) Colossal reversible volume changes in lithium alloys. Electrochem Solid-State Lett 4:A317–A140

Bélanger D, Brousse T, Long JW (2008) Manganese oxides: battery materials make the leap to electrochemical capacitors. Electrochem Soc Interface 17:49–52

Besenhard JO, Yang J, Winter M (1997) Will advanced lithium-alloy anodes have a chance in lithium-ion batteries? J Power Sour 68:87–90

Besenhard JO, Yang J, Winter M (2000) Dispersion of Sn and SnO on carbon anodes. J Power Sour 90:70–75

Bianchini C, Shen PK (2009) Palladium-based electrocatalysts for alcohol oxidation in half cells and in direct alcohol fuel cells. Chem Rev 109:4183–4206

Biener J, Staderman M, Suss M, Worsley MA, Biener MM, Rose KA, Baumann TF (2011) Advanced carbon aerogels for energy applications. Energy Environ Sci 4:656–667

Bierman MJ, Jin S (2009) Potential applications of hierarchical branching nanowires in solar energy conversion. Energy Environ Sci 2:1050–1059

Bolton JR, Strickler SJ, Connolly JS (1985) Limiting and realizable efficiencies of solar photolysis of water. Nature 316:495–500
Brandon NP, Brett DJ (2006) Engineering porous materials for fuel cell applications. Phil Trans R Soc A 15:147–159
Brenner DW, Shenderova OA, Harrison JA, Stuart SJ, Ni B, Sinnott SB (2002) A second-generation reactive empirical bond order (REBO) potential energy expression for hydrocarbons. J Phys Condens Matter 14:783
Bridel JS, Azais T, Morcrette M, Tarascon J-M, Larcher D (2009) Key parameters governing the reversibility of si/carbon/CMC electrodes for li-ion batteries. Chem Mater 22:1229–1241
Bruce PG, Freunberger SA, Hardwick LJ, Tarascon J-M (2012) Li-O_2 and Li-S batteries with high energy storage. Nat Mater 11:19–29
Chan CK, Peng H, Liu G, McIlwrath K, Zhang XF, Huggins RA, Cui Y (2008a) High-performance lithium battery anodes using silicon nanowires. Nat Nanotechnol 3:31–35
Chan CK, Zhang XF, Cui Y (2008b) High capacity Li ion battery anodes using Ge nanowires. Nano Lett 8:307–309
Chen XF, Wang XC, Fu XZ (2009) Hierarchical macro/mesoporous TiO_2/SiO_2 and TiO_2/ZrO_2 nanocomposites for environmental photocatalysis. Energy Environ Sci 2:872–877
Chen P-C, Shen G, Shi Y, Chen H, Zhou C (2010a) Preparation and characterization of flexible asymmetric supercapacitors based on transition-metal-oxide nanowire/single-walled carbon nanotube hybrid thin-film electrodes. ACS Nano 4:4403–4411
Chen S, Zhu J, Wu X, Han Q, Wang X (2010b) Graphene oxide–MnO_2 nanocomposites for supercapacitors. ACS Nano 4:2822–2830
Chen L-H, Li X-J, Tian G, Li Y, Rooke JC, Zhu G-S, Qiu S-L, Yang X-Y, Su B-L (2011) Highly stable and reusable multimodal zeolite TS-1 based catalysts with hierarchically interconnected three-level micro-meso-macroporous structure. Angew Chem Int Ed Engl 50:11156–11161
Chen M, Wu B, Yang J, Zheng N (2012) Small-adsorbate assisted shape control of Pd and Pt nanocrystals. Adv Mater 24:862–79
Cherepy NJ, Krueger R, Fiet KJ, Jankowski AF, Cooper JF (2005) Direct conversion of carbon fuels in a molten carbonate fuel cell. J Electrochem Soc 152:A80–A87
Chmiola J, Largeot C, Taberna P-L, Simon P, Gogotsi Y (2010) Monolithic carbide-derived carbon films for micro-supercapacitors. Science 328:480–483
Cho IS, Chen Z, Forman AJ, Kim DR, Rao PM, Jaramillo TF, Zheng X (2011) Branched TiO_2 nanorods for photoelectrochemical hydrogen production. Nano Lett 11:4978–4984
Coleman JN, Khan U, Gun'ko YK (2006) Mechanical reinforcement of polymers using carbon nanotubes. Adv Mater 18:689–706
Conway BE (1999) Electrochemical supercapacitors: scientific fundamentals and technological applications. Plenum Press, New York.
Cottineau T, Toupin M, Delahaye T, Brousse T, Bélanger D (2006) Nanostructured transition metal oxides for aqueous hybrid electrochemical supercapacitors. Appl Phys A 82:599–606
Dahn JR, McKinnon WR (1987) Structure and electrochemistry of Li_xMoO_2. Solid State Ion 23:1–7
Davies A, Audette P, Farrow B, Hassan F, Chen Z, Choi J-Y, Yu A (2011) Graphene-based flexible supercapacitors: pulse-electropolymerization of polypyrrole on free-standing graphene films. J Phys Chem C 115:17612–17620
Dey AN (1971) Electrochemical alloying of lithium in organic electrolytes. J Electrochem Soc 118:1547–1549
Dhand C, Das M, Datta M, Malhotra BD (2011) Recent advances in polyaniline based biosensors. Biosens Bioelectron 26:2811–2821
Ding Y-S, Shen X-F, Gomez S, Luo H, Aindow M, Suib SL (2006) Hydrothermal growth of manganese dioxide into three-dimensional hierarchical nanoarchitectures. Adv Funct Mater 16:549–555
Ding L-X, Wang A-L, Ou Y-N, Li Q, Guo R, Zhao W-X, Tong Y-X and Li G-R (2013) Hierarchical Pd-Sn alloy nanosheet dendrites: an economical and highly active catalyst for ethanol electrooxidation. Sci Rep 3:1181

Dunn B, Kamath H, Tarascon J-M (2011) Electrical energy storage for the grid: a battery of choices. Science 334:928–935
Fan Z, Yan J, Zhi L, Zhang Q, Wei T, Feng J, Zhang M, Qian W, Wei F (2010) A three-dimensional carbon nanotube/graphene sandwich and its application as electrode in supercapacitors. Adv Mater 22:3723–3728
Fang B, Kim JH, Kim M, Yu J-S (2009) Ordered hierarchical nanostructured carbon as a highly efficient cathode catalyst support in proton exchange membrane fuel cell. Chem Mater 21:789–796
Fang X, Guo B, Shi Y, Li B, Hua C, Yao C, Zhang Y, Hu YS, Wang Z, Stucky GD, Chen L (2012) Enhanced Li storage performance of ordered mesoporous MoO_2 via tungsten doping. Nanoscale 4:1541–1544
Flandrois S, Simon B (1999) Carbon materials for lithium-ion rechargeable batteries. Carbon 37:165–180
Forzani ES, Zhang H, Nagahara LA, Amlani I, Tsui R, Tao N (2004) A conducting polymer nanojunction sensor for glucose detection. Nano Lett 4:1785–1788
Fukishima A, Honda K (1972) Electrochemical photolysis of water at a semiconductor electrode. Nature 238:37–38
Galli G, Martin RM, Car R, Parrinello M (1989) Structural and electronic properties of amorphous carbon. Phys Rev Lett 62:555–558
Galli G, Martin RM, Car R, Parrinello M (1990) *Ab initio* calculation of properties of carbon in the amorphous and liquid states. Phys Rev B Condens Matter 42:7470–7482
Gao P, Wang ZL (2002) Self-assembled nanowire–nanoribbon junction arrays of ZnO. J Phys Chem B 106:12653–12658
Gao M-R, Gao Q, Jiang JW, Cui C-H, Yao W-T, Yu S-H (2011a) A methanol tolerant cathode catalyst Pt/$CoSe_2$ nanobelts for direct methanol fuel cell. Angew Chem Int Ed Engl 50:4905–4908
Gao QA, Guo YY, Zhang WY, Qi HL, Zhang CX (2011b) An amperometric glucose biosensor based on layer-by-layer GOx-SWCNT conjugate/redox polymer multilayer on a screen-printed carbon electrode. Sens Actuators B 153:219–225
Gao PX, Shimpi P, Gao H, Liu C, Cai W, Liao KT, Wrobel G, Zhang Z, Ren Z, Lin HJ (2012) Hierarchical assembly of multifunctional oxide-based composite nanostructures for energy and environmental applications. Int J Mol Sci 13:7393–7423
Giannelis EP (1996) Polymer layered silicate nanocomposites. Adv Mater 8:29–35
Gülzow E (1996) Alkaline fuel cells: a critical view. J Power Sour 61:99–104
Guo S, Dong S, Wang E (2009) Polyaniline/Pt hybrid nanofibers: high-efficiency nanoelectrocatalysts for electrochemical devices. Small 5:1869–1876
Haile SM (2003) Fuel cell materials and components. Acta Mater. 51:5981–6000
Hartmann M, Vinu A (2002) Mechanical stability and porosity analysis of large-pore SBA-15 mesoporous molecular sieves by mercury porosimetry and organics adsorption. Langmuir 18:8010–8016
He Z, Liu J, Qiao Y, Li CM, Tan TTY (2012) Architecture engineering of hierarchically porous chitosan/vacuum-stripped graphene scaffold as bioanode for high performance microbial fuel cell. Nano Lett 12:4738–4741
Heller A, Feldman B (2008) Electrochemical glucose sensors and their applications in diabetes management. Chem Rev 108:2482–2505
Heller A, Feldman B (2010) Electrochemistry in diabetes management. Acc Chem Res 43:963–973
Heon M, lofland S, Applegate J, Nolte R, Cortes E, Hettinger JD, Taberna P-L, Simin P, Huang P, Brunet M, Gototsi Y (2011) Continuous carbide-derived carbon films with high volumetric capacitance. Energy Environ Sci 4:135–138
Hochbaum AI, Gargas D, Hwang YJ, Yang P (2009) Single crystalline mesoporous silicon nanowires. Nano Lett 9:3550–3554
Hoffert MI, Caldeira K, Jain AK, Haites EF, Harvey LDD, Potter SD, Schlesinger ME, Schneider SH, Watts RG, Wigley TML, Wuebbles DJ (1998) Energy implications of future stabilization of atmospheric CO_2 content. Nature 395:881–884

Hong JW, Kim D, Lee YW, Kang SW, Han SW (2011) Atomic-distribution dependent electrocatalytic activity of Au-Pd bimetallic nanocrystals. Angew Chem Int Ed 50:8876–8880
Hou Y, Cheng Y, Hobson T, Liu J (2010) Design and synthesis of hierarchical MnO_2 nanospheres/carbon nanotubes/conducting polymer ternary composite for high performance electrochemical electrodes. Nano Lett 10:2727–2733
Hu C-C, Chang K-H, Lin M-C, Wu Y-T (2006) Design and tailoring of the nanotubular arrayed architecture of hydrous RuO_2 for next generation supercapacitors. Nano Lett 6:2690–2695
Hu L, Wu HB, Hong SS, Cui L, McDonough JR, Bohy S, Cui Y (2011) Si nanoparticle-decorated Si nanowire networks for Li-ion battery anodes. Chem Commun (Camb) 47:367–369
Hu L, Chen M, Fang X, Wu L (2012) Oil-water interfacial self-assembly: a novel strategy for nanofilm and nanodevice fabrication. Chem Soc Rev 41:1350–1362
Hwang YJ, Wu CH, Hahn C, Jeong HE, Yang P (2012) Si/InGaN core/shell hierarchical nanowire arrays and their photoelectrochemical properties. Nano Lett 12:1678–1682
Hyder MN, Lee SW, Cebeci FC, Schmidt DJ, Yang S-H, Hammond PT (2011) Layer-by-layer assembled polyaniline nanofiber/multiwall carbon nanotube thin film electrodes for high-power and high-energy storage applications. ACS Nano 5:8552–8561
Jiang Y-M, Wang K-X, Zhang H-J, Wang J-F, Chen J-S (2013) Hierarchical $Li_4Ti^5O_{12}/TiO_2$ composite tubes with regular structural imperfection for lithium ion storage. Sci Rep 3:3490
Kay A, Cesar I, Graȉtzel M (2006) New benchmark for water photooxidation by nanostructured α-Fe_2O_3 films. J Am Chem Soc 128:15714–15721
Kennedy T, Mullane E, Geaney H, Osiak M, O'Dwyer C, Ryan KM (2014) High-performance germanium nanowire-based lithium-ion battery anodes extending over 1000 cycles through in situ formation of a continuous porous network. Nano Lett 14:716–723
Kim HJ, Han B, Choo J, Cho J (2008) Three-dimensional porous silicon particles for use in high-performance lithium secondary batteries. Angew Chem Int Ed Engl 47:10151–10154
Kim H, Lee I, Kwon Y, Kim BC, Ha S, Lee JH, Kim J (2011) Immobilization of glucose oxidase into polyaniline nanofiber matrix for biofuel cell applications. Biosens Bioelectron 26:3908–3913
Kleitz F, Choi SH, Ryoo R (2003) Cubic *Ia*3*d* large mesoporous silica: synthesis and replication to platinum nanowires, carbon nanorods and carbon nanotubes. Chem Commun (Camb) 2136–2137
Kovalenko I, Zdyrko B, Magasinski A, Hertzberg B, Milicev Z, Burtovyy R, Luzinov I, Yushin G (2011) A major constituent of brown algae for use in high-capacity li-ion batteries. Science 333:75–79
Lao JY, Wen JG, Ren ZF (2002) Hierarchical ZnO nanostructures. Nano Lett 2:1287–1291
Lee SH, Lee DH, Lee WJ, Kim SO (2011) Tailored assembly of carbon nanotubes and graphene. Adv Funct Mater 21:1338–1354
Lee SW, McDowell MT, Berla LA, Nix WD, Cui Y (2012) Fracture of crystalline silicon nanopillars during electrochemical lithium insertion. Proc Natl Acad Sci U S A 109:4080–4085
Lee S-H, Sridhar V, Jung J-H, Karthikeyan K, Lee Y-S, Mukherjee R, Koratkar N, Oh I-K (2013) Graphene-nanotube-iron hierarchical nanostructure as lithium ion battery anode. ACS Nano 7:4242–4251
Leroux F, Goward GR, Power WP, Nazar LF (1998) Understanding the nature of low-potential Li uptake into high volumetric capacity molybdenum oxides. Electrochem Solid-State Lett 1:255–258
Li X, Bohna PW (2000) Metal-assisted chemical etching in HF/H_2O_2 produces porous silicon. Appl Phys Lett 77:2572–2574
Li ZF, Kang ET, Neoh KG, Tan KL (1998) Covalent immobilization of glucose oxidase on the surface of polyaniline films graft copolymerized with acrylic acid. Biomaterials 19:45–53
Li G-R, Feng Z-P, Ou Y-N, Wu D, Fu R, Tong Y-X (2010) Mesoporous MnO_2/carbon aerogel composites as promising electrode materials for high-performance supercapacitors. Langmuir 26:2209–2213

Li S, Luo Y, Lv W, Yu W, Wu S, Hou P, Yang Q, Meng Q, Liu C, Cheng H-M (2011) Vertically aligned carbon nanotubes grown on graphene paper as electrodes in lithium-ion batteries and dye-sensitized solar cells. Adv Energy Mater 1:486–490

Li Y, Fu Z-Y, Su B-L (2012) Hierarchically structured porous materials for energy conversion and storage. Adv Funct Mater 22:4634–4667

Lim B, Kobayashi H, Yu T, Wang J, Kim MJ, Li Z-Y, Rycenga M, Xia Y (2010) Synthesis of Pd-Au bimetallic nanocrystals via controlled overgrowth. J Am Chem Soc 132:2506–2507

Lin YH, Wei TY, Chien HC, Lu SY (2011) Manganese oxide/carbon aerogel composite: an outstanding supercapacitor electrode material. Adv Energy Mater 1:901–907

Liu C, Yu Z, Neff D, Zhamu A, Jang BZ (2010a) Graphene-based supercapacitor with an ultrahigh energy density. Nano Lett 10:4863–4868

Liu J, Essner J, Li J (2010b) Hybrid supercapacitor based on coaxially coated manganese oxide on vertically aligned carbon nanofiber arrays. Chem Mater 22:5022–5030

Liu G, Xun S, Vukmirovic N, Song X, Olalde-Velasco P, Zheng H, Battaglia VS, Wang L, Yang W (2011a) Polymers with tailored electronic structure for high capacity lithium battery electrodes. Adv Mater 23:4679–4683

Liu LQ, Ma WJ, Zhang Z (2011b) Macroscopic carbon nanotube assemblies: preparation, properties, and potential applications. Small 7:1504–1520

Liu B, Zhang JT, Wang X, Chen G, Chen D, Zhou C, Shen G (2012a) Hierarchical three-dimensional $ZnCo_2O_4$ nanowire arrays/carbon cloth anodes for a novel class of high-performance flexible lithium-ion batteries. Nano Lett 12:3005–3011

Liu S, Jia H, Han L, Wang J, Gao P, Xu D, Yang J, Che S (2012b) Nanosheet-constructed porous TiO_2-B for advanced lithium ion batteries. Adv Mater 24:3201–3204

Liu B, Soares P, Checkles C, Zhao Y, Yu G (2013a) Three-dimensional hierarchical ternary nanostructures for high-performance Li-ion battery anodes. Nano Lett 13:3414–3419

Liu C, Tang J, Chen HM, Liu B, Yang P (2013b) A fully integrated nanosystem of semiconductor nanowires for direct solar water splitting. Nano Lett 13:2989–2992

Liu N, Li W, Pasta M, Cui Y (2014) Nanomaterials for electrochemical energy storage. Front Phys 9:323–350

Logan BE (2009) Exoelectrogenic bacteria that power microbial fuel cells. Nat Rev Microbiol 7:735–781

Long JW, Bélanger D, Brousse T, Sugimoto W, Sassin MB, Crosnier O (2011) Asymmetric electrochemical capacitors—stretching the limits of aqueous electrolytes. MRS Bull 36:513–522

Los JH, Fasolino A (2003) Intrinsic long-range bond-order potential for carbon: performance in monte carlo simulations of graphitization. Phys Rev B 68:024107

Lovley DR (2006) Bug juice: harvesting electricity with microorganisms. Nat Rev Microbiol 4:497–508

Lu A-H, Schmidt W, Spliethoff B, Schuth F (2003) Synthesis of ordered mesoporous carbon with bimodal pore system and high pore volume. Adv Mater 15:1602–1606

Maeda K, Domen K (2011) Oxynitride materials for solar water splitting. MRS Bull 36:25–31

Magasinski A, Dixon P, Hertzberg B, Kvit A, Ayala J, Yushin G (2010) High-performance lithium-ion anodes using a hierarchical bottom-up approach. Nat Mater 9:353–358

Marks NA (2000) Generalizing the environment-dependent interaction potential for carbon. Phys Rev B 63:035401

Marks NA, McKenzie DR, Pailthorpe BA, Bernasconi M, Parrinello M (1996a) *Ab initio* simulations of tetrahedral amorphous carbon. Phys Rev B Condens Matter 54:9703–9714

Marks NA, McKenzie DR, Pailthorpe BA, Bernasconi M, Parrinello M (1996b) Microscopic structure of tetrahedral amorphous carbon. Phys Rev Lett 76:768–771

Merchant AR, McKenzie DR, McCulloch DG (2001) *Ab* initio simulations of amorphous carbon nitrides. Phys Rev B 65:024208

Nam KT, Kim D-W, Yoo PJ, Chiang C-Y, Meethong N, Hammond PT, Chiang Y-M, Belcher AM (2006) Virus-enabled synthesis and assembly of nanowires for lithium ion battery electrodes. Science 312:885–888

Naoi K, Morita M (2008) Advanced polymers as active materials and electrolytes for electrochemical capacitors and hybrid capacitor systems. Electrochem Soc Interface 17:44–48

Oezaslan M, Heggen M, Strasser P (2012) Size-dependent morphology of dealloyed bimetallic catalysts: linking the nano to the macro scale. J Am Chem Soc 134:514–524

Ohzuku T, Ueda A, Yamamoto N (1995) Zero-strain insertion material of $Li[Li_{1/3}Ti_{5/3}]O_4$ for rechargeable lithium cells. J Electrochem Soc 142:1431–1435

Ormerod RM (2003) Solid oxide fuel cells. Chem Soc Rev 32:17–28

Pan L, Qiu H, Dou C, Li Y, Pu L, Xu J, Shi Y (2010) Conducting polymer nanostructures: template synthesis and applications in energy storage. Int J Mol Sci 11:2636–2657

Pan L, Yu G, Zhai D, Lee HR, Zhao W-X, Liu N, Wang H, Tee BC-K, Shi Y, Cui Y, Bao Z (2012) Hierarchical nanostructured conducting polymer hydrogel with high electrochemical activity. Proc Natl Acad Sci U S A 109:9287–9292

Park C-M, Kim J-H, Kim H, Sohn H-J (2010) Li-alloy based anode materials for Li secondary batteries. Chem Soc Rev 39:3115–3141

Pech D, Brunet M, Durou H, Huang P, Mochalin V, Gototsi Y, Taberna P-L, Simon P (2010) Ultrahigh-power micrometre-sized supercapacitors based on onion-like carbon. Nat Nanotechnol 5:651–654

Peng K-Q, Wang X, Wu X-L, Lee S-T (2009) Platinum nanoparticle decorated silicon nanowires for efficient solar energy conversion. Nano Lett 9:3704–3709

Peng Y, Chen Z, Wen JG, Xiao Q, Weng D, He S, Geng H, Lu Y (2011) Hierarchical manganese oxide/carbon nanocomposites for supercapacitor electrodes. Nano Res 4:216–225

Poizot P, Laruelle S, Grugeon S, Dupont L, Tarascon J-M (2000) Nano-sized transition-metal oxides as negative-electrode materials for lithium-ion batteries. Nature 407:496–499

Portet C, Yushin G, Gogotsi Y (2007) Electrochemical performance of carbon onions, nanodiamonds, carbon black and multiwalled nanotubes in electrical double layer capacitors. Carbon 45:2511–2518

Qian L, Zhang HF (2011) Controlled freezing and freeze drying: a versatile route for porous and micro-/nano-structured materials. J Chem Technol Biotechnol 86:172–184

Reddy ALM, Ramaprabhu S (2007) Nanocrystalline metal oxides dispersed multiwalled carbon nanotubes as supercapacitor electrodes. J Phys Chem C 111:7727–7734

Ryu JH, Kim JW, Sung YE, Oh SM (2004) Failure modes of silicon powder negative electrode in lithium secondary batteries. Electrochem Solid-State Lett 7:A306–A309

Salimi A, Compton RG, Hallaj R (2004) Glucose biosensor prepared by glucose oxidase encapsulated sol-gel and carbon-nanotube-modified basal plane pyrolytic graphite electrode. Anal Biochem 333:49–56

Shi J, Hara Y, Sun C, Anderson MA, Wang X (2011) Three-dimensional high-density hierarchical nanowire architecture for high-performance photoelectrochemical electrodes. Nano Lett 11:3413–3419

Simon P, Gogotsi Y (2008) Materials for electrochemical capacitors. Nat Mater 7:845–854

Sivakkumar SR, Kim WJ, Choi J-A, MacFarlane DR, Forsyth M, Kim D-W (2007) Electrochemical performance of polyaniline nanofibres and polyaniline/multi-walled carbon nanotube composite as an electrode material for aqueous redox supercapacitors. J Power Sour 171:1062–1068

Sleigh AK, McKinnon WR (1991) Structure and electrochemistry of Li_xWO_2. Solid State Ionics 45:67–75

Springer TE, Zawodzinski TA, Gottesfeld S (1991) Polymer electrolyte fuel cell model. J Electrochem Soc 138:2334–2342

Stoller MD, Park S, Zhu Y, An J, Ruoff RS (2008) Graphene-based ultracapacitors. Nano Lett 8:3498–3502

Su BL, Sanchez C, Yang XY (2011) Hierarchically structured porous materials: from nanoscience to catalysis, biomedicine, optics and energy. Wiley-VCH Verlag & Co. KGaA, Weinheim, Germany.

Subramanian V, Hall SC, Smith PH, Rambabu B (2004) Mesoporous anhydrous RuO_2 as a supercapacitor electrode materia. Solid State Ionics 175:511–515

Tersoff J (1988) Empirical interatomic potential for carbon, with applications to amorphous carbon. Phys Rev Lett 61:2879–2882

Toda T, Igarashi H, Uchida H, Watanabe M (1999) Enhancement of the electroreduction of oxygen on Pt alloys with Fe, Ni, and Co. J Electrochem Soc 146:3750–3756

Uang YM, Chou TC (2003) Fabrication of glucose oxidase/polypyrrole biosensor by galvanostatic method in various pH aqueous solutions. Biosens Bioelectron 19:141–147

van Duin ACT, Dasgupta S, Lorant F, Gooddard III WA (2001) ReaxFF: a reactive force field for hydrocarbons. J Phys Chem A 105:9936–409

Veedu VP, Cao A, Li X-J, Ma K, Soldano C, Kar S, Ajayan PM, Ghasemi-Nejhad MN (2006) Multifunctional composites using reinforced laminae with carbon-nanotube forests. Nat Mater 5:457–462

Verbrugge MW, Cheng Y-T (2008) Stress distribution within spherical particles undergoing electrochemical insertion and extraction. Electrochem Soc Trans 16:127–139

Verbrugge MW, Cheng Y-T (2009) Stress and strain-energy distributions within diffusion-controlled insertion-electrode particles subjected to periodic potential excitations. J Electrochem Soc 156:A927–A937

Vickery JL, Patil AJ, Mann S (2009) Fabrication of graphene–polymer nanocomposites with higher-order three-dimensional architectures. Adv Mater 21:2180–2184

Vinu A, Hartmann M (2005) Characterization and microporosity analysis of mesoporous carbon molecular sieves by nitrogen and organics adsorption. Catal Today 102–103:189–196

Walter MG, Warren EL, McKone JR, Boettcher SW, Mi Q, Santori EA, Lewis NS (2010) Solar water splitting cells. Chem Rev 110:6446–6473

Wang J (2008) Electrochemical glucose biosensors. Chem Rev 108:814–825

Wang YG, Li HQ, Xia YY (2006) Ordered whiskerlike polyaniline grown on the surface of mesoporous carbon and its electrochemical capacitance performance. Adv Mater 18:2619–2623

Wang D-W, Li F, Cheng H-M (2008) Hierarchical porous nickel oxide and carbon as electrode materials for asymmetric supercapacitor. J Power Sour 185:1563–1568

Wang Z, Liu S, Wu P, Cai C (2009) Detection of glucose based on direct electron transfer reaction of glucose oxidase immobilized on highly ordered polyaniline nanotubes. Anal Chem 81:1638–1645

Wang D, Pierre A, Kibria MG, Cui K, Han X, Bevan KH, Guo H, Paradis S, Hakima A-R, Mi Z (2011a) Wafer-level photocatalytic water splitting on GaN nanowire arrays grown by molecular beam epitaxy. Nano Lett 11:2353–2357

Wang W, Epur R, Kumta PN (2011b) Vertically aligned silicon/carbon nanotube (VASCNT) arrays: hierarchical anodes for lithium-ion battery. Electrochem Commun 13:429–432

Wang Y, Chen K, Mishler J, Cho SC, Adroher XC (2011c) A review of polymer electrolyte membrane fuel cells: technology, applications, and needs on fundamental research. Appl Energy 88:981–1007

Wang X, Peng K-Q, Hu Y, Zhang F-Q, Hu B, Li L, Wang M, Meng X-M, Lee S-T (2014) Silicon/hematite core/shell nanowire array decorated with gold nanoparticles for unbiased solar water oxidation. Nano Lett 14:18–23

Wei T, Kumar D, Chen MS, Luo K, Axnanda S, Lundwall M, Goodman DW (2008) Vinyl acetate synthesis over model Pd_2Sn bimetallic catalysts. J Phys Chem C 112:8332–8337

Wen Z, Li J (2009) Hierarchically structured carbon nanocomposites as electrode materials for electrochemical energy storage, conversion and biosensor systems. J Mater Chem 19:8707–8713

Whang D, Jin S, Wu Y, Lieber CM (2003) Large-scale hierarchical organization of nanowire arrays for integrated nanosystems. Nano Lett 3:1255–1259

Winter M, Brodd RJ (2004) What are batteries, fuel cells, and supercapacitors? Chem Rev 104:4245–4269

Wu J, Yin L (2011) Platinum nanoparticle modified polyaniline-functionalized boron nitride nanotubes for amperometric glucose enzyme biosensor. ACS Appl Mater 3:4354–4362

Wu J, Zhang JT, Peng Z, Yang S, Wagner FT, Yang H (2010a) Truncated octahedral Pt3Ni oxygen reduction reaction electrocatalysts. J Am Chem Soc 132:4984–4985

Wu MS, Guo ZS, Jow JJ (2010b) Highly regulated electrodeposition of needle-like manganese oxide nanofibers on carbon fiber fabric for electrochemical capacitors. J Phys Chem C 114:21861–21867

Wu Z-S, Ren W, Wang D-W, Li F, Liu B, Cheng H-M (2010c) High-energy MnO_2 nanowire/graphene and graphene asymmetric electrochemical capacitors. ACS Nano 4:5835–5842

Wu Z, Li W, Xia Y, Webley P, Zhao D (2012) Ordered mesoporous graphitized pyrolytic carbon materials: synthesis, graphitization, and electrochemical properties. J Mater Chem 22:8835–8845

Wu H, Yu G, Pan L, Liu N, McDowell MT, Bao Z, Cui Y (2013a) Stable Li-ion battery anodes by in-situ polymerization of conducting hydrogel to conformally coat silicon nanoparticles. Nat Commun 4:1943

Wu X-L, Wen T, Guo H-L, Yang S, Wang XC, Xu A-W (2013b) Biomass-derived sponge-like carbonaceous hydrogels and aerogels for supercapacitors. ACS Nano 7:3589–3597

Xia BY, Ng W, Wu HB, Wang X, Lou XW (2012) Self-supported interconnected Pt nanoarchitectures as highly stable electrocatalysts for low temperature fuel cells. Angew Chem Int Ed Engl 51:7213–7216

Xian Y, Hu Y, Liu F, Xian Y, Wang H, Jin L (2006) Glucose biosensor based on Au nanoparticles-conductive polyaniline nanocomposite. Biosens Bioelectron 21:1996–2000

Xu C, Wang H, Shen PK, Jiang SP (2007) Highly ordered Pd nanowire arrays as effective electrocatalysts for ethanol oxidation in direct alcohol fuel cells. Adv Mater 19:4256–4259

Yang XY, Li Y, van Tendeloo G, Xiao FS Su BL (2009) One-pot synthesis of catalytically stable and active nanoreactors: encapsulation of size-controlled nanoparticles within a hierarchically macroporous core@ordered mesoporous shell system. Adv Mater 21:1368–1372

Yao Y, McDowell MT, Ryu I, Wu H, Liu N, Hu L, Nix WD, Cui Y (2011) Interconnected silicon hollow nanospheres for lithium-ion battery anodes with long cycle life. Nano Lett 11:2949–2954

Ye J-S, Cui HF, Liu X, Lim TM, Zhang W-D, Sheu F-S (2005) Preparation and characterization of aligned carbon nanotube-ruthenium oxide nanocomposites for supercapacitors. Small 1:560–565

Yu G, Hu L, Liu N, Wang H, Vosgueritchian M, Yang Y, Cui Y, Bao Z (2011a) Enhancing the supercapacitor performance of graphene/MnO_2 nanostructured electrodes by conductive wrapping. Nano Lett 11:4438–4442

Yu G, Hu L, Vosgueritchian M, Wang H, Xie X, McDonough JR, Cui X, Cui Y, Bao Z (2011b) Solution-processed graphene/MnO2 nanostructured textiles for high-performance electrochemical capacitors. Nano Lett 11:2905–2911

Yu G, Xie X, Pan L, Bao Z, Cui Y (2013) Hybrid nanostructured materials for high-performance electrochemical capacitors. Nano Energy 2:213–234

Zeng QH, Yu AB, Lu GQ, Paul DR (2005) Clay-based polymer nanocomposites: research and commercial development. J Nanosci Nanotechnol 5:1574–1592

Zhai D, Liu B, Shi Y, Pan L, Wang Y, Li W, Zhang R, Yu G (2013) Highly sensitive glucose sensor based on Pt nanoparticle/polyaniline hydrogel heterostructures. ACS Nano 7:3540–3546

Zhang WJ (2011) A review of the electrochemical performance of alloy anodes for lithium-ion batteries. J Power Sour 196:13–24

Zhang H, Cao G, Wang ZL, Yang Y, Shi Z, Gu Z (2008) Growth of manganese oxide nanoflowers on vertically-aligned carbon nanotube arrays for high-rate electrochemical capacitive energy storage. Nano Lett 8:2664–2668

Zhang LL, Xiong ZG, Zhao XS (2010) Pillaring chemically exfoliated graphene oxide with carbon nanotubes for photocatalytic degradation of dyes under visible light irradiation. ACS Nano 4:7030–7036

Zhang JT, Jiang JW, Zhao XS (2011) Synthesis and capacitive properties of manganese oxide nanosheets dispersed on functionalized graphene sheets. J Phys Chem C 115:6488–54

Zhang L, Zhang G, Wu HB, Yu L, Lou XW (2013) Hierarchical tubular structures constructed by carbon-coated SnO_2 nanoplates for highly reversible lithium storage. Adv Mater 25;2589–2593

Zhao Y, Bai H, Hu Y, Li Y, Qu L, Zhang S, Shi G (2011) Electrochemical deposition of polyaniline nanosheets mediated by sulfonated polyaniline functionalized graphenes. J Mater Chem 21:13978–13983

Zhao Y, Liu X, Li H, Zhai T, Zhou H (2012) Hierarchical micro/nano porous silicon Li-ion battery anodes. Chem Commun (Camb) 48:5079–5081

Zhao Y, Liu B, Pan L, Yu G (2013) 3D nanostructured conductive polymer hydrogels for high-performance electrochemical devices. Energy Environ Sci 6:2856–2870

Zhou H, Chen H, Luo S, Chen J, Wei W, Kuang Y (2005) Glucose biosensor based on platinum microparticles dispersed in nano-fibrous polyaniline. Biosens Bioelectron 20:1305–1311

Zhou Z-Y, Huang Z-Z, Chen D-J, Wang Q, Tian N, Sun S-G (2010) High index faceted platinum nanocrystals supported on carbon black as highly efficient catalysts for ethanol electrooxidation. Angew Chem Int Ed 49:411–414

Zhou G-M, Wang D-W, Li F, Zhang L-L, Weng Z, Cheng H-M (2011) The effect of carbon particle morphology on the electrochemical properties of nanocarbon/polyaniline composites in supercapacitors. New Carbon Mater 26:180–186

Zhu Y, Murali S, Stoller MD, Ganesh KJ, Cai W, Ferreira PJ, Pirkle A, Wallace RM, Cychosz KA, Thommes M, Su D, Stach EA, Ruoff RS (2011) Carbon-based supercapacitors produced by activation of graphene. Science 332:1537–1541

Zhu Z, Wang S, Du J, Jin Q, Zhang T, Cheng F, Chen J (2014) Ultrasmall Sn nanoparticles embedded in nitrogen-doped porous carbon as high-performance anode for lithium-ion batteries. Nano Lett 14:153–157

Zhuo K, Jeong MG, Chung CH (2013) Highly porous dendritic Ni-Sn anodes for lithium-ion batteries. J Power Sour 244:601–605

Chapter 8
Bioinspired Hierarchical Composites

André R. Studart, Randall M. Erb and Rafael Libanori

Abstract The structural design of composite materials at multiple length scales is a widespread strategy found in biological materials to optimize opposing properties or to combine multiple functional properties in a unique material system. The combination of this hierarchical structuring approach with the vast chemical repertoire available in synthetic systems is expected to lead to man-made composites with unprecedented functionalities. Alternatively, hierarchical materials can potentially achieve sufficient strength and toughness even if made out of weaker environmentally-friendly or bioresorbable building blocks. Replicating the hierarchical design principle of biological systems in synthetic materials is an exciting challenge that has been tackled by researchers across different scientific communities. In this chapter, we present state-of-the-art examples on attempts to identify fundamental design principles of hierarchical natural materials and to then mimic these bioinspired concepts in man-made materials. Three selected structural features that can be independently designed at multiple length scales in biological materials are described as examples: (i) mechanical reinforcement, (ii) porosity, and (iii) topography. By comparing biological and man-made materials exhibiting these hierarchical features, we provide an overview on the limitations of currently exploited top-down and bottom-up manufacturing technologies and on the opportunities for the future development of hierarchical composites inspired by the unique multiscale structure of biological materials.

A. R. Studart (✉) · R. Libanori
Complex Materials, Department of Materials,
ETH Zurich, 8093 Zurich, Switzerland
e-mail: andre.studart@mat.ethz.ch

R. Libanori
e-mail: libanori@mat.ethz.ch

R. M. Erb
Department of Mechanical and Industrial Engineering,
Northeastern University, Boston, MA 02115, USA
e-mail: r.erb@neu.edu

C.-S. Kim et al. (eds.), *Hybrid and Hierarchical Composite Materials*,
DOI 10.1007/978-3-319-12868-9_8

8.1 Introduction

Building hierarchical structures is a powerful approach to create composite materials with outstanding properties and functionalities. Along with other general design principles (Studart 2012; Studart et al. 2013b), nature has extensively exploited this concept of hierarchy to generate remarkable materials from abundant and naturally occurring building blocks. As opposed to the variety of synthetically processed chemical substances available for the fabrication of man-made materials, the chemistry of most biological organisms is limited to elements like carbon, nitrogen, oxygen, hydrogen, calcium, phosphorus, sulfur, and silicon. Constructing materials at multiple length scales is an effective strategy that enables living organisms to compensate for the scarcity of chemical species readily accessible in their natural environments (Fratzl and Weinkamer 2007).

Structural features that may exhibit hierarchy in a natural composite material include, for example, mechanical reinforcing elements (particles, fiber, platelets), porosity (pores, canals, porous networks), and surface topography (roughness, wrinkles, gratings). By organizing these features in a hierarchical fashion, living organisms are able to produce biological materials with properties and functionalities that would not be achievable otherwise using the same set of building blocks implemented at only one particular length scale. Examples in nature are abundant: strong and lightweight bones (Fratzl and Weinkamer 2007; Fratzl 2008a), stiff and tough seashells (Barthelat and Espinosa 2007; Barthelat et al. 2007; Espinosa et al. 2009), fatigue-resistant tendon–bone interfaces (Genin et al. 2009), fast-filtering marine sponges (Tompkins-MacDonald and Leys 2008; Leys et al. 2011), impact-absorbing fruits (Thielen et al. 2013b), self-cleaning plant leaves, sticky insect footpads, and colorful butterflies, among many others.

The wide range of functionalities from mechanics to fluidics to optics illustrates the plethora of applications that can benefit if the hierarchical design principle exploited in biological composites is translated into synthetic systems. The rationale of such a bioinspired approach is that novel high-performance composites can potentially be produced if the rich chemistry of synthetic systems is combined with the hierarchical organization of biological materials (Studart 2012, 2013; Studart et al. 2013a). Alternatively, the introduction of hierarchy in man-made composites might allow for the use of weaker but more environmentally-friendly, abundant, and bioresorbable building blocks without compromising the ultimate properties of existing synthetic materials.

Despite this great potential, the structure of man-made materials remains remarkably poor if compared to the elaborate hierarchical architectures of natural systems. Considering that hierarchy in biological systems is built through cell-mediated complex assembly processes, implementing hierarchical design principles in synthetic systems constitutes an enormous challenge. Therefore, significant research efforts to recreate these bioinspired architectures have been focused on exploiting available fabrication technologies or devising new top-down or bottom-up self- and directed-assembly methods to build hierarchical features on surfaces or bulk synthetic materials.

In this chapter, we showcase exemplary materials that exhibit hierarchical organization of: (i) mechanical reinforcement, (ii) porosity, and (iii) topography. In each section, we first present inspirational biological materials and highlight the design principles and structural features underlying the properties of such hierarchical architectures. Next, we discuss synthetic approaches that have exploited these structural features within man-made systems. Finally, a comparative analysis is drawn between biological and artificial materials to highlight the potential and current challenges in incorporating hierarchy in synthetic systems.

8.2 Hierarchical Mechanical Reinforcement

The wide range of synthetic materials available enables the engineer to tailor the mechanical performance of load-bearing structures simply by selecting a single material according to its mechanical response and compatibility with the fabrication process. In contrast, nature can only start from limited chemical elements to build structural, long-lasting systems with mechanical performance capable of withstanding unexpected loading conditions and significant numbers of loading cycles. This limited availability of chemical elements in the biological realm has driven the introduction of microstructural features at multiple length scales as a means to enhance the overall mechanical performance of natural materials. On the basis of biologically mediated self-assembly principles, natural structures like bones, seashells, and trees have their microstructures designed and controlled at different levels of hierarchy to efficiently harness the load-bearing ability of their organic and inorganic constituents. These natural systems have inspired scientific investigations into the development of synthetic materials that recreate intricate-reinforced microstructures with multiple levels of hierarchy. In this section, we present examples of such natural composites and discuss selected hierarchical synthetic materials whose mechanical performance is greatly improved by implementing specific biological design principles.

8.2.1 Biological Materials with Hierarchical Mechanical Reinforcement

Hierarchical structuring is a design strategy widely used in biological materials to achieve unusual sets of mechanical properties through the implementation of different microstructural features at multiple length scales. A prime example of a hierarchical structure in the biological realm is the mollusk shell, which elegantly combines strength, stiffness, and toughness in a highly mineralized hybrid material (Jackson et al. 1988). At the macroscopic length scale (Fig. 8.1a), these shells are typically composed of a bilayer material exhibiting a hard outer layer reinforced with prismatic calcite crystals and a strong and tough inner layer (nacre) consisting mainly of aragonite platelets (Espinosa et al. 2009). Such structural design works as a strong armor system and offers effective protection to the soft tissue of the mollusks against the attack of predators. Interestingly, while the outer layer presumably increases the hardness and wear resistances of the shell, the strong and tough inner layer is hierarchically designed to arrest cracks and increase the energy needed for a complete shell fracture.

Despite the seemingly straightforward yet crucial role of the calcite prismatic layer for mechanical protection and integrity, the presence of multiple toughening mechanisms acting at different length scales has made the nacreous layer one of the most extensively studied mineralized biological materials. The combination of these toughening mechanisms seems to allow the somewhat unimpressive building blocks of nacre to work together synergistically to produce a robust material that is far tougher than the monolithic material derived from the same building block ($CaCO_3$). Starting from the microscale (Fig. 8.1b, right), the nacreous layer comprises micron-sized aragonite platelets (5–10 μm in length and 0.4–0.5 μm in thickness) arranged in a brick-and-mortar structure and separated by a very thin proteinaceous matrix (20–30 nm in thickness). Seasonal growth lines consisting mainly of organic matter and with thickness between 15 and 70 μm separate the brick-and-mortar material into mesolayers (Fig. 8.1b, left) and work together with the individual platelets as effective crack deflectors to further enhance the ability of the seashell to dissipate energy during fracture (Su et al. 2002).

Interestingly, the surface topography of the aragonite platelets in nacre is also intricately structured at smaller length scales to provide extrinsic toughening mechanisms that operate behind the crack tip to inhibit crack propagation and increase the material's resistance against failure. Nanoscale features including mineral bridges, nanoasperities, and complementary surface waviness (Fig. 8.1c) between adjacent platelets act as a mechanical interlocking system, which requires progressively higher tensile stresses for effective platelet sliding (Barthelat et al. 2006; Barthelat and Espinosa 2007; Espinosa et al. 2009). This mechanical interlocking promotes significant strain delocalization and dilation effects throughout the whole material, which ultimately increases the total energy absorbed during its fracture. Finally, the protein-rich organic matrix of nacre is also believed to contribute to its striking overall toughness (Fig. 8.1d). In situ mechanical tests performed with atomic force microscopy on freshly cleaved nacre surfaces reveal that a significant amount of energy is dissipated through unfolding of modular protein domains during mechanical deformation (Smith et al. 1999). Upon stretching, these modular proteins undergo a series of sequential reversible unfolding events by breaking secondary sacrificial bonds that are programmed in their tertiary structure. Overall, these intrinsic and extrinsic toughening mechanisms operating in a cooperative manner at multiple length scales confer the unique combination of high fracture toughness, mechanical strength, and stiffness for a material that is essentially assembled from 95 vol% of brittle constituents.

In addition to the incorporation of toughening mechanisms in natural materials that are mainly composed of inherently brittle building blocks, hierarchical constructs can also provide a platform to tailor the mechanical properties of load-bearing

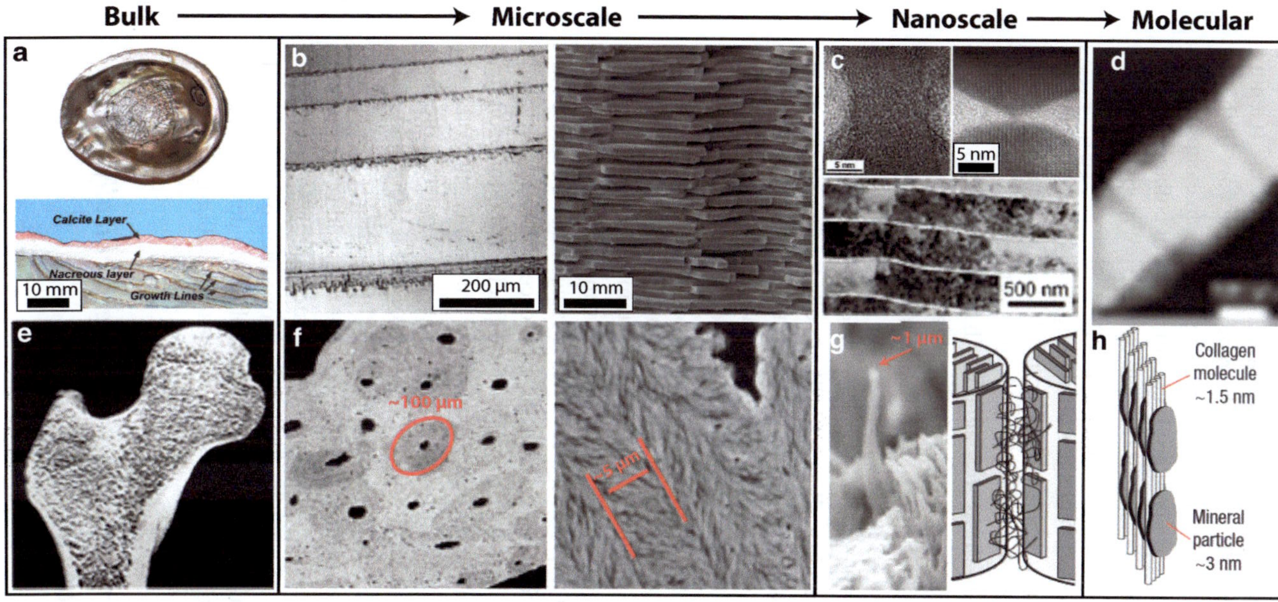

Fig. 8.1 Hierarchical structure of biological materials: **a**–**d** Abalone shell and **e**–**h** human femoral bone. **a** Pictures of abalone shell depicting the inner part (*top*) and its cross section (*bottom*). The cross section reveals the bilayer structure and the seasonal growth lines. (Adapted from Barthelat et al. (2007) with permission from Elsevier.) **b** Optical micrograph (*left*) of a polished surface of the nacreous layer showing the brick-and-mortar material (*right*) separated by the seasonal growth lines. (Adapted from Barthelat et al. (2007) and Menig et al. (2000) with permission from Elsevier.) **c** Nanoscale structures used for geometrical interlocking between adjacent aragonite platelets: mineral bridge (*upper left*), nanoasperities (*upper right*), and complementary surface waviness (*bottom*). (Adapted from Barthelat et al. (2007) with permission from Elsevier.) **d** Fractured surface of the nacreous layer showing the fibrillation mechanism of the proteinaceous matrix achieved through unfolding of modular domains. (Adapted from Sumitomo et al. (2008) with permission from Cambridge University Press.) **e** Picture of the human femoral bone showing its macroscopic structure comprising cortical (*outer layer*) and trabecular (*spongy inner part*) bones. (Adapted from Fratzl et al. (2004) with permission of The Royal Society of Chemistry.) **f** Scanning electron microscopy image of cortical bone highlighting the osteon (*left*) and the lamellar arrangement of collagen fibrils surrounding the osteons (*right*). (Adapted from Fratzl et al. (2004) with permission of The Royal Society of Chemistry.) **g** Mineralized collagen fibrils (*left*) that are held together by an extracellular organic matrix comprising noncollagenous proteins and lipids (*right*). (Adapted from Fratzl et al. (2004, Fratzl and Weinkamer 2007) with permissions of The Royal Society of Chemistry and Pergamon Press.) **h** Molecular structure of the mineralized collagen fibrils depicting the staggered arrangement of the mineral crystals and the collagen molecules. (Adapted from Fratzl et al. (2004) with permission of The Royal Society of Chemistry)

structures within a broad range using a limited number of constituent materials. Collagen-based materials, such as bones and tendons, are fascinating examples of natural materials with a similar hierarchical construct whose mechanical response can be adjusted by controlling the degree of mineralization of the structure (Genin et al. 2009; Fratzl and Weinkamer 2007). The basic hierarchical framework of these collagen-based materials is illustrated in Fig. 8.1e–h using bone as an example. At the molecular level, these collagen-based materials are composed of collagen molecules (300 nm in length and 1.5 nm in diameter) formed by three polypeptide chains assembled in a triple helix. These molecules are aligned parallel to each other and arranged in a staggered manner to form collagen fibrils (Fig. 8.1h). The mineralization degree of such biological constructs is controlled by the local concentration of hydroxyapatite nanoplatelets with a length of 15–200 nm and thickness of 2–7 nm within and/or between the collagen fibrils. Collagen fibrils with controlled level of mineralization are joined together by an amorphous extracellular matrix slightly cross-linked by reversible sacrificial inter- and intramolecular bonds (Fig. 8.1g, Fratzl and Weinkamer 2007).

The morphology of the next hierarchical level depends on the function of the collagen-based material. In bones, the collagen fibrils are often arranged into concentric lamellae around blood vessel canals (Harvesian canals) to form the fundamental functional unit of the bone, the osteon (Fig. 8.1f). The orientation of fibrils within such lamellae is tuned to build a concentric plywood structure (Wagermaier et al. 2006). In contrast to the lamellar structure of the bone, unmineralized collagen fibrils can also be assembled into large fibers and fiber bundles to build tendons (Fratzl 2008b). By adjusting the mineralization degree within collagen fibrils, the elastic modulus of these hierarchically reinforced materials can be tuned from about 400 MPa in tendons up to about 20,000 MPa in bones (Maganaris and Paul 1999; Genin et al. 2009). Bone is further designed at the macroscopic structural level, where the shape and the relative density of the tissue are finely tuned to match its respective function. As illustrated in Fig. 8.1e, bones that are mainly subjected to flexural stresses, such as our femur and humerus, are usually elongated and composed of a dense material (cortical bone) in the outer surface and a spongelike material (trabecular bone) in the interior (see also Sect. 3.1). In contrast, regions that experience mainly compressive loads, such as in the femoral head and vertebrae, are mainly structured with trabecular bone (Fratzl and Weinkamer 2007). Such a multiple hierarchical reinforcement allows for a cooperative deformation of the material and provides an effective protection of the brittle mineral phase by delocalizing the mechanical strain across the whole biological tissue (Gupta et al. 2006).

8.2.2 Bioinspired Synthetic Materials with Hierarchically Reinforced Microstructures

The remarkable mechanical performance exhibited by biological systems has stimulated significant amount of research in the development and fabrication of bioinspired, synthetic materials with unusual mechanical response. In one of the research fronts, efforts have been focused on the fabrication of ceramic–polymer (Munch et al. 2008), ceramic–metal (Prielipp et al. 1995), and all-ceramic composite materials (Bouville et al. 2014) that can potentially reach the unique combination of high strength and toughness in one single load-bearing component. Such properties are often mutually exclusive in conventional synthetic materials because intrinsic toughening mechanisms are intimately related to the ability of the material to undergo mechanical deformation while mechanical strength is correlated to the material resistance to such plastic deformation (Ritchie 2011). For example, ceramics are inherently strong and brittle due to the strong nature of their chemical bonds and rather limited slip planes available for plastic deformation in their crystalline structure. Thus, replicating the design principles found in the microstructure of natural composites into ceramic-based materials can potentially lead to a combination of the ceramic's intrinsically high mechanical strength and stiffness with an unusual ability to dissipate energy that is provided by extrinsic toughening mechanisms programmed in the biological microstructures.

Indeed, strong and tough layered ceramic-based materials have been fabricated using, for example, a freeze-casting technique combined with uniaxial pressing and sintering (Munch et al. 2009; Bouville et al. 2014). Long-range order is obtained by freeze-casting a suspension of nanoparticles, which induces anisotropic growth of ice crystals and expels the particles from the suspension into confined spaces (lamellae). Nacre-like brick-and-mortar or lamellar structures can be created upon pressing and sintering of freeze-cast/dried constructs containing isotropic and/or platelet-like particles. Pressing and sintering convert some of the nanoparticles into ceramic bridges and surface nanoasperities, leading to an effective interlocking system that delocalizes mechanical deformation over a large volume of material. The same approach has been used to produce all-ceramic materials as well as polymer–ceramic layered composites with high volume fractions of ceramic phase. As illustrated in Fig. 8.2a–d for the case of an all-ceramic alumina-based system, the ultimate microstructure exhibits at least four levels of the hierarchical design that is also present in nacre, as follows: (i) long-range structural order with ceramic platelets preferentially aligned along the freezing direction (Fig. 8.2b); (ii) closely packed ceramic platelets (Fig. 8.2c); (iii) surface nanostructures for mechanical interlocking (Fig. 8.2c and d); and (iv) a more compliant secondary phase to allow for load redistribution, crack deflection, platelet pull-out, and crack bridging, which is present in the material as very thin layers in between adjacent platelets (Fig. 8.2c).

The outstanding ability of the nacre-like alumina to dissipate energy during fracture is revealed by testing notched beams under flexural mode. As shown in Fig. 8.2e, the layered ceramic material displays a strikingly stable crack propagation followed by a pronounced strain hardening effect when the asperity- and bridge-forming nanoparticles are introduced in the system Bouville et al. (2014). Such strong inelastic deformation and strain hardening is rather unusual for a material composed of only brittle ceramic constituents. The surprisingly high toughness exhibited by the nacre-like alumina is certainly a result of the extrinsic toughening mechanisms that have been introduced within its hierarchically reinforced microstructure. In fact, removing the ceramic bridges and nanoasperities of the alumina

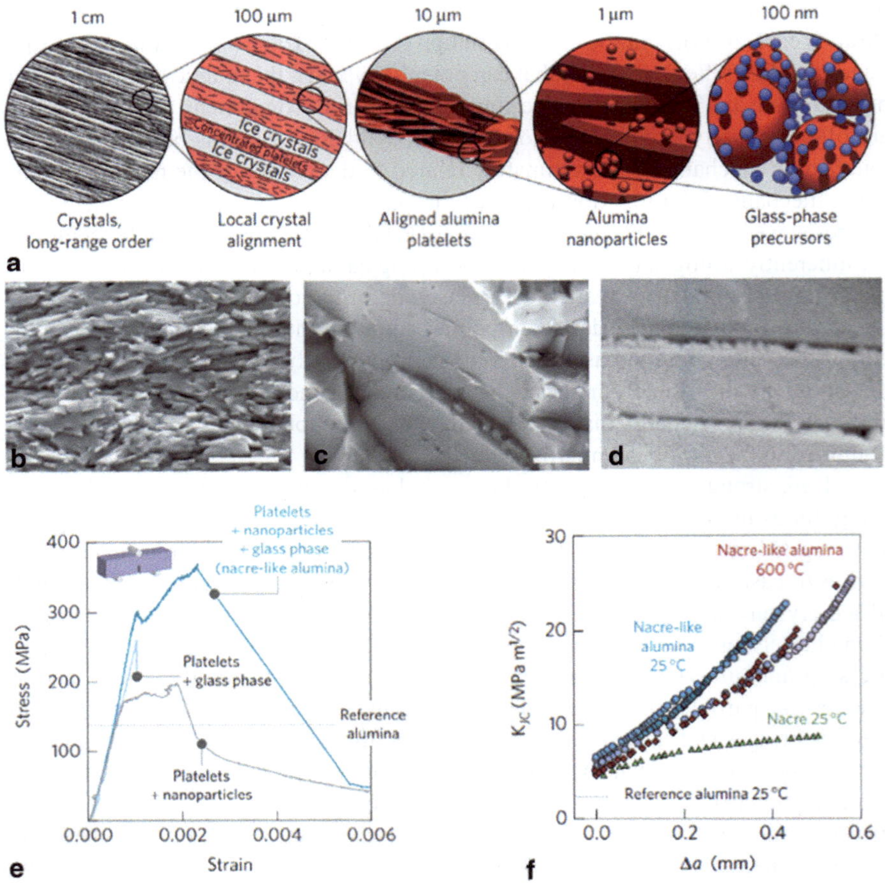

Fig. 8.2 Strong and tough bioinspired ceramics assembled from brittle constituents. **a** Design strategy depicting the control over the microstructural features at multiple length scales. **b–d** Scanning electronic microscopy images of the nacre-like alumina showing **b** the long-range structural order of the microstructure exhibiting platelets preferentially aligned along the freezing direction, **c** local stacking of alumina platelets, and **d** surface of the alumina platelets modified with ceramic bridges and nanoasperities. **e** Stress–strain curves of notched beams of hierarchical nacre-like alumina (*blue curve*), of a structure containing platelets and a glass phase (*gray curve* exhibiting brittle fracture), and of samples comprising platelets and alumina spherical nanoparticles (*gray curve* exhibiting stable crack propagation). **f** Crack-growth resistance curves (R-curves) for the nacre-like alumina samples measured at 25 and 600 °C in comparison to that of the biological material nacre. (Reprinted by permission from Bouville et al. (2014), Macmillan Publishers Ltd, copyright, 2014)

surface leads to the typical brittle failure exhibited by ceramic materials (Fig. 8.2e). The hierarchical nacre-like alumina combines flexural strength as high as 470 MPa and toughness values that increase from 6.2 to astonishingly 22 MPa m$^{1/2}$ as the cracks propagate across the material (Fig. 8.2f). As the material is only composed of ceramics, the resistance against crack propagation is reduced by only 5 % when the temperature is increased from 25 to 600 °C (Fig. 8.2f). These values are unexpectedly high and exceed by far those of a conventional "nonlamellar" polycrystalline alumina and of the biological material nacre (Fig. 8.2f).

Another remarkable characteristic of hierarchically reinforced natural systems is the broad range of mechanical response that can be attained by controlling the mineralization degree of a common continuous organic phase. Using this strategy, nature is able to grow functionally graded structures that are capable of minimizing mechanical stresses at the interfaces between materials with a large mismatch in mechanical properties (Genin et al. 2009; Harrington et al. 2010; Yao et al. 2010). Unlike in nature, the strategy of using mechanical gradients is far less exploited in polymer-based artificial composites, mainly because the level of reinforcement achieved in synthetic systems comprising a common organic phase is often limited by the increasing concentration of processing defects and confinement of the polymer matrix at high volume fraction of reinforcing elements (Bonderer et al. 2008, Kaushik et al. 2009; Bonderer et al. 2010). Developing processing techniques that allow for a tunable reinforcement of a common polymer matrix within a wide range of values can potentially lead to graded polymer-based composites that further extend the functionality of currently available metal/ceramic graded materials (Mortensen and Suresh 1995).

Polyurethane-based composites with elastic moduli that can be adjusted over several orders of magnitude have been recently created using a solvent-based hierarchical reinforcement approach (Libanori et al. 2012a). As shown in Fig. 8.3a, this reinforcement strategy relies on the control of the relative concentration of polyurethane molecular hard domains, nano-sized clay particles, and micron-sized inorganic platelets that are dispersed within a polyurethane soft phase. Remarkably, large-area polyurethane-based films exhibiting elastic moduli spanning from 4 to 7000 MPa can be fabricated by combining this hierarchical reinforcement concept with a standard tape-casting technique. The high chemical compatibility and thermoplastic nature of the continuous polymeric matrix enables the creation of multilayered graded composites by solvent-welding individual layers of polyurethane with different reinforcement levels (Libanori et al. 2012b). As an illustrative example of the potential of such graded materials, this hierarchical reinforcement approach was recently exploited to create polymer-based composites for stretchable electronics. In this application, a highly stretchable polymer substrate is combined with a locally stiff island (patch) that protects brittle electronic devices through local strain minimization. To this end, prismatic patches displaying out-of-plane gradient were solvent-welded on a flexible and elastomeric polyurethane substrate, as illustrated in Fig. 8.3b. Scanning electron microscopy images of the cross section of the multilayered material (Fig. 8.3b) reveal that the interfaces are completely free of macroscopic defects, which is likely enabled by effective polymer entanglement and the formation of an interpenetrating polymer network across the multilayered structure.

To demonstrate the unique combination of global stretchability and local strain minimization, the mechanical response of a model patched substrate was assessed by optical microscopy while being subjected to a global tensile strain. Interestingly, local compressive strains lower than 1% were detected on the top surface of the

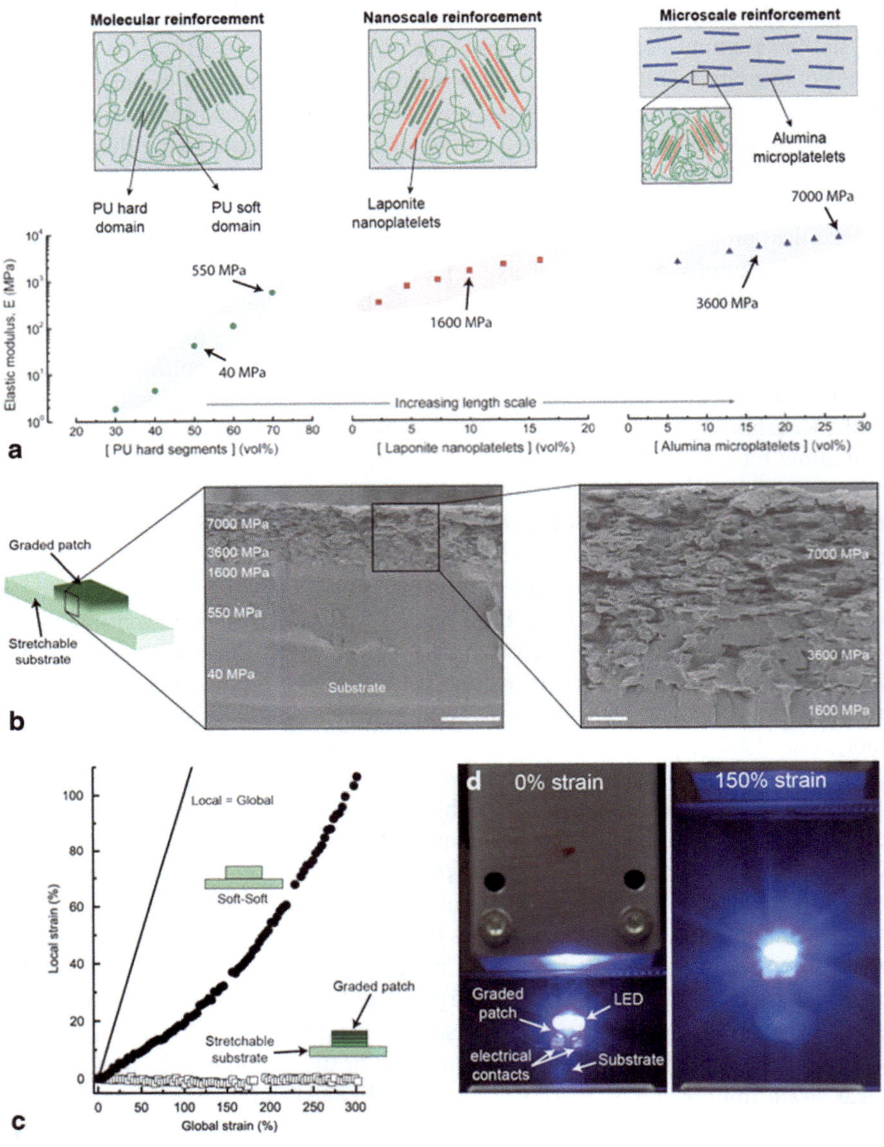

Fig. 8.3 Hierarchically reinforced polyurethane composites with tunable mechanical properties and their assembly into multilayered composites with extreme mechanical gradients. **a** Elastic modulus of polyurethane composites containing different concentrations of reinforcing elements at increasing length scales: molecular hard domains, laponite nanoplatelets, and alumina microplatelets. **b** Cross section of the multilayered composite assembled by solvent-welding a graded patch onto the surface of a stretchable and soft polyurethane substrate. **c** Local strain on the top surface of the patch as a function of increasing global deformations applied on the underlying substrate. **d** Combination of local strain minimization of the graded patch with global stretchability of the underlying substrate allows for enhanced failure resistance of fragile electronic components. (Reprinted from Libanori et al. (2012b) by permission from Macmillan Publishers Ltd, copyright, 2012)

graded patch for global strains as high as 300% (Fig. 8.3c). Conversely, multilayered composites comprised of a patch and substrate exhibiting the same elastic modulus (soft–soft interface) displayed a steady increase in local tensile strain on the top of the patch for increasing global strains. Eventually, the combination of global stretchability and local stiffness enabled the fabrication of substrates for stretchable electronics that can effectively prevent failure of fragile light-emitting diode devices (LEDs) even when the underlying substrate is globally stretched up to 150% (Fig. 8.3d).

Despite the increasing number of successful man-made bioinspired materials, the design and fabrication of artificial composites whose microstructures resemble the intricate hierarchically reinforced architectures of natural materials still remains an exciting challenge for material scientists and engineers. In contrast to the more elaborated cell-mediated assembly of natural materials, state-of-the-art processing techniques that are available for the fabrication of artificial materials usually lack the ability to provide full control of the microstructures over several length scales. The scarcity of experimental studies on the structure–properties relationship of hierarchically reinforced bioinspired composites is partially supplemented by the advanced development of a multitude of computational models that contribute to a more complete understanding of the role of hierarchy on the mechanical properties of natural and bioinspired materials (Buehler 2010; Sen and Buehler 2010; Nova et al. 2010; Garcia et al. 2011). Overall, the examples discussed above illustrate the huge potential of exploring the hierarchical design of natural microstructures to achieve unusual combinations of mechanical properties in artificial composites.

8.3 Hierarchical Porosity

Hierarchically porous systems offer the possibility of combining properties that would not be achievable using pores of one size alone. This strategy is widely exploited by biological and man-made materials to achieve the functionalities required for survival in their natural environment or to fulfill specific engineering targets, respectively. In this section, we first discuss the underlying structure and functionalities of selected multiscale porous materials in biology. Next, we describe processing methods that have been utilized in implementing similar hierarchical porosity in synthetic systems. A comparative analysis between structure–function relationships in natural and synthetic systems is finally drawn to highlight research opportunities to deepen our understanding of biological materials and further advance the performance of synthetic systems through designed porosity at multiple length scales.

8.3.1 Hierarchically Porous Materials in Nature

Hierarchical porous materials can be found across different kingdoms and phyla in nature. Examples range from the vasculature system of plants to aquiferous networks in marine sponges to the skeletal structure of vertebrates (Fig. 2.1). The porous architecture of these materials can exert a wide variety of functions, including for instance shock absorption in plant systems, nutrient uptake in food-deficient marine environments, and weight reduction in load-bearing skeleton parts. Here, we describe the underlying structure and function of the peel of the pomelo fruit, the walls of a marine sponge, and the core–shell architecture of bone tissue to illustrate important design concepts found in hierarchically porous materials in biology.

In the pomelo fruit, a hierarchical structure of pores is found in the thick peel that presumably protects the fruit from impact damage when falling from the high heights of citrus trees (Thielen et al. 2013b). The main part of the peel (mesocarp) consists of a foam-like open structure with pores in the 100–400 μm range (Fig. 8.4a). The struts of the foam are made of liquid-filled plant cells, which define another set of pores within the system. This hierarchical structure is permeated by millimeter-thick vascular bundles responsible for the transport of water and nutrients throughout the peel. The ability of the peel to dissipate up to 90 % of the fruit's kinetic energy during impact has been attributed to the high compressibility and absolute thickness of this hierarchical porous architecture. High compressibility is expected to arise from bending of the struts during compaction and burst of the liquid-filled cells. Vascular bundles presumably also increase energy dissipation during impact by redistributing stresses over a large volume of the peel (Thielen et al. 2013a).

Besides the mechanical function exemplified above, hierarchical porous structures are also used in nature for solid–liquid separation through an energy-efficient filtration process. This is the function exerted by the walls of certain marine sponges, whose hierarchical aquiferous system allows for the filtration of quantities of water equal to several times the volume of the sponge itself, thus enabling the capture of scarce particulate nutrients in the deep sea. The hierarchical porous structure is comprised of channels, chambers, and porous membranes at progressively smaller length scales, as depicted in Fig. 8.4b. Broad inhalant canals bring large quantities of water from the surrounding environment into the internal porosity of the sponge (Fig. 8.4b(i)). Water within inhalant canals is transported to so-called choanocyte chambers (Fig. 8.4b(ii)), which are lined with flagellated cells that can actively capture the particulate matter. The cell flagella presumably contribute to the active water pumping from inhalant to exhalant canals (Leys et al. 2011). Nutrient capture occurs within the 2–5 μm holes that perforate the walls of the choanocyte chamber (Fig. 8.4b(iii)). Such holes are decorated with microvilli coated with a protein layer whose nanoscale mesh size establishes the finest length scale of pores within the system. This unique hierarchical architecture combines a locally high density of capture sites enabled by the smallest pores with the minimum pressure drop provided by the large inhalant and exhalant canals. Such a hierarchy maximizes nutrient capture efficiency while minimizing the energetic cost of pumping.

Bone tissue is an example of a hierarchical biological material whose multiscale porosity is designed to enable load-bearing capability while still being sufficiently light to minimize the energy cost for locomotion. The tissue often displays a core–shell structure with the core made of a foam-like material named trabecular bone and the shell exhibiting a denser structure permeated by isolated pores and

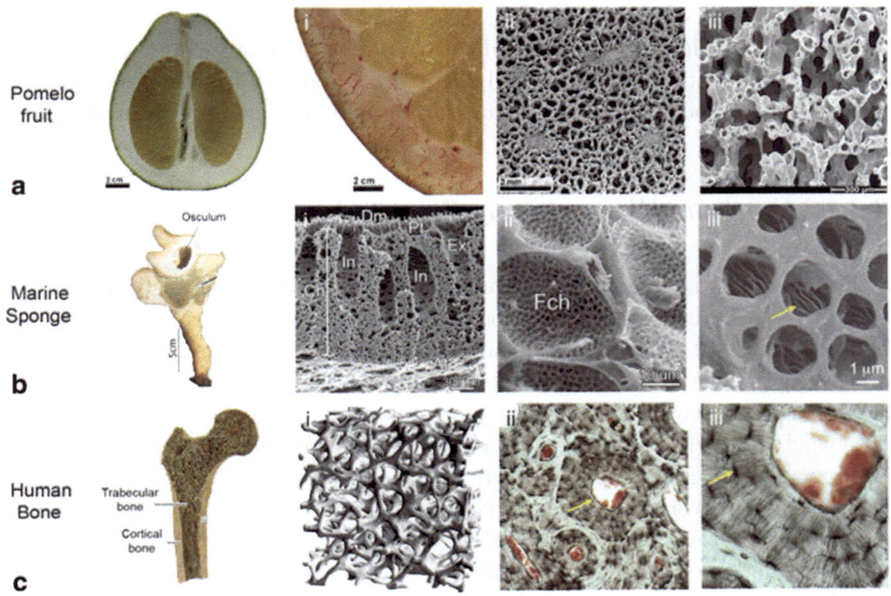

Fig. 8.4 Examples of hierarchical porous structures in natural systems, highlighting the material's microstructure at progressively smaller length scales (from *left* to *right*). **a** Pomelo fruit (*Citrus maxima*): (i) Cross section of the fruit peel, showing the vascular bundles (in red). (ii) Foam-like architecture of the peel. (iii) Detailed view of the foam, revealing the liquid-filled pores (cells) that form the struts of the foam. (Adapted from Thielen et al. (2013b and Fischer et al. 2014) with kind permission from Springer Science+Business Media.) **b** Marine sponge *Aphrocallistes vastus*: (Tompkins-MacDonald and Leys 2008; Leys et al. 2011): (i) Cross section of the wall of the sponge body showing the juxtaposed incurrent (*In*) and excurrent (*Ex*) canals (*Dm*: incurrent dermal surface, *At*: excurrent atrial surface, *Pt*: peripheral trabecular space, *Ch*: choanosome). (ii) Detailed view of a perforated choanocyte chamber (*Fch*). (iii) Holes and collar microvilli (*arrow*) on the wall of the choanocyte chamber. (Adapted from Tompkins-MacDonald and Leys (2008), and Leys et al. (2011) with kind permissions from Public Library of Science and Springer Science and Business Media.) **c** Femoral human bone: (i) Foam-like open structure of trabecular bone. (ii) Cross section of cortical bone, highlighting vascular canals within the so-called osteons (*arrow*). (iii) Closer view of an osteon, indicating the lacunae that host osteocytes (arrow) and the interconnecting canaliculi network. (Picture of femoral bone was modified from Launey et al. (2010) with permission from Annual Reviews of Materials Research,Vol. 40 by Annual Reviews; image (i) from Müller and Rüegsegger (1997) with permission from IOS Press Amsterdam, and pictures (ii) and (iii) were kindly provided by Prof. Tim Arnett, University College London)

interconnected canals known as cortical bone (Fig. 2.1c). The foam-like trabecular architecture reduces the overall weight of the structure and provides room for blood vessels and the marrow (Fig. 8.4c(i)). Long canals run along the outer shell to enable vascularization of the tissue also in the denser cortical bone (Fig. 8.4c(ii)). The highly mineralized extracellular matrix between such canals contains isolated pores (lacunae) that host encapsulated bone cells (osteocytes, Fig. 8.4c(iii)). In turn, osteocytes are connected to one another and to vascular canals through a complex network of canaliculi. The network of canals, lacunae, and canaliculi that permeate the denser shell allows for transport of cells, nutrients, and ions within the material. This enables redistribution of mineral phase throughout the material (homeostasis) and constant remodeling of the structure in response to external mechanical load.

8.3.2 Synthetic Hierarchically Porous Materials

While the creation of hierarchical porous materials in nature relies on the biological activity of living cells, man-made counterparts are mostly produced using assembly methods in the absence of cellular control. Despite these significantly different processing pathways, a common feature of biological and synthetic approaches is the utilization of templates for the creation of pores. In bone, for example, cartilage tissue and bone cells can work as templates for the deposition of the mineral phase hydroxyapatite and are later disintegrated or incorporated into the bone material. Likewise, the network of pores and channels in plants results from the deposition of organic matter around cells and vessels of the plant's vasculature system.

The synthesis or deposition of material around templating structures or building blocks is by far the most widely used approach to generate hierarchical porosity in man-made materials. Such templating approaches include the replica technique, sacrificial template method, and direct foaming, as schematically illustrated in Fig. 3.2a–c. Processing methods that rely on the template-free direct deposition of materials, such as three-dimensional (3D) printing, extrusion, and spinning processes complete the most important set of techniques available for the creation of macroporous materials (Fig. 2.2d). The macroporous structures obtained via these approaches can be further modified to incorporate additional pores at smaller length scales, leading to truly hierarchical porous architectures. Some of the possible strategies to generate this second hierarchical level are schematically depicted in Fig. 2.2e and explained in more detail in the examples discussed below.

One of the most widespread routes to create synthetic macroporous materials is the replica technique. In this method, a preexisting porous structure is used as a template for the deposition of the synthetic material of interest (Fig. 8.5a) (Studart et al. 2006). In the example shown in Fig. 8.6a, the porous structure of rattan palm is the selected template. Typically, synthetic building blocks in the form of suspended particles, precursor sols, or molten phases are incorporated into the chosen porous template through an infiltration process. Figures 3.3a(i) and (ii) show the cross section of rattan palm after pyrolysis and infiltration with liquid silicon (light-colored phase) at 1550 °C. In this example, the infiltrated silicon partly reacts with carbon from the template to form a silicon–silicon carbide (SiSiC) porous structure. Numerous variations of the replica method are possible by changing the type of template and/or the infiltrating phase (Studart et al. 2006). A major advantage of the replica technique is that it can rely on the wide variety of porous structures already available in synthetic or biological systems. As schematically shown in Fig. 8.5a, this process is often used to generate a family of pores at one particular length scale. One possible strategy to create hierarchical architectures is to coat the walls

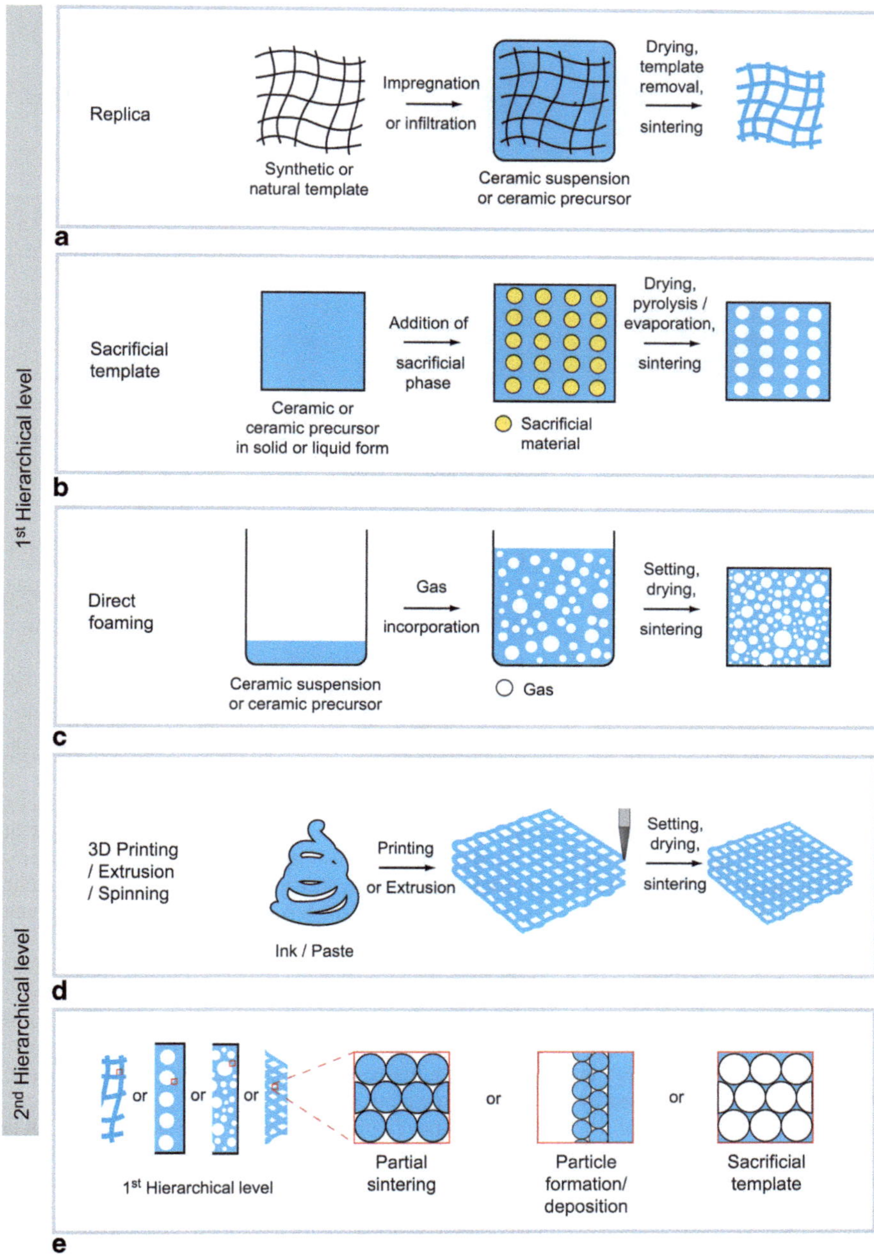

Fig. 8.5 Processing routes to prepare synthetic hierarchical porous structures. **a**–**d** Most common techniques to produce macropores at a first hierarchical level. (Adapted from Studart et al. (2006) with permission of Wiley.) **e** Possible methods to create porosity at finer length scales, thus leading to a second hierarchy of pores

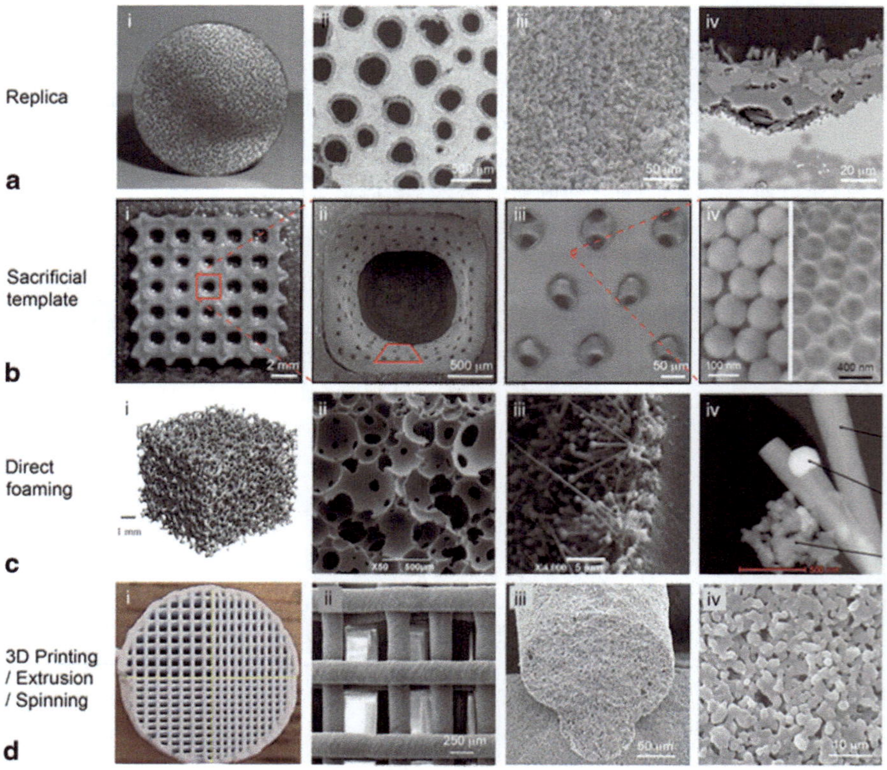

Fig. 8.6 Examples of synthetic hierarchical porous structures. **a** Rattan-derived silicon–silicon carbide (*SiSiC*) porous structure: (i) Transversal cross section of the structure. (ii) Closer view of the macropores generated after infiltration and pyrolysis of the original wood template. (iii) Microporous zeolite particles deposited on the macropore walls. (iv) Detailed lateral view of the zeolite layer on the surface of each macropore. (Adapted from Zampieri et al. (2005) with permission of Wiley.) **b** Hierarchical ordered porous structure generated from sacrificial templates: (i) Glass scaffold with periodic millimeter-sized holes used as substrate for the deposition of suspensions with templating droplets. (ii) Single hole that was coated with particles and oil droplets to generate macropores upon evaporation of the oil phase. (iii) Closer view of the interconnected macropores created after removal of the templating droplets. (iv, *left*) Packed particles between the macropores that are only partially sintered to generate interstitial pores at smaller length scales. (iv, *right*) Fine, ordered porosity created by infiltrating templating polymer particles with sol–gel precursors followed by pyrolysis of the template. (Adapted with permission from Studart et al. (2011) Copyright, American Chemical Society.) **c** Hierarchical porous ceramic made by direct foaming of a poly(siloxane) resin followed by in situ precipitation of nanowires at 1350 °C: (i) Typical 3D macroporous structure of the polysiloxane foam. (ii) Interconnected macropores within the foam. (iii) Detailed view of the macropore wall, depicting the nanowires grown from the wall surface. (iv) Nanowires and nanoparticles that give rise to the second hierarchy of pores. (Adapted from Zeschky et al. (2003) with permission from Elsevier and from Vakifahmetoglu et al. (2010) with permission from Wiley.) **d** Hierarchical cellular structure obtained by 3D printing (direct-write assembly) of hydroxyapatite suspensions followed by drying and partial sintering: (i) Macroscopic sample illustrating the tunable cell sizes and geometries that can be achieved by this process. (ii) Array of ceramic filaments comprising the walls of the cellular structure. (iii) Detailed view of one filament and (iv) of the fine porosity arising from the interstices between partially sintered particles. (Adapted from Simon et al. (2007) with permission from Wiley)

of the macroporous structure with an additional layer of material that generates a second hierarchy of pores. In Figs. 2.3a(iii) and (iv), such a porous layer is generated through the formation of microporous zeolite particles on the SiSiC structure through a dissolution/precipitation process at high temperature.

Another common approach to create porous materials is the sacrificial template method. In this case, a sacrificial phase typically in the form of particles or droplets is mixed with the material of interest and used as temporary template for the formation of macropores (Fig. 8.5b). Figure 8.6b depicts an example of a porous structure created using oil droplets as templates in a suspension of silica particles. Interconnected macropores are generated upon drying of the suspension and evaporation of the oil droplets. The solid particles of the initial suspension form the walls between the macropores (Fig. 2.3b(iii), (iv)). A wide variety of templates can be exploited using this technique, ranging from ice crystals to immiscible phases to solid particles (Studart et al. 2006). The possibility to predefine the level of porosity and pore size/morphology of the final structure by changing, respectively, the initial concentration and size/shape of the template phase is a major advantage of this processing route. Figure 8.6b(iv) shows two possible approaches to create pores at smaller length scales within this structure. These can arise from the interstitial openings between partially sintered packed particles (Fig. 8.6b(iv), left) or from the nanopores formed upon infiltration and removal of packed templating particles (Fig. 8.6b(iv), right).

The need to remove the templating phase is a drawback of the sacrificial approach. This issue can be overcome using gas bubbles as templates in the so-called direct foaming technique (Fig. 8.5c) (Studart et al. 2006). The main challenge in the direct foaming method is to prevent excessive coalescence of the foam before the continuous phase is consolidated through solidification, gelation, or polymerization. This can be accomplished by matching the kinetics of the foaming and consolidation processes or by effectively stabilizing the foam through the interfacial adsorption of surfactants or surface-active particles. Figure 8.6c shows an example of a macroporous structure generated by direct foaming of a polysiloxane resin. In this case, foaming is driven by the evaporation of a physical blowing agent (azodicarbonamide) from the resin while the mixture is heated up to 250 °C. Simultaneous cross-linking of the resin during heating enables consolidation of the foam structure before coalescence of the bubbles (Fig. 8.6c(i), (ii)). The obtained foam can then be pyrolyzed under argon above 1350 °C to promote the in situ formation of SiC nanowires (Fig. 8.6c(iii), (iv)). Together with macropores resulting from the foaming process, the interstitials created between the nanowires ultimately lead to a SiOC-based hierarchical porous ceramic.

In addition to the template-based techniques described above, hierarchical porous structures have also been produced using template-free, top-down fabrication technologies such as 3D printing (direct writing) and electrospinning (Simon et al. 2007; Chen et al. 2011). These processing routes typically involve the formation and deposition of filaments that are assembled into ordered or random cellular structures, as depicted schematically in Fig. 8.5d. Proper formulation of the ink or precursor solution is key to reach the rheological behavior required during the deposition process and to enable consolidation of the cellular structure. Figure 8.6d shows an example of a hierarchical porous hydroxyapatite structure produced by a 3D printing technique known as direct-write assembly. The ink in this case is a viscoelastic paste obtained by initially dispersing hydroxyapatite microparticles in water, followed by gelation of particles into a 3D network using a Ca-based salt. The solid-like nature of this paste in the absence of shear allows for the creation of patterned cellular structures with deliberate window/pore sizes mostly in the millimeter range (Fig. 8.6d(i), (ii)). Pores at much smaller length scales arise from the interstices between the hydroxyapatite particles after partial sintering of the cellular structure at 1200 °C (Fig. 8.6d(iii), (iv)). Although hierarchical porosity is often unexploited in these approaches, 3D printing and other additive manufacturing techniques have been widely used to produce porous materials for tissue regeneration (Cesarano et al. 2005; Billiet et al. 2012), autonomous crack healing (Toohey et al. 2007), energy storage (Sun et al. 2013b), and several other emerging applications (Symes et al. 2012).

The hierarchically porous materials produced with the methods outlined above enable the combination of distinct properties and functionalities due to the presence of pores at different length scales. Although most of such synthetic materials were originally not inspired by natural systems, many of the obtained man-made structures share similar functionalities with those of biological materials. For example, the high permeability and enhanced surface area of the hierarchical porous systems shown in Fig. 2.3a–c have been considered for catalytic and gas adsorption applications that require high density of reactive sites combined with fast mass transport (Zampieri et al. 2005; Vakifahmetoglu et al. 2010). In this case, enhanced fluid transport can be achieved with large macropores, while the high density of reactive sites arises from the pores at smaller length scales. This design strategy is also used by marine sponges to efficiently capture nutrients from seawater while minimizing the energy required to pump water through their porous structure (Fig. 8.4b). In another type of application, the 3D printed hierarchically porous structure shown in Fig. 8.6d has been designed to improve osteointegration of bone substitutes by providing the large pores needed for vascularization with the high surface area required for the adsorption of growth factors and osteoinductive ions. This resembles the multiscale porous architecture of bone tissue itself, which combines an extensively vascularized macroporous trabecular structure with a finer network of lacunae and canaliculi to promote the exchange of ions and molecules between cells and the mineralized extracellular matrix (Fig. 8.4c). Besides catalytic and biomedical applications, attempts have also been made to produce synthetic porous structures that replicate the mechanical impact-absorbing properties of the pomelo fruit (Fig. 8.4a, Fischer et al. 2014).

Since many of the functionalities envisioned in synthetic systems already exist in natural structures, further insights into the design of advanced hierarchical porous materials can be gained by studying structure–property relationships in biological porous systems, as those discussed above. Likewise, the increasing availability of synthetic materials with tunable porous architecture opens the possibility to use them as model systems to better understand the design principles of natural porous materials.

8.4 Hierarchical Topography

Many living organisms and biological composites rely upon topography to attain certain surface characteristics that enhance their performance in at least one regard as material systems. Here, topography pertains to the distribution of features across the surface of an organism or biological material. Specifically, natural biological organisms have seemingly evolved surface topography to increase their competitiveness as a species and their chance for survival. One recurring attribute found in the most effective systems is the predominance of hierarchy within the topography. Hierarchical topography exhibits feature sizes at different registries of scale that can optimize the interaction between the patterned surfaces and foreign material systems. This section will mention some selected hierarchical topographies that exist in natural organisms and then detail exemplary synthetic systems that have recreated this topographical hierarchy to obtain similar material properties and function.

8.4.1 Biological Materials with Hierarchical Topography

Within the biological realm, many *flora* and *fauna* employ hierarchical topography to tune their interaction with foreign materials and organisms. The use of hierarchical topography in nature produces a wide range of results including external protection, adhesion, and coloring. In some species, hierarchical topography works to minimize interaction with foreign species and effectively creates an external protection barrier. In many cases, these species tune their topography to avoid the buildup of dirt or biofilms on their surfaces. In sea creatures, biofilms contaminate the skin of marine animals and can cause recurring infections (Bers and Wahl 2004). Likewise, in certain types of flora, the buildup of debris on a plant surface can impede photosynthesis. Two interesting solutions have been employed by natural systems to address each of these issues. The skin of sharks shows a remarkable antifouling characteristic due in part to the hierarchical topography shown in Fig. 8.7a. In addition to the continuous motion of a shark that works to reduce the contact time of fouling organisms with the skin, the hierarchical texture of the sharkskin not only seems to help for faster swimming (Oeffner and Lauder 2012; Wen et al. 2014) but also exhibits an inconsistent surface area for adhering organisms (Schumacher et al. 2007a, b, 2008). This microscale texturing of the sharkskin realigns and flexes over time constantly changing the characteristics of the targeted surface for the incoming microbe. Meanwhile back on land, the lotus leaf and the rose petal show very similar design concepts to remove both debris from their surfaces that would otherwise work against their photosynthetic processes and pathogens that lead to infection (Barthlott and Neinhuis 1997). In both cases, these plant surfaces show a clear hierarchical topographic design that creates very small pinning points that work to keep a droplet of water beaded up as shown in Fig. 8.7b. Complete wetting of the topography would require the creation of significant surface area that is energetically prohibited (Patankar 2004). Thus, these hierarchical topographies

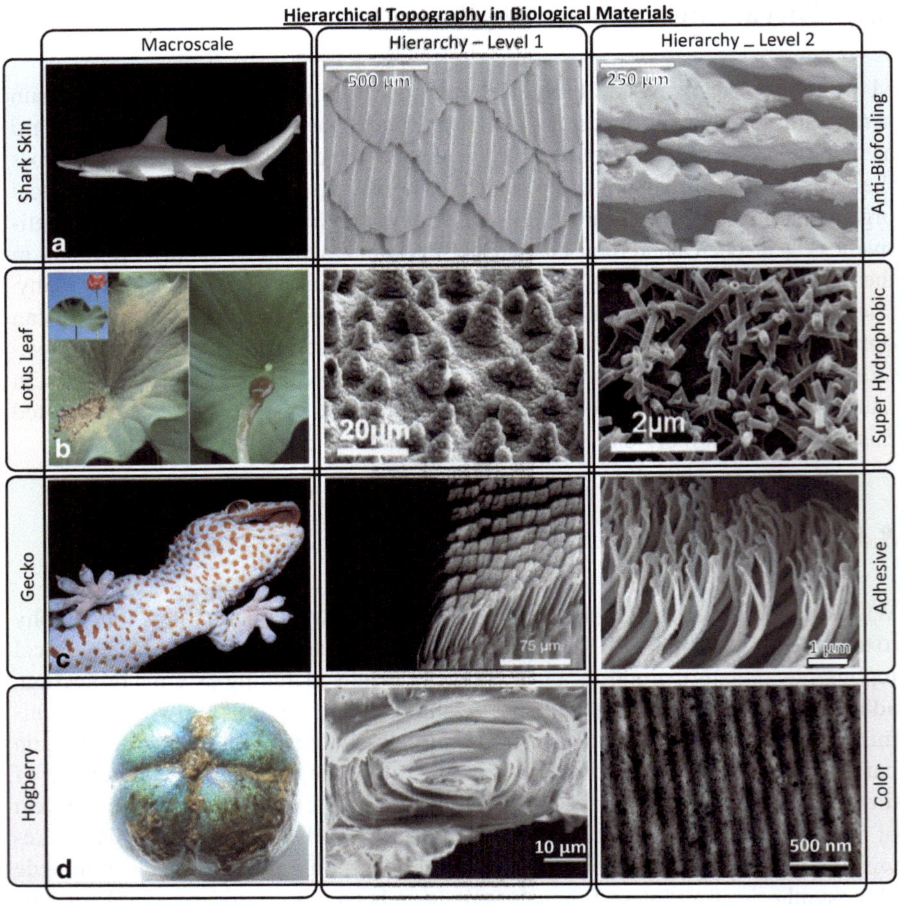

Fig. 8.7 Biological organisms that employ hierarchical topography to seemingly enhance material performances from **a** enhancing antibiofouling in the sharkskin, **b** to being superhydrophobic as in the lotus leaf, **c** to enabling extreme adhesion as in the gecko toepad, **d** to giving off beautiful coloring as in the seed coat of the hogberry. (Image **a** adapted from Wen et al. (2014) with permission of Journal of Experimental Biology and from Callow and Callow (2011) by permission from Macmillan Publishers Ltd, copyright, 2011. Image **b** adapted from Koch et al. (2009) with permission from Elsevier. Image **c** adapted from Autumn and Hansen (2006) and Autumn et al. (2006) with kind permission from Springer Business and Science and Journal of Experimental Biology. Image **d** adapted from Kolle et al. (2013) with permission from Wiley)

showcase superhydrophobic properties. Beaded water droplets on these surfaces are known to roll around and gather debris, effectively cleaning the surfaces to maximize photosynthetic processes (Barthlott and Neinhuis 1997).

Interestingly, hierarchical topography can also work to maximize interactions with foreign bodies. In many systems, the hierarchical nature of the topography leads to an accumulation of attractive interactions such as in the setae-covered feet of climbing *animalia*. For example, beetles, flies, spiders, and geckos all use hierarchical topography in their footpads to increase their adhesion to arbitrary surfaces (Arzt et al. 2013). Figure 8.7c highlights how the footpad of a gecko appears smooth at the macroscale but is in fact composed of many "clumps of macro-setae" at the microscale and many "individual setae" at the sub-microscale (Autumn et al. 2000, 2006). Splitting the number of contact points into so many individual setae eventually increases the total linear length of adhesive points, which in turn enhances the exerted molecular adhesion force, as expected from contact mechanics theory (Arzt et al. 2013). Moreover, the individual setae have also some flexibility that works to maximize their contact points with a foreign rough surface (Hiller 1968; Autumn et al. 2000). Interactions at this scale are dominated by an attractive van der Waals force, which effectively provides sufficient adhesion for the gecko to hold its body weight on a glass wall (Autumn et al. 2002). Without the hierarchical topography, the gecko could not establish as many contact points and the cumulative van der Waals attraction would remain very insignificant similar to when a human touches a glass wall. Such a design strategy goes beyond geckos to beetles, flies, and spiders. In fact, it was found that the setae areal density increases with increasing animal weight (more contact points are required to handle the larger payload) (Arzt et al. 2003). The design principles found in such species could lead us toward the next level of material independent, nonfouling adhesive.

Finally, hierarchical topography is also used throughout nature to create color (Sun et al. 2013a). Color serves a relatively clear process in the biological world: influencing natural selection. A bird tends toward brighter seeds (Cazetta et al. 2009; Vignolini et al. 2012), and brighter butterflies and birds tend to successfully mate (Young 1971; McGraw et al. 2001; Henderson et al. 2013). Hierarchical topography produces some of the most vivid coloring known in nature. The species that employ this "structural" coloring include the abalone, the golden stag beetle, the tachinid fly, the blue morpho butterfly, and many birds (e.g., peacock (Yoshioka and Kinoshita 2002), blue jay, mountain bluebird (Gower 1936), Dodge bunting, hummingbird). In these listed species, the colors that originate from their surfaces are derived from the reflection and interference of incoming sunlight from the textured surfaces. Some of these structures including the abalone and the butterfly are iridescent which means that their colors appear to change with different viewing angles (Sun et al. 2013a). Interestingly, even some plants are iridescent due to their structural topography interfering with incoming light. For example, the bastard hogberry plant of South America has a complex surface structure that produces the bright bluish color shown in Fig. 8.7d (Kolle et al. 2013). Similar structural coloration is found also in the Pollia fruit (Vignolini et al. 2012). Though vivid colors in species of flora and fauna are often seen as attractants or deterrents for other natural organisms, the use of hierarchical topography to produce colorful synthetic materials has very clear benefits including enhanced aesthetics, lower susceptibility to chemical bleaching, and tunability by mechanical deformation.

8.4.2 *Synthetic Materials with Hierarchical Topography: Antifouling*

Due to the outstanding antifouling properties of the sharkskin, lotus leaves, rose petals, and other natural species, significant research effort has been focused on recreating these types of hierarchical topographies to minimize surface fouling (Emily and Geoffrey 2009). Specifically, biofouling is responsible for significant contamination of many of our systems including everything from plumbing infrastructure and marine crafts that result in wasted energy all the way to medical equipment and implants that result in infection and even death (Callow and Callow 2011; Epstein et al. 2013). One of the leading re-creations of the sharkskin topography is demonstrated by the Sharklet™ pattern developed by Brennan and coworkers (Schumacher et al. 2007a, b). These researchers investigated the biofouling properties of different topographies patterned onto a surface with photolithographic techniques. The patterns consisted of repetitive microstructures approximately 25 μm in size made up of sub-elements of roughly 2 μm in size shown in Fig. 8.8a. This investigation found that creating a hierarchical topography with the same sizing as the sharkskin produced significant antifouling properties over flat substrates or even with differently sized topography. The structure that most closely mimicked the sharkskin exhibited topographical features having characteristic dimensions smaller than the settling or exploring microorganisms that initiate biofouling (Schumacher et al. 2008). This bioinspired technology has since been integrated into the aerospace industry and works to reduce biofouling as well as air drag on the surface of certain airplane components. A second example of a hierarchical topographic system is found in wrinkled thin films as studied by Genzer and coworkers (Efimenko et al. 2005, 2009). In these systems, flexible substrates of poly(dimethylsiloxane) (PDMS) are stretched, treated with ultraviolet and ozone, and released to produce the hierarchical wrinkling across five length scales shown in Fig. 8.8b. These substrates were subjected to biofouling environments in open seawater. Flat PDMS surfaces experienced extensive biofouling after 1 month while hierarchically wrinkled PDMS surfaces remained relatively nonfouled (Efimenko et al. 2009). Additionally, Aizenberg and coworkers have looked at combining topography with a dynamic surface to better imitate the sharkskin (Epstein et al. 2013).

Even compared to the attempts to mimic sharkskin, there has been an enormous effort to recreate the superhydrophobic properties of the lotus leaf and rose petal in synthetic materials. Generating superhydrophobic surfaces will prevent the buildup and spreading of aqueous solutions containing pathogens and bacteria on surfaces. These liquids will simply remain beaded up and will roll off of the surface along with the unwanted contaminants (Barthlott and Neinhuis 1997). There is a tremendous market expected for lotus-inspired technologies and to date, a few have been developed including StoLotusan Lotus-Effect®: a topographic paint coating that works to remove residue from buildings during rainstorms. Initially, the lotus effect was recreated in synthetic systems without incorporating hierarchical topography. Preliminary systems were constructed with photolithography on silicon substrates and produced contact angles of about 155° (Yoshimitsu et al. 2002). However,

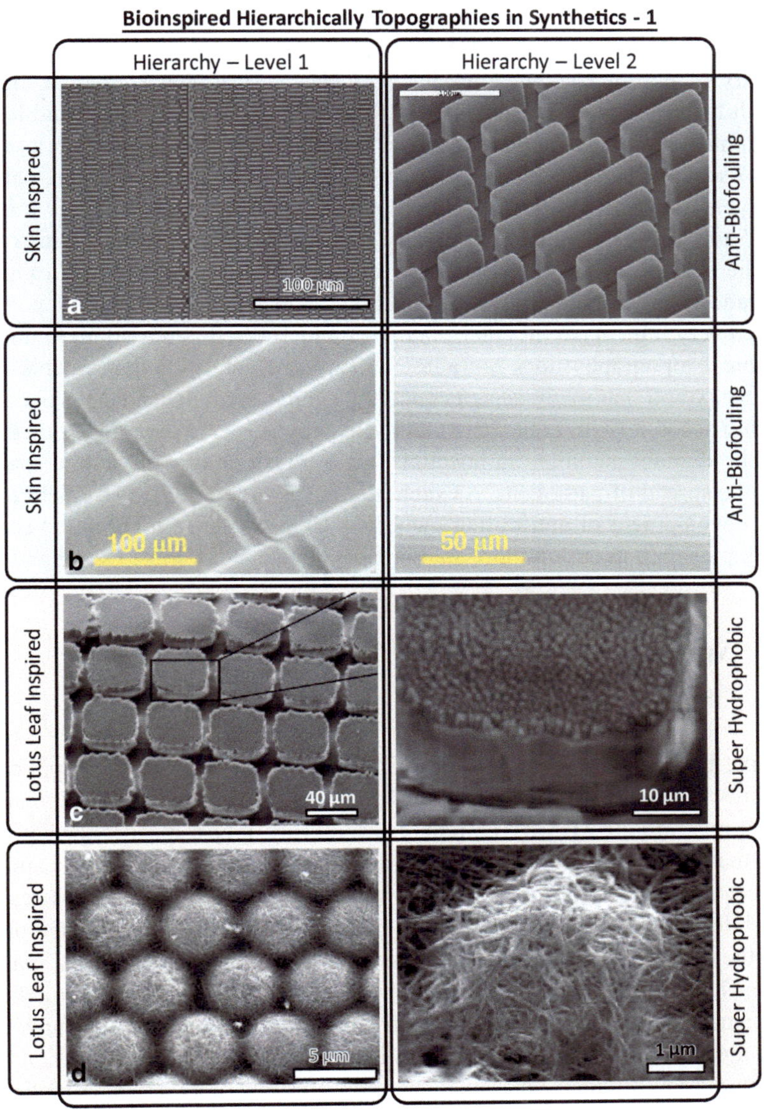

Fig. 8.8 Synthetic materials of hierarchical topography that recreate **a**, **b** the antibiofouling properties of certain skins and **c**, **d** the superhydrophobicity of the lotus leaf. (Image **a** adapted from Schumacher et al. (2007a, b) with permission from Taylor and Francis Ltd. and courtesy of Anthony B. Brennen, University of Florida, USA. Image **b** adapted from Efimenko et al. (2005) by permission from Macmillan Publishers Ltd. Copyright, 2005. Image **c** adapted with permission from Cortese et al. (2008). Copyright, 2008, American Chemical Society. Image **d** adapted with permission from Li et al. (2006). Copyright, 2006, American Chemical Society)

Patanker modeled the double hierarchy nature of the lotus leaf and found that it was responsible for the ~170° apparent contact angle of the natural lotus leaf (Patankar 2004). Soon after, researchers started incorporating this second level of hierarchy in the surface roughness to produce superhydrophobicity. As shown in Fig. 8.8c, Cortese and coworkers patterned a PDMS surface with pillars of roughly 40 μm with a secondary submicron roughness produced by plasma treatment to produce topography with feature sizes of a few 100 nm (Cortese et al. 2008). This second level of topographical hierarchy produced surface contact angles up to 170° (Cortese et al. 2008). A second example of this was envisioned by Li et al. (2006) using colloidal monolayers and single-wall carbon nanotubes. The colloidal monolayer made up of 1.3–5 μm polystyrene particles created a uniform topography at the microscale while the subsequent coating of carbon nanotubes created submicron roughness at the nanoscale. Combined with surface treatments, such hierarchical topography produced very high apparent contact angles up to 160°. It is important to note that hierarchical topography does not automatically translate to enhancements in hydrophobicity and that many design constraints and system dependences need to be taken into account (Kim et al. 2013). Self-cleaning surfaces have huge potential in the public domain including in hospitals, mass transit, and many public spaces to reduce the spread of pathogens. As such, there is a huge effort in this field toward further development of antifouling materials. Here, we have highlighted some of the examples that, in our opinion, are the most developed or most exciting.

8.4.3 *Synthetic Materials with Hierarchical Topography: Adhesion*

The drive to make a new generation of adhesive materials includes applications in mobile robotics, automatic manipulation technology, and medical bandages (Mengüç and Sitti 2013). Most current adhesives rely upon a specific surface chemistry to maximize the chemical interaction between two surfaces, but this approach is subject to damage and dirt, exhibits material specificity, and leaves residues. Meanwhile, geckos, spiders, beetles, and flies can adhere to just about any material due to the hierarchical topography of the setae on their footpads. Thus, many researchers have changed their tune from chemistry to geometry for creating the next generation of adhesives (Mengüç and Sitti 2013). For example, Northen and Turner created hierarchical substrates shown in Fig. 8.9a using photolithography and etching techniques to produce a massively parallel substrate of complex, compliant pillars on the microscale with etched topography on the submicron scale (Michael and Kimberly 2005; Northen and Turner 2006). These substrates showed adhesive properties similar to, but weaker than, geckos. In another embodiment, Dhinojwala and coworkers grew carbon nanotube arrays in a hierarchical fashion by seeding catalysts onto a silicon substrate pre-patterned using photolithographic techniques (Ge et al. 2007). To create flexibility, the structure was then transferred over to a flexible tape substrate as shown in Fig. 8.9b. The resulting structured surfaces showed adhesive properties that were four fold stronger than the normalized adhesion of the gecko's footpad. A similar approach by Wang and coworkers yielded a synthetic adhesive based on carbon nanotubes that exhibited a strong binding strength under

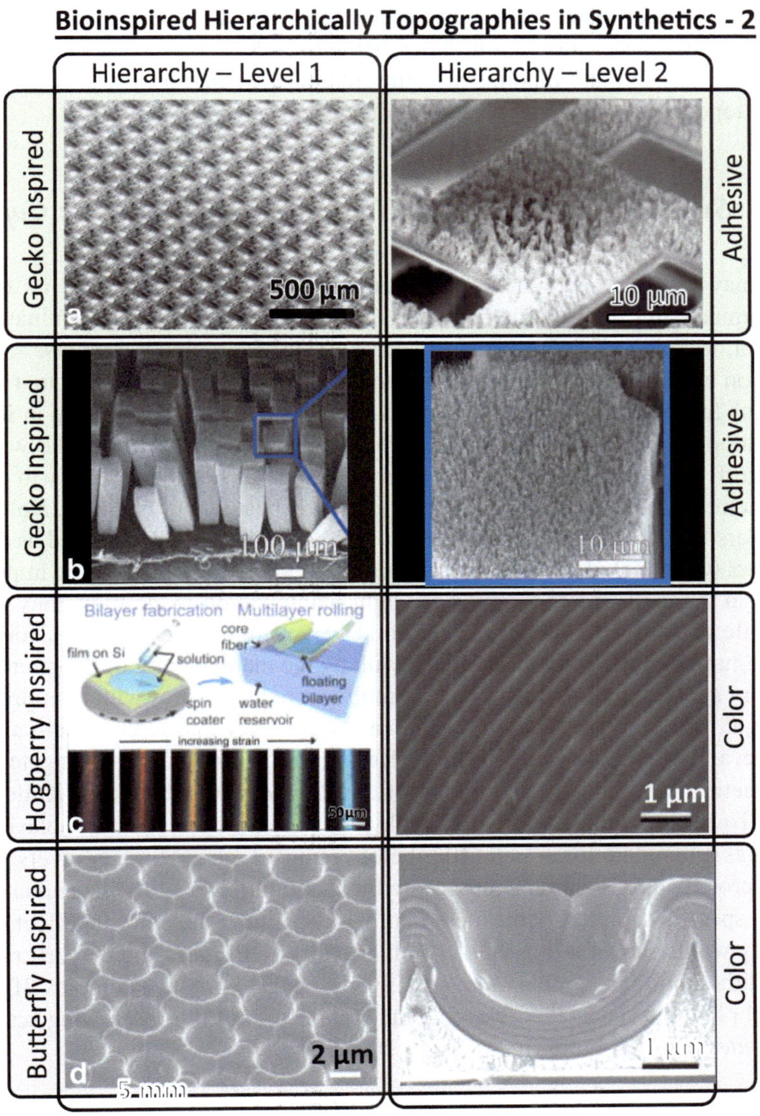

Fig. 8.9 Synthetic materials of hierarchical topography that recreate **a**, **b** the adhesive properties of the gecko footpad and **c**, **d** the color properties of the hogberry and butterfly. (Image **a** adapted from Northen and Turner (2006) with permission from Elsevier. Image **b** adapted from Ge et al. (2007) with permission from National Academy of Sciences. Image **c** adapted from Kolle et al. (2013) with permission from Wiley. Image **d** adapted from Kolle et al. (2010) by permission from Macmillan Publishers Ltd. Copyright, 2010)

shear and a weak binding strength under normal lift. This detail is what allows geckos to remove their footpad from a substrate while they run up walls (Qu et al. 2008). Overall, the added surface area characteristic of hierarchical topography can be a strong enabler of adhesion if the surface forces are attractive. Similar adhesive properties have been observed in other fields including osteointegration where it is believed that implant binding strength dramatically improves with hierarchical surface topography (Zhang et al. 2013; Wang and Carrier 2011).

8.4.4 Synthetic Materials with Hierarchical Topography: Color

Color may influence evolutionary processes in the natural world, but its effects also dominate our daily synthetic materials. Color enables faster dissemination of visual information allowing certain features to stand out above the noise. Such a definition can be applied to everything from graphical images of charts to a lady's earrings. To get to the brightest most vivid colors, nature has relied upon hierarchical topography such as in the seed coat of the hogberry plant or the wing of the butterfly. Recently, these structures have been generated synthetically to obtain manufactured materials with comparable photonic beauty. For example, Kolle and coworkers fabricated polymer bilayers in both 2D and in 3D that served as Bragg reflectors (Guldin et al. 2011; Kamita et al. 2012; Kolle et al. 2013). The important feature in these structures required for light manipulation is the monotony of the nanoscale structure that is on the order of the wavelength of visible light. However, the fine-tuning of the color can be dealt with by the manipulation of the microscale geometry such as in the actuation of the multilayered structures with pressure (Kamita et al. 2012) or the winding of the nanoscale layer into microstructural coils (Kolle et al. 2013) as shown in Fig. 8.9c. Kolle and coworkers also investigated how to synthetically mimic the bright color of the *Papilio blumei* butterfly (Kolle et al. 2010). To form the microscale topography, they create a colloidal monolayer that is used as a template for a metal film before removal, generating an inverse opal. The microstructure is subsequently coated with a multilayer of titania and alumina of very specific submicron thickness that work to mimic the coloring effect of the butterfly wings shown in Fig. 8.9d. Another effort by Burgess and coworkers used planar lithography in combination with inverse opals to produce complex 3D hierarchical photonic stacks that resemble the architecture of the green wing scales of the *Parides sesostris* butterfly (Ian et al. 2013).

8.5 Conclusions and Outlook

The use of hierarchy provides an additional degree of freedom in the vast design space available for the fabrication of biological and man-made composites. Biological systems exhibiting hierarchical reinforcement, porosity, or surface topography clearly illustrate the potential of this design tool as a means to boost the functional properties of composite materials.

Biological composites that exploit hierarchical reinforcement are inspiring novel synthetic composites exhibiting enhanced mechanical performance. As exemplary biological architectures that utilize reinforcing elements at multiple length scales, bone and nacre display fracture toughness levels that are up to two orders of magnitude higher than that expected for "unstructured" synthetic materials of same chemical composition. The implementation of hierarchically structured reinforcement in synthetic systems with chemically stronger building blocks has enabled the fabrication of nacre-inspired artificial composites that are at least two fold tougher than the biological counterpart. Inspired by the graded architecture of tendons and bone tissue, hierarchical reinforcement has also been exploited to produce polymer-based composites with local elastic modulus that can be tuned over an unprecedented four orders of magnitude.

Similarly, building porosity at multiple length scales is a common strategy in nature to obtain composites with outstanding or unusual sets of properties that are highly desirable in a wide range of synthetic systems. The unusual properties of hierarchical porous materials in biology include, for example, the fast fluid transport and high surface area of marine sponges, the high strength-to-weight ratio of bones, and the increased impact absorption properties of the protective peel of certain fruits. Incorporating similar hierarchical designs in synthetic systems has resulted in efficient porous structures for catalysis and tissue regeneration that combine the enhanced permeability associated with large pores with the high surface area arising from finer porosity.

Finally, hierarchical surface topography has also evolved in natural systems providing them with functionalities that researchers are now trying to recreate in man-made materials. The feature of hierarchical surface topography has been widely implemented in nature, generating insect and animal footpads with outstanding adhesive properties, animal skin, and plant surfaces with nonfouling behavior, as well as plant seeds and butterfly wings with robust coloring effects. Using mostly lithographic techniques developed for microelectronics, hierarchical surface topographies have also been synthetically recreated to obtain unprecedented artificial materials exhibiting adhesive, nonfouling, and coloring properties that are determined primarily by structural features rather than chemical composition.

The examples discussed above illustrate the diversity in structure and function of biological materials and the encouraging results of ongoing attempts to harness the potential of hierarchical structures in synthetic systems. The increasing availability of analytical tools to characterize the structure of biological materials at several length scales will likely help unveil many other hierarchical architectures beyond the examples currently understood. Combined experimental and computational work on such hierarchical composites should allow for a deeper understanding of structure–property relationships in these unique materials' systems. Ultimately, progress in the understanding of biomineralization processes and the development of novel self- and directed-assembly routes is expected to provide the synthetic tools required for the fabrication of synthetic hierarchical composites that will outperform today's best synthetic materials.

References

Arzt E, Gorb S, Spolenak R (2003) From micro to nano contacts in biological attachment devices. Proc Natl Acad Sci U S A 100(19):10603–10606. doi:10.1073/pnas.1534701100

Autumn K, Hansen W (2006) Ultrahydrophobicity indicates a non-adhesive default state in gecko setae. J Comp Physiol A Neuroethol Sens Neural Behav Physiol 192(11):1205–1212. doi:10.1007/s00359-006-0149-y

Autumn K, Liang YA, Hsieh ST, Zesch W, Chan WP, Kenny TW, Fearing R, Full RJ (2000) Adhesive force of a single gecko foot-hair. Nature 405(6787):681–685

Autumn K, Sitti M, Liang YA, Peattie AM, Hansen WR, Sponberg S, Kenny TW, Fearing R, Israelachvili JN, Full RJ (2002) Evidence for van der Waals adhesion in gecko setae. Proc Natl Acad Sci U S A 99(19):12252–12256. doi:10.1073/pnas.192252799

Autumn K, Dittmore A, Santos D, Spenko M, Cutkosky M (2006) Frictional adhesion: a new angle on gecko attachment. J Exp Biol 209(18):3569–3579. doi:10.1242/jeb.02486

Barthlott W, Neinhuis C (1997) Purity of the sacred lotus, or escape from contamination in biological surfaces. Planta 202(1):1–8

Barthelat F, Espinosa H (2007) An experimental investigation of deformation and fracture of Nacre–Mother of Pearl. Exp Mech 47(3):311–324. doi:10.1007/s11340-007-9040-1

Barthelat F, Li CM, Comi C, Espinosa HD (2006) Mechanical properties of nacre constituents and their impact on mechanical performance. J Mater Res 21(8):1977–1986. doi:10.1557/jmr.2006.0239

Barthelat F, Tang H, Zavattieri PD, Li CM, Espinosa HD (2007) On the mechanics of mother-of-pearl: a key feature in the material hierarchical structure. J Mech Phys Solids 55(2):306–337. doi:10.1016/j.jmps.2006.07.007

Bers AV, Wahl M (2004) The influence of natural surface microtopographies on fouling. Biofouling 20(1):43–51. doi:10.1080/08927010410001655533

Billiet T, Vandenhaute M, Schelfhout J, Vlierberghe VS, Dubruel P (2012) A review of trends and limitations in hydrogel-rapid prototyping for tissue engineering. Biomaterials 33(26):6020–6041

Bonderer LJ, Studart AR, Gauckler LJ (2008) Bioinspired design and assembly of platelet reinforced polymer films. Science 319(5866):1069–1073. doi:10.1126/science.1148726

Bonderer LJ, Feldman K, Gauckler LJ (2010) Platelet-reinforced polymer matrix composites by combined gel-casting and hot-pressing. Part II: thermoplastic polyurethane matrix composites. Compos Sci Technol 70(13):1966–1972. doi:10.1016/j.compscitech.2010.07.016

Bouville F, Maire E, Meille S, de Moortele VB, Stevenson AJ, Deville S (2014) Strong, tough and stiff bioinspired ceramics from brittle constituents. Nat Mater 13(5):508–514. doi:10.1038/nmat3915

Buehler MJ (2010) Computational and theoretical materiomics: properties of biological and de novo bioinspired materials. J Comput Theor Nanosci 7(7):1203–1209. doi:10.1166/jctn.2010.1474

Callow JA, Callow ME (2011) Trends in the development of environmentally friendly fouling-resistant marine coatings. Nat Commun 2:244

Cazetta E, Schaefer H, Galetti M (2009) Why are fruits colorful? The relative importance of achromatic and chromatic contrasts for detection by birds. Evol Ecol 23(2):233–244. doi:10.1007/s10682-007-9217–1

Cesarano J, Dellinger JG, Saavedra MP, Gill DD, Jamison RD, Grosser BA, Sinn-Hanlon JM, Goldwasser MS (2005) Customization of load-bearing hydroxyapatite lattice scaffolds. Int J Appl Ceram Technol 2(3):212–220. doi:10.1111/j.1744-7402.2005.02026.x

Chen HY, Di JC, Wang N, Dong H, Wu J, Zhao Y, Yu JH, Jiang L (2011) Fabrication of hierarchically porous inorganic nanofibers by a general microemulsion electrospinning approach. Small 7(13):1779–1783. doi:10.1002/smll.201002376

Cortese B, D'Amone S, Manca M, Viola I, Cingolani R, Gigli G (2008) Superhydrophobicity due to the hierarchical scale roughness of PDMS surfaces. Langmuir 24(6):2712–2718. doi:10.1021/la702764x

Efimenko K, Rackaitis M, Manias E, Vaziri A, Mahadevan L, Genzer J (2005) Nested self-similar wrinkling patterns in skins. Nat Mater 4(4):293–297

Efimenko K, Finlay J, Callow ME, Callow JA, Genzer J (2009) Development and testing of hierarchically wrinkled coatings for marine antifouling. ACS Appl Mater Interfaces 1(5):1031–1040. doi:10.1021/am9000562

Emily R, Geoffrey S (2009) Bioinspiration—The solution for biofouling control? Bioinspir Biomim 4(1):015007

Epstein AK, Hong D, Kim P, Aizenberg J (2013) Biofilm attachment reduction on bioinspired, dynamic, micro-wrinkling surfaces. New J Phys 15(9):095018

Espinosa HD, Rim JE, Barthelat F, Buehler MJ (2009) Merger of structure and material in nacre and bone—perspectives on de novo biomimetic materials. Prog Mater Sci 54(8):1059–1100

Fischer SF, Thielen M, Weiss P, Seidel R, Speck T, Buhrig-Polaczek A, Bunck M (2014) Production and properties of a precision-cast bio-inspired composite. J Mater Sci 49(1):43–51. doi:10.1007/s10853-013-7878-4

Fratzl P (2008a) Bone fracture: when the cracks begin to show. Nat Mater 7(8):610–612

Fratzl P (2008b) Collagen: structure and mechanics, an introduction. In: Fratzl P (ed) Collagen: structure and mechanics. Springer, New York

Fratzl P, Weinkamer R (2007) Nature's hierarchical materials. Prog Mater Sci 52(8):1263–1334. doi:10.1016/j.pmatsci.2007.06.001

Fratzl P, Gupta HS, Paschalis EP, Roschger P (2004) Structure and mechanical quality of the collagen-mineral nano-composite in bone. J Mater Chem 14(14):2115–2123. doi:10.1039/B402005G

Garcia AP, Sen D, Buehler MJ (2011) Hierarchical silica nanostructures inspired by diatom algae yield superior deformability, toughness, and strength. Metall Mater Trans A 42A(13):3889–3897. doi:10.1007/s11661-010-0477-y

Ge L, Sethi S, Ci L, Ajayan PM, Dhinojwala A (2007) Carbon nanotube-based synthetic gecko tapes. Proc Natl Acad Sci U S A 104(26):10792–10795. doi:10.1073/pnas.0703505104

Genin GM, Kent A, Birman V, Wopenka B, Pasteris JD, Marquez PJ, Thomopoulos S (2009) Functional grading of mineral and collagen in the attachment of tendon to bone. Biophys J 97(4):976–985. doi:10.1016/j.bpj.2009.05.043

Gower C (1936) The cause of blue color as found in the Bluebird (*Sialia sialis*) and the blue Jay (*Cyanocitta cristata*). Auk 53(2):178–185. doi:10.2307/4077277

Guldin S, Kolle M, Stefik M, Langford R, Eder D, Wiesner U, Steiner U (2011) Tunable mesoporous bragg reflectors based on block-copolymer self-assembly. Adv Mater 23(32):3664–3668. doi:10.1002/adma.201100640

Gupta HS, Seto J, Wagermaier W, Zaslansky P, Boesecke P, Fratzl P (2006) Cooperative deformation of mineral and collagen in bone at the nanoscale. Proc Natl Acad Sci U S A 103(47):17741–17746

Harrington MJ, Masic A, Holten-Andersen N, Waite JH, Fratzl P (2010) Iron-clad fibers: a metal-based biological strategy for hard flexible coatings. Science 328(5975):216–220. doi:10.1126/science.1181044

Henderson LJ, Heidinger BJ, Evans NP, Arnold KE (2013) Ultraviolet crown coloration in female blue tits predicts reproductive success and baseline corticosterone. Behav Ecol doi:10.1093/beheco/art066

Hiller U (1968) Untersuchungen zum Feinbau und zur Funktion der Haftborsten von Reptilien. Z Morph Tiere 62(4):307–362. doi:10.1007/bf00401561

Ian BB, Joanna A, Marko L (2013) Creating bio-inspired hierarchical 3D–2D photonic stacks via planar lithography on self-assembled inverse opals. Bioinspir Biomim 8(4):045004

Jackson AP, Vincent JFV, Turner RM (1988) The mechanical design of nacre. Proc R Soc London Ser B 234(1277):415–440

Kamita G, Kolle M, Huang F, Baumberg JJ, Steiner U (2012) Multilayer mirrored bubbles with spatially-chirped and elastically-tuneable optical bandgaps. Opt Express 20(6):6421–6428. doi:10.1364/oe.20.006421

Kaushik AK, Podsiadlo P, Qin M, Shaw CM, Waas AM, Kotov NA, Arruda EM (2009) The role of nanoparticle layer separation in the finite deformation response of layered polyurethane-clay nanocomposites. Macromolecules 42(17):6588–6595. doi:10.1021/ma901048g

Kim P, Kreder MJ, Alvarenga J, Aizenberg J (2013) Hierarchical or not? Effect of the length scale and hierarchy of the surface roughness on omniphobicity of lubricant-infused substrates. Nano Lett 13(4):1793–1799. doi:10.1021/nl4003969

Koch K, Bhushan B, Barthlott W (2009) Multifunctional surface structures of plants: an inspiration for biomimetics. Prog Mater Sci 54(2):137–178

Kolle M, Salgard-Cunha PM, Scherer MRJ, Huang F, Vukusic P, Mahajan S, Baumberg JJ, Steiner U (2010) Mimicking the colourful wing scale structure of the *Papilio blumei* butterfly. Nat Nano 5(7):511–515

Kolle M, Lethbridge A, Kreysing M, Baumberg JJ, Aizenberg J, Vukusic P (2013) Bio-inspired band-gap tunable elastic optical multilayer fibers. Adv Mater 25(15):2239–2245. doi:10.1002/adma.201203529

Launey ME, Buehler MJ, Ritchie RO (2010) On the mechanistic origins of toughness in bone. In: Clarke DR, Ruhle M, Zok F (eds) Annual review of materials research, vol 40, pp 25–53. doi:10.1146/annurev-matsci-070909-104427

Leys SP, Yahel G, Reidenbach MA, Tunnicliffe V, Shavit U, Reiswig HM (2011) The sponge pump: the role of current induced flow in the design of the sponge body plan. PLoS ONE 6(12). doi:e2778710.1371/journal.pone.0027787

Li Y, Huang XJ, Heo SH, Li CC, Choi YK, Cai WP, Cho SO (2006) Superhydrophobic bionic surfaces with hierarchical microsphere/SWCNT composite arrays. Langmuir 23(4):2169–2174. doi:10.1021/la0620758

Libanori R, Erb RM, Reiser A, Le Ferrand H, Suess MJ, Spolenak R, Studart AR (2012a) Stretchable heterogeneous composites with extreme mechanical gradients. Nat Commun 3:1265. doi:10.1038/ncomms2281

Libanori R, Munch FHL, Montenegro DM, Studart AR (2012b) Hierarchical reinforcement of polyurethane-based composites with inorganic micro- and nanoplatelets. Compos Sci Technol 72(3):435–445. doi:10.1016/j.compscitech.2011.12.005

Maganaris CN, Paul JP (1999) In vivo human tendon mechanical properties. J Physiol 521(1):307–313. doi:10.1111/j.1469-7793.1999.00307.x

McGraw KJ, Stoehr AM, Nolan PM, Hill GE (2001) Plumage redness predicts breeding onset and reproductive success in the House Finch: a validation of Darwin's theory. J Avian Biol 32(1):90–94. doi:10.1034/j.1600-048X.2001.320114.x

Mengüç Y, Sitti M (2013) Gecko-inspired polymer adhesives. In: Polymer adhesion, friction, and lubrication. Wiley, pp 351–389. doi:10.1002/9781118505175.ch9

Menig R, Meyers MH, Meyers MA, Vecchio KS (2000) Quasi-static and dynamic mechanical response of *Haliotis rufescens* (abalone) shells. Acta Materialia 48(9):2383–2398

Michael TN, Kimberly LT (2005) A batch fabricated biomimetic dry adhesive. Nanotechnology 16(8):1159

Mortensen A, Suresh S (1995) Functionally graded metals and metal-ceramic composites.1. Processing. Int Mater Rev 40(6):239–265

Müller R, Rüegsegger P (1997) Micro-tomographic imaging for the nondestructive evaluation of trabecular bone architecture. In: Lowet G, Rüegsegger P, Weinans H, Meunier A (eds) Bone research in biomechanics, vol 40 (Studies in health technology and informatics). IOS Press, Amsterdam, pp 61–79

Munch E, Launey ME, Alsem DH, Saiz E, Tomsia AP, Ritchie RO (2008) Tough, bio-inspired hybrid materials. Science 322(5907):1516–1520. doi:10.1126/science.1164865

Munch E, Saiz E, Tomsia AP, Deville S (2009) Architectural control of freeze-cast ceramics through additives and templating. J Am Ceram Soc 92(7):1534–1539. doi:10.1111/j.1551-2916.2009.03087.x

Northen MT, Turner KL (2006) Meso-scale adhesion testing of integrated micro- and nano-scale structures. Sens Actuators A 130–131(0):583–587

Nova A, Keten S, Pugno NM, Redaelli A, Buehler MJ (2010) Molecular and nanostructural mechanisms of deformation, strength and toughness of spider silk fibrils. Nano Lett 10(7):2626–2634. doi:10.1021/nl101341w

Oeffner J, Lauder GV (2012) The hydrodynamic function of shark skin and two biomimetic applications. J Exp Biol 215(5):785–795. doi:10.1242/jeb.063040

Patankar NA (2004) Mimicking the lotus effect: influence of double roughness structures and slender pillars. Langmuir 20(19):8209–8213. doi:10.1021/la048629t

Prielipp H, Knechtel M, Claussen N, Streiffer SK, Müllejans H, Rühle M, Rödel J (1995) Strength and fracture toughness of aluminum/alumina composites with interpenetrating networks. Mater Sci Eng A 197(1):19–30

Qu L, Dai L, Stone M, Xia Z, Wang ZL (2008) Carbon nanotube arrays with strong shear binding-on and easy normal lifting-off. Science 322(5899):238–242. doi:10.1126/science.1159503

Ritchie RO (2011) The conflicts between strength and toughness. Nat Mater 10(11):817–822

Schumacher JF, Aldred N, Callow ME, Finlay JA, Callow JA, Clare AS, Brennan AB (2007a) Species-specific engineered antifouling topographies: correlations between the settlement of algal zoospores and barnacle cyprids. Biofouling 23(5):307–317. doi:10.1080/08927010701393276

Schumacher JF, Carman ML, Estes TG, Feinberg AW, Wilson LH, Callow ME, Callow JA, Finlay JA, Brennan AB (2007b) Engineered antifouling microtopographies—effect of feature size, geometry, and roughness on settlement of zoospores of the green alga Ulva. Biofouling 23(1):55–62. doi:10.1080/08927010601136957

Schumacher JF, Long CJ, Callow ME, Finlay JA, Callow JA, Brennan AB (2008) Engineered nanoforce gradients for inhibition of settlement (attachment) of swimming algal spores. Langmuir 24(9):4931–4937. doi:10.1021/la703421v

Sen D, Buehler MJ (2010) Atomistically-informed mesoscale model of deformation and failure of bioinspired hierarchical silica nanocomposites. Int J Appl Mech 2(4):699–717. doi:10.1142/s175882511000072x

Simon JL, Michna S, Lewis JA, Rekow ED, Thompson VP, Smay JE, Yampolsky A, Parsons JR, Ricci JL (2007) *In vivo* bone response to 3D periodic hydroxyapatite scaffolds assembled by direct ink writing. J Biomed Mater Res A 83A(3):747–758. doi:10.1002/jbm.a.31329

Smith BL, Schaffer TE, Viani M, Thompson JB, Frederick NA, Kindt J, Belcher A, Stucky GD, Morse DE, Hansma PK (1999) Molecular mechanistic origin of the toughness of natural adhesives, fibres and composites. Nature 399(6738):761–763

Studart AR (2012) Towards high-performance bioinspired composites. Adv Mater 24(37):5024–5044. doi:10.1002/adma.201201471

Studart AR (2013) Biological and bioinspired composites with spatially tunable heterogeneous architectures. Adv Funct Mater doi:10.1002/adfm.201300340

Studart AR, Gonzenbach UT, Tervoort E, Gauckler LJ (2006) Processing routes to macroporous ceramics: a review. J Am Ceram Soc 89(6):1771–1789

Studart AR, Studer J, Xu L, Yoon K, Shum HC, Weitz DA (2011) Hierarchical porous materials made by drying complex suspensions. Langmuir 27(3):955–964. doi:10.1021/la103995 g

Studart AR, Libanori R, Erb RM (2013a) Chap. 15 Replicating biological design principles in synthetic composites. In: Materials design inspired by nature: function through inner architecture. The Royal Society of Chemistry, pp 322–358. doi:10.1039/9781849737555-00322

Studart AR, Libanori R, Erb RM (2013b) Replicating biological design principles in synthetic composites. In: Fratzl P, Dunlop J, Weinkamer R (eds) Materials design inspired by nature: function through inner architecture. RSC Publishing, pp 322–358

Su X, Belcher AM, Zaremba CM, Morse DE, Stucky GD, Heuer AH (2002) Structural and microstructural characterization of the growth lines and prismatic microarchitecture in red abalone shell and the microstructures of abalone "flat pearls". Chem Mater 14(7):3106–3117. doi:10.1021/cm011739q

Sumitomo T, Kakisawa H, Owaki Y, Kagawa Y (2008) In situ transmission electron microscopy observation of reversible deformation in nacre organic matrix. J Mater Res 23(05):1466–1471. doi:10.1557/JMR.2008.0184

Sun J, Bhushan B, Tong J (2013a) Structural coloration in nature. RSC Adv 3(35):14862–14889. doi:10.1039/c3ra41096j

Sun K, Wei T-S, Ahn BY, Seo JY, Dillon SJ, Lewis JA (2013b) 3D printing of interdigitated Li-ion microbattery architectures. Adv Mater 25(33):4539–4543. doi:10.1002/adma.201301036

Symes MD, Kitson PJ, Yan J, Richmond CJ, Cooper GJT, Bowman RW, Vilbrandt T, Cronin L (2012) Integrated 3D-printed reactionware for chemical synthesis and analysis. Nat Chem 4(5):349–354

Thielen M, Schmitt CNZ, Eckert S, Speck T, Seidel R (2013a) Structure-function relationship of the foam-like pomelo peel (*Citrus maxima*)-an inspiration for the development of biomimetic damping materials with high energy dissipation. Bioinspir Biomim 8(2). doi:10.1088/1748-3182/8/2/025001

Thielen M, Speck T, Seidel R (2013b) Viscoelasticity and compaction behaviour of the foam-like pomelo (*Citrus maxima*) peel. J Mater Sci 48(9):3469–3478. doi:10.1007/s10853-013-7137–8

Tompkins-MacDonald GJ, Leys SP (2008) Glass sponges arrest pumping in response to sediment: implications for the physiology of the hexactinellid conduction system. Marine Biol 154(6):973–984. doi:10.1007/s00227-008-0987-y

Toohey KS, Sottos NR, Lewis JA, Moore JS, White SR (2007) Self-healing materials with microvascular networks. Nat Mater 6(8):581–585

Vakifahmetoglu C, Pippel E, Woltersdorf J, Colombo P (2010) Growth of one-dimensional nanostructures in porous polymer-derived ceramics by catalyst-assisted pyrolysis. Part I: iron catalyst. J Am Ceram Soc 93(4):959–968. doi:10.1111/j.1551-2916.2009.03448.x

Vignolini S, Rudall PJ, Rowland AV, Reed A, Moyroud E, Faden RB, Baumberg JJ, Glover BJ, Steiner U (2012) Pointillist structural color in Pollia fruit. Proc Natl Acad Sci U S A 109(39):15712–15715. doi:10.1073/pnas.1210105109

Wagermaier W, Gupta HS, Gourrier A, Burghammer M, Roschger P, Fratzl P (2006) Spiral twisting of fiber orientation inside bone lamellae. Biointerphases 1(1):1–5. doi:10.1116/1.2178386

Wang L, Carrier RL (2011) Biomimetic topography: bioinspired cell culture substrates and scaffolds. In: George A (ed) Advances in biomimetics. (InTech)

Wen L, Weaver JC, Lauder GV (2014) Biomimetic shark skin: design, fabrication and hydrodynamic function. J Exp Biol 217(10):1656–1666. doi:10.1242/jeb.097097

Yao HM, Dao M, Imholt T, Huang JM, Wheeler K, Bonilla A, Suresh S, Ortiz C (2010) Protection mechanisms of the iron-plated armor of a deep-sea hydrothermal vent gastropod. Proc Natl Acad Sci U S A 107(3):987–992. doi:10.1073/pnas.0912988107

Yoshimitsu Z, Nakajima A, Watanabe T, Hashimoto K (2002) Effects of surface structure on the hydrophobicity and sliding behavior of water droplets. Langmuir 18(15):5818–5822. doi:10.1021/la020088p

Yoshioka S, Kinoshita S (2002) Effect of macroscopic structure in iridescent color of the peacock feathers. Forma 17(2):169–181

Young A (1971) Wing coloration and reflectance in Morpho butterflies as related to reproductive behavior and escape from avian predators. Oecologia 7(3):209–222. doi:10.1007/bf00345212

Zampieri A, Sieber H, Selvam T, Mabande GTP, Schwieger W, Scheffler F, Scheffler M, Greil P (2005) Biomorphic cellular SiSiC/zeolite ceramic composites: from Rattan palm to bioinspired structured monoliths for catalysis and sorption. Adv Mater 17(3):344∓. doi:10.1002/adma.200400672

Zeschky J, Goetz-Neunhoeffer F, Neubauer J, Lo SHJ, Kummer B, Scheffler M, Greil P (2003) Preceramic polymer derived cellular ceramics. Compos Sci Technol 63(16):2361–2370. doi:10.1016/s0266-3538(03)00269-0

Zhang W, Wang G, Liu Y, Zhao X, Zou D, Zhu C, Jin Y, Huang Q, Sun J, Liu X, Jiang X, Zreiqat H (2013) The synergistic effect of hierarchical micro/nano-topography and bioactive ions for enhanced osseointegration. Biomaterials 34(13):3184–3195

Chapter 9
Hierarchical Composites Containing Carbon Nanotubes

Meisha L. Shofner

Abstract Hierarchical composites containing carbon nanotubes (CNTs) have the potential to possess improved multifunctional properties as well as unique sensing/active capabilities due to the inherent properties of CNTs (i.e., mechanical, electrical, and thermal). The purpose of this chapter is to review the current state of the art in this research area and highlight opportunities for future research. Specifically, three construction schemes used to produce CNT hierarchical composites are reviewed: dispersed systems, fiber coatings, and CNT structures. In these construction schemes, CNTs are used as a performance additive to reduce matrix mobility, as a sizing to improve adhesion between the matrix and the microscale fiber, and as the building blocks of structures such as fibers to more fully exploit the mechanical properties of CNTs, respectively. To date, research results have indicated that these strategies produce composites with improved properties, and most frequently those improved properties are mechanical properties such as strength, modulus, and fracture toughness. Based on these results, further activities aimed at understanding these property increases in terms of modeling as well as more research activity aimed at producing CNT fibers and exploiting other CNT properties will lead to improved approaches to composite design which merit the routine use of CNTs in structural composites.

9.1 Introduction

Given their unique combination of multifunctional properties and the wide-spread enthusiasm from the academic, government, and industrial research communities surrounding their use in composite applications, the inclusion of carbon nanotubes (CNTs) in hierarchical composites has been a topic of much research. To date, CNTs

---

M. L. Shofner (✉)
School of Materials Science and Engineering, Georgia Institute of Technology,
Atlanta, Georgia 30332, USA
e-mail: meisha.shofner@mse.gatech.edu

C.-S. Kim et al. (eds.), *Hybrid and Hierarchical Composite Materials*,
DOI 10.1007/978-3-319-12868-9_9

Table 9.1. Mechanical, electrical, and thermal properties of CNTs. (Yu et al. 2000a, b; Demczyk et al. 2002; Thostenson et al. 2005; Collins and Avouris 2000; Kim et al. 2001; Han 2004; Ajayan 2002; Fujii et al. 2005; Thess et al. 1996)

Nanotube type	Density (g/cm^3)	Elastic modulus (GPa)	Tensile strength (GPa)	Electrical conductivity (S/m)	Thermal conductivity (W/m K)
SWNT	1.33–1.40	320–1470	13–52	10^6	200–6000
MWNT	1.8–2.6	270–950, 900	11–63, 150	2.7×10^7–8×10^5	2000–3000

SWNT single-walled carbon nanotube, *MWNT* multiwalled nanotube, *CNTs* carbon nanotubes

have been used in various forms in hierarchical composites such as: dispersed fillers in microscale fiber composites, microscale fiber coatings, and microscale structures assembled from individual CNTs. The advantages sought are in areas such as mechanical property improvement (specifically matrix-dominated properties such as interlaminar strength and thermomechanical performance), sensing, fire protection, and electrical properties. Such property improvements are expected based on the inherent nanotube properties, and since they are the motivation behind much of the published research, a short introduction to the structure and properties of CNTs is provided here.

The most basic definition of CNTs is nanoscale carbon fibers. However, CNTs distinguish themselves from other nanoscale carbon fibers such as vapor-grown carbon nanofibers (Tibbetts 1989a, b) in their structure and the properties which result from that structure. CNTs can be divided into two basic categories: single-walled carbon nanotubes (SWNTs; Bethune et al. 1993; Iijima and Ichihashi 1993) and multiwalled carbon nanotubes (MWNTs; Iijima 1991). Typical sizes and properties for SWNTs and MWNTs are given in Table 9.1. Schematic structures of SWNT, MWNTs, as well as a special type of MWNTs, double-walled CNTs (DWNTs), are shown in Fig. 9.1 (Schnorr and Swager 2010). All of the CNT types are formed from graphene sheets which are hexagonally arranged, sp^2-bonded sheets of carbon atoms. SWNTs are composed of a single graphene sheet wrapped to form a tube. The diameters of these fibers typically range from 0.4 to 2 nm (Grady 2011). Specific surface areas of up to 1315 m^2/g are possible, providing opportunities for enhanced interactions with polymer matrices (Peigney et al. 2001). Structurally, SWNTs are quite diverse since the orientation of the hexagons in the graphene sheet with respect to the nanotube axis can vary, leading to SWNTs of different chirality (Yu and Brus 2001). The chirality of the SWNTs is defined by two indices, n and m, that are associated with unit vector in the graphene sheet. These are expressed as (n, m) when describing SWNT chirality. These indices are also shown in Fig. 9.1. If $n = m$, the SWNT is denoted as an armchair SWNT. If $m = 0$, the SWNT is denoted as a zigzag SWNT, and all other values of n and m denote a chiral SWNT. The chirality of SWNTs impacts their electronic properties as well as van der Waals interactions between SWNTs. SWNTs generally assemble into larger rope structures with diameters between 10 and 20 nm and lengths of many micrometers (Thess et al. 1996). The ropes are held together by van der Waals forces which become

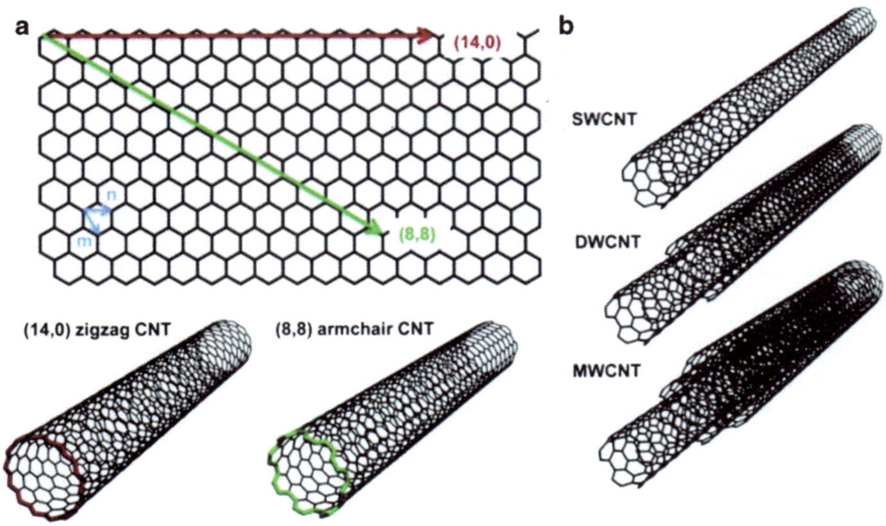

Fig. 9.1 Illustration of carbon nanotube (*CNT*) structural features. CNT chirality relative to a graphene sheet (**a**) and schematic structures of single-walled carbon nanotube (*SWNT*), double-walled carbon nanotube (*DWNT*), and multiwalled nanotube (*MWNT*) (**b**). Abbreviations for CNT types in this chapter omit the "C" used in the abbreviations in this figure. (Reprinted with permission from Chemistry of Materials, 23, J.M. Schnorr and T.M. Swager, Emerging applications of carbon nanotubes, 646–657. Copyright (2010) American Chemical Society.)

sizable due to the long lengths of SWNTs (Szleifer and Yerushalmi-Rozen 2005). In contrast, MWNTs are essentially structured as a nested array of SWNTs with outer diameters typically from 2 to 25 nm (Ajayan 1999). The electronic properties of MWNTs are reported to be influenced by different shells depending on the bias applied (Collins et al. 2001). MWNTs do not form rope structures and thus can be easier to disperse in polymer matrices than SWNTs if the appropriate component compatibility is achieved. The number of CNTs contained in an MWNT is variable, and much recent attention has focused on the properties of DWNT (Cui et al. 2009). These CNTs appear to strike an attractive balance between the benefits of SWNTs and MWNTs (Shen et al. 2011).

The research literature on the topic of CNT/polymer composites is sizable. Over 11,000 journal articles on this general topic have been published as of the writing of this book chapter based on a Web of Science search in April 2014 by the author. The use of CNTs in hierarchical or multiscale composites is a subset of this work which presents one of the likely avenues to exploiting the mechanical properties of CNTs more fully. Additionally, the attractive electrical and thermal properties of CNTs are of interest for designing the next generation of structural composites with smart/active capabilities using a hierarchical design. Therefore, the aim of this chapter is to review the current state of the art in this research area and highlight opportunities for future research. To achieve this aim, the chapter is structured to provide the reader with information about the following topics:

- Salient features of each construction scheme used for CNT hierarchical composites
- Matrix modification by CNTs in hierarchical composites
- Mechanical properties in these materials and how CNT addition affects interfacial and macroscopic properties
- Initial efforts regarding the use of CNTs as sensors in hierarchical composites
- Other properties enabled by CNTs such as enhanced electrical and thermal properties
- Modeling work done to understand the properties of CNT hierarchical composites

The chapter ends with a section summarizing these results as well as presenting an outlook for the future. Overall, the progress to date in this research area has shown that CNTs have been used successfully in multiple construction schemes in hierarchical composites and that their addition provides a combination of value-added properties which would be difficult to achieve in other ways. Ultimately, this application of CNTs has the opportunity to profoundly impact the design and properties of future structural composites for aerospace, infrastructure, and transportation applications.

9.2 Incorporation Schemes

The different construction schemes found in the literature can be grouped into three general categories: dispersed systems, fiber coatings, and CNT structures. These different types of hierarchical composites are shown schematically in Fig. 9.2. The impetus to pursue one of these schemes over another is often tied to the type of property improvement sought. For dispersed systems, the property improvements desired are related to matrix-dominated properties and nanotube-inherent properties such as electrical conductivity. For hierarchical composites containing CNT-coated fibers, interfacial property improvements are often sought, and for hierarchical composites containing CNT structures, fiber-dominated property improvements are targeted. Since each of these types of property improvements are important to improving overall composite performance, much research has been conducted concerning each of these schemes. In this section, the general structure of these types of hierarchical composites is presented. In all of the systems, the levels of structural hierarchy are similar as shown schematically in Fig. 9.3. Individual CNTs have diameters in the nanoscale size range, the continuous fibers have diameters in the microscale size range, and the assembled composite has features at the macroscale. What distinguishes the hierarchical structure in these three types of composite systems is the arrangement of CNTs, i.e., in the matrix as dispersed (individual or aggregated) CNTs, attached to a fiber surface as a coating, or assembled into a structure. This arrangement influences the type of interaction that the CNTs have with the matrix and the continuous fiber and their available mechanisms for property improvement. Discussion about the properties achieved is contained in subsequent sections.

Fig. 9.2 Schematic representation of hierarchical composite construction schemes is described in this chapter. Images are not drawn to scale

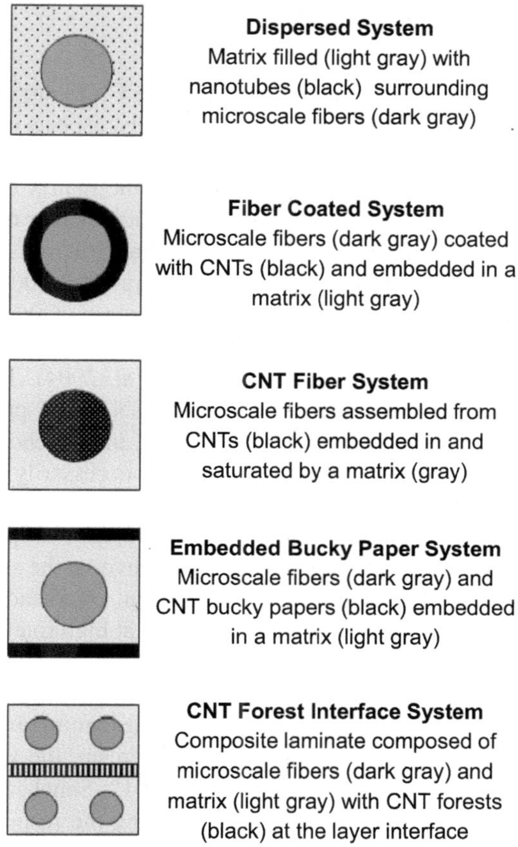

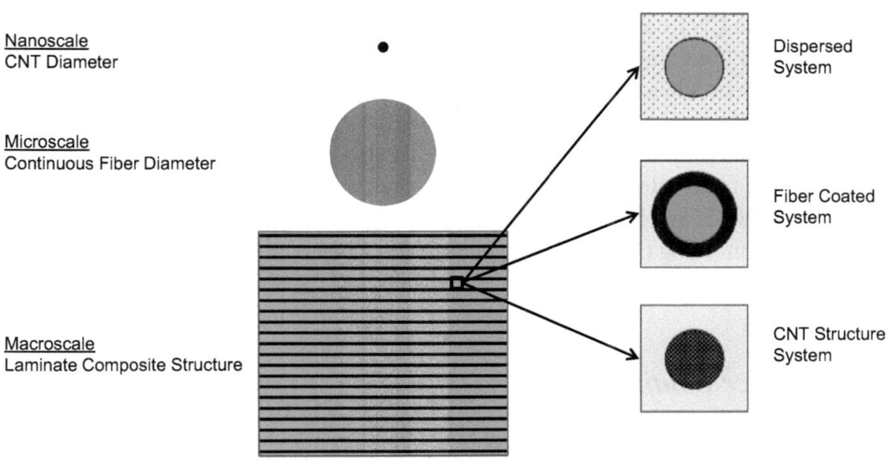

Fig. 9.3 Schematic representation of the different length scales and how they are related in the three general hierarchical composite construction schemes are described in this chapter. Images are not drawn to scale

9.2.1 Dispersed Systems

In dispersed composite systems, the CNTs are contained in the matrix in order to modify matrix properties. Specifically, aspects of thermal stability, thermal service life, and improved through thickness mechanical properties may be improved with CNT addition. CNTs are added to the matrix or its prepolymer precursors using techniques such as sonication in liquid resin or resin–solvent solutions (Wang et al. 2013c; Fan et al. 2004), mixing in three roll mills (Seyhan et al. 2007; Romhany and Szebenyi 2009), calendaring (Godara et al. 2009; Thostenson et al. 2009b; Gojny et al. 2005), batch mixing of thermoplastic (Ashrafi et al. 2012), and thermosetting (Schlea et al. 2010) matrices, extrusion of thermoplastic matrices (Salinier et al. 2013), or overhead stirring (Fan et al. 2004). These processing methods are also used with two-phase mixtures of CNTs and polymers to promote homogeneous dispersion and distribution of CNTs, and the choice of processing method is largely related to the matrix used and not necessarily to a desired property in the final hierarchical composite. In sonication, pressure waves from the cavitation of microbubbles in a liquid-based CNT suspension disrupt CNT aggregates. Three roll mills, calendaring, mixing, and extrusion involve the application of shear forces and elongational forces to improve dispersion. Overhead stirring is generally performed on relatively low viscosity suspensions at high rotational speeds. Often, the concentration of CNTs in dispersed systems is low, in the order of a few percent or less (Qian et al. 2010d). The decision to use low concentrations of CNTs is motivated largely by the processing steps to produce the continuous fiber reinforced composite. When CNTs are added to a low viscosity resin, the viscosity can increase substantially (Wang et al. 2013b; Schlea et al. 2012; Sanchez et al. 2013; Jimenez-Suarez et al. 2012; da Costa et al. 2012), affecting the resin's ability to infiltrate fiber fabrics. Additionally, the fiber fabrics may filter the resin leading to a gradient in CNT concentration through the thickness of the composite (da Costa et al. 2012). In this circumstance, the CNT concentration will be higher on the composite surface with the resin inlet and lower on the opposite surface. As shown in Fig. 9.4, the initial level of CNT dispersion has been found to influence the degree of filtering that occurs. As the level of initial dispersion is increased, the amount of filtering is decreased (Fan et al. 2004). Therefore, to use composite processing techniques that involve infiltration of resin into fiber that preforms such as resin transfer molding (RTM; Gojny et al. 2005; Wichmann et al. 2006; Wang et al. 2013a) and vacuum assisted resin transfer molding (VARTM; Fan et al. 2008; Zhou et al. 2008), the resin viscosity should be impacted as minimally as possible.

9.2.2 Fiber Coatings

CNTs have been used widely as a fiber coating. The typical concentration of CNTs when used in this manner is also low since the CNTs are mainly restricted to the fiber surface; however, concentrations greater than 10 vol.% are possible if desired (An et al. 2013). This construction scheme circumvents the filtering issue with

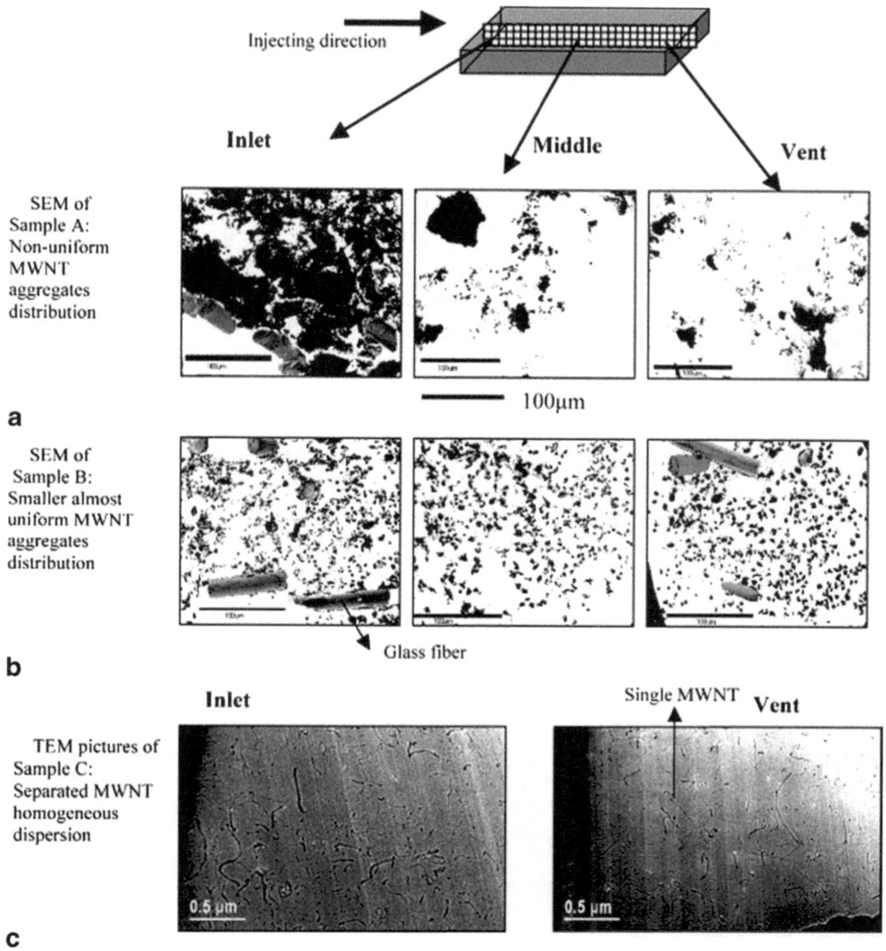

Fig. 9.4. Electron microscopy images of carbon nanotube (*CNT*) dispersion at various locations through multiwalled nanotube (*MWNT*)/glass fiber/vinyl ester composites. The filtering behavior is related to the initial level of dispersion attained in the MWNT/vinyl ester mixture. (Reprinted from Carbon, 42, Zhihang Fan, Kuang-Ting Hsiao, and Suresh G Advani, Experimental investigation of dispersion during flow of multi-walled carbon nanotube/polymer suspension in fibrous porous media, 871–876, Copyright (2004), with permission from Elsevier.)

resin infiltration processing methods seen in some dispersed composite systems (Bekyarova et al. 2007; Zhu et al. 2007). Generally, the character and structure of the CNT coating are characterized through measurements of surface tension and electron microscopy, respectively.

Multiple methods for attaching CNTs to fibers have been investigated. The most common method reported in the literature is in situ growth of CNTs on fibers using chemical vapor deposition (CVD Yamamoto et al. 2009; Jo et al. 2005; Zhao et al. 2008; Makris et al. 2005). The density of the coating produced by this method is quite variable as shown in Fig. 9.5 (Steiner et al. 2013). In some instances, the coating is relatively sparse. CNTs lay on the fiber surface, providing a conformal coating, and the fiber may have uncoated areas. However, dense coatings may also be produced such that the fiber has a bottle-brush structure where the CNTs are aligned approximately perpendicularly to the fiber surface (Cao et al. 2005). These coatings can cover the entire fiber surface. In both cases, control over the CVD process parameters allows for variation in the structure of the CNT coating (Zhang et al. 2009b), and the structural perfection of the CNTs has been found to vary along the length of the CNT coating (i.e., base to tip) with the material at the base showing a higher degree of crystallinity (Qian et al. 2010c). The CVD process often degrades the inherent properties of carbon fibers and fibers of other types (Sager et al. 2009; Qian et al. 2010a, b). Thus, trade-offs must often be made in order to use the CVD process to produce a CNT coating. However, tensioning of carbon fibers during CVD of CNTs or performing the CVD operation at lower temperatures can preserve fiber properties (Steiner et al. 2013). In the case of tensioning, losses in fiber properties are thought to be mitigated through guiding or restricting mechanochemical changes such as thermally activated graphitic reorganization. For lower temperature processing, the temperature used is below the threshold temperature for activating these processes.

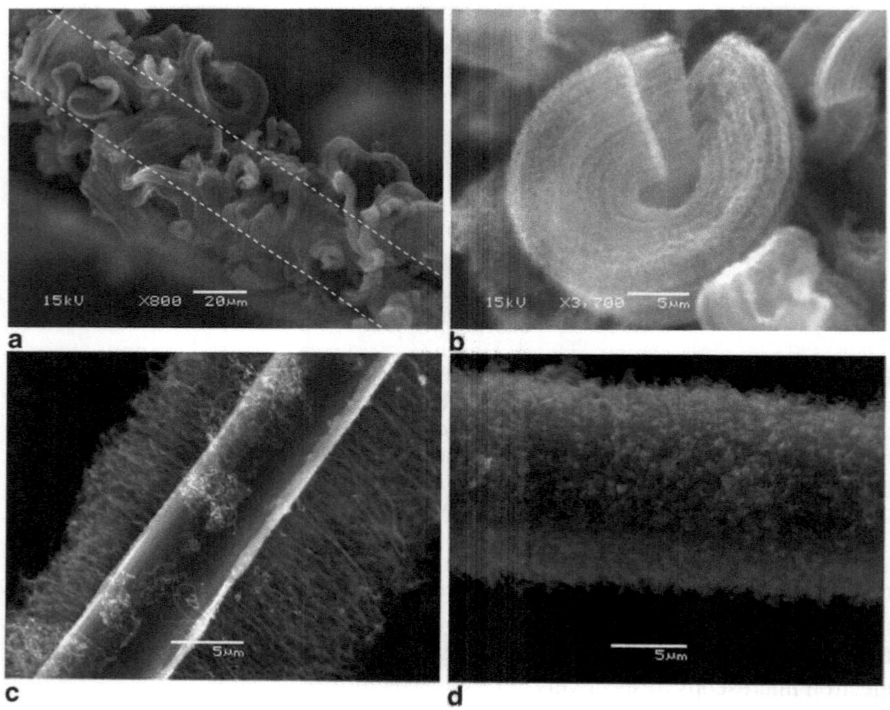

Fig. 9.5 Scanning electron microscope (*SEM*) images of carbon nanotube (*CNT*)-coated fibers demonstrating variations in coating structure: brush structures at different magnifications (**a**) and (**b**), sparse CNT coating (**c**), and conformal coating with curled CNTs (**d**). (Reprinted with permission from ACS Applied Materials and Interfaces, 5, Stephen A. Steiner, III, Richard Li, and Brian L. Wardle, Circumventing the mechanochemical origins of strength loss in the synthesis of hierarchical carbon fibers, 4892–4903. Copyright (2013) American Chemical Society.)

Other techniques to add CNTs to fiber surfaces which typically do not involve elevated temperatures include electrophoretic deposition (Bekyarova et al. 2007; Wang et al. 2010), spray deposition (Zhu et al. 2007), coating as a sizing (Barber et al. 2004; Siddiqui et al. 2009), covalent grafting (He et al. 2007; Laachachi et al. 2008), and layer-by-layer (LbL) deposition (Loyola et al. 2010). When electrophoresis is used, an applied voltage is used to drive CNT deposition from a CNT-liquid suspension onto a carbon fabric immersed in the suspension. This method has shown the ability to attain dense coatings (An et al. 2013) and may be applied continuously (Guo and Lu 2012). Spray deposition and other coating methods offer the possibility of even coatings using well-developed methods which can be scaled to be continuous processes. In these processes, a CNT-liquid suspension is prepared and applied to the fabric. In the case of spray deposition, the viscosity of the suspension and its rheological properties at a range of shear rates must be carefully considered. Chemical grafting methods involve the formation of a chemical bond between the CNT and the microscale fiber. This coating method leverages research efforts to functionalize the surfaces of CNTs and provides opportunities to strongly bind CNTs to fiber surfaces. In the LbL process, two coating solutions are used: a polyelectrolyte solution and a CNT dispersion, and a coating is built up one nanoscale layer at a time. Since an additional coating solution is required to balance the charge, the coating produced is a layered nanocomposite instead of the neat CNT coating produced by most of the other methods. In this method, the fabric is first dipped into the polyelectrolyte solution then washed to remove excess coating material. Next, the fabric is dipped into the CNT dispersion and then washed to remove the excess coating material. The process is repeated until the desired coating thickness is achieved. All of these alternative methods generally produce conformal coatings as opposed to brush coatings.

9.2.3 *CNT Structures*

Beyond their use as a discontinuous matrix filler and fiber coating, structures consisting of CNTs are used in hierarchical composites. Since the structures are themselves hierarchical with microscale/macroscale overall dimensions and nanoscale building blocks and surface topology, composites incorporating most of these structures can be considered hierarchical with or without the inclusion of additional microscale fibers. Three main types of structures are used in these applications: CNT fibers, CNT sheets or tapes, and CNT forests. While CNT structures are used less frequently than the other two types of hierarchical composites, these types of composites hold promise for fully exploiting the desirable properties of CNTs and leading to disruptive changes in composite technology.

9.2.3.1 CNTs Assembled into Fibers

Microscale fibers consisting only of CNTs have been pursued for some time through three main routes, and a brief description of these major methods are described

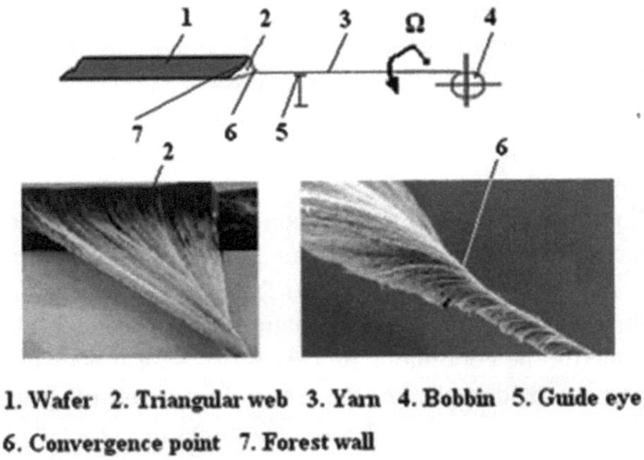

Fig. 9.6 Schematic of fiber spinning process from a carbon nanotube (*CNT*) forest. (Reprinted from Carbon, 47, C.D. Tran, W. Humphries, S.M. Smith, C. Huynh, and S. Lucas, Improving the tensile strength of carbon nanotube spun yarns using a modified spinning process, 2662–2670, Copyright (2009), with permission from Elsevier.)

here. Readers interested in a more detailed discussion of CNT fiber production and properties are directed to review articles on this topic (Lu et al. 2012; Wu and Chou 2012; Behabtu et al. 2008). The first route involves spinning a fiber from a CNT dispersion. This methodology is similar to that used to produce aramid fibers. Specifically, it has been suggested that some CNT dispersions behave as liquid crystals facilitating assembly and alignment during fiber spinning (Davis et al. 2004, 2009; Ramesh et al. 2004; Parra-Vasquez et al. 2010; Ericson et al. 2004; Zhang et al. 2008; Zhang and Kumar 2008). SWNT fibers with diameters of approximately 50 μm following annealing and continuous lengths on the order of 30 m have been produced from a SWNT/sulfuric acid dispersion (Ericson et al. 2004). These fibers have tensile strength and modulus values of 160 MPa and 120 GPa, respectively. Neat fibers of MWNTs have also been produced following a similar methodology with MWNT/ethylene glycol dispersion (Zhang et al. 2008). These fibers had diameters in the range of 10–80 μm. The reported tensile modulus value for these fibers, 69 GPa, is lower than that reported for fibers made from SWNTs, but the tensile strength value is similar, 150 MPa. These values are less than the properties of an individual CNT; however, these fibers, and CNT fibers made using the other methods described here, offer benefits over other construction methods. Specifically, resin infusion into the CNT fiber may occur since the structure has some porosity, providing an additional mechanical reinforcement method. This feature of these composites is described in the context of the mechanical properties later in the chapter.

The second route reported in the literature is dry spinning of fibers from CNT forests (Jiang et al. 2002; Zhang et al. 2004; Tran et al. 2009; Lepro et al. 2010; Liu et al. 2010). In this method, an adhesive tape is applied to the edge of the forest and pulled. This process is shown in Fig. 9.6. Additional CNTs are removed from the forest and pulled into the developing fiber due to the highly entangled nature of CNTs in the forest. Complex CNT yarns may also be produced from multiple fibers (Bogdanovich et al. 2007). It has been shown that the properties of the CNT fibers increase as CNTs with fewer walls are used to construct them (Jia et al. 2011). Neat CNT fibers produced by this method have been reported to have moduli up to 330 GPa and tensile strengths less than 5 GPa (Zhang et al. 2007a, b; Jayasinghe et al. 2011).

The third route to producing neat CNT fibers is through drawing a CNT aerogel from a flowing reaction tube (Koziol et al. 2007; Li et al. 2004; Motta et al. 2005; Vilatela and Windle 2010). The CNTs are drawn from the end of the tube and collected as a fiber on a spinning reel. The combination of drawing and application of an acetone spray consolidate the aerogel into a fiber with a diameter of approximately 10 μm (Vilatela and Windle 2010). The fiber consists solely of CNTs held together by physical entanglements. Neat CNT fibers produced by this method had modulus and tensile strength values up to 357 and 8.8 GPa, respectively (Koziol et al. 2007). Process optimization studies have led to the production of fibers enriched with metallic CNTs (Sundaram et al. 2011). Additionally, chemical treatment with a UV-curable monomer has increased the strength, elongation, and toughness of these fibers (Boncel et al. 2011).

Overall, the properties of CNT fibers depend on several factors including the aspect ratio of the nanotubes used to construct the fiber. As the aspect ratio of the nanotubes increases, the factor governing the macroscale fiber strength changes. Computational studies have shown that there are three failure modes for aligned bundles of CNTs: surface tension from capillary forces at the low aspect ratio values, internanotube friction at moderate aspect ratio values, and individual CNT failure at high aspect ratio values (Yakobson et al. 2000). This relationship is shown in Fig. 9.7. The prediction indicates that an aspect ratio of 10^6 is needed to fully exploit the inherent tensile strength of CNTs in a bundle, and CNT fibers made to date have not attained the tensile strength value of an individual CNT. Likely explanations for this discrepancy include low levels of individual CNT alignment, void space in the fiber, and/or limited CNT cohesion in the fiber (Behabtu et al. 2008).

9.2.3.2 Embedded Sheets and Tapes

One of the earliest CNT structures to be produced was a CNT paper, commonly referred to as a bucky paper (Rinzler et al. 1998). These papers are produced by microfiltration of a CNT/solvent dispersion. As the dispersion is filtered through the microfiltration membrane, the CNTs are excluded from the filter and accumulate at the filter surface until the solvent is completely removed. The thickness of the paper depends on the quantity and concentration of the CNT/solvent dispersion used, though papers generally have thicknesses on the order of tens of microns. These materials have been employed in composite applications with and without other fibers and some bucky papers containing SWNTs and MWNTs have been shown to have a negative Poisson's ratio (Hall et al. 2008). When used without microscale fibers, the papers are infiltrated with polymer in order to achieve high CNT concentrations

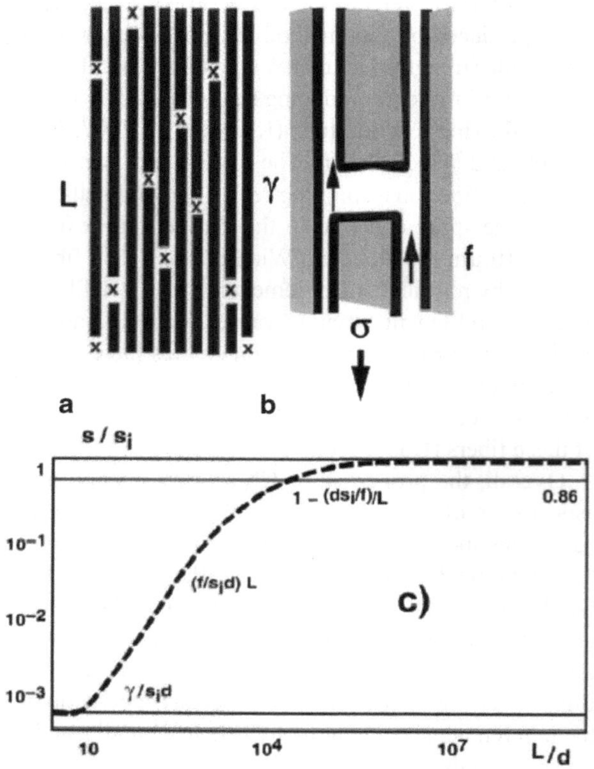

Fig. 9.7 Schematic representation of the effects of carbon nanotube (*CNT*) length (*L*) (**a**) and lateral friction (*f*) and surface energy (γ) (**b**) on the strength of a nanotube bundle (*s*) normalized by the strength of an individual CNT (s_i) (**c**). Based on the computational results presented here, an aspect ratio (*L/d*) of approximately 10^6 is needed to achieve the strength of an individual CNT. (Reprinted from Carbon, 38, B.I. Yakobson, G. Samsonidze, and G.G. Samsonidze, Atomistic theory of mechanical relaxation in fullerene nanotubes, 1675–1680, Copyright (2000), with permission from Elsevier.)

(Pham et al. 2008; Song et al. 2007). When used with microscale fibers, the papers may be applied at surfaces to impart functional properties (Wu et al. 2010, 2011).

9.2.3.3 CNT Forests

The use of CNT forests is not as common as the use of the previous two types of CNT structures. CNT forests have been investigated as interlaminar bonding agents (Garcia et al. 2008a). As shown in Fig. 9.8, these forests can be delaminated from the substrate they were grown on using a roll coated with prepreg. When an additional layer is placed on top of the CNT forest and the structure is cured, the resin is able to infiltrate the forest such that part of the forest is embedded in each composite layer. The length of the CNTs in the forest may be changed so that different embedded lengths are possible though excessive CNT lengths may affect in-plane properties negatively by increasing the thickness of the ply interlayer.

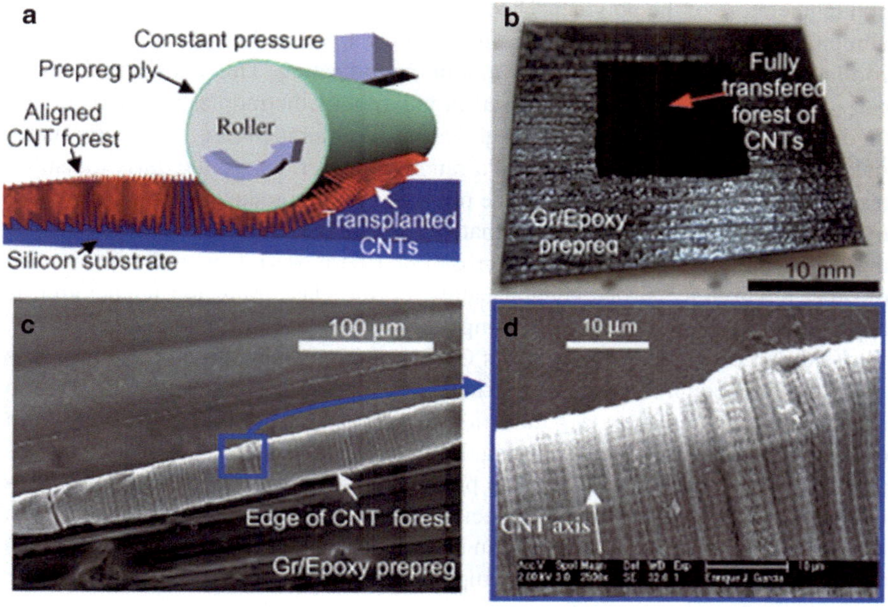

Fig. 9.8 The application of carbon nanotube (CNT) forests to interlaminar interfaces can improve properties limited by interlaminar adhesion: illustration of transfer printing processes (**a**), CNT forest on the prepreg surface (**b**), Scanning electron microscope (*SEM*) images of the CNT forest on the prepreg surface (**c** and **d**). (Reprinted from Composites Part A: Applied Science and Manufacturing, 39, Enrique J. Garcia, Brian L. Wardle, and A. John Hart, Joining prepreg composite interfaces with aligned carbon nanotubes, 1065–1070, Copyright (2008), with permission from Elsevier.)

9.3 Alteration of Matrix Properties

One route to improving the properties of continuous fiber reinforced composites through hierarchical design is improving the properties of the matrix. In composite applications, the matrix properties can limit the performance of the composite by limiting the upper limit service temperature and limiting the interlaminar properties. Adding CNTs to the matrix as a dispersed filler or as a CNT structure at an interface may improve this aspect of composite performance. The former option is presented in this section to highlight how CNTs impact polymer mobility and degradation behavior. The latter option is presented in Sect. 9.4.2.

The types of polymers most often used in continuous fiber reinforced composites are thermosets such as epoxy, polyimide, vinyl ester, phenolic, and bismaleimide resins. These polymers are formed in the presence of the fibers through chain extension and cross-linking processes, ultimately leading to the formation of a chemically bonded network structure with very high molecular weight. The mobility of the polymer chains in the network changes as a function of temperature and is influenced by the density of cross-links in the network as well as structural features of the polymer. Physically, this change in mobility is represented by the glass transition temperature of the polymer. The glass transition is associated with cooperative motion of polymer chain segments with either the arrest of this type of motion upon cooling or the onset of this type of motion upon heating. This transition dramatically changes the state of the material in amorphous thermoplastic polymers and thermosetting polymers. Below the glass transition temperature, polymeric materials are rigid with moduli in the GPa range; however, above the glass transition temperature, chain segments can slide past one another, reducing the macroscopic modulus value. The modulus of the material in this temperature range is related to the entanglement (thermoplastic) or cross-link (thermoset) density. In the absence of other environmental degradation pathways, the glass transition temperature is approximately equal to the service temperature of the composite (Campbell 2010). The addition of nanoparticles such as CNTs has been shown to increase the glass transition temperature of polymers. There are two possible pathways for CNTs to cause an increase in the glass transition temperature. First, the CNTs may form chemical bonds with the matrix during the cross-linking process, acting as additional cross-links. Second, the CNTs may physically associate with the matrix, impeding molecular mobility through interfacial interactions. Since both methods result in the same bulk physical response, it can be difficult to identify which mechanism(s) are responsible for glass transition temperature increases in a given system (Chen et al. 2009). At adequate CNT concentrations, infrared or Raman spectroscopy can provide guidance in determining the type of CNT–polymer interaction present.

Thermal degradation of the matrix is also an important factor which can limit the performance of continuous fiber reinforced polymer composites. Often, this degradation is assessed using thermogravimetric analysis (TGA). The data obtained from this method are the weight loss as a function of temperature or time in a prescribed environment, allowing different degradation pathways to be observed. Decreased flammability of CNT composites is also an area of active research in light of the need to replace halogenated flame retardants and increased use of composite materials in aviation. CNTs generally improve the thermal stability and can decrease the flammability of polymers through component interactions at the CNT–polymer interface (Seo and Park 2004; Jana and Cho 2008; Barus et al. 2010) as well as, at sufficient concentrations, network formation (Kashiwagi et al. 2005a, b; Moniruzzaman and Winey 2006; Cipiriano et al. 2007).

In matrices for hierarchical composites containing dispersed CNTs, the effect of CNT addition on the value of the glass transition temperature is variable and dependent on the component level interactions and the degree of dispersion attained. Generally, as the level of dispersion increases and/or the interactions between the CNTs and the polymer become stronger, the value of the glass transition temperature increases. An impressive increase in the glass transition temperature of 29 °C was observed in MWNT/carbon fabric/epoxy composite at an MWNT concentration of 0.025 wt.% (Wang et al. 2013b). Additionally, an increase in the glass transition temperature of approximately 6 °C was observed at a CNT concentration of 0.5 wt.% in an MWNT/carbon fiber/epoxy composite (Siddiqui et al. 2013). At higher MWNT concentrations, the value of the glass transition temperature was approximately equal to that of the carbon fiber/epoxy composite, supporting results

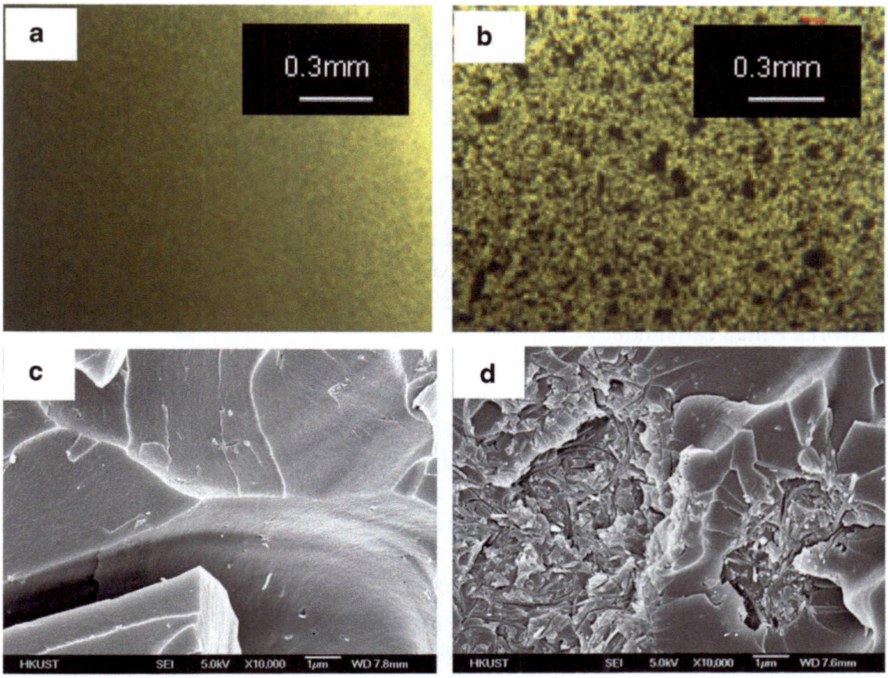

Fig. 9.9 Optical (**a**, **b**) and scanning electron microscopy (**c**, **d**) images illustrated decreased carbon nanotube (*CNT*) dispersion as the concentration is increased from 0.5 wt.% multiwalled nanotubes (*MWNTs*) (**a**, **c**) to 1.0 wt.% MWNTs (**b**, **d**). In the optical microscopy images, MWNT agglomerates appear as black objects. (Reprinted from Composite Structures, 104, Naveed A. Siddiqui, Shafi Ullah Khan, and Jang-Kyo Kim, Experimental torsional shear properties of carbon fiber reinforced epoxy composites containing carbon nanotubes, 230–238, Copyright (2013), with permission from Elsevier.)

from optical and electron microscopy images shown in Fig. 9.9 that indicated a diminished level of dispersion at MWNT concentrations greater than 0.5 wt.% with their processing method. Alternatively, some systems have shown decreased values of glass transition temperature due to the addition of certain functional groups on the CNT surface which are thought to inhibit the curing process of an MWNT/glass fiber/epoxy composite (Qiu et al. 2007). Specifically, while pristine MWNTs decreased the value of the matrix's glass transition temperature slightly, the addition of MWNTs functionalized through treatment with acid decreased the value of the matrix's glass transition temperature by approximately 13 °C at a concentration of 1 wt.% MWNT.

Changes in the coefficient of thermal expansion (CTE) with CNT addition are also related to changes in polymer mobility. Measurements of CTE perpendicular to the fiber direction in CNT/carbon fiber/epoxy composites showed that smaller CNTs and functionalized CNTs led to decreased values of CTE when CNT concentration was kept at 0.5 wt.% (Godara et al. 2009). In this system, the functionalization moiety chosen, amine ($-NH_2$), interacted favorably with the matrix and was thought to participate in the curing reaction since its chemistry is similar to typical curing agents for epoxies. These changes were attributed to an increased ability for smaller CNTs and functionalized CNTs to interact with polymer chains physically and chemically, respectively. Additionally in composites containing CNT fibers, CTE measured in the axial direction was observed to decrease, suggesting effective load transfer and coupling between the components (Vilatela et al. 2012). Considering these results together, the impacts of processing and CNT concentration should also be considered. Increased values of the glass transition temperature were associated with quite low concentrations of CNTs. These concentrations would be easier to disperse in the matrix and have a lesser tendency to form entangled CNT networks which can hinder infiltration processes.

Overall, hierarchical composite systems with CNTs as a dispersed filler in the matrix have been shown to modify polymer matrices by restricting molecular mobility. This restriction in polymer chain mobility is observed experimentally through increased glass transition temperature and decreased CTE in composite structures. These property changes have been shown to be related to the level of CNT dispersion achieved and chemical functionalization, similar to results for two-phase mixtures of CNTs and polymers. Generally, the level of dispersion achieved is more homogeneous at low concentrations of CNTs, though improvements in incorporation techniques and functionalization schemes can aid in achieving homogeneous dispersion at somewhat higher CNT concentrations. When designing a system to achieve improvements in service temperature as well as improved flame resistance, it may be necessary to choose the CNT concentration judiciously, since improvements in each of these areas can be related to different CNT arrangements, i.e., well dispersed and networked, respectively.

9.4 Mechanical Properties of Hierarchical Composites

The types of mechanical properties measured for hierarchical composites are generally the same as those measured for conventional continuous fiber reinforced composites and their components. In this section, aspects of mechanical performance are presented with separate sections for areas of interest to different hierarchical composite construction schemes. The properties of CNT structures outside of composites were presented in a previous section and are not repeated here.

9.4.1 Interfacial Properties

Two main types of interfacial properties are measured in hierarchical CNT composites: interlaminar shear strength (ILSS) and interfacial shear strength (IFSS). These properties describe the macroscopic interfaces in these composite materials where ILSS describes the adhesion between individual composite lamina and IFSS describes the adhesion between components (matrix and fiber). These properties are often measured for dispersed systems (ILSS) and coated fiber composites (ILSS and IFSS). In dispersed systems, improvement in ILSS is often sought, overcoming a critical shortcoming of continuous fiber reinforced composites (Sanchez et al. 2013; Siddiqui et al. 2013; Wang et al. 2013a, c; Ashrafi et al. 2012; Godara et al. 2009; Qiu et al. 2007). CNTs are thought to improve ILSS by bridging composite layers. The dispersion level achieved and the interaction between the CNTs and the matrix are key to achieving improvements in ILSS (Godara et al. 2009; Wang et al. 2013c). In MWNT/glass fiber/epoxy composites, a maximum in ILSS was observed at MWNT concentrations less than 0.1 wt.% (Wang et al. 2013a). In these composites, the MWNTs were not chemically functionalized, but the dispersion was assisted through the use of a surfactant. Qualitatively similar results have been observed in MWNT/carbon fiber/epoxy composites and have been reported where a maximum in ILSS values was observed at MWNT concentrations less than 1 wt.% (Siddiqui et al. 2013). Similar trends were observed in the value of the glass transition temperature and storage modulus with CNT concentration. In these studies, the decrease in ILSS values at higher CNT concentrations was attributed to decreased levels of dispersion as CNT concentration increased. Given these results, chemical functionalization has the potential to improve component interactions and improve dispersion with respect to pristine CNTs, similar to results for nanocomposites containing only a polymer matrix and CNTs (Mitchell et al. 2002). Both effects should improve ILSS values. As described in the previous section, amine functionalization has the potential to participate in the curing of epoxy polymers; therefore, it is a commonly explored chemical functionalization scheme. Amine functionalized CNTs have shown to favorably interact with an epoxy matrix and improve ILSS in a CNT/carbon fiber/epoxy composite by more than 30% at a CNT concentration of 1 wt.% (Wang et al. 2013c). Similarly, a composite system consisting of amine functionalized DWNT, glass fiber, and epoxy showed a 20 % improvement in ILSS at a DWNT concentration of 0.3 wt.%, and this composite had an ILSS value greater than a comparison hierarchical composite containing carbon black (Gojny et al. 2005). ILSS values were also observed to decrease in this system at higher CNT concentrations due increased aggregation. The trend in flexural strength was qualitatively the same with CNT concentration. When compared directly in the same hierarchical composite system, amine functionalized CNTs provided greater ILSS improvements than pristine CNTs, reinforcing the need to select components that interact favorably when seeking ILSS improvements. Though not strictly a comparative ILSS measurement due to the test results obtained, short-beam bending strength in CNT/glass fiber fabric/polyetheretherketone (PEEK) composites was observed to increase by 64 % when a compatibilizing agent was used with 1 wt.% CNTs as a dispersed filler (Ashrafi et al. 2012). Additionally, as shown in Fig. 9.10, the failure mode changed from interlaminar failure of the outer laminae in the glass fiber/PEEK composite to interlaminar failure of the interior laminae in the composite structure in the compatibilized CNT/glass fiber fabric/PEEK composite. Among the contributing factors for the change in failure mode, it was suggested that the impregnation of the polymer was more homogeneous through the thickness of the

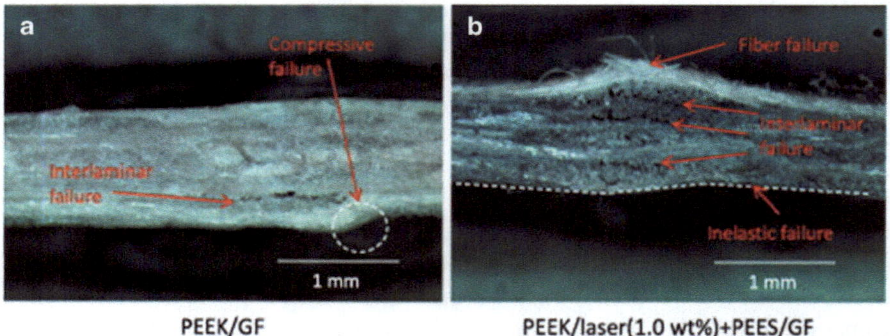

Fig. 9.10 Cross-sectional images of samples following short-beam shear tests. The inclusion of compatibilized carbon nanotubes (*CNTs*) changed the failure mode in the composite system. (Reprinted from Composites Part A: Applied Science and Manufacturing, 43, Behnam Ashrafi, Ana M. Díez-Pascual, Leon Johnson, Marc Genest, Simon Hind, Yadienka Martinez-Rubi, Jose M. González-Domínguez, M. Teresa Martínez, Benoit Simard, Marián A. Gómez-Fatou, and Andrew Johnston, Processing and properties of PEEK/glass fiber laminates: Effect of addition of single-walled carbon nanotubes, 1267–1279, Copyright (2012), with permission from Elsevier)

composite structure due to enhanced thermal conductivity and reduced temperature variations through the thickness of the hierarchical composite. In the comparison of glass fiber/PEEK composite, voids present at the inner surfaces of the outer laminae reduced the strength and dictated the failure mode.

The ILSS values for hierarchical composites made with CNT-coated fibers have also been measured in some systems. These systems generally do not show the same trends in ILSS with CNT concentration as dispersed systems. Since the CNTs are restricted to the fiber surfaces and for the most part not incorporated into the resin, reductions in ILSS due to CNT aggregation do not occur. In these systems, surface area for component interactions increases with increasing CNT concentration. Even small CNT concentrations can produce measureable increases. At a concentration of 0.25 wt. % MWNT with respect to the carbon fabric, a 27 % improvement in ILSS was seen in an MWNT/carbon fabric/epoxy composite (Bekyarova et al. 2007). In MWNT/carbon fiber/epoxy composites, ILSS values were observed to increase with MWNT content up to the highest concentration used in the study, 3 vol. % (Singh et al. 2012). At the highest MWNT concentration, the ILSS value was approximately doubled. Using densely covered alumina fiber fabrics, a 69 % improvement in ILSS was observed (Garcia et al. 2008b). This result compared favorably with other mechanisms to improve ILSS such as stitching. SWNT spray coated glass fabrics in a vinyl ester matrix have shown improvements in ILSS between 20 and 45 % at SWNT concentration less than or equal to 1 wt. % (Zhu et al. 2007). The improvement in ILSS values were shown to be dependent on the surface functionality of the CNTs with pristine SWNTs coating producing the largest improvements. In this construction scheme, functional groups were found to interfere with the curing of the matrix, decreasing the composite properties and suggesting that the effects of chemical functionalization are not uniform in CNT hierarchical composites.

IFSS measurements on CNT-coated fibers generally show an improvement which is attributed to the increased amount of surface area available for component

Table 9.2 Comparison of IFSS values for CNT-coated fibers

Composite system	Deposition method	Coating structure	Testing method	IFSS improvement (%)	Reference
MWNT/T300 carbon fiber	Chemical grafting	Conformal	Microdroplet	150	Zhang et al. (2009a)
MWNT/T300B-3000–40B carbon fiber	Chemical grafting	Conformal	Microdroplet	111	Peng et al. (2012)
MWNT/Silfa silica fiber	CVD	Aligned brush	Fiber fragmentation	150	Qian et al. (2010b)
MWNT/T650 carbon fiber	CVD	Aligned brush	Fiber fragmentation	11	Sager et al. (2009)
MWNT/T650 carbon fiber	CVD	Brush	Fiber fragmentation	71	Sager et al. (2009)
MWNT/carbon fiber	CVD	Brush	Microdroplet	108	An et al. (2011)
MWNT/silica	CVD	Brush	Fiber fragmentation	107	Lachman et al. (2012)
MWNT/alumina	CVD	Brush	Fiber fragmentation	55	Lachman et al. (2012)
MWNT/Granoc YS-50–30S carbon fiber	CVD	Conformal	Fiber fragmentation	15	Thostenson et al. (2002)
MWNT/C320.00A carbon fiber	CVD	Conformal	Single fiber pullout	60	Qian et al. (2008)
MWNT/IM7 carbon fiber	CVD	Conformal	Fiber fragmentation	26	Qian et al. (2010a)
MWNT/T300 carbon fiber	CVD	Conformal	Microdroplet	94	An et al. (2012)
MWNT/T300 carbon fiber	CVD	Conformal	Fiber fragmentation	23	Lachman et al. (2013)
MWNT/PAN carbon fiber	Electrophoresis	Conformal	Microdroplet	68.8	Guo et al. (2012)
MWNT/carbon fiber	Electrophoresis	Conformal	Microdroplet	35.6	Zhang et al. (2013)

IFSS interfacial shear strength, *CNT* carbon nanotube, *MWNT* multiwalled nanotube, *CVD* chemical vapor deposition

interactions. Table 9.2 gives some results from the literature concerning IFSS values for CNT-coated fibers. Table 9.2 includes information as available from the source papers concerning the type of microscale fiber used in the composite, the method for applying the CNT coating to the microscale fiber, the morphology of the coating, how the IFSS value was measured, and the percent improvement in IFSS of the hierarchical composite with respect to a composite containing an uncoated fiber. The coating morphologies were not always given so the descriptions are based on electron microscopy images provided in the papers. These categories were defined as conformal, where the CNTs appeared to coat the fibers and brush where the CNTs protruded from the microscale fiber. In the brush coatings, the CNTs in the coating were aligned perpendicularly to the microscale fiber axis or showed little preferential alignment. These data suggest some general trends with respect to coating method and structure. CNT coatings that are chemically grafted onto fiber surfaces have shown high values of IFSS (Peng et al. 2012; Zhang et al. 2009a). Appropriate chemical functionalization of the CNTs in the coating may allow for chemical bonds to form between the CNTs and matrix, producing even larger improvements in IFSS (Zhang et al. 2009a). For coatings grown by CVD processes, the structure of the coatings influenced the value of IFSS obtained. Generally, coatings structured as dense brushes (An et al. 2011; Lachman et al. 2012; Sager et al. 2009; Qian et al. 2010b) showed higher IFSS values than those which were conformal (An et al. 2012; Lachman et al. 2013; Qian et al. 2008, 2010a; Thostenson et al. 2002), coating the surface with curly CNTs. The orientation of the CNTs in the coating has also been shown to have an effect on the value of IFSS. In companion studies with a radially aligned CNT brush coating and randomly aligned CNT brush coating, the IFSS value was found to increase 11 and 71 %, respectively, when compared to a composite containing unsized fibers (Sager et al. 2009). The randomly oriented CNT brush coating was assumed to perform at a higher level since it was possible for some of the CNTs to be aligned with the principle tensile stress direction (±45° with respect to the microscale fiber axis). Conformal coatings applied by electrophoresis have shown similar IFSS improvement percentages to similarly structured coatings produced by CVD. IFSS testing for CNT-coated fibers has also been shown to be more complicated to interpret than for uncoated microscale fibers. In some cases, pull-out tests have given different results than push-out tests due to differences in the failure mode under these different loading conditions (Qian et al. 2008). Specifically, the fiber was loaded from the outside surface in pull-out tests and from the interior in push-out tests, so the CNT coating behaved differently in these different loading conditions. When loaded from the outside surface, failure occurred gradually in the region containing CNTs and epoxy, but when loaded from the interior of the microscale fiber, the failure occurred abruptly in the fiber.

Understanding the properties of the composite in the vicinity of CNT-coated fibers is also an important area of research. Nanoindentation has been used to understand the properties of the interfacial material (Qian et al. 2011). Nanoindentation data indicated that as expected the hardness and modulus were increased in the vicinity of the fiber. These values decayed to those of the neat (unfilled) epoxy at sufficient distances from the fiber surface. Such a methodology could be used to map the interfacial zone and understand the efficacy of the CNT coating in bridging the fiber and matrix properties. Beyond these types of interfacial properties, CNT–epoxy coatings have also been shown to provide healing capabilities when used on glass fibers, effectively filling in defects prior to and during mechanical loading (Siddiqui et al. 2009). This effect was distinct from property changes resulting from a neat epoxy coating on the fibers.

9.4.2 Macroscopic Properties

Macroscopic mechanical properties are commonly measured for hierarchical CNT composites to understand how CNTs can influence or preserve fiber-dominated properties. In dispersed composite systems, the changes to macroscale mechanical properties may be seen as modest; however, it should be considered when interpreting these properties that only small CNT concentrations are used in this hierarchical composite design. Incorporation of 1 wt.% functionalized MWNTs increased the tensile strength and modulus of a glass fabric/epoxy composite by 16 and 27%, respectively (Qiu et al. 2007). Companion composites produced with pristine MWNTs showed a lesser degree of improvement in modulus and a decreased tensile strength, emphasizing the importance of obtaining good adhesion between components and homogeneous CNT dispersion. Aggregation of CNTs has been shown to influence tensile properties differently based on the direction of testing relative to the microscale fiber direction. When testing parallel to the fiber direction, aggregated CNTs decreased the tensile strength of composites while impacting modulus to a lesser degree, as expected (Godara et al. 2009). However, results for tensile testing at an angle of 45° relative to the fiber direction and perpendicular to the fiber direction showed limited change in tensile strength, indicating improvement to matrix-dominated properties. Hierarchical dispersed composites produced by VARTM processing have been shown to possess mechanical properties beyond those predicted by micromechanical models due to processing-induced changes in CNT orientation and distribution (Chen et al. 2009). In some instances, strain at break has also been improved by the addition of CNTs. In an MWNT/glass fiber/poly(etherimide) composite system, the strain at break in the hierarchical composite was slightly improved with the addition of 5 wt.% MWNTs (Salinier et al. 2013). This result was observed in tensile and flexural loading. The improved property was attributed to a higher local concentration of MWNTs at the glass fiber surface, leading to better adhesion. These fibers were also preferentially oriented, with the MWNT arranged with their long axes perpendicular to the fiber surface.

While the fiber properties are often somewhat compromised when a CNT coating is applied, the fiber-dominated properties in a composite can show improvement. In MWNT/carbon fiber/phenolic resin and MWNT/carbon fiber fabric/phenolic resin composites, tensile strength and modulus increased with increasing MWNT concentration, though MWNT concentrations of less than approximately 9 and 2 wt.% had lower values than those of the composites without MWNTs for composites containing unidirectional carbon fibers and carbon fiber fabric, respectively (Mathur et al. 2008). Flexural testing results for an MWNT/silicon carbide fabric/epoxy composite showed improvements of 140, 5, and 424% with respect to a similar composite without CNTs in strength, modulus, and toughness, respectively (Veedu et al. 2006).

Fracture toughness is also often improved in hierarchical composites due to crack bridging of CNTs. Measurements of fracture toughness have shown that CNT addition as a dispersed filler increased resistance to crack initiation and propagation (Godara et al. 2009; Ashrafi et al. 2011; Gorbatikh et al. 2011). Increased CNT–

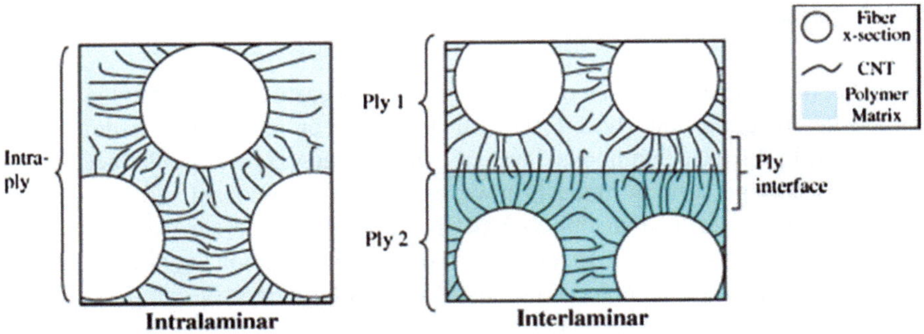

Fig. 9.11 Carbon nanotube (*CNT*) coatings on fibers have the ability to improve intralaminar and interlaminar properties due to bridging by the CNTs. (Reprinted from Composites Science and Technology, 68, Enrique J. Garcia, Brian L. Wardle, A. John Hart, and Namiko Yamamoto, Fabrication and multifunctional properties of a hybrid laminate with aligned carbon nanotubes grown *in situ*, 2034–2041, Copyright (2008), with permission from Elsevier.)

matrix interaction through decreased CNT diameter and chemical functionalization of the CNT surfaces have produced greater increases in the fracture toughness with respect to pristine MWNTs (Godara et al. 2009). Hierarchical composites containing CNT-coated fibers have also shown improved fracture toughness. At sufficient CNT lengths and coating densities, the CNT coating may bridge fibers in a within and across lamina leading to improved intralaminar and interlaminar properties. This ability is shown schematically in Fig. 9.11. An improvement of 348 % in Mode I fracture toughness measured at initiation was observed for an MWNT/silicon carbide fabric/epoxy composite system (Veedu et al. 2006). In an MWNT/alumina fiber fabric/epoxy composite system, the addition of a CNT coating to the fabric improved the Mode I steady-state fracture toughness by 76 % (Wicks et al. 2010). Additionally, initiation fracture toughness was improved substantially. Fracture toughness was also observed to increase with CNT coating in carbon fiber/epoxy composites by 46 and 23 % in samples with warp and fill orientation, respectively (Kepple et al. 2008).

Modification of laminate interfaces through the incorporation of vertically aligned CNT (VACNT) forests has shown that fracture toughness of the composite structure may be enhanced through crack bridging and interlayer toughening (Garcia et al. 2008a). The results of Modes I and II fracture toughness tests showed increases in toughness by factors of 1.5 and 3, respectively, for the same prepreg system. The VACNT interface penetrates the plies to a depth of 10–20 μm, acting as "interlaminar nanostitches."

Composites containing microscale fibers of assembled CNTs have also shown attractive macroscale properties in tension and compression. In composites containing CNT fibers drawn from an aerogel in an epoxy matrix, the properties measured in these tests were quite similar to rule of mixtures predictions, suggesting that the reinforcing potential of these CNT structures may be effectively harnessed in this hierarchical composite design (Mora et al. 2009). The tensile strength and modulus of composites containing CNT yarns consolidated from forests of CNTs and epoxy matrices exceeded the predictions of a rule of mixtures model significantly

(Bogdanovich and Bradford 2010). The properties of the composites exceeded those of the CNT yarns by factors of at least six. Additionally, the strain at failure values for the composites were similar and not a function of the strain at failure for the neat epoxy. In composites containing an epoxy matrix and microscale CNT yarns consolidated from forests and processed into composites using a method similar to pultrusion, large increases in mechanical properties were achieved with maximum improvements in modulus and tensile strength of 15 and 7 times, respectively (Shimamura et al. 2014). Composites produced by this technique were mechanically reinforced to levels that were greater than a rule of mixtures prediction for the matrix and CNT yarn. The nature of the reinforcement is, however, somewhat uncertain since in these composites, the epoxy matrix prepolymer or precursor may infiltrate the fiber during processing and then cure inside and outside of the fiber. The increased surface area available for component interaction provides a reinforcement mechanism similar to CNTs used in filled polymers and the CNT fiber macroscopic properties provide reinforcement in a similar way that microscale fibers reinforce conventional composites. Further study is needed to ascertain which mechanism(s) are appropriate for these composites.

9.4.3 Summary

Overall, research to date has shown that the addition of CNTs is a promising route to improving the mechanical performance of composite materials. With regard to interfacial mechanical properties assessed by values of ILSS and IFSS, the addition of CNTs can increase the values of ILSS and IFSS over composites containing only a matrix and microscale fiber. The trends in ILSS values for hierarchical composites containing dispersed CNTs follow those observed for glass transition temperature values. Specifically, homogeneous CNT dispersion and good compatibility between the CNTs and the matrix are needed to maximize ILSS improvement. CNT-coated fiber composites have shown improvements in ILSS and IFSS though the mechanism for ILSS improvement is different than the mechanism in dispersed composite systems. In these composites, interactions between the CNTs and the matrix as well as the CNTs and the fiber contribute to the improvements.

Improvements in macroscopic mechanical properties under different loading conditions were observed though the mechanisms for these property improvements are also likely different due to differences in the way that the CNTs interact with the matrix and fibers (where applicable). Dispersed systems show improvements in matrix-dominated properties and fracture toughness with the dispersion level achieved again playing a significant role in the properties of the composite. CNT-coated fibers also show improvements in macroscale properties, including those which are fiber dominated. However, in these composite systems, consideration of the CNT coating method is needed to prevent or mitigate loss of fiber properties which impact the final composite properties. Fracture toughness is also improved in CNT-coated fiber composites, though the mechanism is through CNT-mediated microscale fiber bridging. Finally, hierarchical composites containing CNT structures show improved macroscale properties and benefit from an additional reinforcement mechanism related to infiltration of the resin into the CNT fiber. This is a unique design aspect to hierarchical composites containing CNT fibers and potentially other CNT structure and is a likely avenue for further property enhancement through control of CNT structure porosity.

9.5 Other Properties

Among the other types of properties sought through CNT incorporation in hierarchical composites are thermal properties and electrical properties. As mentioned earlier in this chapter, the inherent thermal and electrical properties of CNTs are exceptional, and it is intuitive to attempt to harness these properties in composite applications, producing the next generation of multifunctional materials.

9.5.1 Thermal Properties

Improved thermal conductivity can be difficult to achieve in CNT composites since the interfacial structure developed in the composite plays an important role in the efficacy of thermal transport between components. In a CNT/glass fiber/PEEK composite, thermal conductivity was found to increase two times when an appropriate compatibilizing agent was used with 1 wt.% SWNTs (Ashrafi et al. 2012). In this case, poly(1,4-phenylene ether-ether sulfone) was used to improve CNT–matrix interactions. Thermally treated MWNTs were found to improve thermal conductivity in an MWNT/carbon fiber/phenolic composite by acting as bridges which connected the microscale carbon fibers (Kim et al. 2007). Similar composites containing MWNTs in the as-synthesized state did not produce this result, suggesting that the inherent properties of the CNTs as well as interfacial interactions can have significant effects on the thermal conductivity of a hierarchical composite. In composites containing CNT-coated fibers, out-of-plane improvement in thermal conductivity was larger than the in-plane improvement (Hu et al. 2013). In the case of hierarchical composites containing CNT fibers, thermal conductivity was also found to improve (Vilatela et al. 2012). In that system, thermal conductivity improved over two orders of magnitude with respect to the neat epoxy at a CNT fiber concentration of 10 wt.%. This value was approximately one order of magnitude higher than a comparison composite system containing T300 carbon fiber from Toray and epoxy. The improved thermal conductivity in the hierarchical composite was attributed to the improved contact between components which was enabled by resin infiltration.

Beyond thermal conductivity, fire protection of composite structures has also been seen through the application of CNT structures at surfaces. CNT bucky papers have been shown to act as protective coatings for carbon fiber/epoxy and carbon fiber/bismaleimide composites (Wu et al. 2011). In cone calorimetry tests, bucky papers made of MWNTs were found to decrease the heat release rate and smoke generation. The performance of bucky papers made from different types of CNTs was found to be tied to the pore structure of the paper and the purity of the CNTs (Wu et al. 2010).

9.5.2 Electrical Properties

Electrical conductivity has been effectively increased in CNT composites of many different designs, including hierarchical. For a dispersed composite system consisting of MWNT, glass fabric, and an epoxy matrix, the electrical conductivity of the hierarchical composite was found to be similar to the electrical conductivity of the MWNT/epoxy mixture (Qiu et al. 2007). Similar results were obtained for an MWNT/glass fiber/poly(etherimide) composite system (Salinier et al. 2013). Such a result indicates that the hierarchical composite design does not necessarily hinder the capabilities of the MWNTs to impart this property. Additionally, CNT length has been shown to influence network formation as would be expected. Longer MWNTs led to higher values of electrical conductivity when compared to similar hierarchical composites containing shorter MWNTs (Zimmer et al. 2012). In hierarchical composites containing CNT-coated fibers, out of plane changes in electrical conductivity were found to be preferentially improved with respect to in-plane properties (Bekyarova et al. 2007; Garcia et al. 2008b). These trends with regard to direction in the composite are similar to those seen for thermal conductivity for CNT-coated fiber composites. In CNT fiber/epoxy composites, electrical conductivity was observed to increase to a value up to 3600 S/m, a change of more than ten orders of magnitude with respect to the neat epoxy (Vilatela et al. 2012).

9.5.3 Summary

Improvements in thermal and electrical properties through the addition of CNTs have been achieved in hierarchical composites. The guidelines which govern improvement in these areas seem to be similar to those which govern improvement in two-phase mixtures of CNTs and polymers. Notably for electrical property improvement, similar gains may be achieved to those seen in two-phase mixtures, suggesting that property screening may be done prior to the construction of a hierarchical composite. When considering composite design for multifunctional improvement, aspects of CNT concentration and functionalization which produce maximum improvements in mechanical properties may not lead to the maximum improvements in electrical or thermal properties, since electrical and thermal properties are enhanced most notably when CNT networks are present. Therefore, individual property tradeoffs may be necessary to achieve the desired multifunctional improvement for a given application.

9.6 Sensing Capabilities of Hierarchical Composites

The increasing use of composite materials in structural applications such as aerospace and infrastructure has motivated research into the topic of structural health monitoring of composites. Hierarchical composites filled with CNTs have the potential to provide structural health monitoring capabilities through monitoring a change in a specific CNT property, most often electrical conductivity or resistivity. In this methodology, damage to the composite structure would result in the damage to a percolated CNT network in a specific area of the composite. If the CNT network is altered or destroyed, the electrical conductivity would decrease (electrical resistivity would increase). Given the appropriate calibration of electrical properties to damage, these changes could be used for structural health monitoring.

In hierarchical composites where CNTs were used as a dispersed filler in glass fabric/epoxy composites, changes in electrical resistance were observed under different loading conditions, indicating that disruptions in the CNT network structure could be used for sensing (Böger et al. 2008). Additionally, a change in the resistance was observed for residual strain in the composite after unloading, suggesting that CNTs also serve as health monitors during the life of the composite. The efficacy of this method in evaluating damage due to impact are not as clear (Arronche et al. 2013). Monitoring changes in the CNT network dispersed in the matrix using electrical resistivity measurements also helps to support other nondestructive evaluation techniques such as acoustic emission by providing data about existing damage in the structure which can be coupled to measurements of accumulating new damage and provide a comprehensive view of the composite's health (Gao et al. 2009). Other hierarchical composite constructions have also shown capabilities for structural health monitoring. The electrical resistivity of composites containing CNT-coated fibers has been shown to change with the application of uniaxial tension (Loyola et al. 2010), shear strain (An et al. 2013), cyclic and fatigue loading (Loyola et al. 2011), as well as temperature and humidity (Loyola et al. 2013). Though many of these applications use two point or four point measurements to assess the state of the CNT network with applied strain, monitoring with electrical impedance spectroscopy has been shown to allow for a greater understanding of the types of damage that is occurring through effective circuit modeling of the response (Loyola et al. 2010, 2011).

The Raman spectra for CNTs have also been shown to be sensitive to the local strain experienced and this change has been used in filled polymers to assess load transfer (Schadler et al. 1998). This concept is extended in hierarchical composites to sense strain around a microscale fiber (Zhao et al. 2001). Additionally, CNT coatings on glass fiber have been shown to be effective sensors for interfacial adhesion (Barber et al. 2004). In this application, the Raman shift of the CNTs was found to correspond with the results of fragmentation tests, suggesting another characterization method for assessing composite structure health.

9.7 Modeling

Developing improved predictive capabilities for designing hierarchical CNT composites is a critical need in this research area. Accounting for the different structures as well as the changes in the component properties due to the incorporation of CNTs complicates the model development; thus to date, few studies have been conducted. Micromechanical approaches have also been attempted using the rule of mixtures model and the semiempirical Halpin–Tsai model (Sharma and Shukla 2012). The rule of mixtures models (parallel and series arrangement of components) represent the upper and lower bound of expected behavior in a two-phase system, and the Halpin–Tsai models fill in the property space bounded by the rule of mixtures expressions through the use of a reinforcement efficiency factor. Generally, these models are applied to two-phase systems; therefore, an additional modeling step was used to adapt them to the three-phase hierarchical system. Specifically, the matrix modulus was replaced with the CNT/matrix modulus in the expressions. The modulus of the CNT/matrix mixture was calculated using the Tsai–Pagano equations. However, the model predictions from this approach were shown to overestimate the modulus in CNT/carbon fiber/epoxy composites at CNT concentrations up to 1.5 wt.%. Combining Halpin–Tsai equations with micromechanical expressions for woven fiber composites also resulted in an overestimation of composite modulus for a CNT/carbon fiber fabric/epoxy composite (Kim et al. 2009).

Multiscale modeling approaches have been employed to capture the interactions between components in hierarchical CNT composites (Nie et al. 2011). The composite was modeled by beginning with a representative volume element for a CNT-coated fiber in an epoxy matrix and then applying a cohesive zone model to account for load transfer. In this model, the CNT coating on the fiber is considered part of the matrix instead of part of the fiber. The results of this model were consistent with general experimental observations that matrix-dominated properties are improved as CNT–epoxy interfacial strength is increased.

Molecular modeling methods have also been used to understand differences in component interactions relevant to load transfer in hierarchical composites (Mei et al. 2011). Molecular dynamics was used to understand how adding a polymer grafted CNT coating to carbon fiber influences the IFSS of a hierarchical composite. Simulations were conducted on two materials: a composite containing the grafted CNT fiber coating and a composite without CNTs consisting of the carbon fiber and epoxy only. The simulation results indicated that the addition of the CNT fiber coating increased the interaction energy 2.7 times. The authors also conducted IFSS measurements on microdroplet composites to validate the simulation results. The experimental and simulation results were qualitatively consistent as the IFSS of the hierarchical composites increased by approximately 170% with respect to the conventional composite. In this system, improved chemical adhesion and mechanical interlocking between components improved composite properties and were attributed to the addition of CNTs and the polymer grafting agent.

Electrical properties have also been studied using modeling approaches. Electrical percolation behavior in CNT/glass fiber/vinyl ester composite have been analyzed using Monte Carlo simulations (Thostenson et al. 2009a). These simulations have shown that CNT network formation in unidirectional composites is influenced by the orientation of the microscale fiber relative to resin flow direction during VARTM processing. Assuming that the CNTs form two-dimensional networks in the composites, the simulation results suggest that percolating networks are more likely to form when the fiber orientation is parallel to the resin flow direction. These simulation results were qualitatively consistent with experimental results.

9.8 Summary and Outlook

While not as widely researched as other types of CNT/polymer composites such as two-phase mixtures of CNTs and polymers, CNT hierarchical composites represent important application areas for CNTs such as aerospace, infrastructure, and transportation. Research results to date suggest that these composites provide opportunities for harnessing the inherent mechanical properties of CNTs to a greater extent than some other composite designs. In this chapter, three hierarchical composite construction schemes were presented. These schemes have all been shown to improve the macroscopic mechanical properties of fiber reinforced polymers; however, the design features of each of these schemes differentiate them from each other within the application space. The salient features of composites produced with each of these construction schemes are given below.

Composites containing CNTs in the matrix as a dispersed filler have been extensively investigated. The incorporation of CNTs in this way has generally improved ILSS values in composites and additionally increased matrix performance through higher glass transition temperatures and reductions in CTE. Other construction schemes discussed in this chapter would not be expected to impact the resin properties as greatly since the CNTs are largely located in localized areas, i.e., fiber surfaces and inside of CNT structures. Additionally, these composites can be manufactured by established processes if the resin viscosity is not impacted greatly, facilitating their implementation outside of the laboratory. CNT dispersion in the matrix as well as adhesion to the matrix remain important considerations with these types of composites, and this issue is not fully resolved, though appropriate chemical functionalization of CNTs has shown to improve properties relative to those obtained with pristine CNTs. This composite construction scheme is intuitive, builds upon research results in the nanocomposite literature, and is likely to be useful in applications where matrix properties are limiting the application or where sensing capabilities are desired. However, the effective design of a composite material using this construction scheme requires an understanding and prioritization of the application requirements. Sensing properties generally require monitoring a change in electrical conductivity, so a CNT concentration large enough to form a percolating network is required. However, obtaining a maximum improvement in ILSS may require that the CNTs remain isolated from one another, indicating no one CNT arrangement in the matrix provides ideal values of both properties simultaneously.

CNT hierarchical composites containing CNT-coated fibers overcome the key limitation of attaining homogeneous dispersion and distribution in the matrix since the CNTs are purposefully segregated to fiber surfaces, allowing the use of industrially relevant composite processing methods. Many methods, some of which are scalable to continuous processes, are available to attach CNTs to fiber surfaces such as CVD, coating processes, electrophoresis, and chemical grafting. Generally, composites containing these coated fibers have shown improved ILSS values, and measurements of IFSS have also indicated that the CNT coating improves fiber–matrix adhesion. The extent to which IFSS is improved is related to the structure of the coating and the coating method. Relative to the other construction schemes, these types of composites show particular advantages in the area of processing. This type of hierarchical composite can be designed for improvements in macroscopic properties through improvement of interfacial bonding between the matrix and the CNT-coated microscale fiber. Additionally, this type of composite could be simultaneously designed to have suitable electrical properties for sensing using conformal coatings or brush coatings. A lesser degree of trade-off would be expected for this type of system if fiber properties can be preserved through the coating process.

Finally, hierarchical composites containing CNT fibers are a promising route to incorporating high concentrations of CNTs into composites. In some cases, these systems have shown higher mechanical properties than those predicted by the rule of mixtures, suggesting that another reinforcement mechanism is available. Specifically, the additional mechanical reinforcement has been attributed to resin infusion into the CNT fiber during processing, leading to additional chemical and physical interactions between the components. Based on the mechanical property results available, research directed at improving the properties of the CNT fibers is likely to improve the properties further, since the CNT fiber properties are somewhat lower than the properties of an individual CNT. A hierarchical composite containing CNT fibers would likely be designed to improve the macroscopic mechanical properties for an application requiring high specific mechanical properties by more fully harnessing the inherent properties of CNTs. As the field progresses, fabric architectures could be envisioned that allow electrical transport or dispersed CNTs could be added to the matrix to improve the abilities of this construction scheme for applications requiring sensing and active properties.

Beyond the opportunities and challenges presented by each hierarchical composite construction scheme, further research into the properties which clearly distinguishes these materials from conventional fiber reinforced polymers will provide an impetus for their application. A few of those research areas are listed here.

- Sensing capabilities are an important area to explore since they impact the long-term cost incurred by using composite materials, not only in terms of material costs but also in terms of maintenance and inspection costs.
- Thermal property improvements have been particularly difficult to attain in all kinds of polymer composites containing CNTs, but improvements in this area

may allow for the use of polymer composites in new applications which require higher service temperatures.

- Developing modeling methods which enable a greater understanding of the phenomena responsible for their multifunctional properties and allow for rapid materials design is critically needed to support wider scale implementation of these materials.

Overall, research in this area has shown that the use of CNT in hierarchical composites is a method to improve properties beyond those of conventional fiber reinforced polymers. These materials may be versatile since they can be produced using multiple construction schemes which can be tailored for existing applications as well as potentially for new applications which have not used polymer composites traditionally. Finally, more research activities aimed at modeling to improve materials design capabilities, multifunctional property improvement, and differentiating these materials from conventional composites are needed to merit the routine use of CNTs in structural composites.

References

Ajayan PM (1999) Nanotubes from carbon. Chem Rev 99(7):1787–1799. doi:10.1021/cr970102g

Ajayan PM (2002) Carbon nanotubes. In: Nalwa HS (ed) Nanostructured materials and nanotechnology. Academic, San Diego, pp 329–360

An F, Lu CX, Guo JH, He SQ, Lu HB, Yang Y (2011) Preparation of vertically aligned carbon nanotube arrays grown onto carbon fiber fabric and evaluating its wettability on effect of composite. Appl Surface Sci 258(3):1069–1076. doi:10.1016/j.apsusc.2011.09.003

An F, Lu CX, Li YH, Guo JH, Lu XX, Lu HB, He SQ, Yang Y (2012) Preparation and characterization of carbon nanotube-hybridized carbon fiber to reinforce epoxy composite. Mater Des 33:197–202. doi:10.1016/j.matdes.2011.07.027

An Q, Rider AN, Thostenson ET (2013) Hierarchical composite structures prepared by electrophoretic deposition of carbon nanotubes onto glass fibers. ACS Appl Mater Interfaces 5(6):2022–2032. doi:10.1021/am3028734

Arronche L, La Saponara V, Yesil S, Bayram G (2013) Impact damage sensing of multiscale composites through epoxy matrix containing carbon nanotubes. J Appl Polym Sci 128(5):2797–2806. doi:10.1002/app.38448

Ashrafi B, Guan JW, Mirjalili V, Zhang YF, Chun L, Hubert P, Simard B, Kingston CT, Bourne O, Johnston A (2011) Enhancement of mechanical performance of epoxy/carbon fiber laminate composites using single-walled carbon nanotubes. Compos Sci Technol 71(13):1569–1578. doi:10.1016/j.compscitech.2011.06.015

Ashrafi B, Diez-Pascual AM, Johnson L, Genest M, Hind S, Martinez-Rubi Y, Gonzalez-Dominguez JM, Martinez MT, Simard B, Gomez-Fatou MA, Johnston A (2012) Processing and properties of PEEK/glass fiber laminates: effect of addition of single-walled carbon nanotubes. Compos Part A Appl Sci Manuf 43(8):1267–1279. doi:10.1016/j.compositesa.2012.02.022

Barber AH, Zhao Q, Wagner HD, Baillie CA (2004) Characterization of E-glass-polypropylene interfaces using carbon nanotubes as strain sensors. Compos Sci Technol 64(13–14):1915–1919. doi:10.1016/j.compscitech.2004.02.004

Barus S, Zanetti M, Bracco P, Musso S, Chiodoni A, Tagliaferro A (2010) Influence of MWCNT morphology on dispersion and thermal properties of polyethylene nanocomposites. Polym Degrad Stab 95(5):756–762. doi:10.1016/j.polymdegradstab.2010.02.013

Behabtu N, Green MJ, Pasquali M (2008) Carbon nanotube-based neat fibers. Nano Today 3(5–6):24–34. doi:10.1016/S1748–0132(08)70062–8

Bekyarova E, Thostenson ET, Yu A, Kim H, Gao J, Tang J, Hahn HT, Chou TW, Itkis ME, Haddon RC (2007) Multiscale carbon nanotube-carbon fiber reinforcement for advanced epoxy composites. Langmuir 23(7):3970–3974. doi:10.1021/la062743p

Bethune DS, Kiang CH, deVries MS, Gorman G, Savoy R, Vasquez J, Beyers R (1993) Cobalt-catalysed growth of carbon nanotubes with single-atomic-layer walls. Nature 363(6430):605–607. doi:10.1038/363605a0

Bogdanovich AE, Bradford PD (2010) Carbon nanotube yarn and 3-D braid composites. Part I: tensile testing and mechanical properties analysis. Compos Part A Appl Sci Manuf 41(2):230–237. doi:10.1016/j.compositesa.2009.10.002

Bogdanovich A, Bradford P, Mungalov D, Fang SL, Zhang M, Baughman RH, Hudson S (2007) Fabrication and mechanical characterization of carbon nanotube yarns, 3-D braids, and their composites. SAMPE J 43(1):6–19

Böger L, Wichmann MHG, Meyer LO, Schulte K (2008) Load and health monitoring in glass fibre reinforced composites with an electrically conductive nanocomposite epoxy matrix. Compos Sci Technol 68(7–8):1886–1894. doi:10.1016/j.compscitech.2008.01.001

Boncel S, Sundaram RM, Windle AH, Koziol KKK (2011) Enhancement of the mechanical properties of directly spun CNT fibers by chemical treatment. ACS Nano 5(12):9339–9344. doi:10.1021/nn202685x

Campbell FC (2010) Structural composite materials ASM international. Materials Park, Ohio

Cao AY, Veedu VP, Li XS, Yao ZL, Ghasemi-Nejhad MN, Ajayan PM (2005) Multifunctional brushes made from carbon nanotubes. Nat Mater 4(7):540–545. doi:10.1038/nmat1415

Chen W, Shen HB, Auad ML, Huang CZ, Nutt S (2009) Basalt fiber-epoxy laminates with functionalized multi-walled carbon nanotubes. Composites Part A Appl Sci Manuf 40(8):1082–1089. doi:10.1016/j.compositesa.2009.04.027

Cipiriano BH, Kashiwagi T, Raghavan SR, Yang Y, Grulke EA, Yamamoto K, Shields JR, Douglas JF (2007) Effects of aspect ratio of MWNT on the flammability properties of polymer nanocomposites. Polymer 48(20):6086–6096. doi:10.1016/j.polymer.2007.07.070

Collins PG, Avouris P (2000) Nanotubes for electronics. Sci Am 283(6):62–69

Collins PG, Arnold MS, Avouris P (2001) Engineering carbon nanotubes and nanotube circuits using electrical breakdown. Science 292(5517):706–709. doi:10.1126/science.1058782

Cui S, Kinloch IA, Young RJ, Noe L, Monthioux M (2009) The effect of stress transfer within double-walled carbon nanotubes upon their ability to reinforce composites. Adv Mater 21(35):3591–3595. doi:10.1002/adma.200803683

da Costa EFR, Skordos AA, Partridge IK, Rezai A (2012) RTM processing and electrical performance of carbon nanotube modified epoxy/fibre composites. Compos Part A Appl Sci Manuf 43(4):593–602. doi:10.1016/j.compositesa.2011.12.019

Davis VA, Ericson LM, Parra-Vasquez ANG, Fan H, Wang YH, Prieto V, Longoria JA, Ramesh S, Saini RK, Kittrell C, Billups WE, Adams WW, Hauge RH, Smalley RE, Pasquali M (2004) Phase behavior and rheology of SWNTs in superacids. Macromolecules 37(1):154–160. doi:10.1021/ma0352328

Davis VA, Parra-Vasquez ANG, Green MJ, Rai PK, Behabtu N, Prieto V, Booker RD, Schmidt J, Kesselman E, Zhou W, Fan H, Adams WW, Hauge RH, Fischer JE, Cohen Y, Talmon Y, Smalley RE, Pasquali M (2009) True solutions of single-walled carbon nanotubes for assembly into macroscopic materials. Nat Nanotechnol 4(12):830–834. doi:10.1038/nnano.2009.302

Demczyk BG, Wang YM, Cumings J, Hetman M, Han W, Zettl A, Ritchie RO (2002) Direct mechanical measurement of the tensile strength and elastic modulus of multiwalled carbon nanotubes. Mat Sci Eng A Struct 334(1–2):173–178. doi:10.1016/S0921–5093(01)01807-X

Ericson LM, Fan H, Peng HQ, Davis VA, Zhou W, Sulpizio J, Wang YH, Booker R, Vavro J, Guthy C, Parra-Vasquez ANG, Kim MJ, Ramesh S, Saini RK, Kittrell C, Lavin G, Schmidt H, Adams WW, Billups WE, Pasquali M, Hwang WF, Hauge RH, Fischer JE, Smalley RE (2004) Macroscopic, neat, single-walled carbon nanotube fibers. Science 305(5689):1447–1450. doi:10.1126/science.1101398

Fan ZH, Hsiao KT, Advani SG (2004) Experimental investigation of dispersion during flow of multi-walled carbon nanotube/polymer suspension in fibrous porous media. Carbon 42(4):871–876. doi:10.1016/j.carbon.2004.01.067

Fan ZH, Santare MH, Advani SG (2008) Interlaminar shear strength of glass fiber reinforced epoxy composites enhanced with multi-walled carbon nanotubes. Compos Part A Appl Sci Manuf 39(3):540–554. doi:10.1016/j.compositesa.2007.11.013

Fujii M, Zhang X, Xie HQ, Ago H, Takahashi K, Ikuta T, Abe H, Shimizu T (2005) Measuring the thermal conductivity of a single carbon nanotube. Phys Rev Lett 95(6):065502. doi:10.1103/PhysRevLett.95.065502

Gao L, Thostenson ET, Zhang Z, Chou T-W (2009) Coupled carbon nanotube network and acoustic emission monitoring for sensing of damage development in composites. Carbon 47(5):1381–1388. doi:10.1016/j.carbon.2009.01.030

Garcia EJ, Wardle BL, John Hart A (2008a) Joining prepreg composite interfaces with aligned carbon nanotubes. Compos Part A Appl Sci Manuf 39(6):1065–1070. doi:10.1016/j.compositesa.2008.03.011

Garcia EJ, Wardle BL, John Hart A, Yamamoto N (2008b) Fabrication and multifunctional properties of a hybrid laminate with aligned carbon nanotubes grown In Situ. Compos Sci Technol 68(9):2034–2041. doi:10.1016/j.compscitech.2008.02.028

Godara A, Mezzo L, Luizi F, Warrier A, Lomov SV, van Vuure AW, Gorbatikh L, Moldenaers P, Verpoest I (2009) Influence of carbon nanotube reinforcement on the processing and the mechanical behaviour of carbon fiber/epoxy composites. Carbon 47(12):2914–2923. doi:10.1016/j.carbon.2009.06.039

Gojny FH, Wichmann MHG, Fiedler B, Bauhofer W, Schulte K (2005) Influence of nano-modification on the mechanical and electrical properties of conventional fibre-reinforced composites. Compos Part A Appl Sci Manuf 36(11):1525–1535. doi:10.1016/j.compositesa.2005.02.007

Gorbatikh L, Lomov SV, Verpoest I (2011) Nano-engineered composites: a multiscale approach for adding toughness to fibre reinforced composites. In: Guagliano M, Vergani L (eds) 11th international conference on the mechanical behavior of materials, vol 10. Procedia Engineering. pp 3252–3258. doi:10.1016/j.proeng.2011.04.537

Grady BP (2011) Carbon nanotube-polymer composites. Wiley, Hoboken

Guo JH, Lu CX (2012) Continuous preparation of multiscale reinforcement by electrophoretic deposition of carbon nanotubes onto carbon fiber tows. Carbon 50(8):3101–3103. doi:10.1016/j.carbon.2012.02.044

Guo JH, Lu CX, An F (2012) Effect of electrophoretically deposited carbon nanotubes on the interface of carbon fiber reinforced epoxy composite. J Mat Sci 47(6):2831–2836. doi:10.1007/s10853–011-6112–5

Hall LJ, Coluci VR, Galvao DS, Kozlov ME, Zhang M, Dantas SO, Baughman RH (2008) Sign change of Poisson's ratio for carbon nanotube sheets. Science 320(5875):504–507. doi:10.1126/science.1149815

Han J (2004) Structures and properties of carbon nanotubes. In: Meyyappan M (ed) Carbon nanotubes: science and applications. CRC, Boca Raton

He XD, Zhang FH, Wang RG, Liu WB (2007) Preparation of a carbon nanotube/carbon fiber multi-scale reinforcement by grafting multi-walled carbon nanotubes onto the fibers. Carbon 45(13):2559–2563. doi:10.1016/j.carbon.2007.08.018

Hu JB, Dong SM, Wu B, Zhang XY, Wang Z, Zhou HJ, He P, Yang JS, Li QG (2013) Mechanical and thermal properties of C_f/SiC composites reinforced with carbon nanotube grown in situ. Ceram Int 39(3):3387–3391. doi:10.1016/j.ceramint.2012.08.072

Iijima S (1991) Helical microtubules of graphitic carbon. Nature 354(6348):56–58. doi:10.1038/354056a0

Iijima S, Ichihashi T (1993) Single-shell carbon nanotubes of 1-nm diameter. Nature 363(6430):603–605. doi:10.1038/363603a0

Jana RN, Cho JW (2008) Thermal stability and molecular interaction of polyurethane nanocomposites prepared by in situ polymerization with functionalized multiwalled carbon nanotubes. J Appl Polym Sci 108(5):2857–2864. doi:10.1002/app.27895

Jayasinghe C, Chakrabarti S, Schulz MJ, Shanov V (2011) Spinning yarn from long carbon nanotube arrays. J Mat Res 26(05):645–651. doi:10.1557/jmr.2010.91

Jia JJ, Zhao JN, Xu G, Di JT, Yong ZZ, Tao YY, Fang CO, Zhang ZG, Zhang XH, Zheng LX, Li QW (2011) A comparison of the mechanical properties of fibers spun from different carbon nanotubes. Carbon 49(4):1333–1339. doi:10.1016/j.carbon.2010.11.054

Jiang K, Li Q, Fan S (2002) Nanotechnology: spinning continuous carbon nanotube yarns. Nature 419(6909):801. doi:10.1038/419801a

Jimenez-Suarez A, Campo M, Sanchez M, Romon C, Urena A (2012) Influence of the functionalization of carbon nanotubes on calendering dispersion effectiveness in a low viscosity resin for VARIM processes. Compos Part B Eng 43(8):3482–3490. doi:10.1016/j.compositesb.2011.12.009

Jo SH, Huang JY, Chen S, Xiong GY, Wang DZ, Ren ZF (2005) Field emission of carbon nanotubes grown on carbon cloth. J Vac Sci Technol B 23(6):2363–2368. doi:10.1116/1.2102970

Kashiwagi T, Du FM, Douglas JF, Winey KI, Harris RH, Shields JR (2005a) Nanoparticle networks reduce the flammability of polymer nanocomposites. Nat Mat 4(12):928–933. doi:10.1038/nmat1502

Kashiwagi T, Du FM, Winey KI, Groth KA, Shields JR, Bellayer SP, Kim H, Douglas JF (2005b) Flammability properties of polymer nanocomposites with single-walled carbon nanotubes: effects of nanotube dispersion and concentration. Polymer 46(2):471–481. doi:10.1016/j.polymer.2004.10.087

Kepple KL, Sanborn GP, Lacasse PA, Gruenberg KM, Ready WJ (2008) Improved fracture toughness of carbon fiber composite functionalized with multi walled carbon nanotubes. Carbon 46(15):2026–2033. doi:10.1016/j.carbon.2008.08.010

Kim P, Shi L, Majumdar A, McEuen PL (2001) Thermal transport measurements of individual multiwalled nanotubes. Phys Rev Lett 87(21):215502. doi:10.1103/PhysRevLett.87.215502

Kim YA, Kamio S, Tajiri T, Hayashi T, Song SM, Endo M, Terrones M, Dresselhaus MS (2007) Enhanced thermal conductivity of carbon fiber/phenolic resin composites by the introduction of carbon nanotubes. Appl Phys Lett 90(9):093125. doi:10.1063/1.2710778

Kim M, Park YB, Okoli OI, Zhang C (2009) Processing, characterization, and modeling of carbon nanotube-reinforced multiscale composites. Compos Sci Technol 69(3–4):335–342. doi:10.1016/j.compscitech.2008.10.019

Koziol K, Vilatela J, Moisala A, Motta M, Cunniff P, Sennett M, Windle A (2007) High-performance carbon nanotube fiber. Science 318(5858):1892–1895. doi:10.1126/science.1147635

Laachachi A, Vivet A, Nouet G, Ben Doudou B, Poilane C, Chen J, Bai JB, Ayachi M (2008) A chemical method to graft carbon nanotubes onto a carbon fiber. Mat Lett 62(3):394–397. doi:10.1016/j.matlet.2007.05.044

Lachman N, Wiesel E, de Villoria RG, Wardle BL, Wagner HD (2012) Interfacial load transfer in carbon nanotube/ceramic microfiber hybrid polymer composites. Compos Sci Technol 72(12):1416–1422. doi:10.1016/j.compscitech.2012.05.015

Lachman N, Qian H, Houlle M, Amadou J, Shaffer MSP, Wagner HD (2013) Fracture behavior of carbon nanotube/carbon microfiber hybrid polymer composites. J Mat Sci 48(16):5590–5595. doi:10.1007/s10853-013-7353-2

Lepro X, Lima MD, Baughman RH (2010) Spinnable carbon nanotube forests grown on thin, flexible metallic substrates. Carbon 48(12):3621–3627. doi:10.1016/j.carbon.2010.06.016

Li YL, Kinloch IA, Windle AH (2004) Direct spinning of carbon nanotube fibers from chemical vapor deposition synthesis. Science 304(5668):276–278. doi:10.1126/science.1094982

Liu K, Sun YH, Zhou RF, Zhu HY, Wang JP, Liu L, Fan SS, Jiang KL (2010) Carbon nanotube yarns with high tensile strength made by a twisting and shrinking method. Nanotechnology 21(4):045708. doi:10.1088/0957-4484/21/4/045708

Loyola BR, La Saponara V, Loh KJ (2010) In situ strain monitoring of fiber-reinforced polymers using embedded piezoresistive nanocomposites. J Mat Sci 45(24):6786–6798. doi:10.1007/s10853-010-4775-y

Loyola BR, Loh KJ, La Saponara V (2011) Static and dynamic strain monitoring of GFRP composites using carbon nanotube thin films. In: Tomizuka M, Yun CB, Giurgiutiu V, Lynch JP (eds) Sensors and smart structures technologies for civil, Mmchanical, and aerospace systems 2011, vol 7981. Proceedings of SPIE. p 798108. doi:10.1117/12.881006

Loyola BR, Zhao YJ, Loh KJ, La Saponara V (2013) The electrical response of carbon nanotube-based thin film sensors subjected to mechanical and environmental effects. Smart Mat Struct 22(2):025010. doi:10.1088/0964-1726/22/2/025010

Lu W, Zu M, Byun J-H, Kim B-S, Chou T-W (2012) State of the art of carbon nanotube fibers: opportunities and challenges. Adv Mat 24(14):1805–1833. doi:10.1002/adma.201104672

Makris TD, Giorgi R, Lisi N, Pilloni L, Salernitano E, De Riccardis MF, Carbone D (2005) Carbon nanotube growth on PAN- and pitch-based carbon fibres by HFCVD. Fullerenes Nanotubes and Carbon Nanostruct 13:383–392. doi:10.1081/fst-200039380

Mathur RB, Chatterjee S, Singh BP (2008) Growth of carbon nanotubes on carbon fibre substrates to produce hybrid/phenolic composites with improved mechanical properties. Compos Sci Technol 68(7–8):1608–1615. doi:10.1016/j.compscitech.2008.02.020

Mei L, Li YB, Wang RG, Wang C, Peng QY, He XD (2011) Multiscale carbon nanotube-carbon fiber reinforcement for advanced epoxy composites with high interfacial strength. Polym Polym Compos 19(2–3):107–112

Mitchell CA, Bahr JL, Arepalli S, Tour JM, Krishnamoorti R (2002) Dispersion of functionalized carbon nanotubes in polystyrene. Macromolecules 35(23):8825–8830. doi:10.1021/ma020890y

Moniruzzaman M, Winey KI (2006) Polymer nanocomposites containing carbon nanotubes. Macromolecules 39(16):5194–5205. doi:10.1021/ma060733p

Mora RJ, Vilatela JJ, Windle AH (2009) Properties of composites of carbon nanotube fibres. Compos Sci Technol 69(10):1558–1563. doi:10.1016/j.compscitech.2008.11.038

Motta M, Li YL, Kinloch I, Windle A (2005) Mechanical properties of continuously spun fibers of carbon nanotubes. Nano Lett 5(8):1529–1533. doi:10.1021/nl050634 +

Nie JQ, Jia YX, Qu P, Shi QR (2011) Carbon nanotube/carbon fiber multiscale composite: influence of interfacial strength on mechanical properties. J Inorg Organomet Polym Mater 21(4):937–940. doi:10.1007/s10904-011-9565-8

Parra-Vasquez ANG, Behabtu N, Green MJ, Pint CL, Young CC, Schmidt J, Kesselman E, Goyal A, Ajayan PM, Cohen Y, Talmon Y, Hauge RH, Pasquali M (2010) Spontaneous dissolution of ultralong single- and multiwalled carbon nanotubes. ACS Nano 4(7):3969–3978. doi:10.1021/nn100864v

Peigney A, Laurent C, Flahaut E, Bacsa RR, Rousset A (2001) Specific surface area of carbon nanotubes and bundles of carbon nanotubes. Carbon 39(4):507–514. doi:10.1016/S0008-6223(00)00155-X

Peng QY, He XD, Li YB, Wang C, Wang RG, Hu PA, Yan YD, Sritharan T (2012) Chemically and uniformly grafting carbon nanotubes onto carbon fibers by poly(amidoamine) for enhancing interfacial strength in carbon fiber composites. J Mat Chem 22(13):5928–5931. doi:10.1039/c2jm16723a

Pham GT, Park YB, Wang SR, Liang ZY, Wang B, Zhang C, Funchess P, Kramer L (2008) Mechanical and electrical properties of polycarbonate nanotube buckypaper composite sheets. Nanotechnology 19(32):325705. doi:10.1088/0957–4484/19/32/325705

Qian H, Bismarck A, Greenhalgh ES, Kalinka G, Shaffer MSP (2008) Hierarchical composites reinforced with carbon nanotube grafted fibers: the potential assessed at the single fiber level. Chem Mat 20(5):1862–1869. doi:10.1021/cm702782j

Qian H, Bismarck A, Greenhalgh ES, Shaffer MSP (2010a) Carbon nanotube grafted carbon fibres: a study of wetting and fibre fragmentation. Compos Part A Appl Sci Manuf 41(9):1107–1114. doi:10.1016/j.compositesa.2010.04.004

Qian H, Bismarck A, Greenhalgh ES, Shaffer MSP (2010b) Carbon nanotube grafted silica fibres: characterising the interface at the single fibre level. Compos Sci Technol 70(2):393–399. doi:10.1016/j.compscitech.2009.11.014

Qian H, Bismarck A, Greenhalgh ES, Shaffer MSP (2010c) Synthesis and characterisation of carbon nanotubes grown on silica fibres by injection CVD. Carbon 48(1):277–286. doi:10.1016/j.carbon.2009.09.029

Qian H, Greenhalgh ES, Shaffer MSP, Bismarck A (2010d) Carbon nanotube-based hierarchical composites: a review. J Mat Chem 20(23):4751–4762. doi:10.1039/c000041h

Qian H, Kalinka G, Chan KLA, Kazarian SG, Greenhalgh ES, Bismarck A, Shaffer MSP (2011) Mapping local microstructure and mechanical performance around carbon nanotube grafted silica fibres: methodologies for hierarchical composites. Nanoscale 3(11):4759–4767. doi:10.1039/c1nr10497g

Qiu JJ, Zhang C, Wang B, Liang R (2007) Carbon nanotube integrated multifunctional multiscale composites. Nanotechnology 18(27):275708. doi:10.1088/0957-4484/18/27/275708

Ramesh S, Ericson LM, Davis VA, Saini RK, Kittrell C, Pasquali M, Billups WE, Adams WW, Hauge RH, Smalley RE (2004) Dissolution of pristine single walled carbon nanotubes in superacids by direct protonation. J Phys Chem B 108(26):8794–8798. doi:10.1021/jp036971t

Rinzler AG, Liu J, Dai H, Nikolaev P, Huffman CB, Rodriguez-Macias FJ, Boul PJ, Lu AH, Heymann D, Colbert DT, Lee RS, Fischer JE, Rao AM, Eklund PC, Smalley RE (1998) Large-scale purification of single-wall carbon nanotubes: process, product, and characterization. Appl Phys a Mat Sci Proc 67(1):29–37. doi:10.1007/s003390050734

Romhany G, Szebenyi G (2009) Interlaminar crack propagation in MWCNT/fiber reinforced hybrid composites. Exp Poly Lett 3(3):145–151. doi:10.3144/expresspolymlett.2009.19

Sager RJ, Klein PJ, Lagoudas DC, Zhang Q, Liu J, Dai L, Baur JW (2009) Effect of carbon nanotubes on the interfacial shear strength of T650 carbon fiber in an epoxy matrix. Compos Sci Technol 69(7–8):898–904. doi:10.1016/j.compscitech.2008.12.021

Salinier A, Dagreou S, Leonardi F, Derail C, Navascues N (2013) Electrical, rheological and mechanical characterization of multiscale composite materials based on poly(etherimide)/short glass fibers/multiwalled carbon nanotubes. Compos Struct 102:81–89. doi:10.1016/j.compstruct.2013.02.025

Sanchez M, Campo M, Jimenez-Suarez A, Urena A (2013) Effect of the carbon nanotube functionalization on flexural properties of multiscale carbon fiber/epoxy composites manufactured by VARIM. Compos Part B Eng 45(1):1613–1619. doi:10.1016/j.compositesb.2012.09.063

Schadler LS, Giannaris SC, Ajayan PM (1998) Load transfer in carbon nanotube epoxy composites. Appl Phys Lett 73(26):3842–3844. doi:10.1063/1.122911

Schlea MR, Brown TR, Bush JR, Criss JM, Mintz EA, Shofner ML (2010) Dispersion control and characterization in multiwalled carbon nanotube and phenylethynyl-terminated imide composites. Compos Sci Technol 70(5):822–828. doi:10.1016/j.compscitech.2010.01.019

Schlea MR, Meree CE, Gerhardt RA, Mintz EA, Shofner ML (2012) Network behavior of thermosetting polyimide/multiwalled carbon nanotube composites. Polymer 53(4):1020–1027. doi:10.1016/j.polymer.2011.12.050

Schnorr JM, Swager TM (2010) Emerging applications of carbon nanotubes. Chem Mater 23(3):646–657. doi:10.1021/cm102406h

Seo MK, Park SJ (2004) A kinetic study on the thermal degradation of multi-walled carbon nanotubes-reinforced poly(propylene) composites. Macromol Mat Eng 289(4):368–374. doi:10.1002/mame.200300303

Seyhan AT, Gojny FH, Tanoglu M, Schulte K (2007) Critical aspects related to processing of carbon nano tube/unsaturated thermoset polyester nanocomposites. Eur Polym J 43(2):374–379. doi:10.1016/j.eurpolymj.2006.11.018

Sharma K, Shukla M (2012) Experimental study of mechanical properties of multiscale carbon fiber-epoxy-CNT composites. In: Fan W (ed) Manufacturing science and technology, Pts 1–8, vol 383–390. Advanced Materials Research. pp 2723–2727. doi:10.4028/www.scientific.net/AMR.383-390.2723

Shen C, Brozena AH, Wang Y (2011) Double-walled carbon nanotubes: challenges and opportunities. Nanoscale 3(2):503–518. doi:10.1039/C0NR00620C

Shimamura Y, Oshima K, Tohgo K, Fujii T, Shirasu K, Yamamoto G, Hashida T, Goto K, Ogasawara T, Naito K, Nakano T, Inoue Y (2014) Tensile mechanical properties of carbon nanotube/epoxy composite fabricated by pultrusion of carbon nanotube spun yarn preform. Compos Part A Appl Sci Manuf 62(0):32–38. doi:10.1016/j.compositesa.2014.03.011

Siddiqui NA, Sham ML, Tang BZ, Munir A, Kim JK (2009) Tensile strength of glass fibres with carbon nanotube-epoxy nanocomposite coating. Compos Part A Appl Sci Manuf 40(10):1606–1614. doi:10.1016/j.compositesa.2009.07.005

Siddiqui NA, Khan SU, Kim JK (2013) Experimental torsional shear properties of carbon fiber reinforced epoxy composites containing carbon nanotubes. Compos Struct 104:230–238. doi:10.1016/j.compstruct.2013.04.033

Singh BP, Choudhary V, Saini P, Mathur RB (2012) Designing of epoxy composites reinforced with carbon nanotubes grown carbon fiber fabric for improved electromagnetic interference shielding. AIP Adv 2(2):022151. doi:10.1063/1.4730043

Song L, Zhang H, Zhang Z, Xie SS (2007) Processing and performance improvements of SWNT paper reinforced PEEK nanocomposites. Compos Part A Appl Sci Manuf 38(2):388–392. doi:10.1016/j.compositesa.2006.03.007

Steiner SA, Li R, Wardle BL (2013) Circumventing the mechanochemical origins of strength loss in the synthesis of hierarchical carbon fibers. ACS Appl Mater Interf 5(11):4892–4903. doi:10.1021/am4006385

Sundaram RM, Koziol KKK, Windle AH (2011) Continuous direct spinning of fibers of single-walled carbon nanotubes with metallic chirality. Adv Mat 23(43):5064–5068. doi:10.1002/adma.201102754

Szleifer I, Yerushalmi-Rozen R (2005) Polymers and carbon nanotubes—dimensionality, interactions and nanotechnology. Polymer 46(19):7803–7818. doi:10.1016/j.polymer.2005.05.104

Thess A, Lee R, Nikolaev P, Dai H, Petit P, Robert J, Xu C, Lee YH, Kim SG, Rinzler AG, Colbert DT, Scuseria GE, Tomanek D, Fischer JE, Smalley RE (1996) Crystalline ropes of metallic carbon nanotubes. Science 273(5274):483–487. doi:10.1126/science.273.5274.483

Thostenson ET, Li WZ, Wang DZ, Ren ZF, Chou TW (2002) Carbon nanotube/carbon fiber hybrid multiscale composites. J Appl Phys 91(9):6034–6037. doi:10.1063/1.1466880

Thostenson ET, Li C, Chou T-W (2005) Nanocomposites in context. Compos Sci Technol 65(3–4):491–516. doi:10.1016/j.compscitech.2004.11.003

Thostenson ET, Gangloff JJ, Li CY, Byun JH (2009a) Electrical anisotropy in multiscale nanotube/fiber hybrid composites. Appl Phys Lett 95(7):073111. doi:10.1063/1.3202788

Thostenson ET, Ziaee S, Chou TW (2009b) Processing and electrical properties of carbon nanotube/vinyl ester nanocomposites. Compos Sci Technol 69(6):801–804. doi:10.1016/j.compscitech.2008.06.023

Tibbetts GG (1989a) Vapor-grown carbon fibers. In: Figueiredo JL, Bernardo CA, Baker RTK, Huttinger KJ (eds) Carbon fibers filaments and composites. Kluwer, Dordrecht

Tibbetts GG (1989b) Vapor-grown carbon fibers: status and prospects. Carbon 27(5):745–747. doi:10.1016/0008-6223(89)90208-X

Tran CD, Humphries W, Smith SM, Huynh C, Lucas S (2009) Improving the tensile strength of carbon nanotube spun yarns using a modified spinning process. Carbon 47(11):2662–2670. doi:10.1016/j.carbon.2009.05.020

Veedu VP, Cao AY, Li XS, Ma KG, Soldano C, Kar S, Ajayan PM, Ghasemi-Nejhad MN (2006) Multifunctional composites using reinforced laminae with carbon-nanotube forests. Nat Mat 5(6):457–462. doi:10.1038/nmat1650

Vilatela JJ, Windle AH (2010) Yarn-like carbon nanotube fibers. Adv Mat 22(44):4959–4963. doi:10.1002/adma.201002131

Vilatela JJ, Khare R, Windle AH (2012) The hierarchical structure and properties of multifunctional carbon nanotube fibre composites. Carbon 50(3):1227–1234. doi:10.1016/j.carbon.2011.10.040

Wang BC, Yin JS, Wang L (2013a) Structure and properties of aeronautical composites using carbon nanotubes/epoxy dispersion as nanocomposite matrix. Polym Compos 34(10):1690–1697. doi:10.1002/pc.22570

Wang BC, Zhou X, Ma KM (2013b) Fabrication and properties of CNTs/carbon fabric hybrid multiscale composites processed via resin transfer molding technique. Compos Part B Eng 46:123–129. doi:10.1016/j.compositesb.2012.10.022

Wang BC, Zhou X, Yin JS, Wang L (2013c) Investigation on some matrix-dominated properties of hybrid multiscale composites based on carbon fiber/carbon nanotube modified epoxy. J Appl Polym Sci 128(2):990–996. doi:10.1002/app.38191

Wang YQ, Ku HK, Byun JH, Song JI (2010) Optimization of deposition conditions for CNTs/carbon fiber hybrid multiscale composites via Taguchi method. In: Multi-functional materials and structures III, Pts 1 and 2, vol 123–125. Advanced Materials Research. pp 1179–1182. doi:10.4028/www.scientific.net/AMR.123-125.1179

Wichmann MHG, Sumfleth J, Gojny FH, Quaresimin M, Fiedler B, Schulte K (2006) Glass-fibre-reinforced composites with enhanced mechanical and electrical properties—Benefits and limitations of a nanoparticle modified matrix. Eng Fract Mech 73(16):2346–2359. doi:10.1016/j.engfracmech.2006.05.015

Wicks SS, de Villoria RG, Wardle BL (2010) Interlaminar and intralaminar reinforcement of composite laminates with aligned carbon nanotubes. Compos Sci Technol 70(1):20–28. doi:10.1016/j.compscitech.2009.09.001

Wu AS, Chou T-W (2012) Carbon nanotube fibers for advanced composites. Mat Today 15(7–8):302–310. doi:10.1016/S1369-7021(12)70135-9

Wu Q, Zhu W, Zhang C, Liang ZY, Wang B (2010) Study of fire retardant behavior of carbon nanotube membranes and carbon nanofiber paper in carbon fiber reinforced epoxy composites. Carbon 48(6):1799–1806. doi:10.1016/j.carbon.2010.01.023

Wu QA, Bao JW, Zhang C, Liang RC, Wang B (2011) The effect of thermal stability of carbon nanotubes on the flame retardancy of epoxy and bismaleimide/carbon fiber/buckypaper composites. J Therm Anal Calorim 103(1):237–242. doi:10.1007/s10973-010-0960-0

Yakobson BI, Samsonidze G, Samsonidze GG (2000) Atomistic theory of mechanical relaxation in fullerene nanotubes. Carbon 38(11–12):1675–1680. doi:10.1016/S0008-6223(00)00093-2

Yamamoto N, Hart AJ, Garcia EJ, Wicks SS, Duong HM, Slocum AH, Wardle BL (2009) High-yield growth and morphology control of aligned carbon nanotubes on ceramic fibers for multifunctional enhancement of structural composites. Carbon 47(3):551–560. doi:10.1016/j.carbon.2008.10.030

Yu M-F, Lourie O, Dyer MJ, Moloni K, Kelly TF, Ruoff RS (2000a) Strength and breaking mechanism of multiwalled carbon nanotubes under tensile load. Science 287(5453):637–640. doi:10.1126/science.287.5453.637

Yu MF, Files BS, Arepalli S, Ruoff RS (2000b) Tensile loading of ropes of single wall carbon nanotubes and their mechanical properties. Phys Rev Lett 84(24):5552–5555. doi:10.1103/PhysRevLett.84.5552

Yu Z, Brus LE (2001) (n, m) Structural assignments and chirality dependence in single-wall carbon nanotube Raman scattering. J Phys Chem B 105(29):6831–6837. doi:10.1021/jp010853t

Zhang FH, Wang RG, He XD, Wang C, Ren LN (2009a) Interfacial shearing strength and reinforcing mechanisms of an epoxy composite reinforced using a carbon nanotube/carbon fiber hybrid. J Mat Sci 44(13):3574–3577. doi:10.1007/s10853-009-3484-x

Zhang M, Atkinson KR, Baughman RH (2004) Multifunctional carbon nanotube yarns by downsizing an ancient technology. Science 306(5700):1358–1361. doi:10.1126/science.1104276

Zhang Q, Liu J, Sager R, Dai L, Baur J (2009b) Hierarchical composites of carbon nanotubes on carbon fiber: influence of growth condition on fiber tensile properties. Compos Sci Technol 69(5):594–601. doi:10.1016/j.compscitech.2008.12.002

Zhang S, Kumar S (2008) Carbon nanotubes as liquid crystals. Small 4(9):1270–1283. doi:10.1002/smll.200700082

Zhang S, Koziol KK, Kinloch IA, Windle AH (2008) Macroscopic fibers of well-aligned carbon nanotubes by wet spinning. Small 4(8):1217–1222. doi:10.1002/smll.200700998

Zhang S, Liu WB, Hao LF, Jiao WC, Yang F, Wang RG (2013) Preparation of carbon nanotube/carbon fiber hybrid fiber by combining electrophoretic deposition and sizing process for enhancing interfacial strength in carbon fiber composites. Compos Sci Technol 88:120–125. doi:10.1016/j.compscitech.2013.08.035

Zhang XF, Li QW, Holesinger TG, Arendt PN, Huang JY, Kirven PD, Clapp TG, DePaula RF, Liao XZ, Zhao YH, Zheng LX, Peterson DE, Zhu YT (2007a) Ultrastrong, stiff, and lightweight carbon-nanotube fibers. Adv Mat 19(23):4198–4201. doi:10.1002/adma.200700776

Zhang XF, Li QW, Tu Y, Li YA, Coulter JY, Zheng LX, Zhao YH, Jia QX, Peterson DE, Zhu YT (2007b) Strong carbon-nanotube fibers spun from long carbon-nanotube arrays. Small 3(2):244–248. doi:10.1002/smll.200600368

Zhao JO, Liu L, Guo QG, Shi JL, Zhai GT, Song JR, Liu ZJ (2008) Growth of carbon nanotubes on the surface of carbon fibers. Carbon 46(2):380–383. doi:10.1016/j.carbon.2007.11.021

Zhao Q, Frogley MD, Wagner HD (2001) The use of carbon nanotubes to sense matrix stresses around a single glass fiber. Compos Sci Technol 61(14):2139–2143. doi:10.1016/S0266-3538(01)00166-X

Zhou YX, Pervin F, Lewis L, Jeelani S (2008) Fabrication and characterization of carbon/epoxy composites mixed with multi-walled carbon nanotubes. Mater Sci Eng A Struct Mater Proper Microstruct Proc 475(1–2):157–165. doi:10.1016/j.msea.2007.04.043

Zhu J, Imam A, Crane R, Lozano K, Khabashesku VN, Barrera EV (2007) Processing a glass fiber reinforced vinyl ester composite with nanotube enhancement of interlaminar shear strength. Compos Sci Technol 67(7–8):1509–1517. doi:10.1016/j.compscitech.2006.07.018

Zimmer M, Cheng QF, Li S, Brooks J, Liang R, Wang B, Zhang C (2012) Comparative characterization of multiscale carbon fiber composite with long and short MWCNTs at higher weight fractions. J Nanomater 2012:532080. doi:10.1155/2012/532080

MIX
Papier aus verantwortungsvollen Quellen
Paper from responsible sources
FSC® C105338
FSC
www.fsc.org

If you have any concerns about our products,
you can contact us on
ProductSafety@springernature.com

In case Publisher is established outside the EU,
the EU authorized representative is:
Springer Nature Customer Service Center GmbH
Europaplatz 3, 69115 Heidelberg, Germany

Printed by Libri Plureos GmbH
in Hamburg, Germany